500.F.5

GB
330
BEA
1981

3009388985

D1165854

The inland waters of tropical Africa

L. C. Beadle

Honorary Senior Research Fellow, University of Newcastle upon Tyne. Formerly Professor of Zoology and Wellcome Research Professor, Makerere University, Uganda

The inland waters of tropical Africa
An introduction to tropical limnology

Second edition

Longman
London and New York

Longman Group Limited
Longman House
Burnt Mill
Harlow
Essex

*Published in the United States of America
by Longman Inc., New York*

© Longman Group Limited 1974, 1981

First published 1974

Second edition 1981

British Library Cataloguing in Publication Data
Beadle, Leonard Clayton
    The inland waters of tropical Africa. – 2nd ed.
    1. Fresh-water ecology – Africa
    I. Title
    574.5'2632'0967   QH541.5.F7   80-41376

    ISBN 0-582-46341-6

Printed in Great Britain by William Clowes (Beccles) Ltd
Beccles and London

Contents

Acknowledgements and note on geographical names

I have much benefited from conversations and correspondence with many of the people who have contributed to this subject. Their names are recorded with gratitude by means of asterisks in the reference list of publications. Among them are some friends with long experience in Africa and other tropical regions with whom I have discussed these matters at greater length and, with some, over a number of years. They have provided me with facts and ideas and have stimulated me to produce ideas of my own, with some of which they may not agree. I must especially mention W. W. Bishop, G. S. Carter, C. W. Coulter, B. Dussart, D. H. Eccles, L. K. H. Goma, P. H. Greenwood, A. T. Grove, A. J. Hopson, M. Kalk, E. M. Lind, G. Marlier, R. H. Lowe-McConnell, A. J. McLachlan, T. R. Milburn, M. Poll, H. Sioli, J. F. Talling, E. Trewavas, S. A. Visser, E. B. Worthington.

Théodore Monod has both honoured me with his foreword and introduced me to much of the French literature and has given me good advice from his long experience in the Sahara and francophone Tropical Africa.

My wife has helped me very much at many stages in the field and in the final proof reading and making of indices.

The production of this book was assisted by a generous grant from the Wellcome Trust.

All the diagrams and sketchmaps have been redrawn and the photographs reprocessed by the Graphics Section of the Photographic Department in the University of Newcastle upon Tyne.

Photographs whose source is not acknowledged were taken by the author.

Since the independence of the African countries during the 1960s many of the names of the countries, towns, lakes and rivers have been changed. Most of the new names have been adopted, though of course the literature before 1960 refers to them by the old names (e.g. Lake Nyasa, now Lake Malawi). Very recent changes, however, made while this book was in its final stages are noted, but, to avoid confusion, are not adopted through the text, since they do not yet appear in any of the published literature.

Preface to second edition

It was inevitable that the first attempt to review and discuss such a wide ranging subject would soon have to be revised in response to the reactions of specialists and in the light of more recent work. Apart from correcting errors, many alterations have been made in the text to improve the presentation and in some places to modify the assessments previously made. The major change is, however, the introduction of new material into nearly every chapter, especially in relation to Lake Turkana, the Ethiopian Lakes, Lake Chad, the Okavango Delta and the manmade lakes. More than 260 new references have been added.

Several of those whose special help I had previously acknowledged (above) have again assisted with information and comments. To these I must now add K. E. Banister, H. J. Dumont, G. Fryer, J. J. Gaudet, J. Lemoalle, H. W. Lissmann, D. A. Livingstone, J. M. Melack, T. Petr, J. Rzóska and F. A. Street.

Foreword
by Professor Théodore Monod

Il n'est pas surprenant que la limnologie africaine ait connu des progrès plus lents que l'océanographie: si les côtes de l'Afrique étaient connues de l'Europe dès la fin du XVe siècle, l'intérieur d'un continent exceptionellement massif se trouvait efficacement défendu par des obstacles naturels sérieux, fleuves souvent peu navigables, coupés de rapides, vastes zones de marécages, déserts. L'histoire de la découverte des sources du Nil reste, à cet égard, exemplaire puisqu'il a fallu attendre le milieu du XIXe siècle pour résoudre un problème posé plus de 2000 ans plus tôt.

L'approche marine était plus aisée et l'on connaîtra donc les Copépodes du Golfe de Guinée avant ceux du Lac Victoria on du Tchad. De plus, les premiers voyageurs occidentaux dans l'intérieur de l'Afrique, même naturalistes, se sont livrés en priorité à l'étude des groupes les plus 'remarquables' – au sens étymologique du mot: Mammifères, Oiseaux, Papillons, Coléoptères, etc. Peu d'entre eux se souciaient de récolter des Collemboles, des Oligochètes ou des Ostracodes . . .

Les choses ont, depuis, et très heureusement, changé. Et maintenant que la totalité du continent est ouvert à la recherche scientifique, maintenant qu'il existe toute une série d'universités, de services ou laboratoires spécialisés (des pêches, en particulier), l'étude des eaux douces africaines a pu faire, enfin, au cours des cinquante dernières années, des progrès très considérables.

Sans doute l'étendue de nos ignorances demeure-t-elle immense et le champ de travail ouvert aux chercheurs reste-t-il pratiquement sans limites. Mais on pouvait, au stade où nous parvenons, penser que l'heure était tout de même venue de préparer une première synthèse de nos connaissances.

Cette synthèse, rendue exceptionnellement difficile par l'extrême dispersion de la documentation, le Professeur L. C. Beadle a eu le courage de la tenter et il faut lui en savoir gré, car il a de la sorte mis entre les mains de tous les intéressés, hydrobiologistes, zoologistes, écologistes, biogéographes, responsables des pêches ou de la santé publique, etc, un véritable traité – le tout premier du genre, bien entendu – de limnologie africaine. Celui-ci sera consultè avec fruit, non seulement pour y découvrir les données acquises sur tel ou tel sujet mais aussi, et ce n'est pas moins important, pour y trouver des thèmes de recherche sur les problèmes qui attendent encore leur solution, et il n'en manque pas...

Limnologie *africaine*? Oui, mais en même temps limnologie *tropicale*, et le Professeur Beadle a bien raison d'insister sur l'individualité du monde tropical qui possède, à tant d'égards, ses caractères propres et qui n'ont peut-être pas été suffisamment reconnus jusqu'ici, sauf sans doute par les botanistes. On appréciera aussi, j'en suis sûr, la rigueur et al prudence d'un auteur qui sans cesse refuse d'en dire plus qu'on n'en sait, insiste sur le caractère incertain ou provisoire de bien des conclusions, se méfie des classifications trop ambitieuses et trop précises, risquant de masquer parfois sous une nomenclature prétentieuse les transitions ménagées, les nuances, les complexités du réel.

Certains seront peut-être surpris par l'importance accordée dans cet ouvrage d'une part aux développements *historiques*, de l'autre aux questions *géologiques* et *paléoclimatiques*. Mais comment ne pas sentir la nécessité pour expliquer le présent de connaître le passé, récent (celui de l'histoire humaine) ou plus lointain (évolution des climate et de la morphologie), si l'on veut pouvoir tenter d'expliquer les faits actuels, par exemple la géographie des bassins fluviaux.

Sans doute, et l'auteur sera le premier à le reconnaître, le livre est plus une *limnologie* (étymologiquement: science des *lacs*) qu'une *hydrobiologie* africaine générale. Et par conséquent un ouvrage ou l'Afrique des grands lacs, l'Afrique *orientale* se verra l'objet d'un traitement privilégié. Mais pouvait-il en être autrement, étant données et l'abondance de la documentation concernant les grands lacs orientaux et l'expérience personnelle de l'auteur? On notera d'ailleurs que ce dernier, soucieux d'équilibrer le plus possible son texte, a su faire leur place à d'autres régions, au Tchad par exemple, au Niger, et même aux eaux sahariennes, à tant d'égards si intéressantes, à la fois par la faune de leurs mares temporaires que par le caractère relictuel de tant d'éléments de leurs 'gueltas' pérennes ou de leurs eaux artésiennes.

Alors que jusqu'ici toute approche d'un des aspects de la limnologie africaine exigeait, d'abord, de patientes et laborieuses recherches bibliographiques, l'ouvrage du Professeur Beadle va en permettre un accès grandement facilité. L'étudiant disposera enfin de la sorte, avec le premier traité de limnologie africaine jamais rédigé, d'une riche mine d'informations solides, appuyées soit sur une vaste expérience personnelle, soit sur une connaissance inégalée des sources disponibles. Le chercheur confirmé, de son côté, viendra ici non seulement compléter sa documentation personnelle mais découvrir de nouveaux sujets d'étude: à voir le volume des lacunes de notre savoir, qui ne souhaitera, à son tour, rejoindre la trop petite équipe des hydrobiologistes au travail en Afrique et contribuer, sur le terrain, au progrès d'une discipline d'une aussi haute importance, tant théorique que pratique, d'ailleurs, si l'on songe aux développements de l'industrie des pêches sur les lacs ou les grands fleuves soudaniens, à ceux aussi de la pisciculture?

L'ouvrage du Professeur Beadle, et l'auteur nous en a très honnêtement averti lui-même, ne dit pas *tout* sur *tout*: le domaine des eaux saumâtres littorales, lagunes et mangroves, n'est pas traité, sans doute comme relevant plutôt de l'océanographie que de la limnologie, certaines faunes aquatiques spécialisées, interstitielles, souterraines ou cavernicoles ne sont pas décrites non plus, malgré leur prodigieux intérêt biogéographique ou évolutif.

Mais tel qu'il est, enrichi d'une très importante bibliographie, d'un précieux index et d'une illustration bien choisie, ce volume fera date dans l'histoire de la recherche scientifique africaine. Et, de plus, il suscitera, j'en ai la conviction, des vocations parmi les jeunes biologistes qui se forment aujourd'hui en Afrique: quel plus noble et plus utile rôle pourrait-on lui promettre?

Théodore Monod
Membre de l'Institut de France
(Académie des Sciences)
Professeur honoraire au Muséum National d'Histoire Naturelle
(Pêches Outre-Mer)
Directeur honoraire de l'Institut Français d'Afrique Noire

1

Introduction

Limnology (from the Greek *limne*, a pond or lake) is the study of inland waters –
lakes, rivers, swamps, etc. It is concerned not only with discovering what they
are composed of but, more especially, with understanding the complex interrelations between the physical, chemical and biological events which maintain them
and link them with the outside world and have moulded them in the past. The
subject has its main practical applications in fisheries, public health and water
pollution. Hydrology, on the other hand, is confined to the study of the geological, chemical, physical and climatic aspects of water resources, and aims at
finding, mobilising and regulating water for domestic, irrigation and hydroelectric purposes. But geological factors, and the chemical and physical conditions,
decide the environment for aquatic organisms, and much of hydrology is necessarily included in limnology.

The subject developed in Europe and North America during the first quarter
of this century, and has spread to most regions of the world where there is active
research on the environment. A number of rather rigid concepts and systems of
classification of lake types and of seasonal cycles were originally developed on the
basis of studies in temperate climates. Further work even in temperate regions
has loosened the structure of the subject, but recent investigations on tropical
lakes have made it clear that some of the older concepts are of limited application
and that rigid classification of the subject-matter is far from helpful. The German expedition of Thienemann and Ruttner to the tropical lakes of Indonesia in
1928–29 first showed the limited value of the contemporary classification of lake
types based on studies in Europe and North America (Thienemann, 1932; Ruttner, 1931; and a short review in English by Rohde, 1974). This too, I hope, will
emerge from the book, and not all readers will be convinced that limnology in the
tropics is sufficiently different to be regarded as a separate compartment of the
subject. But in certain respects, notably in relation to seasonal cycles of production and to speciation, profitable comparisons can be made between the course of
events in temperate and in equatorial regions. Such comparisons are made here,
and the extent to which principles previously formulated from work in temperate
regions are applicable to the tropics is discussed.

It must be emphasised that no very precise meaning can be attached to the word 'tropical' that can have any ecological value. Latitude, as such, is clearly irrelevant, and the climatic conditions of any region within the tropics depend greatly on altitude, local topography, position relative to the main continental air-streams, etc. The best we can do is to regard a typical 'tropical' climate as one involving a relatively high temperature with small seasonal fluctuations. We must, however, face the fact that there is no clearcut division between 'tropical' and 'temperate' climates, which grade one into the other with many intermediates. There are also parts of the tropics that do not have a 'tropical' climate. Apart from the regions of high altitude the 'equatorial' region (closed to the Equator) tends to have a more uniform climate – high rainfall and humidity with relatively constant high temperature, and from an ecological standpoint is rather easier to define.

There are many curious unexplained facts and unsolved problems presented by the inland waters of tropical Africa, and I shall be at pains to emphasise the extent of our ignorance. I am tempted to take this opportunity to warn the student of a tendency, that is certainly not confined to limnologists, to propose a simple explanation for a puzzling phenomenon and to allow this to become generally accepted though, in fact, it has never been put to the final test. There are, of course, degrees of uncertainty for explanations that cannot strictly speaking be directly tested. For example, the presence of Nile ('soudanian') fish in the now closed basin of Lake Turkana is good evidence of a recent connection with the Nile, because it is consistent with what is known of the distribution and behaviour of African fish and with some palaeoclimatic and geological evidence (Ch. 10). No other explanation is at present conceivable. At the other extreme, there are phenomena for which few people have had the temerity to suggest an explanation. Such are the scarcity of papyrus in West Africa (p. 312), the absence from tropical Africa to Urodele Amphibia (newts, etc.), of green *Hydra* and of free-living freshwater Isopod and Amphipod Crustacea, except for some rare cave dwellers (p. 149), though brown *Hydra*, Anura (frogs and toads) and other freshwater Crustacea (crabs, prawns, Cladocera, Copepoda and Ostracoda) are abundant. Between are a mass of phenomena for which the suggested explanations present a gradient of probability.

It is very tempting to associate some peculiar feature in the fauna or flora with an unusual chemical composition of the water that happens to have been noted, and some of these suggestions have been in danger of general acceptance though they have never been put to the test of experiment, and some are inherently unlikely. However close the correlation between the occurrence of two phenomena, there is no certainty that one is the 'cause' of the other. For example, there is probably a high correlation between the occurrences of hippopotamus and of fish in tropical African lakes though no one could reasonably suggest a direct causal connection. It is not always certain that the phenomenon to be explained is a real one. For example, repeated collections over a long period are needed to establish with certainty the absence of certain species, especially of invertebrate animals which are apt to fluctuate greatly for reasons which we do not yet understand, and there have been cases of mistaken identity. Unbalanced estimates of the presence and abundance of some species have been made by collectors whose interests were restricted to certain groups of organisms. For example the bacteria and protozoa, that are now known to occupy a very important position in the plankton of many lakes, were often neglected. The difficulties of preservation and identification are partly responsible. Green (1976) reported that the relative abundance and species composition of the zooplankton in three of the volcanic barrier lakes in western Uganda – Bunyoni, Mutanda and Mulehe – changed considerably between 1962 and 1975. Some species had disappeared. Alterations in the water following an increase in the surrounding agricultural population was suggested as a possible cause of these changes. Occasional reversion to resting stages in the sediment might also be suggested. Whatever the cause, such cases serve as a warning against conclusions drawn from infrequent sampling. These considerations apply with great force to the subject of production. In order to establish the reality of recurrent events, such as seasonal fluctuations in productivity, observations must be spread over a long period, even of years. This is especially important in the tropics where seasonal changes are apt to be less well marked. Another point worth emphasising is that we are limited to measuring those features of the environment for which we have methods that can easily be applied in the field or to preserved samples brought back to a laboratory. Features that are of great importance in determining the character of the ecosystem may be, and in some cases have been, missed. Modern techniques have much extended the range of our methods, but there is still a long way to go.

The reader should adopt a critical attitude to all numerical data. There is a great deal of quantitative information concerning the chemical composition of waters, the occurrence and abundance of certain species and on other matters. In some cases, examination of the literature reveals that the figures were calculated from observations made at one point on a single day by an expedition many years ago. Such data are undoubtedly valuable and the early work was mostly of this kind, but, apart from possible errors in the methods used, and possible changes during storage of samples, the figures represent only the situation at the time and are not necessarily permanently characteristic. With the more recent establishment of permanent bases for research we now have data relating to certain water systems that have been collected over several years and from which general conclusions may reasonably be drawn. I hope that not all of the interpretations of data to be found in this book will accepted by the reader as reasonable without reference to the original literature, from which he may well form another opinion.

Apart from differences of temperament and outlook, lack of decisive evidence is the main cause of disagreement among experts. A student of the literature on the present subject will find some opposing opinions expressed with great conviction. The topics within the scope of this book that are especially subject to controversy are (a) the origins and history of the water systems and (b) the origins and evolution of the species, particularly the fish, that inhabit them. Decisions on the latter subject are the more difficult, because it cannot yet be said with confidence that we fully understand all of the causes and mechanisms of speciation (Ch. 8). As between archaeologists, the debates are sometimes heated, but in most cases the uncommitted reader will find it difficult to agree that any final conclusion is justified on the evidence so far available.

These warnings must be countered by a declaration of faith, shared now by most scientists, in the supreme value of imaginative guesswork provided only that the guess is based on a background of knowledge and is not accepted without supporting evidence as a certain step towards the truth. An excellent example of an inspired and valuable, but wrong, guess was the marine relict theory of the origin of the fauna of Lake Tanganyika propounded at the end of the last century. It was at one time regarded by several reputable biologists as a near-certainty, but was later quite descredited. Nevertheless, there was at the time much apparently good evidence to support it and the great interest in this remarkable lake and the vigorous investigations inspired by the theory, contributed much to the study of evolution in relation to geographical distribution. It was also the beginning of tropical limnology, though more than twenty years elapsed before any further serious work was done on the African lakes.

Limnology is potentially of much greater economic importance in tropical Africa than in western Europe or North America. In the arid deserts of the northern and southern tropics, water is the great limiting factor, and the study of temporary and saline waters is therefore important. In much of the equatorial region, on the other hand, surface water is so abundant as to be a major natural resource, as important as the sea to the countries of western Europe, and one that has as yet been only partially exploited.

There are eight great lakes in tropical Africa ranging in size from 2 300 to 75 000 km^2, of which all but Lake Chad are in the eastern half of the continent. The largest, Lake Victoria, has about the same area as the Aegean Sea between Greece and Turkey. There are innumerable smaller lakes that in Europe would be considered rather large. In addition, there are four of the world's largest river systems, the Nile, Zaïre, Niger and Zambezi, as well as very great areas of swamp covering many thousands of square kilometres. Most of the large lakes and rivers were, until recently, the most convenient means of communication across great expanses of difficult country. Even now, with railways, motor roads and air-routes, ships provide transport for men and freight on Lakes Victoria, Tanganyika and Malawi, and on long stretches of the rivers Nile, Zaïre and Niger. But the future lies in the exploitation of lakes and rivers for fisheries, irrigation, hydroelectric power and recreation. One of the objects of this book is to provide a scientific background to the fisheries and other potential biological resources of the inland waters of tropical Africa. It is not generally realised how great is this potential reservoir of animal protein for a population for which protein deficiency is a major and widespread dietary problem. Most of the continent is remote from the sea, and the marine fisheries present greater technical and economic problems for their successful exploitation. The introduction of the gill-net and of modern methods of preservation and transport have caused a great expansion of the inland fisheries previously conducted by some very ingenious, but less productive, native methods. Even now, most of the fishing is done on a small scale by private fishermen, mostly in the shallow inshore waters of lakes and in rivers. Here the 'small man' will probably remain an efficient exploiter, provided that the methods of preservation, transport and marketing are improved.

In recent years the more technically advanced countries, either directly or through United Nations agencies, have been assisting the developing countries to exploit their natural resources more efficiently. Such aid is usually in the form of 'development schemes', involving a preliminary investigation or research programme by specialist scientists armed with appropriate equipment and funds. On the basis of their findings, new industries may be launched with further funds and technical assistance. These may be concerned with the extraction of minerals or with manufactures based on the country's organic and inorganic natural resources. Considerable improvements have also been made in native techniques of agriculture and fisheries and in methods of transporting and marketing of the produce. These are fundamental human activities which in tropical Africa, as elsewhere, have determined the way of life and outlook of the people, and provide the majority with an occupation, as well as a means of subsistence. The social effects of suddenly introducing large-scale mechanical exploitation are at least as important as the overall economic benefits that are expected to result therefrom. When a government or commercial firm plans to operate a fishery using large, powered boats and mechanically operated dredges, ring-nets, etc., attention must be paid to the possible social, as well as to the economic and ecological, consequences. For example, further expansion of the large-scale 'sardine' fishering of Lake Tanganyika (p. 292) and the proposed trawler-dredging of cichlid fishes in the deep water of Lake Victoria (p. 265) must be planned and controlled with these possible consequences in mind (see Jackson, 1971). Another scheme that would have grave social consequences is the proposed Jonglei canal to short-circuit some of the Nile water that enters the Sudd swamps (p. 331).

Important tropical diseases of man and domestic animals are caused by organisms that can live in water (e.g. water-borne bacterial infections) or are carried by animals that are partly or wholly aquatic (e.g. malaria, bilharzia, liver-fluke, onchocerciasis, filariasis and several virus infections). Owing to the continuously high temperature, tropical waters provide favourable conditions for the maintenance of reservoirs of these organisms, and the climate is conducive to frequent direct contact between men and water. The relevance of limnology to tropical public health is thus particularly close.

This book is intended in the first place as a background for students in Africa destined to become fishery or public health officers, members of research establishments concerned with inland waters or university teachers and research workers specialising in some aspect of this subject. I would hope also to interest visiting scientists intending to work in this or some related field and even an inquiring layman. Almost all of the research has been published in French or English, and until recently, there has been too little contact between the workers in the franco- and anglophone regions of Africa. I therefore hope to give the student an overall view of research in both regions.

For many countries in tropical Africa, having in mind the abundance of their lakes and rivers, limnology provides great opportunities for research and teaching in some basic biological subjects, such as evolution of faunas and floras in the light of geological and climatic history, biogeography, ecological problems in a wide range of ecosystems, energy and nutrient cycles, photosynthesis, the seasonal cycles of primary and secondary production under tropical conditions, and the physiological and biochemical adaptation of organisms to peculiar conditions (high salinity, periodic desiccation, lack of oxygen, etc.).

Most of the these subjects could in principle be studied in any of the major water systems of tropical Africa. But our knowledge is, as yet, fragmentary, and we have perforce to illustrate a particular subject with examples from places in which it has been most effectively studied. This is particularly obvious with productivity and production cycles.

The study of aquatic ecosystems is advancing rapidly, and within another ten years there will be many new ideas concerning the circulation of matter and energy relevant to tropical African inland waters and a much greater understanding of their ecology and past history. The reader will no doubt be impressed with the present extent of our ignorance and with the number of doubtful interpretations.

It is to be hoped that neither the student nor the general reader will find it necessary to turn at intervals to other books to inform himself or to refresh his memory of basic principles. With this in mind some elementary discussion of such matters as water movements, water chemistry, ecosystems, photosynthesis, productivity and the evolution of species have been included in the first seven chapters. The glossary should help those who are not already familiar with the jargon of the subject.

This is not a comprehensive treatise, and, indeed, it would be beyond the author's knowledge to make it so. It is an introduction to some important topics in limnology and a survey of the main water systems in tropical Africa and of their peculiar problems. The references will enable anyone to study the original work in greater detail.

Though I have made some effort to maintain a balance, I cannot pretend to have avoided giving relatively more prominence to certain subjects and regional studies than a dispassionate reader might think that they deserve. I can only reply that it is better for a student to hear about subjects of which the author has a direct knowledge and in which he is especially interested. It is impossible to disguise the fact that I know more from direct experience about eastern than about western tropical Africa.

2
Historical background to scientific exploration

The roots of science are, no doubt, more diverse and widespread than the surviving records would lead us to suppose, but during the last five hundred years B.C. the countries bordering the eastern Mediterranean saw the beginnings of a new attitude to the material world that became the basis of modern scientific thought. This attitude assumed that the causes of events could, in principle, be discovered by reasoning, and that the world could be explained in rational terms from established premises. This revolutionary concept was at first entirely theoretical and confined to relatively few people, but it spread gradually across western Europe during the following centuries, and began to be applied to the practical art and crafts which flourished greatly during the Middle Ages. From this mating of the theoretical and the practical came the birth of experimental science during the sixteenth and seventeenth centuries. It inevitably led to the dramatic advances in science and technology of the present day, and for good and ill will continue to lead us much further. In view of the proximity of these momentous developments, why were the practical arts and skills of the Tropical African peoples, from which Europe too might then have learnt much, almost totally unaffected by them until well on into the nineteenth century? Why, during that crucial period of the sixteenth to eighteenth centuries were cultural communications between the Mediterranean countries and tropical Africa more firmly blocked than during any previous period in history? These are clearly the most important and basic questions relating to the history of modern tropical Africa. It is because water and water systems played an important part in this history that these questions are worth some discussion at this point.

It is, however, not true that during the first 1700 years A.D. no European set foot in the interior of tropical Africa. There is even rather good evidence that during the first century A.D. some Roman soldiers penetrated far into the Sahara and may even have reached the Niger (discussion in Hallett, 1965a, 46–7). There is even better evidence that in the year 61 A.D. the Emperor Nero sent an expedition up the Nile from the then southern Roman frontier a hundred miles south of Aswan. A thousand miles of travel brought them to Malakal where they were held up by the Sudd Swamps. They eventually returned to Rome (Kirwan, 1957).

From the fifteenth to the eighteenth centuries, Portuguese, French, British and Germans penetrated varying distances, mainly from bases on the Senegal and Gambia rivers. But these contributed little to the solution of the major geographical problems of the interior, and until the late eighteenth century, accurate knowledge of tropical Africa, other than the coastal regions, had made very little progress in Europe over the previous thousand years, though closer contacts with the much more remote tropics of the Americas and the Far East had been established for several centuries. For an introduction to the history of Africa, with some discussion of this subject, see Oliver and Fage (1966), de la Roncière (1924–27), Hallett (1965a), Bovill (1968a), Fage (1978), and Davidson (1959) for a stimulating discussion and interpretation of Iron Age archaeology in Africa.

For a long time, up to about 5000 B.C., the western Sahara and the entire southern edge of the present desert from the River Senegal through Chad to the Upper Nile had a higher rainfall than now.* Lake Chad was about five times its present size, and there were other lakes in the now complete desert, including the very large Lake Araouane north-west of Timbuktu. These conditions allowed movements of fauna and flora between North Africa and the humid tropics (see Ch. 11). It is therefore not surprising that at that time the west central Sahara supported a considerable population of pastoralists, and there were at least two well-worn trade routes crossing it from north to south. Evidence for these facts comes from lake sediments, fossils, human artifacts (e.g. fishing harpoons) and innumerable rock-paintings of men with domestic and wild animals (Lhote, 1958, 1959). From the paintings we also learn that horse-drawn vehicles were used on the two main trade routes towards the end of the period before the climate had reached its present state of extreme desiccation. The camel was introduced into the Sahara from the Near East during the first two centuries A.D. As in later historic times, the goods exchanged were mainly gold, slaves and ivory from the south, for salt, metal weapons, cloth and ornaments from the north. The middlemen and transporters in this trans-Saharan trade were at that time probably the ancestors of those Berber peoples (Touareg) who remained in the western Sahara when it became drier and managed to adapt to a nomadic desert life. The negro pastoralists retreated southwards, though leaving a few remnant groups that have survived to this day in the Tibesti Mountains and in some of the desert oases.

During the last one thousand years B.C. the western Sahara became progressively more difficult to traverse by any but people specially experienced in desert life, and the desert barrier must surely have contributed to the cultural separation of tropical Africa from the Mediterranean. Nevertheless, the trans-Saharan trade continued and was, in fact, greatly expanded by the Arabs who invaded and occupied the whole of the North African coast during the seventh and eighth centuries A.D. Already experienced in desert life and they rapidly monopolised the trade. Between then and the eighteenth century there were more than six north–

---

* The present savanna belt across the continent between the desert and the tropical forests (Fig. 11.3) is often referred to in English as the 'Sudan', and sometimes as the 'Sahel' (the 'border land' i.e. between the desert and the humid tropics). To avoid confusion with the Republic of Sudan at the eastern edge of the belt, the French spelling 'Soudan' is used in this book (see also p. 144).

south trade routes between the Senegal and the Nile (Fage, 1958, Map 13; Bovill, 1968a, Fig. IX). During the middle ages there arose in the Soudan, south of the desert, a number of highly organised negro states and empires whose prosperity depended upon the trans-Saharan trade. Though they eventually adopted Islam, it seems that some basic elements in their culture were derived from the ancient Kush civilisation centred on Meroe on the middle Nile during the last 500 years B.C., which, in turn, had inherited some of the traditions of ancient Egypt. These states waxed and waned, but at times reached a level of technical skill, culture and sophistication not surpassed by any contemporary European country. For example, the kingdom of Mali in the fourteenth century became an important world centre of Islamic culture and scholarship, especially at Timbuktu and Djenne. Their main source of riches was the gold found between them and the Guinea Coast, for which they acted as intermediaries between the negro miners and the Arab traders from the north.

From A.D. 1000 to 1500 there was therfore a flourishing trans-Saharan trade between well-organised and sophisticated states in northern tropical Africa and the countries of Europe. Moreover, it was conducted by the Arabs who were noted for their men of learning and skill. That this commerce provided practically no cultural contacts between Europe and the interior of Africa is at least partly explained by the fact that the Arabs not only were exclusive specialists in the difficult art of desert travel, but also saw to it that no one else, and especially the Christian infidels of Europe, disturbed their monopoly of the trade. Under such circumstances, commercial and cultural relations are not necessarily linked. The Soudanic states had much closer cultural relations with North Africa, Egypt and the Near East. The Arab traders during this period learnt much of the geography of the Soudan, especially of the upper Niger basin and Lake Chad and, indeed, several Arabic accounts of the medieval Soudanic states have survived from what must originally have been a considerable mass of literature. But these were slow to become known in Europe, and for the most part were not recognised much before the late fifteenth century when exploration of the New World began and the attention of Europe became deflected to the apparently greater and more easily obtainable riches of the Americas. This was another major reason for the further deterioration of the already tenuous cultural links with tropical Africa, and until the end of the eighteenth century, information reaching Europe concerning the geography of the African interior was of the vaguest, and was largely incorrect. The Soudanic states themselves, after flourishing for several centuries, eventually broke down as powerful political entities, partly through mutual dissension, but mainly from external invasion particularly from Morocco armed with the newly acquired firearms.

In the meantime, however, the opinion of some Arab geographers that Africa was surrounded by sea had become known, and Portugal was the first European nation to act on this supposition in order to open a new route to the Far East and to exclude the Arab middlemen by approaching western tropical Africa from the sea. The remarkable and successful voyages by Portuguese seamen along the Mauritanian and Guinea coasts during the fifteenth century were followed by other European nations, and thus began a new phase in African–European trade that ultimately reduced the relative importance of the trans-Saharan routes.

Though the west coast trade was more direct, in that Europeans dealt immediately with the negro peoples of the coast, the latter were in fact middlemen who obtained the goods (mainly gold, ivory and slaves) from further inland. Apart from some praiseworthy, but shortlived, work by Christian missionaries during the early stages of the Portuguese presence along the coast, contacts with the people were strictly concerned with trade and confined to the immediate coast. Slavery was well established in Africa, Europe and Asia long before the dawn of recorded history. European slaves were very common in North Africa in the seventeenth century, and some are known to have been at Timbuktu; others were in the service of the negro State of Bornu west of Lake Chad during the same period (Hallett, 1965a, p. 101). Nevertheless, the appalling intensification of the slave trade during the seventeenth and eighteenth centuries, to supply the plantations of tropical America, further ensured that no proper cultural relations were established between the Europeans and the people of the interior. Consequently Europe received even less reliable information about the geography and human affairs of the Soudan, which from 1500 onwards lay within 800 km of their West African trading posts, than had reached them across the 1 500 km of desert and the Mediterranean.

The idea that Timbuktu lay on the shore of a large lake from which one river flowed westward to the Atlantic and another, the Nile, flowed to the east, seems to have originated from secondhand information got by the Portuguese on the west coast before 1500, and from the writings of Leo Africanus, an Arab who travelled over much of the Soudan in the early sixteenth century. Some traces of this story, e.g. a west-flowing River Niger and a branch of the Nile rising from a lake or lakes in the central Soudan (Lake Chad?), survived in some minds until the final explorations in the nineteenth century (see Fig. 2.1). One wonders whether the many bizarre ideas that have been expressed in the past concerning the disposition of the African rivers and lakes (see Langlands, 1962) have owned something to ancient legends from prehistoric times, when, as we know, the water systems in the Sahara and Soudan were very different from now. Even during the final millenium B.C. they were still settling down to their present drier condition and news from tropical Africa could probably spread to Europe more easily then than at any subsequent time before 1800. Most of the ideas, expressed before the final explorations that settled these questions, originated in Africa, and were no doubt distorted by repeated telling. But there is no evidence that the inhabitants of the interior of the continent had any more accurate knowledge of the wider relations of the water systems in their own immediate neighbourhood than did the armchair geographers in Europe at the end of the eighteenth century.

The history of European exploration of Africa has, for obvious reasons, been dominated by the great mystery of the sources of the Nile, a river that played a predominant part in the early history of Mediterranean civilisation. For this very reason, the fact that its sources were unknown until the nineteenth century is in need of explanation. From about 3000 B.C. the civilisation of ancient dynastic Egypt owed its existence to the Nile, or, more particularly, to the annual floods and the sediments by which, as now, the crops and grazing were nourished. There is no evidence that Egyptians ever reached the lake region of the Upper

Nile basin, but they certainly went south in search of ivory, ebony and slaves, and the tropical savanna at that time extended further north than at present. As already mentioned, there is reason to suppose that some of the culture of ancient Egypt ultimately spread right across the Soudan as well as southwards into eastern Africa during the last few centuries B.C. from the upper Nile kingdom of Kush, presumably by migration of peoples (for a recent authoritative discussion of this controversial subject see Fage, 1978). Subsequently, however, there does not seem to have been any fruitful contact between the Nile valley and the immediate south, and no trade route was developed comparable with those across the Sahara. Eastern equatorial Africa was presumably less attractive as a source of valuable materials and the Sudd swamps of the Upper Nile made access by the Nile valley very difficult.

In any case, the gold- and iron-producing region still further south in southeast Africa had had trading relations across the Indian Ocean with Arabia, India and China at least from the beginning of the Christian era. Egyptian, Kushite, Chinese, Arab and, later, Portuguese ships took part in this. There was thus little incentive to establish an overland route. Both Roman and Alexandrian ships based on Red Sea ports visited the East African coast during the first few centuries A.D. It is possible that the well-known map of Claudius Ptolemy of Alexandria (A.D. second century) showing the sources of the Nile as lakes near the equator fed from a range of mountains ('Mountains of the Moon') was based on information collected on the East Coast by these traders. On the other hand, Crawford (1949) made a careful study of maps published in Europe between the fifteenth and seventeenth centuries, traced the pedigrees of names and consulted the available literary sources. His conclusion was that the myth of the lakes fed by the 'Mountains of the Moon' probably originated in Ethiopia but was applied to the Rwenzoris and the Great Lakes as soon as they were discovered. The fact that the interior of Eastern Tropical Africa was practically unknown in Europe until the late eighteenth century but that information about Ethiopia (Abyssinia) had been steadily accumulating since before the Christian Era, would seem to give some support to this idea. Here again, as on the Saharan routes and along the Guinea Coast, the men who transported the goods to the outside world did not themselves go all the way into the interior to fetch them, at least not until the early nineteenth century when Arab traders from Zanzibar penetrated inland as far as the eastern Congo in search of slaves and ivory. In other directions during the heyday of Islamic expansion (8th–11th centuries) the Arabs had direct trade and cultural relations with most of the world then known to them – China, India, South Russia and Europe (Lombard, 1975).

As a brief summary, it may be said that, though the archaeology of Africa is still in its infancy, we now have enough information, together with such written records as have survived, to give at least some of the answers to the question raised at the beginning of this chapter. There were, clearly, several reasons for the cultural separation of Europe and tropical Africa from about the fifth to the late eighteenth century, during which period Europe learnt very little indeed of the geography of the interior of the continent. We may reasonably conclude that the following factors among others contributed to this situation, though it must be added that opinions differ on their relative importance:

1. The progressive desiccation of the Sahara and the difficult obstacle presented by the swamps of the Upper Nile.
2. Though the Saharan and East Coast trade continued, and the West Coast trade was started during this period, there was no direct contact between the final customer and the initial producer in the interior.
3. When the Arabs became the principal middlemen for the Saharan trade, the economic rivalry and the religious antagonisms between Christendom and Islam virtually closed the door to cultural exchanges between Europe and Africa.
4. Arab scholars, who had previously played a great part in the early science and technology, did not contribute to or share in the Scientific Revolution of the seventeenth and eighteenth centuries.
5. The discovery of the New World in the fifteenth century began to deflect Europe's attention from Africa as a source of precious metals.
6. The disastrous expansion of the slave trade in West Africa by the Europeans and their African middlemen in the seventeenth and eighteenth centuries, and on the East Coast by the Arabs in the late eighteenth century, made effective cultural contacts still more difficult.
7. Diseases such as malaria and dysentery, against which Europeans had little resistance, the causes of which were unknown and for which there were no effective remedies. The reputation of parts of the West Coast as 'the white man's grave' was indeed well earned.

This situation began to change towards the end of the eighteenth century.

It is not easy to decide on the relative influence of the various motives behind European exploration in Africa during the nineteenth century. Economic profits, national and personal prestige, and the lure of adventure, had always played their part in such undertakings. In this respect, a distinction must be made between the persons or organisations sponsoring the expeditions and the explorers themselves who could never have endured the hardships without a consuming passion for discovery and adventure. But during the first half of the nineteenth century, it can truly be said, both humanitarian motives and scientific curiosity had a greater influence on the promotors of expeditions than at any previous period in the history of exploration. The former was marked by a great resurgence of missionary enterprise which had been much reduced since the early Portuguese days on the West Coast four hundred years previously. It was, of course, closely linked with the general revulsion against the excesses of the slave trade.

The rising prosperity of western Europe in the eighteenth century produced an educated class of people with the time and opportunities to indulge their curiosity about the world around them.* The outlook and techniques of the rapidly developing sciences began to be applied to geographical exploration, and a very important event was the founding in London in 1788 of the Association for the

---

* During the late eighteenth century, three freed West African slaves (Equiano, Sancho and Cugoane), all living in England, wrote autobiographical books in English, which were popular at the time, and which disclosed much about the country and the African character and life. They also included some shrewd observations on European life (Hallett, 1965a, 147–8; Edwards, 1967). The present spate of novels and poetry in French and English from the newly independent African states has had its forerunners.

Discovery of the Interior Parts of Africa, or 'African Association', by a group of men whose interests were partly scientific and humanitarian (Hallett, 1964). Since, however, its membership included some very influential people, notably the treasurer Sir Joseph Banks, they were able to obtain some financial support from the British Government. It was the Association that sponsored the four expeditions to the Niger which are discussed below. In 1831, it was merged into the newly formed Royal Geographical Society which supported the major expeditions to the Nile Basin in the middle of the century, as well as David Livingstone's Zambezi expedition of 1859–64. Missionary societies also began to support men who contributed much to the exploration of these regions. These facts have been mentioned primarily to emphasise that scientific and humanitarian motives were the main forces behind many of these ventures. It is true that the expansion of trade was also in mind, but even this was considered by some, notably by David Livingstone, chiefly as a means of benefiting the African peoples. Political motives were less important until after 1870, by which time the major geographical problems had been solved.

European ignorance of the geography of Africa at the beginning of the nineteenth century is well illustrated by the map (Fig. 2.1) published in London in 1821. Particularly striking is the contrast between the accuracy of the entire coastline and the small amount of information, much of it incorrect, relating to the interior. A large proportion of the tropical region is devoid even of tentative geographical suggestions. The Scottish doctor Mungo Park had recently, on two expeditions from the Gambia River in 1795 and 1805, sponsored by the African Association, reached the Upper Niger at Bamako and followed it for more than 2 000 km to Bussa. This finally disposed of the idea, got by the Portuguese from west-coast Africans in the fifteenth century, that it flowed westwards into the Atlantic.

The map in Fig. 2.1 does not represent all of the features that had actually been established at the time of publication, nor does it properly distinguish between the known and the conjectured, particularly with regard to the courses of the Niger and Nile. Mungo Park had already, on his second expedition in 1805, established that the Niger turns to the south about 300 km east of Timbuktu, which he did not attempt to visit. This did not, however, deter some influential geographers from continuing to favour a final junction with the Nile, and it is not surprising that the revolutionary notion of a separate outlet for the Niger to the south, though it had in fact been suggested some years before, was not even tentatively placed on this map. In the face of such exciting mysteries, it was inevitable that some less reputable theories were put forward by enthusiasts. One of these involved the Niger turning northwards and flowing under the desert to reach the Mediterranean. It is probable that the two lakes, depicted in Fig. 2.1 in the central Soudan on the courses of the Nile and Niger, were both based on rumours of the existence of Lake Chad, which had not yet been found.

A year after the publication of this map (1822) Laing located the sources of the Niger within less than 300 km of the Guinea Coast on the border of the present Sierra Leone. In the same year, Denham and Clapperton, members of the so-called 'Bornu Mission' from London, crossed the Sahara from Tripoli and reached Lake Chad in the following year and, after crossing the inflowing

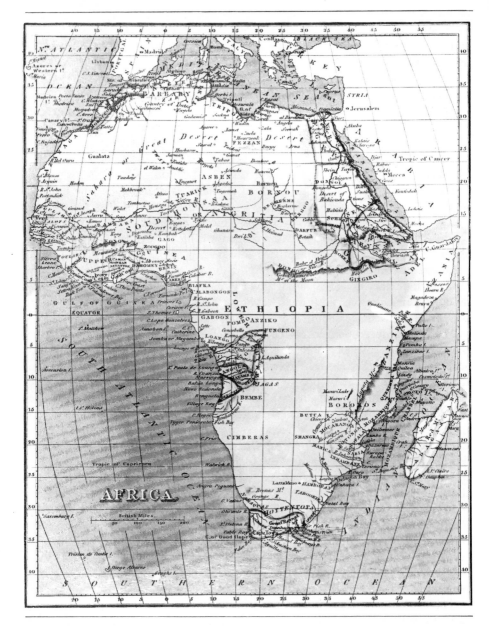

Fig. 2.1 Map of Africa published by Sherwood, Neely and Jones, London, 1821

Komodugu-Yobe and Shari Rivers, followed most of the northeast shore of the lake (see Fig. 12.1). They thus finally disposed of the idea that this lake was connected with either the Niger or the Nile. Even then, still less justification, some geographers continued to believe in a Niger–Nile connection. But a southern outlet into the Bight of Benin, previously a minority opinion (and not represented in Fig. 2.1), gained increased support and, with this in mind, Clapperton, with his servant Richard Lander, embarked on a second expedition in 1825, starting from Badagri on the Guinea Coast. This failed in its main objective and Clapperton died two years later at Sokoto – one of the many who lost their lives in pursuit of the Niger mystery.

In 1827, the Frenchman Caillié, disguised as an Arab, and starting from Sierra Leone, reached the Niger at Djenne and travelled 500 km by canoe to Timbuktu, where he spent some weeks. He then crossed the Sahara with a camel caravan to reach the Atlantic near Fez. This great feat, which was not supported by any official organisation, not only provided information on the swampy lake region of the Middle Niger, but marked the beginning of modern European exploration of the Sahara.

The subsequent major explorations of the central Sahara and its southern fringe, which have particular importance in relation of the history of the tropical African water systems, were made during the second half of the nineteenth century by the Germans Barth and Nachtigal. Barth crossed the desert from Tripoli and spent six years (1849–55) on the Middle Niger and the country around Lake Chad and discovered the River Benue, the main tributary of the Niger. Nachtigal (1870–74) visited the Tibesti Mountains and was the first European to cross the desert from Lake Chad to the Nile near Khartoum. From an introductory account of European exploration of the Sahara see Eydoux (1938).

The final episode was the most remarkable. Lander, who had had no education nor training suitable for an explorer, determined to finish his master's work and, surprisingly in those days of rigid social stratification, was officially encouraged, though very ungenerously financed, by the British Government (a suggestion to include a 'gentleman' on the expedition was fortunately not adopted). He and his brother, who had never previously left his native England, landed at Badagri on the Bight of Benin early in 1830 and, within the year, reached the mouth of the river, having descended by canoe about a thousand kilometres from Bussa, suffering extreme hardships, dangers and disease (Hallett, 1965b).

Thus ended the first major episode in the history of nineteenth-century European exploration of tropical Africa. 'The greatest problem that Africa had to offer the scientific world, around which bitter controversy had raged since Mungo Park's discovery of the Niger in 1796, had, after all those years of almost continuous endeavour and at immense cost in human lives, at last been solved' (Bovill, 1968b, p. 244). The sources of the Nile were surely an equally great and well-publicised problem at that time, but Bovill's book provides an excellent introduction to this story and to the original literature.

The explorations which contributed to the solution of the Nile problem are best studied from the books of the explorers themselves (for a general introduction see Moorhead, 1960, 1962). A few episodes of particular relevance to our subject will be mentioned here.

In 1848–49, the German missionaries Krapf and Rebmann, employed on the East Coast by the Church Missionary Society of London, first saw the snow-capped mountains Kenya and Kilimanjaro, though they were not at first believed. They also recorded local reports of an immense lake, 'Ujiji', in the interior. The Royal Geographical Society became interested and supported an expedition by Burton and Speke in 1857–58, the objects of which included an investigation of this lake and its relation with the Nile which was assumed by some to flow out of it. Lake Tanganyika was thus found in 1858. They did not actually visit the north end, but got local information that the river (Ruzizi) flowed into and not out of the lake. Stories of other large lakes were heard on the course of this journey, and Speke parted from Burton at Tabora and reached the southeast corner of Lake Victoria. His conviction that this was the main reservoir of the Nile was, at the time, premature and based on very little evidence, but it ultimately proved to be true. He returned with Grant in 1860, again with support from the Royal Geographical Society, and, in 1862, they found the very copious outflow from the lake at Jinja. They did not follow the river in its diversion to Lake Albert, but went due north to reach the White Nile at Gondokoro. Speke was, however, now well justified in his claim to have solved the problem. The missing link was soon provided by Baker who, coming from Egypt, met Speke and Grant at Gondokoro and heard from them a rumour of another large lake fed by the river from Lake Victoria. Lake Albert and its connections with both Lake Victoria and the White Nile were discovered by Baker in 1864.

David Livingstone's first expedition (1853–56) was from the south. He discovered much of the hitherto unknown upper Zambezi River, including the Victoria Falls, turned west across the southern edge of the Zaïre Basin to reach Loanda on the Atlantic. Returning by the same route, he followed the Zambezi to its mouth through country that had been occupied by the Portuguese for two centuries. His second expedition (1859–64) included Lake Nyasa (now Lake Malawi): the existence of this lake had been known to the Portuguese for more than a century (see Fig. 2.1), but its northern extent and most of its detailed features were first found by Livingstone. During his last expedition (1867–73), he travelled in the region of Lake Tanganyika and the upper south-east Zaïre Basin and discovered Lakes Mweru and Bangweulu. He later reached the Lualaba River from the north end of Lake Tanganyika and followed it to Nyangwe (Fig. 9.3). It was here that on 11 July 1871 he noted, 'I bought the different species of fish brought to the market in order to sketch eight of them, and compare them with those of the Nile lower down: most are the same as Nyasa'. He, at that time, thought that the Lualaba might flow into the Nile, and this very enlightened attempt to apply the principles of zoogeography to the Nile problem might have supported him (though falsely) in his belief, had he been armed with the knowledge of African fish accumulated during the following fifty years (Bòulenger, 1909–16) (see pp. 166–167).

It was at Nyangwe that Livingstone witnessed an appalling massacre by Arab traders, after which he struggled back in very bad health to Ujiji on Lake Tanganyika, where Stanley finally found him on that famous occasion on 10 November 1871. Though Speke and Grant had already found the outlet from Lake Victoria, which Livingstone was now prepared to believe was one of the sources of the White Nile, there were still considerable doubts expressed in Europe. Stanley had therefore been asked by the Royal Geographical Society finally to settle the question of an outlet from the north end of Lake Tanganyika which Burton, in spite of previous local information to the contrary, was again suggesting to be an affluent to the Nile. Stanley and Livingstone together visited the Ruzizi River and found that it flows into the lake.

After Livingstone's death near Lake Bangweulu in 1873, Stanley undertook a second expedition (1874–77) with the principal objects of solving the last remaining details concerning the course of the Upper Nile and of following the Zaïre from its eastern watershed to the Atlantic. In the course of this great expedition, he circumnavigated Lake Victoria by canoe, finally confirming Speke's finding of the Nile outlet, discovered Lake Edward and followed the Zaïre to the ocean.

No account of biological exploration in tropical Africa can omit a reference to that curious and eccentric character Emin Pasha, *alias* Eduard Schnitzer, a German medical doctor who adopted Islam and changed his name. He was eventually appointed by General Gordon, then Governor-General of the Egyptian Sudan, as Governor of the so-called Equatorial Province in the largely unknown region of the Upper Nile. This expansion of Egyptian influence under the Kedhive Ismail, and the contemporary exploits by Stanley on behalf of the Belgian King Leopold in the Congo, marked the beginnings of political penetration of tropical Africa, though in the Sudan one of the objects was suppression of the slave trade, in which, indeed, it was partially and temporarily successful. Among his many accomplishments, Emin was a good naturalist and, during his twelve years on the Nile, he worked with meticulous care in collecting and preparing biological specimens with detailed notes, which were sent to museums in Europe. He contributed much to our knowledge of the fauna and flora of the region. He also discovered the mouth of the Semliki River which brings the water of Lake Edward through the edge of the Ituri Forest into Lake Albert and, thus, into the Nile.

I shall not repeat the story of Emin's southward retreat in 1885 from the Sudanese revolutionary forces of the Mahdi, who finally exterminated Gordon in Khartoum, and his disappearance from the world for two years with his 10 000 followers in the then unknown region of the Albert Nile. It was, however, partly to find and rescue Emin from this predicament that Stanley undertook his final and most hazardous expedition from the mouth of the Zaïre to the East Coast. It was during the course of this journey that Stanley followed the course of the Semliki River between Lakes Albert and Edward and was fortunate enough to walk along the foothills of the Rwenzoris at a moment when the cloud cover had dispersed and the glaciers and snowfields were exposed to view. Whether this, the most extensive of the high mountain ranges of Africa, was the origin of the 'Mountains of the Moon' which had figured on maps for the past 1 700 years (Fig. 2.1) we shall never know. But the snowfields and heavy rains over the montane forests which drain into the Edward and Albert basins are a major source of water for the Upper Nile.

In 1888, the Hungarian Teleki with the Austrian von Höhnel reached the lake on the southern border of Ethiopia which they named Lake Rudolf (von Höhnel, 1894, 1938). This, the last of the great lakes to be found by Europeans, lies in a closed basin but, as we now know, previously drained to the north-west into the Nile (Ch. 10).

Thus, within less than a hundred years, after two thousand years of speculation, the georgraphy of the great rivers of tropical Africa – Nile, Niger, Zaïre and Zambezi, and the lakes associated with them, was finally exposed to the outside world. It is of interest to note that from Mungo Park's first expedition to the Niger in 1796 to Stanley's discovery of the Rwenzoris in 1889, the techniques of tropical exploration were progressively improved. From the unbelievably in-adequate equipment and funds provided for the original Niger ventures, Stan-ley's final expedition was equipped as competently and financed as lavishly as was possible at the time. Nevertheless, though more was achieved in a shorter time, with travelling on foot and without the protection of modern tropical medicine, the hardships and dangers were in no way reduced.

From 1890 onwards, exploration in tropical Africa began to assume a different complexion. The main geographical problems had been solved and there re-mained only details to be added. There was a progressive increase of interest in scientific exploration other than purely geographical, particularly in the fields of geology, biology, medicine and anthropology. Specialist scientific societies, which were mostly founded in Europe during the latter half of the nineteenth century, began to support specifically scientific expeditions or to finance scientists attached to expeditions planned for other purposes. The annexation as colonies of most of tropical Africa by European nations during the 1880s ultimately provided addi-tional and more practical incentives for scientific investigations. But for some time, the interest of European scientists was focused on tropical Africa primarily as an exciting virgin field for making important contributions to their subjects.

One of the first purely scientific expeditions was that of the Scottish geologist J. W. Gregory who, in 1893, worked in that section of the Eastern Rift Valley situated in what is now Kenya (also known as the Gregory Rift) as far north as Lake Baringo (Gregory, 1896). This pioneer work was the foundation of geologic-al investigations on the East African Rifts and was of obvious importance for the study of the Rift Valley lakes. J. E. S. Moore's expeditions to Lake Tanganyika in 1894 and 1897 (Moore, 1903) could be regarded as the first designed primarily to tackle a limnological problem: namely, the supposed marine origin of much of the fauna of the lake. This idea was inspired by specimens of the remarkable endemic fauna sent to Europe by Speke and subsequent travellers in that region. Though finally refuted by Cunnington (1920) during an expedition to the lake in 1904 conducted primarily to re-examine the fauna in the light of this theory, Moore's work stimulated subsequent studies in the African Great Lakes and could reasonably be regarded as the foundation of tropical African limnology.

Apart from the presentation of some general principles, the rest of this book is essentially a discussion of the research done since 1920. But the manner of con-ducting these researches, which has progressively changed during this period, will be discussed briefly as a finale to this chapter. Though the meaning of the word 'expedition', like 'safari', has been somewhat debased, it may be defined here as a short-term investigation in some rather remote place where facilities and equipment needed for the work are lacking and have to be transported or impro-vised on the spot. The early expeditions required little special equipment for making their observations and collecting specimens – collecting-gear, preserva-tives, portable meteorological instruments, etc. The increasing sophistication of limnological techniques during the 1930s involving chemical and physical measurements began to complicate the conduct of an expedition. Much of the chemical information at that time came from titrations, pH estimations by indica-tors, etc., made in a small, open boat, in a tent or back of a lorry, or even in the open air exposed to heat, wind and dust. Suitable boats, especially on the large and stormy lakes, were a more serious problem, though a surprising amount of the early work was done from canoes or very small open boats, or, as a last resort, by swimming with a plankton net or sample bottle in tow. The original Fisheries Survey of Lake Victoria in 1928 (Ch. 14) and the Belgian Hydrobiolog-ical Survey of Lake Tanganyika in 1946–47 (Ch. 16) were, however, done from steamboats lent by the respective colonial governments. Both of these lakes have been connected by railway with the East Coast since before the First World War. Motor transport has been available since 1920. A general account of the work and results of biological expeditions to Lakes Victoria, Albert, Rudolf and Edward between 1925 and 1931 is to be found in Worthington and Worthington (1933).

Apart, however, from the growing technical complications, it became clear that short-term expeditions, though still valuable for collecting preliminary or special information from remote regions, cannot provide the kind of knowledge now needed of the processes and cyclical (especially seasonal) changes at work in a water system, which relate to productivity. Prolonged investigations over several years, preferably from a fixed and adequately equipped base, are required. The colonial governments began at an early stage to provide facilities and to employ scientists to discover and research on local problems in medicine, agriculture, veterinary science and geology. The importance of water resources, except in relation to agriculture and stock-raising, was relatively slow in gaining recogni-tion. The Fisheries Departments, originally concerned mainly in regulating fishing and marketing, eventually began to employ experts to investigate the biol-ogy of the commercially important species. After the Second World War, this kind of research was much expanded and, with financial support from overseas, some organisations well provided with laboratories and equipment were set up to research, partly or wholly, on limnology and fisheries science. Most of these have been taken over and some have been augmented by the independent African Governments.

The principal expeditions from Europe, mainly from Belgium, Britain and France and more recently from U.S.A., and the local research institutes on which much of the recent research has been based, will be mentioned in later chapters.

The foundation of universities in the tropical African countries from about 1950 onwards has already had considerable impact on this as on other fields of research. In the long run, apart from their function in education and training, they will surely provide the main enduring focus for discussion, a stimulus for research and opportunities for the kind of interdisciplinary approach that most problems require for their solution. Several of the East and West African univer-sities are engaged in limnological research and in closely related fields. Their success in a very short time is perhaps best demonstrated by the reference list at the end of this book which includes the names of twelve African biologists who have contributed to this subject within the past fifteen years.

3

Geological and climatic history, present climates

A reconstruction of the remote past must necessarily be based upon inference, rarely from systematically collected data, more often from an inadequate number of facts which chance has placed in the way of competent investigators who can recognise their significance. As time goes on some gaps are filled, others are likely to remain for ever empty, but the picture as a whole becomes progressively clearer. The following outline sketch of the origin and history of the African inland waters is based upon what can reasonably be inferred from the facts so far available. This will be amplified in more detail in later chapters, which deal with the separate water systems, and where reference will be made to some of the original work upon which the ideas are based.

The main catchment areas and drainage channels of the African continent have existed for an immense length of time. Except in the extreme north-west (Atlas Mountains) and south (Cape Mountains) the continental crust has been little affected, at least for many millions of years, by the horizontal or orogenic compressions which have folded up the world's major mountain ranges (Himalayas, Alps, Andes and Rockies). Its surface has been shaped mainly by spasmodic earth movements (tectonics) in a vertical direction – uplifting, faulting, subsidence and volcanic outpourings, as well as by the continuous forces of erosion. The presence of marine sediments dating from the Cretaceous (some 100 million years ago) show that much of what is now the Sahara, Ethiopia and Somalia was then inundated by the sea. But the rest of the continent has remained above sea-level since the Precambrian, more than 600 million years ago, though large areas have from time to time been covered by lakes or swamps, most of which have now disappeared.

Uplifting and subsidence have moulded the surface to a pattern of large depressions separated by ridges (basins and swells), and it is this pattern which has determined the outlines of Africa's hydrology (Fig. 3.1). Most of the basins, which are the main catchment areas, now drain into the sea through gaps in their rims, via Zaïre, Niger, Nile, Zambezi and Orange Rivers. Some are 'closed' and have no outlet. Lake Chad, which during part of the Pleistocene was larger than any of the world's present existing lakes, has no surface outlet, but seeps into the

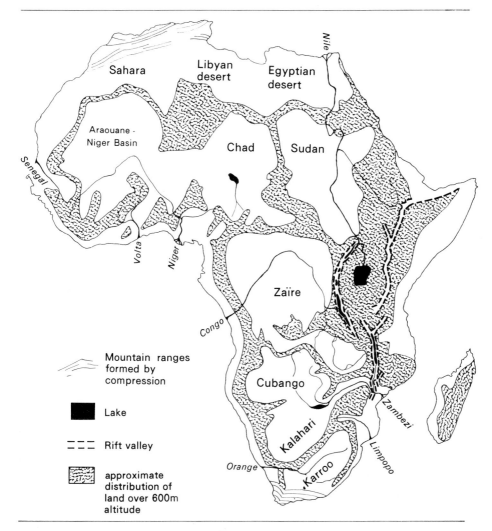

Fig. 3.1 Diagrammatic representation of the basin and swell structure of the African continent (redrawn from Holmes, 1965)

sandy soils to the northeast, rises to the surfaces by capillarity, evaporates and deposits salts. The lake thus remains fresh through the basin is 'closed'.

Though some of the northern edges of the Cubango and Kalahari basins in Angola and Botswana are drained by the Zambezi, the water of most of these areas finds no permanent outlet and collects in the swamps of the Okavango Delta and in the saltpans of the Makgadikgadi basin (p. 328).

With exceptionally heavy rains in the highlands of Angola these swamps occasionally overflow to the north by the Selinda Spillway into the Zambezi (Fig. 9.4). The same situation is to be found in the northwestern Sahara depression some of whose subterranean water from the central highlands (Ahaggar) collects and evaporates in the salt-pans (Chotts) at the foot of the eastern Atlas Mountains (Fig. 11.3).

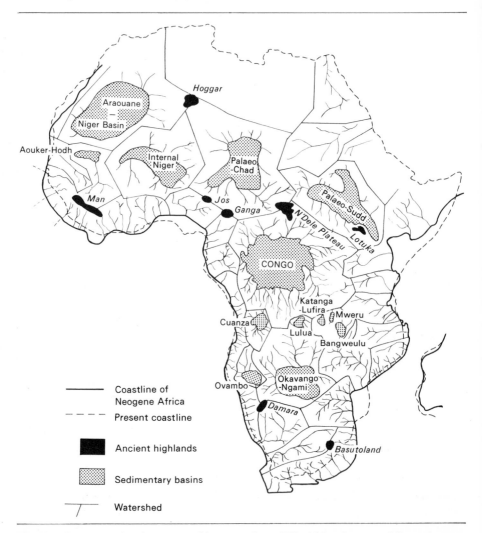

Fig. 3.2 A suggested palaeogeographic map of pre-Rift Africa (e.g. pre-Miocene) compiled by de Heinzelin, in Howell and Bourlière (eds), 1964, p. 650. The small details, as of the water courses, are guesswork, and there is no evidence to indicate the distribution of lakes or swamps within the basins

It seems that in the early Miocene the surface of the continent had been worn down by a long period of erosion with tectonic stability (Fig. 3.2). The watersheds between the basins were generally lower than now and there were probably more frequent faunal connections between the water systems (pp. 143–144).

This ancient pattern of drainage is still the basis of the hydrology of most of the continent, but in East and Central Africa it has been disrupted by dramatic earth movements, mainly since the Miocene, which have had profound hydrological, biological and human consequences.* A new complex of catchment basins

---

* The geological history of Southern Africa, including the Zambezi, is discussed by Wellington (1955).

and drainage channels, including all of the Great Lakes, except Lake Chad, has been superimposed upon the old pattern. The vertical upwarping has been more vigorous and has extended over a much greater area than elsewhere in Africa, and a huge tract of country from Eritrea to the Zambezi, some 500–800 km wide, has been raised a thousand or so metres since the Miocene. The centre of this great ridge in its East African section has sagged to form the enormous, though shallow, basin of Lake Victoria. The two edges have been raised still further and the consequent stretching of the crust has caused it to crack and a central strip to sink along the crests of these two ridges (Fig. 13.1). This is a much oversimplified account of the origin of the two Great Rift Valleys, and the nature of the forces involved has been a subject for lively controversy among geologists for many years. It was part of a complex of earth movements whose effects are apparent over a much wider area including the Red Sea trough and the valley of the Jordan (Holmes, 1965, p. 1059; McConnell, 1967).

The release of internal pressure by cracking of the earth's crust has been accompanied by earthquakes and volcanic eruptions. We can only imagine the intensity of the former, but the innumerable volcanic mountains, craters and lava fields in and near the two Rift Valleys must convince any but the most unobservant of the extreme violence of these forces, even in the very recent past.

The rifting was, in general, at right-angles to the previous east and west drainage from the high ridge of eastern Africa and the trenches so formed, whose floors are in places more than 1 000 m below the tops of the bounding walls, have collected a great amount of water. All of the Great Lakes of eastern Africa have originated in this way with the exception of Lake Victoria which occupies the shallow depression between the two Rifts (Fig. 3.4 and 15.1).

Some lakes have been formed as the direct result of volcanic activity, and many extinct volcanoes have craters whose rims are sufficiently intact to allow water to collect in them. There are several hundreds of such crater lakes in East Africa particularly along the Western Rift on the Zaïre-Uganda border (Fig. 13.1). They are mostly rather small, not exceeding about 2 km in diameter. Their catchment areas also, as would be expected, are usually small, and some have no surface outlet or develop one only during heavy rains. They tend therefore to be isolated from other water systems and to support a poor indigenous fauna. So-called explosion craters, which are common on the floor of the Western Rift, are little more than large shallow depressions on the flat plain and were due to short-lived and violent eruptions. Some of these are flooded with underground water or by surface streams, others form bays to lakes (e.g. Lakes Edward and George) whose waters have inundated them (Fig. 13.1).

In the Virunga volcanic region where Uganda, Zaïre and Rwanda meet (Fig. 15.1) several lakes were formed during the Pleistocene through the damming of river valleys by lava flows. The largest of these, Lake Kivu, is within the Western Rift and originated from the blockage and reversal of the previously north-flowing river system by the chain of Virunga volcanoes. There is evidence of a former smaller lake on the same site which must have drained to the north (p. 271). The valley has been flooded to form a beautiful branching fjora-like lake which overflows to the south into Lake Tanganyika. In the same volcanic region are a number of smaller lakes just outside the Rift which have been formed in the same manner. These include Lakes Bunyoni and Mutanda in Uganda and Bulera and Luhondo in Rwanda (15.1). Volcanic blockage during the Pleistocene was the origin of Lake Tsana, the natural reservoir of the Blue Nile in the highlands of Ethiopia (Fig. 11.3). Finally, there are several small alpine lakes at high altitudes, as on Rwenzori and Mount Kenya, which were formed by glacial excavation and moraine blockage at a time when the glaciers reached a lower level than at present.

It will be repeatedly emphasised in the following chapters that the predominant influence in the biological history of the inland waters of eastern tropical Africa was the appearance after the Miocene (about 20 million years ago) of a great range of lake habitats for a fauna that was previously mainly adapted to rivers. The large number of new species of fish and other animals that were evolved in response to these events will be discussed in later chapters. The history of the eastern African lakes and of their faunas was not one of steady progress in one direction. Spasmodic earth movements and volcanic eruptions, as well as fluctuations of climate during the Pleistocene, caused great changes in the area and volume of the lakes and rivers, some being temporarily reduced to swamps or even obliterated. There is evidence of two especially violent periods, one at the beginning and another at the end of the Pleistocene. Minor earth movements are still common and some faulting has actually been observed during the last fifty years. These signs and the contemporary volcanic activity of Nyamlagira and Nyiro Gongo in the Western and of Oldonyo Lengai in the Eastern Rift, show that complete stability has not yet been reached. The pattern of interconnections between the lakes has thus changed from time to time, and some have been, or still are, quite isolated such as Lake Tanganyika in the recent past or Lakes Turkana, Naivasha and Chad at the present time. Others such as Lake Victoria and Lake Edward are connected by rivers to other water systems, but are isolated by waterfalls and rapids which fish are unable to surmount.

The water courses in the rest of the continent have never been disrupted by earth movements comparable with those which have affected eastern Africa since the Miocene. Though the basic pattern has therefore been comparatively stable, there have been considerable fluctuations in the quantity of water within the basins and in the connections between them through changes of climate, especially during the Pleistocene. The most dramatic of these changes, disclosed by recent palaeontological and archaeological research, affected certain parts of the Sahara Desert which comparatively recently (20 000–5 000 years ago) had a higher rainfall and a much more abundant fauna and flora, and human population, than at present (Ch. 11). Such extreme fluctuations have on the whole been unfavourable to the evolution of rich endemic faunas such as we find in those Rift Valley lakes which have persisted for a long time as large and isolated water masses.

The Zaïre basin, containing the greatest expanse of humid tropical forest in Africa, presents a rather special situation. Of the ancient continental basins (Fig. 3.1) it has, as a whole, been least affected by earth movements or great climatic changes during the Pleistocene. Previous to this however, in the late Pliocene, there was a large lake in the centre of the basin which was later drained into the Atlantic through 'capture' (p. 135) by a coastal river (Cahen, 1954). Soil surveys have shown that there were subsequently at least two incursions into the central basin from the south of wind-blown sand, indicating temporary arid conditions. The last of these was 10 000–15 000 years ago, and the previously dry central area is now again covered with forest. It is evident, however, that the rainfall in peripheral regions of the catchment remained sufficiently high during the dry period in the centre to preserve the river and its main tributaries which must have traversed the arid region on the way to the coast. There have been some recent invasions of 'soudanian' fish (p. 144) from the northwest and west, but the Zaïre basin has been sufficiently isolated for a long enough time to evolve more endemic species than any other African water system (Ch. 9). This abundant speciation was presumably stimulated in part by the previous lacustrine history of the central basin.

There have been some differences of opinion regarding the relative importance of tectonics (earth movements) and of climatic changes in causing the great fluctuations in the amount and distribution of standing water that are known to have occurred, especially during the Pleistocene (the past two to three million years). The doubts mainly concern eastern tropical Africa where earth movements have been so extensive and continuous over this period. Evidence for the previous much higher water levels of many of the lakes and of the existence of lakes in regions where there are now none has been provided by raised beaches and sediments, some of which contain fossils of fish and other aquatic organisms. It must be admitted that beaches and sediments can be raised and lowered by earth movements and lakes can be drained by faulting and created by volcanic action without a change of climate. Nevertheless, the evidence is overwhelming that fluctuations of climate have in fact played an important part in the history of most of the tropical African lakes. They have obviously played the predominant part in the now arid regions of the western Sahara and in the Chad basin which were tectonically relatively stable during the Pleistocene.

Serious study of past climates in tropical Africa, as revealed in exposed alluvial and lake sediments and raised beaches with their fossils and human artifacts, was begun in the 1920s. This led to the formulation by L. S. B. Leakey of a climatic scheme for the Pleistocene in sub-Saharan Africa, based on work in East Africa, involving a series of alternating pluvial (wet) and interpluvial (dry) periods and these were related to glacial and interglacial periods in the northern hemisphere. There were certainly great climatic fluctuations during the Pleistocene, especially in rainfall, but more recent research has disclosed a number of difficulties in the way of finding a universal timetable of climatic events applicable to the whole of tropical Africa. A popular account of East African prehistory, based on the pluvial–interpluvial climate scheme, is to be found in Cole (1954).

It is, however, only the climatic events of the past 20 000 years or so (very late Pleistocene*) that can easily be related to the present condition of the African lakes and their faunas. The introduction during the late 1940s of methods for extracting cores from underwater sediments greatly expanded the scope of the investigations. Much of the original chemical and biological features are preserved, and sediments laid down at successive epochs are arranged in an orderly sequence and can be separated for analysis. Moreover, the invention of radiocar-

---

* I have adopted the attitude of Moreau (1966) in regarding the present as not sufficiently distinguished to be assigned to a separate 'Post-Pleistocene' or 'Holocene' period.

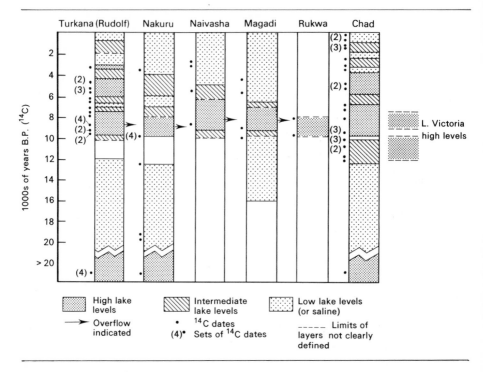

Fig. 3.3 Periods of high, medium and low water-levels during the past 20 000 years in six lakes having at present no surface outlets. Based on recent research on sediment cores with radiocarbon dating. Redrawn from Butzer *et al.* (1972), with additional data on Lake Victoria high levels from Kendall (1969) (B.P. = before present time)

bon (^{14}C) dating has provided an absolute time scale. Butzer (1972) includes an instructive and critical discussion of the techniques of investigation and of the interpretations that have been proposed from the evidence concerning Pleistocene environments, with a chapter (20) devoted to Africa in the late Pleistocene. By these techniques it is possible to gain some knowledge of past fluctuations in water composition and level, in productivity and in the general climatic conditions, from the chemical characteristics and remains of organisms (especially pollen grains and diatom skeletons) at different levels in the cores. Livingstone (1975) has reviewed the most important information got from the study of sediment cores in Africa relating to late Pleistocene climates. There are still large gaps to be filled and some interpretations are uncertain. Some special points will be made on the history of each of the large lakes in the appropriate chapters, but Fig. 3.3 is included here to summarise what these techniques have disclosed of the fluctuations of water levels during the past 20 000 years of some lakes that at present have no surface outlets. It is derived from Butzer *et al.* (1972) with some additional information concerning Lake Victoria from Kendall (1969, discussed in Ch. 14). A glance at Figs. 3.4 and 11.3 will show the immense distances that separate Lakes Chad, Turkana and Rukwa. It is therefore of great significance that all these lakes were at an exceptionally high level between 10 000 and 8 000

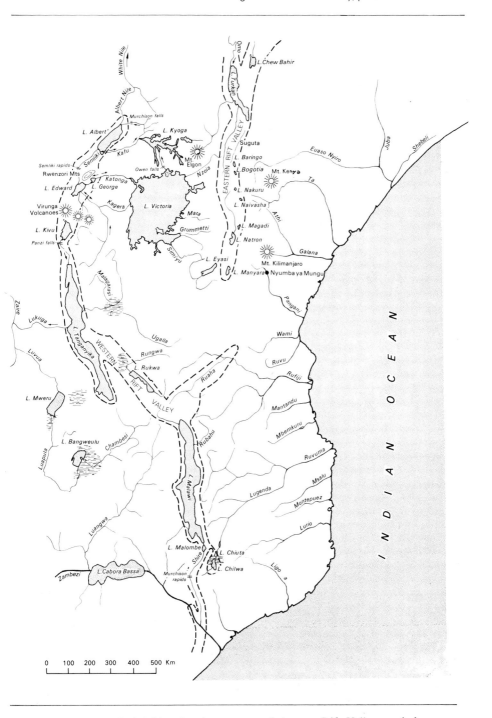

Fig. 3.4 Eastern tropical Africa showing courses of the two Rift Valleys and the water systems

years ago, which was also included in the period of the now extinct Saharan lakes (Ch. 11). The arrows in Fig. 3.3 indicate the points at which there is reasonable evidence of surface outflows. These would, of course, have been in operation for a considerable length of time, perhaps for a thousand years or more. During the past 8 000 years there have been several smaller oscillations of water level in most of the lakes, but with a general lowering towards present levels†. In relation to present limnology and to the distribution of freshwater faunas, especially fish in Tropical Africa, the apparently widespread more humid climate, with a great increase in the amount of standing water, around 8 000–10 000 years ago, followed by a general, though unsteady, decline up to the present time, was of great importance. The consequences will be discussed in future chapters.

It has not until recently been practically possible to raise cores from the African lake sediments long enough to represent more than about 20 000 years of deposition. It is however known that several of the lake basins contain very much deeper sediments dating back several million years. For example, drilling near the shore of Lake Albert showed that the sediments exceeded 1 200 m in depth (Harris et al., 1956), and seismic sounding in Lake Tanganyika disclosed at least 1 500 m of sediment (Degens et al., 1971a). These depths represent more than 10 million years of deposition. It is now technically possible, though very expensive, to raise cores from such depths of sediment. If it were done we should have a detailed and continuous record, reliably dated, both of the chemical and ecological history of the lakes and of the changing climatic conditions in the region since the Miocene. This would contribute to a better understanding of past world climates and would include the period of the evolution of Man, which is now thought to have gone through its critical stages on the African continent. We need to know what were the climatic and ecological conditions during that period.

The African Rift Lakes are much older than most of the north temperate lakes which are post-glacial in age. In the deepest, such as Lake Tanganyika, the sediments are not disturbed by strong water movements nor by benthic animals because of the permanently anoxic bottom water which has probably persisted for a very long time. Besides extending much further into the past, the sedimentary record is therefore likely to be more detailed than that from temperate lakes.

The amount of standing water in a region depends not only on the rate of surface and subsurface inflow and outflow, but also on the rate of evaporation. The water budget of a lake is clearly compounded of direct rainfall and inflow on the positive side and evaporation and outflow on the negative side. But the rate of evaporation is dependent both on the temperature and humidity of the atmosphere, as well as on the wind stress at the water surface. In principle, therefore, a general change in the mean temperature in the past could, by its effect on evaporation, have altered the amount of standing water without any change in rainfall. It has been estimated that the present rate of evaporation from Lake Victoria is about 1 700 mm per annum and is approximately equal to the annual rainfall direct onto the lake, which is about four-fifths of the total annual intake (Gill-

---

† See also Richardson (1969), Richardson and Richardson (1972), Grove and Goudie (1971), Livingstone (1975), Holdship (1976) and Harvey (1976) for discussions and data on changes in level of African lakes during the last 20 000 years.

man, 1933; Walker, 1956). A reduction in the rate of evaporation by a fall in temperature could therefore affect a most important item in the water budget.

Moreau (1963b; 1966, Ch. 3), in outlining the major climatic changes in Africa during the late Pleistocene, was primarily concerned with the fluctuations in the positions and areas of the main vegetation zones which are the most important determinants of the distribution of bird faunas. He summarised the considerable body of evidence to indicate a general lowering of temperature by about 5°C between about 30 000 and 8 000 years ago – a period contemporary with the end of the last European glaciation (Würm). This, through the consequent lowering of the rate of evaporation, might have contributed to the high lake levels 8 000–10 000 years ago (Fig. 3.3). The evidence comes from several directions (chiefly from sediment cores) and the following are examples of events that have been shown to have occurred during this period. The glaciers on the Rwenzoris extended to a maximum of about 2 000 m below the present level and their final recession started about 15 000 years ago (Nilsson, 1963; Livingstone, 1962). The vegetation belts in the Kenya Highlands were about 1 000 m lower than now (Bakker, 1964; Coetzee, 1964), and there was vegetation characteristic of a much cooler and wetter climate at Kalambo near the southern end of Lake Tanganyika (Clark, 1962).

Further evidence and other findings relating to the climate and vegetation of the Sudan during the past 20 000 years has been compiled by Wickens (1975). There is, however, little doubt that the high lake levels were caused more by increased rainfall than by the general drop in temperature. Grove and Pullan (1963) have calculated that, if the temperature of Lake Mega-Chad were 5°C lower than at present, evaporation from the twentyfold greater surface area than now could only have been balanced by an annual inflow from direct rainfall and rivers about sixteen times that at the present time. This calculation neglects the loss from an outflow which surely must have existed when the lake was at that level (Ch. 12).

It might well be asked whether such a general drop in temperature could have had a direct effect on the aquatic fauna and flora of tropical Africa, enough to influence their present distribution. As pointed out by Moreau (1966), a general lowering of temperature by 5°C would have resulted in a vastly greater area of cold montane forest than at present – an important factor in relation to the present distribution of terrestrial plants and animals at high altitudes. There would at the same time have been more water, chiefly streams and small lakes, subject to freezing, as well as very large areas of water with temperatures between 5° and 15°C. On the other hand, the temperature of the Great Lakes, including Lake Victoria which is higher than the others, would not have fallen below about 15°C.

We know very little for certain about the direct effect of temperature in determining the distribution of freshwater organisms because there are so many other factors to be taken into account. We would nevertheless suspect that though the present faunas of waters at mid-altitudes would probably be replaced by those now living at higher levels, it is unlikely that those in the Great Lakes would be subject to temperature conditions outside their normal range of tolerance. We have anyway to face the fact that the organisms living in the Great Lakes show no signs of having suffered from this cause during the late Pleistocene, though other aspects of the past climate, particularly rainfall, have certainly been involved.

In Table 3.1 is set out an approximate timetable for some of the important events since the early Miocene relating to the history of the Tropical African lakes and rivers. The table demonstrates in a striking manner how much of the present disposition and interrelations of the water systems has been determined by events during the past few thousand years.

Table 3.1(a)

Approximate dates. *Years* B.P.	*Geological periods*	
		Expanded in b.
35 000		
	PLEISTOCENE	Large volcanoes – Emi Kusi (Tibesti), Marsabit (N. Kenya). Virunga Volcanoes (Zaïre – Uganda border), final major eruptions. 'Semliki' lake in Albert and Edward basins.
		Start of upwarping of western side of future Victoria basin.
		'Kaiso' lake in Albert and Edward basins.
		Further rifting and formation of more Rift Lakes. Rwenzori rising (continued through remainder of Pleistocene). Drainage of Zaïre lake to west coast.
		Early Man.
2–3 million	PLIOCENE	Large isolated volcanoes – Kilimanjaro, Kenya, Elgon, Cameroon. Large lake in Zaïre basin.
12 million	MIOCENE	Begining of rifting and formation of Rift lake basins. Uplifting of land surface in E. and S.E. Africa by more than 1 000 m, with great extension of montane flora and fauna. Lake Karunga covering site of the N.E. corner of the future Lake Victoria and extending north-eastward.
25 million		

In considering the present climatic pattern of the African tropics, we must concentrate on those features that have most influence on the ecological conditions in inland waters.* They are as follows (starting with the most obvious):

1. Given a suitable geomorphological situation (i.e. a catchment area draining into a basin that holds water), the extent and indeed the very existence of standing water depends upon the relation between the rainfall and evaporation. The rate of evaporation depends on the temperature and thus the vapour-pressure, the humidity of the atmosphere and wind-stress at the surface. But the rate of evaporation per unit volume of a lake, and thus the rate at which it may alter the composition of the water, depends also on the relation between the surface area and the volume of the lake.

* For a short elementary introduction to tropical African climatology see Grove, 1970a, p. 12 ff., and Gourou, 1970, p. 30 ff; for more advanced discussion and extensive data see Thompson, 1965, and Griffiths, 1972.

Table 3.1 Some of the important events relating to the history of the African water systems from the Miocene onwards. The evidence for the approximate dating of some of the events is controversial and of varying reliability, and further research will certainly result in some modifications of the chronological details. For discussion of these and other events, and references, see this Chapter and those concerned with the different lakes and rivers.

(b)

1 000s years B.P.				
5– Temperature generally about 2°C higher than now.	Desiccation of Western Sahara	Lake Victoria falls in steps to present level		
	Level of Mega-Chad finally begins to fall		Lake Rudolf overflows into the Nile.	
	Upper Niger and Lake Araouane overflow to lower Niger.	Lake Victoria rising to 18 m above present level	High levels in Eastern Rift Lakes	
10–	Saharan Great Lakes, and faunal and floral exchanges across Western Sahara	Level of Lake Victoria below outlet		
Rising temperature				
15–	Max. extent of Lake Mega-Chad		Max. extension of glacier on Mountains Rwenzori and Kenya, and beginnings of retreat	
		Lake Victoria falls from previous high level, partly by cutting down of outlet		
20– Temperature generally about 5° lower than now	Lake Chad rising		Major eruptions of Virunga Volcanoes begin formation of present Lake Kivu and other volcanic barrier lakes. Lake Tanganyika rises to the Lukuga outlet	
25–		West-flowing rivers across site of future Victoria Basin reversed by continuous upwarping, the lake fills to a very high level, and eventually overflows to the north into the Nile Basin		
30–				

A major distinguishing feature between the climates of different regions of tropical Africa is in fact the state of the balance between rainfall and evaporation at different seasons and over the whole year. If evaporation is greatly in excess, as in the Sahara Desert and in much of the northern savanna belt (Soudan), standing waters are likely to be very shortlived, perhaps lasting only for a few days or even hours. At the other extreme, with rainfall much in excess through most of the year, as in the central Zaïre basin, standing water is a permanent feature of the region. The obvious point must, however, be made, that a lake or the lower reaches of a river are often subject to a very different climate from that in the upper regions of the catchment. Neither the Nile in Egypt nor Lakes Chad and Turkana would exist if their local very arid climates were spread over their entire catchments. The present distribution of permanent or near-permanent surface water over the continent is shown in Fig. 3.5.

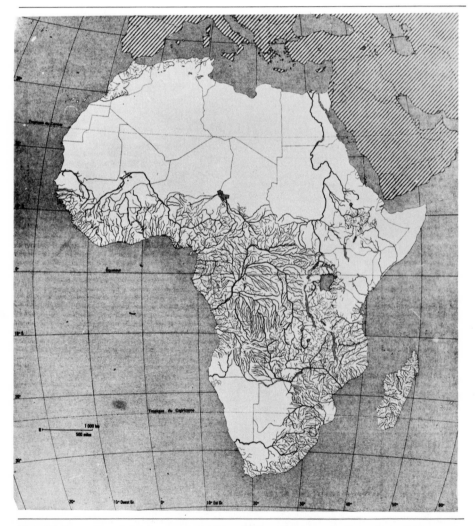

Fig. 3.5 Permanent inland waters of Africa. This omits very small patches of permanent standing water which are found even in the Sahara (from Gourou, 1970)

It is clear from the evidence of past climates that there have been great changes in many parts of tropical Africa in the relation between rainfall and evaporation during the Pleistocene (Table 3.1). An example is illustrated in Fig. 3.6 from the Eastern Rift Valley near Nairobi which shows the extent of the large lake during the late Pleistocene as judged by the distribution of lake sediments. Records of lake levels during the present century have shown that, in addition to seasonal fluctuations, there have been more extensive long-term changes in rainfall/evaporation in the catchments and these have had considerable ecological consequences. They are discussed in the chapters dealing with the separate water systems (see, for example, Fig. 14.3 for changes in level of Lake Victoria).

2. In temperate climates the mean atmospheric temperature and intensity of illumination are the major factors that determine the seasonal change in organic production both on land and in the water. Winter is, by definition, the period when both temperature and the intensity and duration of solar radiation are

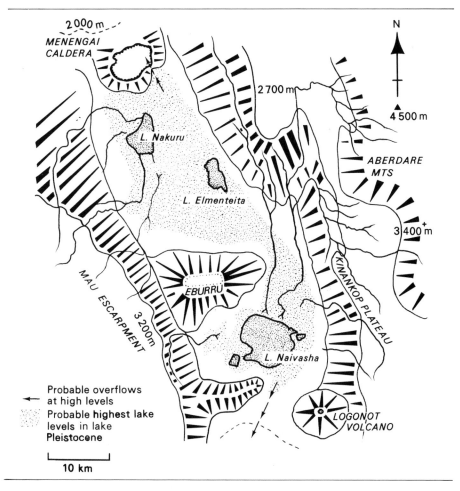

Fig. 3.6 Nakuru–Elmenteita and Naivasha basins in the Eastern Rift (Kenya) showing major relief features, escarpments, volcanics and former late Pleistocene lake levels (after Bishop, 1971)

too low to allow a net production of new organic matter, and life processes are reduced to a low level. In general, it may be said that in the tropics temperature and illumination are continuously adequate and that they do not control seasonal organic production in this manner. Towards the edges of the tropics, however, the seasonal temperature change may be enough to influence directly the rate of production and there is evidence of this in Lakes Chad and Kariba (Chs 12 and 20). The magnitude of the seasonal fluctuations in temperature and sunlight is, of course, a function of latitude. But temperature is also a function of altitude and, in the alpine zones at 4 000–5 000 m on the tropical African high mountains, the terrestrial and aquatic faunas and floras are limited to species that are adapted to low temperatures, even near freezing point. From these high regions down to altitudes of about 2 000 m, it is likely that the background temperature is still one of the factors that determine the type of fauna and flora. But at altitudes lower than 2 000 m, within the tropical belt, which includes all the Great Lakes, there is no evidence that differences in the general level of temperature are responsible for any of the differences that distinguish the biota of one water system from that of another.

The seasonal uniformity in mean monthly temperatures has, however, tended to distract attention from the diurnal changes which, in some parts of the tropics, are very large indeed, especially in dry regions where there is a great loss of heat by radiation at night. Small pools exposed to direct sunlight may reach a temperature of more than 50°C. The shallower the water the greater the diurnal swing in its temperature because a larger proportion of the whole is within the range of penetration of the radiation. The temperature at the surface of deeper water in calm weather can be very high in the afternoon with a consequent very steep vertical temperature gradient. Surface temperatures of 30–35°C, more than 10°C above that at the bottom (2·5 m) have been recorded in Lake George. Apart from the known inhibitory effects on photosynthesis of the high intensity of radiation just below the water surface (Ch. 7), it is possible that this and the associated high temperature may exceed the optimum for some planktonic animals. But in general in tropical lakes and in densely shaded shallow waters such as papyrus swamps temperature is continuously favourable for organic production.

3. The general rate of organic production and its seasonal fluctuations in temperate lakes is determined not only by the direct effects of temperature and illumination on photosynthesis and growth. There are other more subtle and less direct ways in which climate plays a part, and in the tropics these are of special importance. Solar radiation is absorbed by water so that primary (photosynthetic) production of living matter in a lake is confined to a layer near the surface (euphotic zone) to which radiation of the required wavelength and of adequate intensity for photosynthetis can penetrate. On the other hand, the bodies of the organisms ultimately tend to sink and to decompose in the water below the euphotic zone. If the nutrient substances so released were not returned to the surface, production would be greatly reduced and most of the nutrients would be trapped in the lower water and bottom sediments. Since temperature and illumination are, in general, optimal in tropical lakes it is clear that the supply of nutrients in the euphotic zone is a major factor con-

trolling production. Hence the great importance of those features of the climate that influence the vertical movements of water masses in lakes (Ch. 6). The situation is, of course, different in flowing rivers which are continuously stirred, but in very slow-flowing reaches, or in stagnant pools in a river bed during the dry season, conditions in this respect can resemble those in a lake.

As a background to a discussion of water circulation in lakes, we need to know the geographical and seasonal distribution in tropical Africa of atmospheric temperature, humidity and winds. The vertical circulation of water is resisted by density gradients induced mainly by heating at the surface. The water column is made less stable by a reduction of the density gradient through cooling at the surface following a fall in atmospheric temperature. This cooling is sometimes accelerated by evaporation into dry air, especially when augmented by wind. The main source of the energy that moves the water masses is, in fact, wind. The various kinds of wind-induced water movements in lakes and their importance in the circulation of nutrients are discussed in Ch. 6.

In Fig. 3.7 are set out some curves to represent mean monthly rainfall, max-

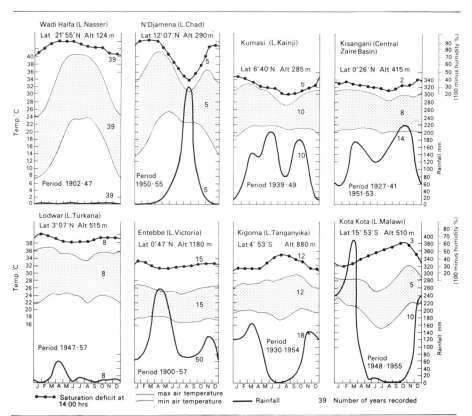

Fig. 3.7 Mean monthly rainfall, maximum and minimum temperature and saturation deficit (100 – humidity %) from eight meteorological stations in tropical Africa. The nearest water-system discussed in this text is shown in parentheses. Drawn from data in *Tables of temperature, humidity and precipitation for the World* with the permission of the Controller of Her Majesty's Stationary Office

imum and minimum air temperatures, and atmospheric saturation deficit at 14·00 hours.* They are plotted from data obtained over a varying number of years at stations selected for their proximity to some of the major water systems discussed in this book. The temperature and humidity measurements are made in the standard screen above the ground, but it is the conditions at the air–water interface that are significant in the present connection. Consequently, the absolute values are of little relevance; attention should be given to the relative levels and to the differences between the seasonal patterns of climate. It must also be borne in mind that each diagram represents conditions at one point in a very large catchment. Those stations that are close to the shore of some lakes, e.g. Kigoma (Lake Tanganyika) and Kota Kota (Lake Malawi), may be taken to indicate the type of climate over much of the lake. But records from the several meteorological stations around Lake Victoria show that its climate is by no means uniform. From Entebbe in the northwest corner the lake spreads into a progressively drier climatic region to the southeast, which contributes to the ecological diversity of this enormous lake (Ch. 14). The climate of N'Djamena represents that of the southern end of Lake Chad. From south to north of the lake, which extends towards the southern edge of the Sahara, the rainfall decreases and the saturation deficit and rate of evaporation increase (Ch. 12).

The general temperature level at each station is influenced by altitude as well as by latitude. The rainfall in some regions is determined by the topography of the surrounding country. For example, the plains between Lakes Edward and George on the floor of the Western Rift have a relatively low rainfall because they lie in the 'rain-shadow' of the Rwenzori Mountains which trap much of the rain coming from the west across the Ituri Forest (Ch. 13).

The value of the information set out in diagrams such as those in Fig. 3.7 is limited also because, being based on monthly mean figures, it gives no indication of the occasional wide departures from those means, which may have considerable ecological consequences. A good example is the long-term climatic fluctuations that are reflected in the changes in level of Lake Victoria during the past century (Fig. 14.3).

Nevertheless, the diagrams illustrate two very important climatic features of the regions, namely, the general level and the pattern of seasonal fluctuations in rainfall, in air temperature and in the saturation deficit of the atmosphere. These features are more relevant to the events in inland waters than any absolute figures obtainable from the meteorological records.

The characteristic two rainfall peaks in the equatorial region and the corresponding changes in the saturation deficit of the atmosphere, due to the back and forth passage of the sun across the tropics, can be clearly seen in the diagrams for Entebbe, Kigoma, Kisangani and Kumasi. The two extreme positions of the equatorial rainbelt are shown in Fig. 3.8, mostly south of the Equator in January and north in July. The single-peak rainfall in the northern and southern tropics is shown by N'Djamena (August) and Kota Kota (March) respectively.

* Measured rates of evaporation from a standard water surface (potential evaporation) are apparently available from only a few stations. Atmospheric saturation deficit (100 minus humidity %) is a factor which, together with the frequency and speed of the winds, determines the rate of evaporation from a water surface. It represents the 'dryness' of the air, which tends to be maximal around 14·00 hours in the afternoon.

Wadi Halfa has an extreme desert climate with its negligible rainfall and very high maximum temperatures and wide seasonal and diurnal swing in temperature. These features are due to the very clear and dry atmosphere which offers little obstruction to solar radiation reaching the ground by day and escaping upwards at night.

Attention should be given to the two stations situated in the humid forests of western equatorial Africa. Kisangani, in the Middle Zaïre basin, has a typical tropical forest climate – an almost continuous rainfall with two well-marked peaks, long-continued cloud cover, and a relatively constant temperature and saturation deficit. Compare these conditions with those at Kumasi situated in the strip of forest inland from the Guinea Coast. In January–February the rainfall drops to a low level, the maximum temperature is high, the difference between maximum and minimum is at its greatest, and the saturation deficit is at its maximum. These peculiar features are due to the position of the Guinea forests relative to the Sahara Desert. December–February is the period during which the northeast 'harmattan' wind blows across this region from the desert via the savanna belt (Fig. 3.8). For a short period each year, these forests are therefore subjected to a rather diluted desert climate – dry and with clear skies which allow a greater gain of heat during the day and loss at night, i.e. an increased difference between maximum and minimum temperatures. The considerable inter-peak fall in rainfall during August is however accompanied by a *drop* in maximum temperature and a reduction in the difference between maximum and minimum, and the saturation deficit falls to the annual minimum. The explanation is that this is the time of the humid southwesterly winds from the Atlantic (Fig. 3.8), with consequent continuous and heavy cloud-cover, which reduces the daytime rise of temperature and brings the saturation deficit to a low level. The effects of these conditions on the hydrology of the Volta Lake and Lake Bosumtwi are discussed later (pp. 93 and 378).

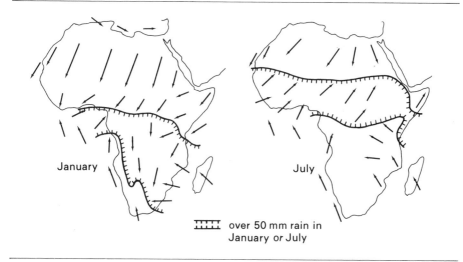

January   July

⫴ over 50 mm rain in January or July

Fig. 3.8 Seasonal changes in the prevailing winds at ground level and in the position of the tropical rainbelt (redrawn from Grove, 1970a)

The principal types of climate and their distribution (Fig. 3.7) are determined mainly by the horizontal and vertical movements of air whose force and direction at different levels and seasons are caused by differences of temperature and density and are modified by the distribution of sea and land, by differences of altitude and by the rotation of the earth. For the present purpose we are particularly interested in the winds at ground level because the circulation of water masses, upon which organic productivity largely depends, is caused mainly by wind-stress at the water surface.

The seasonal changes in direction of the prevailing winds over the African continent are represented in Fig. 3.8, together with the associated shifts of the tropical rainbelt. The two extreme conditions are shown in January when the sun is vertically above the southern tropic, and in July when in its corresponding northern position. These changes are especially important for the seasonal regimes of production in the large and deep lakes and are discussed in Ch. 6. The lakes (except Chad) are all situated in eastern tropical Africa where the prevailing northeast winds at the turn of the year are rather spasmodic and in general not as strong or continuous as the southeast trade winds during June–August. There is therefore a tendency for increased deep stirring of the water and a consequent higher production during the middle of the year.

These are, however, general overall features that are superimposed with varying effect on local climatic conditions. For example, Lake Victoria is large enough to cause regular diurnal land and sea breezes across its coasts, especially noticeable during otherwise calmer periods in the year and in the absence of local storms. Heating of the land during the day induces an often quite powerful wind from the lake during the afternoon and, following an evening calm, a wind may blow from the land during the second half of the night because more heat has been lost from the land than from the water. These breezes must make some contribution to the productivity of the shallow inshore waters by maintaining vertical circulation during otherwise calm periods. The trench-like Western Rift Valley, in which Lakes Tanganyika and Malawi lie, is orientated northwest–southeast. This fortuitous circumstance enhances the effect of the southeast trade winds which are funnelled down the valley, often with great force.

4. Lastly, as will frequently be mentioned in subsequent chapters, sudden violent storms with heavy wind and rain are a characteristic feature of the tropical climate. They are most frequent towards the end of the low-rainfall season. They are shortlived and, though moving, are usually of small area, sometimes only a few km^2. If one happens to hit a lake, the ecological consequences may be considerable. In large lakes the effects will be localised, but in small ones they can be catastrophic. They provide an important, but unpredictable, element in the local climate, and one which will be discussed in more detail in later chapters.

4

Aquatic ecosystems: Productivity

As a background to what follows and to provide a general introduction to the study of organic production in the inland waters of tropical Africa, a discussion of some basic principles is presented in this chapter.

Ecology is the study of the interrelations between living organisms and between them and their inanimate environment. The ultimate object of such studies is, on the one hand, to understand the factors upon which the continued existence of species in nature depends and to find the causes of fluctuations in numbers and in some cases of temporary or permanent extinction. On the other hand, we hope to make some practical use of this knowledge in harvesting species of organisms for food and other purposes in such a manner that the stocks are not permanently diminished, and to control certain species that are especially harmful to our interests, such as some parasites and their vectors or competing species that are of no immediate value. But we are now beginning to realise the urgent necessity for preserving the balance of nature, upon which for many reasons our own survival may well, in the long run, depend. Though we are far from fully understanding the nature of this balance and our position in it, we do know that the wholesale destruction of some natural habitats may be irrevocable. Man is an important object for ecological study.

The popular classification of the natural environment into recognisable types of situation – sea, forest, savanna, desert, river, lake, swamp, etc. – is, in general, accepted by ecologists as a basis for their investigations. A situation, such as the alpine region of one of the African high mountains or Lake Victoria, has recognisable physical and biological features, and at least roughly discernible boundaries. The term *ecosystem* is now used for such a situation. This implies that it is something more than a characteristic assemblage of organisms living together under certain conditions. It emphasises a dynamic aspect. An ecosystem is the site of a complex pattern or system of processes, the basic feature of which is the circulation of the incoming energy and materials between the members of a more or less localised community of organisms and their immediate environment. (For an introduction to the study of energy circulation in ecosystems, see Phillipson, 1966, and Mann, 1970.) It is obvious, however, that no ecosystem is strictly definable in space (its boundaries are often blurred and fluctuating) and that there is always some exchange of energy and materials, and even of organisms, with the surroundings. Some ecosystems, such as a hypersaline pool in the desert, are more self-contained than others, such as a lake fed from inflows draining heavily forested country, but none are completely closed. Yet it is by virtue of a certain tendency to overall stability that we are justified in recognising 'ecosystem' as a useful concept.

Apart from estimations of the fishery resources of some rivers, almost all productivity studies of inland waters in tropical Africa have so far been done on lakes. It would therefore be appropriate to discuss a tropical lake as an example of an ecosystem in order to make these generalities more meaningful, and to introduce the notion of *productivity*. The main routes by which energy enters a lake and is circulated and finally dissipated are shown in Fig. 4.1. The initial trapping of solar energy by green plants, and the photosynthesis of the organic compounds in which their energy is stored, is known as *primary production*. There are three obvious sites at which this occurs – in the open water by the microscopic algae or *phytoplankton*, in the shallow water (usually up to about 5 m deep) by the submerged waterweeds such as *Ceratophyllum*, and in the very shallow water by the emergent plants, such as reeds and papyrus. The green photosynthesising portions of the emergent vegetation are in the air, but ultimately fall into the water, and their storage organs are submerged and contribute to the matter and energy in the lake. Less obvious but in some shallow waters very important sites of primary production are the surface of rocks, stones, sand, and the stems of aquatic plants (in fact, any submerged surfaces exposed to sunlight) which are encrusted with a dense growth of algae. Another source of organic matter is that synthesised by land plants within the catchment area and transported to the lake by the inflows as fragments of plants and animals or their excreta. Still less obvious is the influx of organic matter from the atmosphere as small airborne particles such as seeds, pollen grains and fragments of disintegrated plants and animals, or even whole animals such as insects. This last source is sometimes very important. In Lake Bosumtwi in Ghana (Fig. 11.3) there is a fish *Epiplatys chaperi* (Cyprinodontidae) that feeds almost exclusively on ants that fall into the water (Whyte, 1975). Another species of the same genus (*E. sexfasciatus*) in the West Cameroon crater Lake Barombi Mbo feeds largely on terrestrial insects as well as on emerging lake flies (Trewavas *et al*, 1972). From measurements made on certain lakes in the U.S.A., Gasith and Hasler (1976) have concluded that airborne particles can provide organic matter of carbon content greater than 10% of that synthesised within the lake. The relative amounts of the contributions from these sources vary from one lake to another according to the situation, climate, etc. It is usual to refer to the organic matter synthesised within the ecosystem as *autochthonous* and to that from outside sources as *allochthonous*.

Since the existence of an ecosystem, such as that of a lake, depends upon a continuous supply of energy contained in the organic matter synthesised by plants, it is important to understand the conditions required for a maximum rate of primary production. The mechanism and basic requirements for photosynthesis in all plants can be studied from any textbook of plant physiology (see also Fogg, 1968). We are here concerned with the manner in which these requirements are provided for plants in inland waters and especially for the phytoplankton in tropical lakes. Water is clearly no problem and carbon dioxide is usually abundant as dissolved CO_2 and bicarbonate, though the conditions under which it is available to aquatic plants are discussed in Ch. 7, which is mainly devoted to the measurement and assessment of the rate of primary production by the phytoplankton in lakes and to the factors which actually control it. Apart from the unavoidable restriction of sunlight to the daytime, solar energy is, in general, continuously adequate in the tropics and is not subject to seasonal reduction to a level insufficient for plant growth as in temperate climates in winter. The same can usually be said of the temperature level, though in lakes situated near the limits of the tropics, such as Lake Chad, there is a sufficient winter fall in temperature to retard significantly the general rate of production (Ch. 12).

In addition to these basic requirements, plants are dependent upon an adequate supply of what are by convention called *nutrients* (not represented in Fig. 4.1). They do not themselves provide available energy, but are essential for plant growth. From nitrates, phosphates and sulphates dissolved in the water, the plants incorporate N, P and S into their proteins and other organic substances. The phosphate bonds in adenosine triphosphate (ATP) store the energy immediately available for use in most living organisms though the basic stores are in carbohydrate and fat. Silica, in the form of dissolved silicate, is essential for the skeletons of diatoms which are usually an important component of the phytoplankton and for some of the larger aquatic plants including papyrus. In the present connection, we are interested in the manner in which these substances become available to the phytoplankton. In most large lakes the important immediate source of these substances is from *recycling*. Through excretion of the products of metabolism by both plants and animals, and from the decomposition of dead organisms, simple organic and inorganic substances are continuously being returned to the water and used again by the phytoplankton. The importance to the nutrient cycle of the excretion of soluble organic substances by living planktonic plants and animals (so-called 'extracellular products') has been discovered and investigated quite recently (for reviews see Lund, 1965; Fogg, 1971). To a great extent these processes take place in the upper illuminated water where the algae can immediately absorb the organic substances as well as inorganic nutrients from both living and dead organisms and incorporate them into their own living matter. But, since dead and decomposing organisms tend to sink in water and to form an organic sediment, it is in the bottom water and in the mud that most of the nutrients are released through decomposition. The phytoplankton, however, can make use of them only in the well-illuminated water near the surface. For this reason the movements of water masses, especially in a vertical direction, are of extreme importance for the recycling of nutrients and can have a very great influence on primary production in lakes. The movements of water masses and the factors which control them are discussed in Ch. 6.

The meaning of the word 'nutrient' having been defined in the above manner it would be reasonable to include inorganic ions such as Na, K, Ca and Mg, as well as the organic extracellular products mentioned above, many of which are taken up and used by aquatic plants. But compounds of N, P, S and Si (nitrates, phosphates, sulphates and silicates) are of particular interest because in

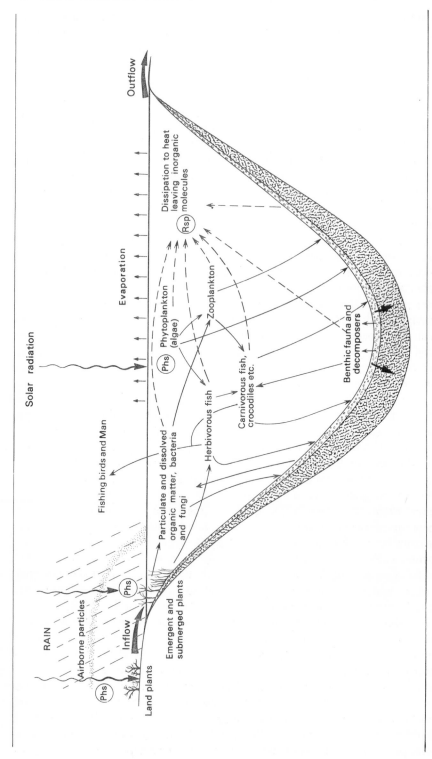

Fig. 4.1 The circulation of organic matter and energy in a tropical lake. Dissolved plant nutrients are not included. Phs: photosynthesis, Rsp: respiration

some situations, and at certain times, the rate of primary production may be retarded by scarcity of one of these because they have been consumed by the plants themselves. In this book it is these substances that are implied by 'nutrient' unless otherwise stated.

But however efficient the recycling of nutrients (N, P, S, and Si) the supply obviously cannot be maintained indefinitely without an adequate influx from outside the system. There is in fact a continuous loss in outflows, by permanent deposition in deep sediments, through removal of organisms, particularly fish, by men and other terrestrial predators, and by other routes. Sulphur may be lost as gaseous H_2S under reducing conditions, as in swamps (Ch. 18).

The ultimate source of most of the nutrients, like that of the inorganic ions (Ch. 5) is through leaching of rocks and soils by rainwater. The composition of the descending waters, both surface and subterranean, thus depends primarily on that of the rocks in the catchment and may be altered by vegetation (live or dead) through which they may flow and by herbivorous animals and their excreta. Prolonged leaching depletes the soluble minerals at a rate depending on the nature of the rocks and rainfall. The resulting erosion may expose other kinds of rocks to leaching, which may also result from earth movements and volcanic action. So the composition of the water flowing into a lake cannot be expected to remain constant for great lengths of time. Indeed in the tropics, where large seasonal fluctuations in rainfall are a characteristic feature of the climate, both the amount and the composition of the drainage waters are subject to large periodic changes during the year. Some lakes have very little or even no inflow during the dry season but receive masses of mineral and nutrient-rich water during the rains. Lake Chad is an outstanding example (Ch. 12). Rain falling direct into lakes may contribute significant quantities of minerals and nutrients (Table 5.1). The main source in rainwater is atmospheric dust and possibly the oxidation of nitrogen gas to nitrate by photochemical action or even by lighting discharges (Visser, 1964b, 1974). In coastal regions winds transport sea-salt to inland waters with detectable effects on their composition. Dry dust particles and several kinds of small organisms (including insects) fall into lakes and rivers, and in some regions probably make an important contribution to the mineral and nutrient content of the water.

The artificial and often excessive influx of nutrients with consequent pollution is discussed on p. 57.

Another natural source of nutrient is the elemental nitrogen in the air which is now known to contribute more to inland waters than was previously thought. The fixation of atmospheric nitrogen through reduction to ammonia by certain bacteria in the root nodules of leguminous plants has long been well known as a source of nitrogenous nutrient for land plants. This process requires energy, but it is not necessarily linked directly with photosynthesis (Fogg, 1959; 1968, pp. 77–8). It has now been found that some aquatic blue-green algae, e.g. species of *Anabaena*, and some bacteria fix atmospheric nitrogen which is thus brought into the nutrient cycle of lakes (reviewed by Fogg *et al.*, 1973). Measurements on Lake George in Uganda suggest that nitrogen-fixation by both blue-green algae and bacteria makes a significant contribution (Horne and Viner, 1971). The importance of this relative to other sources has yet to be determined. Many shallow lakes, some with very saline and alkaline waters, have a phytoplankton dominated by blue-green algae, often in enormous numbers. Under those conditions nitrogen fixation is perhaps a major source of compounds of nitrogen.

In recent years it has become increasingly clear that most natural waters, even the apparently cleanest, contain a significant amount of dissolved organic matter which is recycled through the system before it is fully decomposed. It has been shown that, apart from normal excretion, organic compounds such as polysaccharides, polypeptides, amino acids, glycollic acid, and even active enzymes diffuse out from living algae and zooplankton ('extra cellular products') as well as from the decomposing bodies of dead organisms, and some are directly assimilated by both phyto- and zooplankton (Lund, 1965; Fogg, 1971; Jørgensen, 1976). Even in the deep sea, far from land, there is a surprising amount of dissolved organic matter together with organic particles from partially decomposed organisms. This forms a reservoir of food maintained in dynamic equilibrium between production and consumption amounting to 25–60% of the annual primary production by the marine phytoplankton (Parsons and Seki, 1970). How much larger must be the organic reservoir in the generally shallower *inland* waters where a greater proportion of the bottom sediments are stirred into the water and much organic matter is washed in from the land. This reservoir perhaps provides a 'buffer' against large seasonal and other fluctuations in the rate of primary production.

The relative amounts of organic matter that are photosynthesised within and outside the system vary according to circumstances. In dense swamps, for example, external (allochthonous) production is practically the only source. In assessing the productivity of a water the relative importance of these two sources must be considered.

Why, in contrast to the major ions, is the quantity of available 'nutrients' often sufficiently reduced to retard production, though both are taken up by aquatic plants and the ions by animals for whose existence they are in fact essential? Two main reasons could be suggested,

(1) the range of concentrations of the nutrients found in lake waters is very much lower than that of the major ions. A reasonable basis of quantitative comparison is not easy to decide, but as an indication of the differences Table 4.1 gives the ranges of concentration in African lakes as milliequivalents per litre of nitrate (NO_3^-), orthophosphate (PO_4^{3-}), sodium (Na^+) and potassium (K^+). Seasonal changes in the concentrations of the major ions, due to the activities of organisms, of the same order of magnitude as those often shown by the nutrients would be too small a percentage of the total concentration to be detected by analysis. Such periodic changes in ion content, if they occur, apparently fail to reach a limiting level.

Table 4.1 Concentration ranges of some nutrients compared with those of some major ions measured in African lake waters, in milliequivalents per litre (< indicates that in some samples the nutrients were indetectable).

NO_3^-	<0·0007	–	0·0035
PO_4^{--}	<0·0001	–	0·02
Na^+	1.3	–	22000
K^+	0.05	–	1900

Calculated from the data collected by Talling and Talling (1965)

(2) The major ions in the bodies of aquatic organisms are on the whole freely exchangeable with those in the water, and in fact are continually being actively absorbed during life to compensate for perpetual loss by diffusion. Immediately after death this leakage continues more rapidly and unchecked. On the other hand, though some organic compounds containing N and P are released and reabsorbed by living organisms ('extracellular products',), many of the nutrients are built more firmly into the molecular structure of the organisms, and are not released until final decomposition and then mainly in the lower water and sediments. They are recycled only when conditions favour vertical movements of water (Ch. 6) and some are permanently trapped in the lower sediments. Though a complete answer to this question requires more knowledge than we possess regarding rates of exchange of these substances between the organisms and the water, the above general statements are likely to remain valid.

This brings us to the idea of a *limiting factor*, a term commonly used by ecologists. It is evident that, if any one of the requirements is deficient, then the growth and reproduction of the organisms under consideration will be depressed and in extreme cases may be completely extinguished. The requirement has then become a limiting factor and is putting a brake on production because it is in short supply. The low temperature and low intensity and short duration of sunlight during the winter in north temperate regions can reduce the rate of production in lakes to a very low level. In the tropics we should not expect temperature and sunlight to become limiting, in a very direct manner, but, as will be shown in subsequent chapters, occasional and sometimes seasonal lowering of the rate of production in lakes is usually caused by deficiency of a nutrient because at certain times it is not sufficiently rapidly recycled to sustain the full growth of the phytoplankton. The deficient nutrient is, in principle, identified by finding which of the suspected substances will stimulate photosynthesis of the algae when added to a sample of lake water. Using cultures of the alga *Selenastrum capricornutum* Roberts and Southall (1977) concluded that in seven manmade lakes in Zimbabwe (including Kariba) the main limiting nutrient was phosphate. The concentration of a nutrient in the water at any moment is the result of a balance between the rates of supply and consumption. It is important to realise that the actual concentration measured in the water at a given moment bears no necessary relation to the rate of growth of the plants. A very high rate of growth can be balanced by a correspondingly high rate of recycling so that very little of the nutrient accumulates in the water. Very low or even indetectable quantities of nitrate and phosphate in lakes are commonly associated with a very high rate of primary production. This has been made evident from experiments with added radioisotopes, e.g. phosphate containing ^{32}P, whereby the rate of uptake can be estimated. The size of the balance in the current account at the bank is no necessary indication either of the wealth or of the rate of spending of the customer.

Some of the energy fixed by plants in organic compounds is required for their own immediate metabolism. This is got by oxidising some of the material which they have synthesised. What remains and is available to other organisms is the *net primary production* and is thus less than the original total or *gross production*. The transfer of matter and energy between organisms is effected by the simple process of consuming one another or by assimilating organic matter released by them. Green plants and photosynthetic bacteria are by their very nature independent of an external supply of organic matter, they are *autotrophic*, whereas most other organisms are *heterotrophic*, that is they rely on organic matter that has originally been synthesised by plants. Heterotrophic animals and micro-organisms then use these substances for their own energy and growth requirements by essentially the same processes as in plants. The accumulation of organic matter in the heterotrophs is known as *secondary production*, a term to which, as will be seen, it is more difficult to apply a precise definition.

Some species of blue-green algae (Cyanophyceae) can, in some circumstances, function heterotrophically and assimilate simple organic molecules dissolved in the water, such as glucose and acetate, in low light intensities or even in the dark (Fogg *et al.*, 1973). It is probable that they can thereby survive in situations where light is insufficient for photosynthesis, as in winter or in very turbid water especially in calm weather when larger numbers are below the euphotic zone, as in Lake George (Ganf, 1974).

The algae in the lake plankton are eaten by some of the small invertebrate animals of the zooplankton, such as crustacea and rotifers, and by some of the herbivorous fish, especially of the family Cichlidae. They are consumed alive in the upper layers of water, and in various stages of decomposition, as they fall towards the bottom. It will be seen from Fig. 4.1, which omits many details, that the subsequent pathways, or food chains, are complicated. But, as a general summary, it may be said that the main steps are as follows: primary producers – herbivores – small carnivores – large carnivores. All of the organisms concerned are also subject to death and to decomposition through the activities of bacteria and fungi (the *decomposers*) which incorporate the energy-rich organic matter into their own bodies. The above steps are sometimes called *trophic levels*, but the practical value of this term is limited because of the difficulty of defining them in any particular ecosystem. For example, many animals are herbivorous and carnivorous at the same time or at different stages of their life history, or according to the kind of food available, and decomposers are themselves consumed by most if not all of the animals, and they are probably an essential component of the diet of some of the zooplankton.

Except in lakes of very low productivity, usually due to a scarcity of nutrients, much of the organic matter synthesised at the surface and of the allochthonous material brought in by the inflows reaches the bottom in the not fully decomposed bodies of dead plants and animals and accumulates as an organic sediment. In the upper layer of this mud, provided that the overlying water is sufficiently oxygenated, there live a great variety of organisms collectively known as *benthos*. In bulk, the micro-organisms are probably always the most abundant. These are engaged in absorbing and breaking down organic matter in a series of steps to simpler molecules and eventually to inorganic nutrient salts. At each step, the organism concerned takes some of the released energy for its own metabolic purposes.

Apart from the successive steps involved in the breakdown of organic matter in dead organisms, we now know that bacteria and fungi play their parts at other points in the circulation of matter and energy intakes. Some blue-green algae fix atmospheric nitrogen and thus contribute to the nutrient requirements of photosynthesis. There is even evidence that enzymes released into the water by some bacteria may break down soluble organic compounds derived from other organisms. For example, bacterial alkaline phosphatase can decompose organic phosphorus compounds in solution to yield inorganic phosphate (Kusnezov, 1968; Overbeck, 1974). There is clearly much more to be discovered about the recycling of materials in lakes.

There are a number of animals living in the mud, such as oligochaet and nematode worms, ostracod Crustacea, insect larvae especially of the chironomid midges, snails and bivalve molluscs. They all feed on the mud itself and get their organic matter both from the undecomposed remains and from the bodies of the micro-organisms. These 'mud feeders' are, in turn, consumed by certain fish, for example some of the mormyrids (elephant-snout fish), and it is probable that several other fish get nourishment directly from the mud. The fly larvae eventually migrate to the surface to emerge, and very many of them are consumed on the way by carnivorous fish. By all these means, some of the energy-rich organic matter and nutrients derived from its decomposition are returned to circulate in the ecosystem.

At each step, when there is a transfer of matter and energy, the receptor organism uses some of the energy for its own metabolism, movements, etc., and this is released by oxidation of some of the organic matter, and is finally dissipated as heat. The efficiency of the transfer, that is to say the proportion which is retained and stored as organic matter during reproduction and growth, has been assessed by experiment for several aquatic organisms. It depends, as expected, partly on the nature of the food. The methods employed and some of the results are reviewed by Mann (1969). The whole cycle of processes thus involves the continuous loss and dissipation of energy and the machinery of the ecosystem is kept running by the influx of more solar energy.

The main pathways by which the organic matter synthesised by freshwater planktonic algae is decomposed and 'mineralised', have been defined and studied by Golterman (1960, 1964, 1971). These processes occur at all levels in the water but, since much of the particulate organic matter sinks towards the bottom, the events in the bottom sediments are of particular importance for the turnover of materials in the lake as a whole (Ch. 6). The rate of recycling of nutrients from the lower water and mud, and thus of primary production, depends much upon the rate at which the products of decomposition are conveyed by movements of water to the upper illuminated regions (euphotic zone). The spring outburst of production in lakes of temperate regions is retarded in midsummer by a shortage of nutrients due to temperature stratification which prevents the stirring of the bottom water. In deep tropical lakes there is a tendency for stratification to be more prolonged. The reasons for this and the effects on primary production will be discussed in Chs. 6 and 7.

The bottom mud provides a particularly favourable medium for micro-organisms. These, in releasing energy for their own purposes, consume oxygen rapidly and, if the water above the mud is not sufficiently stirred, as in deep water during periods of stratification, it may become totally deoxygenated. In the deep regions of some large tropical lakes, this condition persists indefinitely, and the mud is for long periods and sometimes permanently an impossible habitat for the normal benthic animals that depend upon oxygen. They are then restricted to the mud under the shallow water round the edge, where the bottom water is near enough to the surface to be frequently stirred by winds. Deoxygenation may occur also in the shallow water of swamps in which there are large quantities of decomposing vegetable matter near the surface (Ch. 18).

The first stages of decomposition in the mud involve aerobic organisms deriving their energy from a chain of reactions for which oxygen is used as the final acceptor of hydrogen. When, however, the oxygen is exhausted, i.e. 'reduced', the process is continued further by various anaerobic micro-organisms that use substances other than oxygen, organic and inorganic, as hydrogen acceptors. As a result a number of reduction products appear of which the commonest are methane (CH_4), ammonium compounds (NH_4^-) and hydrogen sulphide (H_2S), though ethylene (C_2H_2) and even elemental nitrogen and hydrogen may be produced in small quantities. Bubbles of methane commonly emerge from the mud, and samples of water taken from the stagnant lower anoxic water of deep productive lakes usually smell of H_2S. Methane is usually much the most abundant of these gaseous end-products, but not all of it is lost from the ecosystem. There are species of bacteria that gain their energy from aerobic oxidation of methane (and indeed others from H_2S). In dense tropical swamps which are free of oxygen to near the surface (Ch. 18) most of the gas must escape, but in those lakes that have a permanently anoxic lower layer of water, it is all oxidised by bacteria as it diffuses into the upper oxygenated water. This process has been studied in Lake Kivu where an enormous quantity of methane is produced (Deuser et al., 1973; Jannasch, 1975). It was estimated that the total quantity of methane in this lake (confined to the anoxic water and mud) is about 50 km^3! Commercial exploitation of the gas has been tried (p. 129).

Some of the organic matter is more resistant to decomposition, especially in the absence of oxygen, so that it accumulates beneath the upper layer of mud, which is being augmented by decomposition from above, and is preserved in a reduced state together with a varying quantity of inorganic particles that have been washed in from the land. Of all the organic matter that has fallen to the bottom, only a part is recycled through the ecosystem. The remainder passes out of circulation and is preserved in the deep sediments. It is to this kind of preservation that we owe our fossil fuels (coal and oil) from which we release the energy stored within it in the distant past. Though the remains of organisms have been preserved in ice and under layers of volcanic ash in a desiccated state, it is only under water that it has happened on a large scale. The decomposer organisms need water, which also cuts off the sediments from access to oxygen and so maintains the anaerobic conditions produced by the decomposers. The deposition of more material from above further seals the deeper sediments and prevents exchanges with the water above. In some lake basins, especially in those like the Albert basin (p. 174) that have subsided, the old lake sediments may be hundreds of metres deep. Sediments under water may be sampled by coring to several metres under the mud surface. The amount of organic matter at any particular level depends upon the conditions at the time of deposition. Some layers may be composed almost entirely of inorganic particles (sand, silt, etc.) brought in from outside by the inflows at times of heavy rainfall, others are highly organic and may include the fossil remains of plants and animals. Much can obviously be learnt from sediment cores of the history of a lake basin and of past climates by studying the physical structure of the inorganic particles, the chemical composition and the fossil organisms in the various layers (e.g. Lake Windermere: Mackereth, 1966). An approximate age within the past few thousand years can often be assigned to a particular layer by the radiocarbon (^{14}C) dating method. As an example of the information that has been deduced from cores taken from a tropical lake, see Kendall (1969) on the sediments of Lake Victoria (p. 252).

Confining himself to the diatoms Hustedt (1949) studied lakes in and around the Western Rift Valley and concluded that alkalinity (HCO_3^- and CO_3^-) is the main determinant of the species composition. He classified his lakes into three categories of increasing alkalinity in which the dominant diatoms are: (1) acid waters with *Eunotia* and *Pinnularia* as dominant diatoms; (2) (a) more alkaline waters with dominant *Melosira*, and (b) higher alkaline waters with *Nitzschia*. From a wider survey in East and Central Africa and from a study of previous literature Richardson (1968) found many apparent exceptions to Hustedt's classification though alkalinity obviously is a major correlate. But there are other influences, particularly salinity, silica levels and probably organic matter. A reasonable correlation was found by Kilham (1971b) between type of diatom flora and the mean content of silicate in waters in Africa and elsewhere.

Until we know more about their physiology, the chief value in establishing correlations between the diatom flora and the composition of the water is that their skeletons found in lake sediments can help to indicate the conditions in the water at different periods in the past. It is assumed that the pattern of diatom species at a certain depth in the sediment core indicates that the chemical conditions in the water (at least salinity, alkalinity and silicate content) during the period of deposition were similar to those in present existing lakes that have a similar assemblage of diatom species. This adds to the evidence from chemical analysis of cores, while pollen analysis indicates the prevailing climatic conditions which may affect the composition of the water. Diatom analysis has been used, for example, to determine the recent hydrological history of Lake Naivasha by Richardson and Richardson (1972), of Lake Albert by Harvey (1976), of Lake Manyara by Holdship (1976) and of Lake George by Haworth (1977). (pp. 28, 237).

For the ecosystem of a lake, however, the preserved organic sediments represent a permanent loss of materials and energy. Organic matter and nutrients are also lost in the outflow, but the importance of this loss depends upon the rate of flow-through relative to the volume of the lake. From very large tropical lakes much of the water is lost by evaporation and loss of materials from the outflow is insignificant in relation to the total resources of the lake. For small and particularly shallow lakes the outflow is likely to account for more serious deficits. The possibility must be mentioned of significant losses suffered by some lakes from the emergence of immense clouds of lake flies (chironomids and chaoborids) that drift onto the land and perish there (Fig. 4.2). We do not yet have enough data to estimate the quantities involved and to assess the relative importance of losses by this route. Losses are also caused by predators from the land that seize and remove organisms from a lake, and in particular fish. The activities of fish-eating birds and mammals are probably always a relatively small item in the economy of

Fig. 4.2 Lake-fly swarm being blown onto land, Lake Edward, Uganda. Photograph by L. MacGowan.

a lake and, until the introduction of modern methods of fishing, the same could have been said of fish-eating human beings. Now, however, with the introduction of new methods and with increasing populations to be fed, the ecosystems of many lakes and other waters are threatened with serious disruption by indiscriminate overfishing to the detriment of the fisheries themselves. To understand the conditions required to maintain a permanently productive fishery is one of the practical object of limnology.

It will now be understood from what has been said and from a look at Fig. 4.1 that much of the organic matter synthesised by plants goes at first through animals and other heterotrophic organisms in several steps. What is left, after these have satisfied their needs for energy, is decomposed in the surface mud or escapes permanently into the lower sediments.

The herbivorous fish, such as the tilapias, feed directly on plants, and their production is thereby more 'efficient' in the sense that there are no intermediaries to dissipate the energy on the way. Some of them, such as *Tilapia zillii*, feed on large vegetation – water weeds, submerged grasses, etc. Others, such as *Sarotherodon niloticus** and *esculentus* consume the phytoplankton which are separated from the water by a filtering device on the gill arches (Greenwood, 1953). Until recently it was thought from visual observations of the contents of the gut that only the diatoms and the green algae were digested. The blue-green algae, though ingested, were apparently not assimilated, owing, perhaps, to the resistance to enzymes of the polysaccharide layer surrounding the cells of these algae (Fish, 1952, 1955a). In view of the abundance, and in many lakes the predominance, of blue-green algae, especially *Microcystis*, and since the direct consumption of phytoplankton by fish is clearly a very important link in the production cycle of many lakes, it is essential to know for certain whether or not these algae are involved in this way.

* See note on tilapias (p. 150).

Using cultures of algae labelled with radio-carbon (^{14}C), assimilated from bicarbonate, Moriarty showed that both *Sarotherodon niloticus* and *Haplochromis nigripinnis* from Lake George do in fact digest and assimilate up to about 70 per cent of the carbon contained in the blue-green algae *Microcystis* and *Anabaena*, the amount depending upon the time of day during which the fish feed. Experiments showed that the very acid conditions in the stomach (pH about 1·4) are responsible for making the algal polysaccharide coat permeable to the gastric digestive enzymes (Moriarty, 1973; Moriarty and Moriarty, 1973a, b). The confirmation of this direct link between blue-green algae and fish is essential to an understanding of the production cycle in many African lakes of which Lake George is an outstanding example.

The animals of the zooplankton feed on the phytoplankton, organic particles and micro-organisms, and some are predators on the others. Of the carnivorous fish some small species, e.g. of the genera *Aplocheilichthys*, *Alestes* and *Stolothrissa*, consume the zooplankton. The large carnivorous fish, such as *Lates* and *Bagrus*, eat the smaller fish including the herbivores and the zooplankton-feeders. Some of the organic matter that has escaped decomposition in the mud and has been consumed by the benthic animals is taken by certain fish, for example, chironomid larvae and oligochaet worms by mormyrids and snails and bivalve Molluscs by the lungfish *Protopterus*. These are a few of the innumerable pathways by which energy is circulated and dissipated in a mature and productive lake.

The productivity of the fishery of a lake therefore depends partly upon the presence of certain organisms that serve as links in the chains* between the primary producers and the commercially important fishes. Examples will be given of lakes that are not as productive as they might be because, for reasons connected with their origin and history, one or more of these links is missing or is inadequately represented. This can, in principle, be rectified by the introduction of a suitable species from elsewhere. It is important, however, to be aware of the risks that are involved in the transfer of species from one place to another. The workings of most ecosystems are so complex and delicately balanced that we cannot, and perhaps never will, be certain of the consequences of introducing a foreign species. All that we can learn of the biology of the species in its own habitat or from previous introductions elsewhere, or from small-scale experiments in fishponds or tanks, will not necessarily enable us to predict just what will happen in a particular ecosystem, even in one that we have previously investigated as thoroughly as possible by all available methods. There are questions of the genetic make-up, immunity relations, fecundity, conditions for and length of breeding cycle, feeding and breeding behaviour and physiological adaptability of the introduced species compared with those already there, and in relation to the new physical environment. These and other factors can contribute to the result which may be anything from failure to become established to complete dominance of the introduced species with suppression or extinction of native species. The process of reaching a final equilibrium may involve only a temporary catastrophic upset, as with the water fern, *Salvinia auriculata*, on Lake Kariba.

---

* The word 'niche' is sometimes used to denote such a functional position occupied by a species in an ecosystem, though it is more usually defined from the standpoint of the species itself and does not relate only to feeding, e.g. 'its specific way of utilising its environment' (Mayr, 1970).

A general introduction to this important subject may be found in Elton (1958). Among the many examples of the devastating effects of accidental or intentional transfer of species are several involving organisms dependent upon inland waters. One of the most sensational of tropical examples was the accidental transport by ship in 1929 of *Anopheles gambiae* from West Africa to the northeast coast of Brazil. There followed during the early 1930s the most catastrophic and lethal epidemic of malaria ever experienced in Brazil. In some areas the death rate was over 10%. The disease spread with the mosquito about 300 km from the original focus, but was finally arrested by a large-scale operation organised by the Rockefeller Foundation whereby the entire population of adults and larvae of the invading mosquito was destroyed by the end of 1940 – one of the most decisive and successful operations in the history of tropical hygiene. An example of damage done to a purely aquatic ecosystem is the invasion of the Great Lakes of North America by the sea-lamprey *Petromyzon marina* during the 1930s. This followed, after an interval of many years, the construction of the Welland Ship Canal between Lakes Ontario and Erie, bypassing the Niagara Falls which had previously been an effective faunal barrier. The lampreys are vigorous predators on fish, and the result has been a very serious reduction of the commercially important lake fisheries. These, and many other examples, emphasise the need for great caution and for thorough preliminary investigation of the biology of the organism to be transferred and of the ecology of the water system into which it is to be introduced. The risks involved can then be more clearly seen, though never quite certainly defined.

The usual objective is the improvement of the fishery by the introduction of a species of fish more palatable or of a larger size than those already present. Species of fish or of invertebrates are also introduced in order to exploit some source of organic matter not previously contributing significantly to the production of commercial fish. Species of *Sarotherodon* and *Tilapia* have been introduced into Lake Victoria to feed on the large submerged vegetation and proposals to introduce the large and predaceous Nile perch into Lake Victoria (now accidentally accomplished) were based on the assumption that they would feed mainly on the large populations of cichlid fishes too small for commercial exploitation (p. 263). A common deficiency of newly formed and immature lakes is the absence of fish that consume the zooplankton or the benthic fauna. Among the many attempts to rectify this deficiency was the introduction of the zooplankton-feeding 'sardine' from Lake Tanganyika into Lake Kariba (p. 372).

In some small and isolated lakes, such as crater and volcanic-barrier lakes (Ch. 18), the indigenous fish fauna is often very poor and sometimes wholly lacking. In some, such as Lakes Bunyoni and Nkugute in western Uganda, the basic nutrient resources of the ecosystem appear to be deficient and the introduction of fish has met with little success. But in others, such as Lake Naivasha in the Kenya Highlands, a profitable fishery has been created by the introduction over the past forty years of *Sarotherodon niger*, *S. leucostictus* and *Tilapia zillii** and of

---

* *Sarotherodon*, previously *Tilapia* (p. 150) Hyder (1969) showed that the breeding cycle of *S. leucostictus* in Lake Naivasha is adjusted to conditions in its new environment. The gonads develop during the warmer and more sunny period (December–March) and breeding occurs in the rainy period (March–April).

the predatory North American black-bass *Micropterus salmoides* which was originally brought in for sporting fishing (Mann and Ssentongo, 1969). For the same
reason trout have been successfully introduced into many highland streams of
East Africa.

That in past times fish may have been transferred from one water system to
another by the indigenous peoples of tropical Africa is a possibility that has rarely
been discussed. In view of their knowledge of the habits of fish and of their
ingenuity in devising methods of capture, it must be admitted that this rather
simple operation may occasionally have been performed. There is, however, no
certain evidence for it, and it is unlikely to have been done except to improve a
very poor fishery or to introduce fish where there were none before. Few fish
can survive much handling nor can they easily be transported in containers without aeration. Those that are most resistant to temporarily unfavourable conditions
are more likely to have a wide natural distribution.

The word *productivity* has been frequently used in this discussion. It certainly
conveys a generally understandable meaning, but when applied to a particular
situation with a view to measuring and expressing it quantitatively, it becomes
necessary to decide more exactly what is meant by the word. If the *rate* of production is implied, then it must be concerned with a particular organism or group
of organisms. It can then be defined as the rate at which new living matter is
being produced in the bodies of these organisms – either the *gross* rate, or the *net*
rate after the organisms have satisfied their own energy needs. To refer to the
productivity of an ecosystem *as a whole*, without further qualification, is therefore unjustified. If the system is in equilibrium, with a steady biomass, the *net*
productivity is in fact zero however intense the circulation of matter and energy.
It is often actually negative (e.g. at night, in winter and after exhaustion of a
nutrient).

Of all the energy fixed by the planktonic algae, only a fraction, which may be
very small (see Ch. 7) is actually passed on to their consumers. Taking the invertebrate animals of the zooplankton as an example of secondary production,
these rates can, in principle, be estimated from measurements of the rate of
growth and reproduction and of changes in the density of the population from
which the total amount, or *biomass*, of the organism and its fluctuations in the
whole lake can be calculated. The energy requirements as reflected in respiratory
rate, and losses from excretion, can be estimated from measurements in the
laboratory. Each type of organism, e.g. zooplankton, benthic animals and fish,
requires very different methods for getting the required information and it is impossible to review these here. Recent reviews of methods and results from work
on freshwater invertebrate production are to be found in Edmondson and Winberg (1971) and Winberg (1971). It can be imagined that the difficulties involved
in the quantitative sampling of an organism to give figures applicable over a wide
area of the ecosystem are often very great. This is particularly so with the benthic
fauna. There is also the doubt as to how far laboratory measurements of growth
and respiratory rate represent the rate of these processes in the lake. Measurements of primary production by the phytoplankton (see Ch. 7) are much easier
than those of secondary (heterotrophic) production. For that reason techniques
for the latter are much less developed, and only during the past ten years have

any significant advances been made. In the tropics these studies are only just beginning.

As might be expected, until recently more effort has been put into estimating the production of fish* than of any other secondary producer. It is sometimes given the humanly-biased title of 'terminal production'. Recorded fisheries statistics, if collected thoroughly enough, can provide a figure for the total weight of fish caught in a lake per year. This clearly does not necessarily represent 'productivity' because the annual catch depends on the efforts and skill of the fishermen and on the type of equipment used. For the purpose of comparing one lake with another these figures are sometimes converted to give the weight of fish caught per annum per square metre of lake surface. This can have some clear meaning for a lake such as Lake George (Ch. 13) where the fish and fishing are distributed fairly uniformly over the lake. But with most of the large African lakes this is not so. For example, the fishing in Lake Victoria has so far been confined to the shallow inshore waters which cover only a small proportion of the total area. If the proposed deepwater *Haplochromis* fishery is put into operation, it must be expected that the total annual catch per square metre will be much increased. The value of any such figures as an estimate of productivity can thus be assessed only when all the relevant circumstances are known, and when it can be demonstrated that the rate of extraction of organic matter by the fishery is about balanced by the rate of net production on the part of the fish and an increase in the fishing effort will reduce the stocks. A commonsense meaning for 'productivity' in relation to a fishery is 'the maximum sustainable yield'. This is the rate at which fish can be continuously taken without depleting the stocks of the species that are being caught. This figure is clearly related to, but not identical with, the net rate of production. There are predators other than man, as well as decomposers that consume these fish. Nevertheless, it is the most important information needed for the practical regulation of the fishery though prolonged investigation of the fish populations may be required to determine it. The maximum sustainable yield does not necessarily remain constant for long periods, and may well be altered by natural and manmade events as, for example, changes of climate or of agricultural practices in the lake basin that may alter the amount of nutrient-rich water flowing into the lake. A practical objective of a fishery biologist is, however, to find the maximum sustainable yield and to suggest regulations for controlling the fishing so that it is in fact sustained. It is usually less important to limit the total amount caught than to control the size of the fish in the catches by, for instance, insisting on a minimum mesh-size for gill-nets. Fisheries have often suffered from the loss of breeding fish because they are caught before they reach maturity.

The techniques for studying fish production are aimed at obtaining the same kind of information as was outlined above for the invertebrate animals. But, unlike the latter, they are large enough to mark (for which there are several methods) so that individual fish can be recognised on recapture. This can provide reliable information on the movements, development, rate of growth and reproduction

* For an introduction to the ecology of tropical fish see Lowe-McConnell, 1975 and 1977.

in the natural environment (Hickling, 1961; Gerking, 1967; Ricker, 1968; Mann, 1969).

For the practical purpose of rapidly predicting the potential productivity of an, as yet, undeveloped fishery, or assessing the degree to which a fishery is falling short or exceeding its potentialities, Ryder (1965) has proposed a 'morphoedaphic index'. This is defined as the total dissolved solids in mg/litre divided by the mean depth in metres. The former can be sufficiently accurately estimated from the conductivity, which is easily measured. A possible scientific basis for the use of this seemingly arbitrarily chosen index is not suggested. The lack of good evidence for a positive correlation between conductivity (and salinity) and productivity of African freshwaters is discussed in Ch. 5. There is, however, no doubt that in general the more frequently a lake is stirred by winds to the bottom the faster the nutrients are recycled from the mud into the photozone where they may accelerate the rate of primary production (Ch. 7). The depth of the water is thus, in principle, negatively correlated with the rate of production. It could be objected that the mean depth does not necessarily represent the proportion of the mud surface that lies near enough to the water surface to be stirred directly by winds, and that there are equally important variables that determine the degree to which the surface is exposed to the winds. Such are the wind regime itself, the topographical situation and the wind-fetch (Ch. 6). It was admitted that the index can be proportional to production, as represented by annual records of fish catches, only for a group of lakes subjected to the same climate, with no large changes in water-level, with bicarbonate as the major cation, with no significant turbidity due to inorganic particles and provided that all of the lakes are subjected to the same intensity of fishing (Ryder, 1965, and in Regier, 1971). With these rather restrictive provisos it was claimed that, for a series of North American lakes, there is a good enough correlation between the index and fish production to make predictions concerning the potential production of an as yet unexploited lake in the same group. Decisions can then be made on the direction of research and development on the basis of the results of two simple measurements. The expensive and time-consuming investigations of biomasses, population structures, fecundity, growth rates, etc., needed to make a reliable estimate of potential productivity of the fishery and to decide on the best methods of exploitation (see Ricker, 1968) could then be restricted to the more promising lakes.

In attempting to apply the index to African lakes, which differ greatly from each other in many important features, it must be remembered that in some of the larger lakes fishing is restricted to a small proportion of the total area, and the mean depth of the whole lake is not necessarily relevant. If a large proportion of central Lake Victoria were 500 m deep it would not obviously affect the productivity of the shallow inshore waters. The effort and technical skill applied to fishing, as well as to the collection of fishery statistics, has varied greatly from one lake to another. In some cases illegal fishing (poaching) has accounted for a very large proportion of the catches and does not appear in the records.

To a scientist, the validity of this procedure must at first sight appear very doubtful in view of the many variables whose interrelations are not at all obvious. If, however, it can be proved that in some situations practical results of value can be got in this way, it is not for the scientist to object. He must then try to show why it is successful, with a view to improving it. 'The proof of the pudding is in the eating.'*

Assessments of the actual and potential resources of the inland fisheries of tropical Africa were presented at a Symposium at Fort Lamy in 1972 (Okedi (ed.), 1973).

Artificial tropical fish culture in ponds dug for the purpose has been practised for centuries in the Far East and has been introduced to tropical Africa during the past thirty years. The ecosystems so created are relatively very simple, and are usually aimed at producing only one or two species of fish, e.g. a herbivore with or without a predator. Primary production may be sustained by the addition of organic or inorganic fertilisers, or for fish that feed on larger vegetation (e.g. *Tilapia zillii*) the food is supplied as grass or the leaves of other land plants.

Though productive methods of culture suited to local conditions have been evolved by trial and error over the ages in the Far East, there is room for improvement in the light of limnological research. This applies especially when fish culture is introduced to another part of the world where both the conditions and the fish may be very different and the people have no background of experience (Hickling, 1962, 1968). Fishponds are also valuable for some experimental purposes. Conditions can be controlled and simplified more easily than in a natural lake, but they can approach more nearly to natural conditions than a laboratory tank.

In conclusion, we should return once more to some generalities concerning the productivity of inland waters and especially lakes. The rate of primary production depends, among other things, both on the supply of nutrients from outside and on their recycling through the system. Productivity, in whatever sense the word is used, varies greatly from one lake to another and even in the same lake at different periods. It is customary to call a productive water *eutrophic* and conversely one of low productivity *oligotrophic*. There have been attempts to classify lakes on this basis and to use certain quantitative chemical and biological data as criteria for one or other condition. However, it has become clear, and is supported by growing experience with tropical lakes, that these terms can only be relative ones, like 'hot' and 'cold', and cannot be given precise quantitative meaning of any practical value.

Dystrophic is a word applied to a particular kind of water found in upland moorland bogs in temperate regions. It is characterised by a brown colour due to humic substances, an acid reaction, a very low calcium content and often a deficiency of other nutrients. The deficiency is mainly due to the fact that the water is derived from direct rainfall rather than from inflows from the surrounding land. It is usually unproductive and supports only a few specialised plants and animals. This particular combination of features is rare in tropical Africa, though it is found in some high mountain pools and in some volcanic crater lakes at high altitude whose water supply is almost entirely derived from direct rainfall. The sphagnum swamps of Lake Nabugabo (p. 311) might be put into this category, though the very low salinity and calcium content of the inflows

---

* An expression which can be traced in English literature at least as far back as the seventeenth century. It was perhaps a product of the Scientific Revolution.

are due to geological causes. The very extensive tropical swamps, though often coloured with humic substances, are certainly not of this type (Ch. 18).

An interesting and very important feature of aquatic ecosystems, especially of relatively static bodies of water such as lakes, is the self-limiting effect of organic production. This can happen in several ways. When primary production is very rapid, the algal population can become so dense that the depth to which light can penetrate in sufficient intensity for photosynthesis (euphotic zone) is reduced, sometimes to a few centimetres. A limit to primary production is therefore reached, however favourable the other conditions may be (p. 124). A very high rate of production in the upper layers can also result in a correspondingly large amount of organic matter falling towards the bottom. On decomposing this may release sufficient soluble matter to raise the density of the lower water enough to reduce or even to prevent mixing with the water above for long periods or even permanently. In such cases, the rate of recycling of nutrients reaches a limit. Associated with these conditions, as already mentioned, is the reduction or complete exhaustion of dissolved oxygen. This in itself sets a limit to the production of most animals, and secondary production, other than of anaerobic micro-organisms, is virtually brought to a standstill in the lower stagnant water. When a large increase of nutrients encourages the massive growth of floating vegetation, as during the initial stages of Lake Kariba (p. 368), exclusion of light and prevention of stirring by wind may reduce the oxygen and make even the water near to the surface unfavourable for many organisms.

An increase in primary productivity of a water is known as *eutrophication*. This has recently become a matter of great urgency because, though the self-limiting effects of production in natural waters has seldom in the past been regarded as of immediate practical importance, the limits are now being very greatly overstepped through the artificial addition of nutrients (Hynes, 1960; J. R. E. Jones, 1964; Rohlich, 1969). Organic pollution of natural waters from the discharge of sewage and other organic wastes is now one of the great menaces to the environment. Overproduction is also caused by the excessive use of chemical fertilisers on the neighbouring land, combined with an efficient drainage system which transports them into the rivers and lakes.

The course of eutrophication is marked by progressive changes in the chemistry of the water and in the composition of the fauna and flora. Highly polluted waters support few organisms other than bacteria, fungi, blue-green algae, protozoa, nematode and oligochaet worms and a few insect larvae. Fish may be excluded not only through lack of oxygen or of suitable food, but also by poisonous substances produced by some organisms and especially blue-green algae. Some species of alga have been shown to be responsible for deaths of fish and even of cattle that drink the water (Prescott, 1948; Gentile, 1971). The irony of the situation lies in the fact that the controlled addition of these substances could well increase the useful productivity of both land and water. Not only are valuable materials being wasted, they may even cause the extinction of the fishery.* There are also the dangers to public health, which might well be more

---

* Toxic chemicals from industrial wastes and insecticides can be equally devastating to aquatic life and may contaminate fish and make them unfit for food. However, they pose rather different problems that are not so closely related to the normal workings of the ecosystem.

acute in the tropics, and the devastating effects on the amenities. Rivers, lakes and the sea provide enjoyment for more people than any other kind of natural environment. The overproduction of micro-organisms, fungi and blue-green algae with, in extreme cases, a nauseating stench, can turn an attractive and economically valuable river or lake into a useless, repulsive and even dangerous stretch of water. These disasters are now well known in Europe and North America (see Rohlich, 1969). With expanding industries and the urbanisation of a rapidly increasing population they will certainly be repeated in tropical Africa, whose rivers and lakes are of even greater potential value, unless the warning is heeded in time. (The problem has already arisen in the pollution of the artificial Lake McIlwaine, near-Salisbury, Zimbabwe – see p. 390.) This requires not only the application of research now in progress elsewhere; it is even more important that the ecosystems of the African lakes and rivers should be well understood as a background to the planning of counter-measures.

The problems associated with excessive eutrophication are very important and of great scientific interest. But they belong to what might be called an accidental section of applied limnology, which is mainly concerned with the wider and more positive aspects of preservation and efficient exploitation.

In Ch. 21 there is a short discussion on methods of approach to the study of limnology, with the ecosystems of the tropical African lakes especially in mind.

5

The mineral composition of tropical African fresh waters in relation to ecology

There is a great amount of information on the chemical composition of inland waters that could be discussed from several viewpoints – the geological origin and geochemistry of the dissolved constituents, the comparative compositions of different waters, and the chemical processes and equilibria within them. A discussion of those aspects of chemical hydrology that are related to geology and limnology is to be found in Hutchinson (1957), and data on the chemical composition of the inland waters of the world have been compiled by Livingstone (1963). Talling and Talling (1965) record and discuss the comparative ionic composition of the lake waters of eastern Africa and Kilham (1971a) presents a large amount of original and collected data from most of tropical Africa on the basis of which, like the Tallings, he proposes a classification of African inland water (p. 57). Symoens (1968) records chemical analyses of waters in the south-eastern section of the Zaïre basin (p. 64), and Viner (1975) studied the sources of minerals in the lakes of Uganda from analyses of inflowing rivers. In this chapter we shall consider briefly only a few topics of direct biological importance that relate particularly to African inland waters.

The dissolved gases concerned in photosynthesis and respiration (oxygen, and free and combined carbon dioxide) or that are end-products of metabolism (methane and hydrogen sulphide) and the role of compounds of nitrogen, phosphorus and silicon ('nutrients') are all discussed in other chapters (especially Chs 4, 6, 7 and 18). We are here concerned with the dissolved solids and mainly with those inorganic salts that are normally the most abundant of the dissolved substances and are sometimes for this reason known as the 'major' constituents. In fresh waters they are sufficiently diluted to be almost entirely dissociated into their component electrically charged ions – mainly sodium Na^+, potassium K^+, calcium Ca^{++}, magnesium Mg^{++}, bicarbonate HCO_3^-, chloride Cl^-, and sulphate SO_4^{--}. The electrical conductivity of the water is affected both by the total concentration of ions and by other factors such as the mobility of the individual ions, but for most ecological purposes, except for very saline waters, it reflects sufficiently closely the total concentration of the major ions and thus the salinity (concentration of salts by weight), and is particularly easy to measure with suitable equipment. It is usually expressed as the reciprocal of the resistance (^{1}/ohms or mhos) of 1 cm of water at 20°C (K_{20}). For example, the water of Lake Victoria has a conductivity varying between 90 and 100 μmhos (micromhos or mhos $\times$ 10^{-6}).

Since, as already stated, the salts in fresh waters are almost fully dissociated into ions, there is no justification for quoting water analyses, as was at one time customary, in terms of weights of associated salts (e.g. NaCl, $MgSO_4$ etc.). Moreover, it is the number and balance of ions, rather than their weight, in a given quantity of water that is of most biological and chemical importance. Hence it is now accepted limnological practice to express water analyses in terms of milliequivalents of each ion per litre (meq/litre, i.e. weight of ion in mgms/ equivalent wt per litre). This has the additional advantage that any difference between the sum of the equivalents of the positive cations and the negative anions, that cannot be traced to errors in the analytical methods, reveals the presence of other undetected ions in significant quantities.

The *total* concentration of the mineral salts (salinity) is however usually expressed as weight per volume of water and in this book as parts per thousand or $^o/_{oo}$, a sign commonly used in oceanography for salinity of seawater. 'Total dissolved solids' (normally quoted as weight per volume of water) is usually estimated by evaporating samples at 105°C. It may be significantly different from 'salinity' in waters of high organic content. Information on methods of water analysis and on accepted methods of expressing the data are to be found in Golterman and Clymo (1969).

A small selection from the available data has been set out in Table 5.1. In order, however, to discuss these in an ecological context, some general points must be made concerning the known or probable biological importance of these ions. HCO_3^- together with CO_3^- in very alkaline waters are of special importance in relation to the availability of carbon dioxide for photosynthesis, and their relative concentrations are linked with the concentration of hydrogen ions of which the pH is the reciprocal log. These equilibria are discussed in connection with photosynthetic production in Ch. 7. All of the major ions listed above are necessary constituents of the living matter of organisms and in the case of aquatic animals a proportion of them comes from the food. It has long been known that submerged aquatic plants (macrophytes) and planktonic algae obtain them, as would be expected, from solution in the water. It was later shown by Krogh (reviewed in 1939) that freshwater animals too absorb them from the water. In fact there is a continuous flux of ions in and out of freshwater plants and animals, the loss by outward diffusion being compensated by uptake from the water. This accounts for the fact that many freshwater animals, including fish, can be kept for a very long time without food in a healthy and active state. The continuous loss of ions is combated by active reabsorption. Since the concentration of each ion inside the organism is very much greater than that in the surrounding water, uptake involves the movement of ions against a considerable gradient of concentration (Table 5.2). Therefore, uptake of ions from fresh water in principle involves the expenditure of energy, though the amount of energy directly required may be modified by the presence of a gradient of electrical potential along which the

Table 5.1 Inorganic composition of some tropical African lake and rainwater

Lake	Reference	Date of sampling	Conductivity K_{20} (μ mhos)	Approx. salinity % (g/l)	pH range	Na^+	K^+	Ca^{++}	Mg^{++}	CO_3^{2-} + HCO_3	Cl^-	SO_4^{2-}	Cations	Anions
Lungwe	Dubois, 1955	1953	15–17	0.010	6.5–6.7			0.07	0.03	0				
Tumba	Dubois, 1959	1955	24–32	0.016	4.5–5.0			0.03	0.02					
Nabugabo	Beadle & Heron (unpublished)	June 1967	25	0.015	7.0–8.2	0.090	0.028	0.060	0.020	0.140	0.040	0.019	0.198	0.199
Bangweulu	Harding & Heron (unpublished)	1960	35	0.023	7.0–8.3	0.114	0.033	0.075	0.066	0.260	0.009	0.021	0.288	0.290
Victoria	Talling & Talling, 1965	May 1961	96	0.093	7.1–8.5	0.430	0.095	0.280	0.211	0.900	0.112	0.037	1.02	1.05
George	Talling & Talling, 1965	June 1961	200	0.139	8.5–9.8	0.59	0.09	1.00	0.67	1.91	0.25	0.23	2.35	2.39
Chad (Baga Sola)	Maglione, 1969	July 1967	180	0.165	8.0–8.5	0.5	0.2	0.8	0.3	1.8	0	0.1	1.8	1.9
Malawi	Talling & Talling, 1965	Sept. 1961	210	0.192	8.2–8.9	0.91	0.16	0.99	0.39	2.36	0.12	0.11	2.46	2.59
Tanganyika	Talling & Talling, 1965	Jan. 1961	610	0.530	8.0–9.0	2.47	0.90	0.49	3.60	6.71	0.76	0.15	7.46	7.62
Albert	Talling & Talling, 1965	Feb. 1961	735	0.597	8.9–9.5	3.96	1.67	0.49	2.69	7.33	0.94	0.76	8.81	9.0?
Edward	Talling & Talling, 1965	June 1961	925	0.789	8.8–9.1	4.78	2.32	0.57	3.98	9.85	1.03	0.89	11.65	11.77
Kivu	van der Ben, 1959	Feb. 1954	1 240	1.115	9.1–9.5	5.70	2.17	1.06	7.00	16.40	0.89	0.33	15.93	17.62
Turkana	Talling & Talling, 1965	Jan. 1961	3 300	2.482	9.5–9.7	35.30	0.54	0.28	0.25	24.50	13.50	1.40	36.37	39.40
Rainwater														
Kampala, Uganda	Visser, 1961	1960			7.7–8.1	0.28	0.10	0.005			0.05	0.05		
Gambia 9 km from coast	Thornton, 1965	1963				0.026–0.36	0.01–0.80	0–1.10						
136 km	Thornton, 1965	1963				0.021–0.066	0.010–0.015	0.05–0.25						

Each set of figures (except pH) was obtained by analysis of a single sample of *surface* water. In all respects the waters of these lakes are subject to variations according to location of sampling and to seasonal and longer-term fluctuations. This is especially so with Lake Chad (Ch. 12). The composition of rainwater is very variable but it may have a significant influence on the mineral composition of freshwaters, especially those of low salinity. The salinities of Lungwe and Tumba, for which analyses are incomplete, were calculated from the mean conductivity using the curves provided by Maglione (1969) for Lake Chad. The other salinities are the sum of weights in gms of the major inorganic ions assuming the approximate proportions of HCO_3 and CO_3 theoretically corresponding to the prevailing pH (Fig. 7.1).

Table 5.2 Minimum external equilibrium concentrations of Na^+, K^+ and Ca^{++} for some freshwater invertebrate animals (meq/l)

		Na^+	K^+	Ca^{++}	Reference
*Potamon	Crustacea				
niloticus	Decapoda	0·05	0·07†	—	Shaw, 1959a
Astacus	Crustacea	0·04	—	—	Shaw, 1959b
pallipes	Decapoda	—	—	0·18 (intermoult)	Chaisemartin, 1965
Gammarus	Crustacea	—	—	0·14 (from hard	Vincent, 1963, 1969
pulex				stream)	
	Amphipoda	—	—	0·05 (soft stream)	
Lymnaea	Mollusca	0·025	—	0·10	van der Borght, 1962
stagnalis	Pulmonata	—	—	0·12	Greenaway, 1970, 1971
*Aedes aegypti	Insecta				
(larva)	Diptera	0·005	—	—	Stobbart, 1965

* Tropical African species
† The sample from Lake Nabugabo, from which *P. niloticus* is absent (Table 5.1), contained 0·028 meq/l K^+

charged ions are moved. These regulatory activities of freshwater animals in which both external membranes and the excretory organs play a part, are discussed in Potts and Parry (1964, Ch. 5).

An important conclusion from these facts is that the dissolved inorganic salts in fresh waters, though present in relatively low concentrations and sometimes very low indeed, are nevertheless vitally important for the existence of aquatic animals as well as plants. The question arises as to whether the occurrence and distribution of organisms in inland waters are determined to any extent by the concentration of all or of some ions – in other words by the inorganic composition of the water. Evidence on this matter can, in principle, be got from two sources; first, from simultaneous observations in the field on the fauna and flora and on the composition of the water, and, second, from laboratory studies on the physiology of adaptation to artificial waters of controlled composition. There are limitations to both of these approaches, but considered together they can provide some useful indications. It is often very difficult to decide whether the absence of certain organisms from a water of peculiar composition is due to the mineral composition of the water and not to other circumstances such as temperature, light, oxygen supply, toxic organic substances, inadequate food, predators, or simply to the inaccessibility of the water to the organisms in question. The ecological evidence, however, is overwhelming at least on one point – that above a certain range of total salinity most of the typical 'freshwater' plants and animals are eliminated and only a few species can survive that have special powers of preventing loss of their internal water by diffusion into the external saline water (see Ch. 19). Though there can be no sharp distinction between 'fresh' and 'saline' waters, changes in fauna and flora are obvious when the total salinity reaches $5-10^o/_{oo}$ (5–10 g/l). Above about $20^o/_{oo}$ there remain only those few species that are specially adapted. Within the 'saline' water range there are some indications of differences in relative proportions of certain ions that are reflected in faunal and floral differences. This is demonstrated by a comparison between the neutral and alkaline saline waters of Africa (p. 339).

In this chapter we are concerned with waters that harbour a typical 'freshwater' fauna, that is those with a salinity lower than $5^o/_{oo}$. Of the large African lakes which have such a fauna, Lake Turkana is the most saline (c. $2 \cdot 5^o/_{oo}$). The surface oxygenated water of Lake Kivu is around $1^o/_{oo}$ salinity, but the remainder, as most freshwater lakes elsewhere, are much less saline than these (Table 5.1). As a general conclusion from studies of the faunas and floras of these lakes it could be said that such differences as there are bear little obvious relation to the differences in either salinity or inorganic composition of their waters, except for the few waters of exceptionally low salinity (below $0 \cdot 03^o/_{oo}$), and these will be discussed below. Within this very wide range of waters ($0 \cdot 03 - 2 \cdot 5^o/_{oo}$) there are great differences in the proportions of the various ions (Table 5.1 and discussion in Talling and Talling, 1965). Lund (1965) has reviewed, among many other aspects of algal ecology, the evidence so far available concerning the relations between freshwater algae and the total quantity and proportions of the major ions in the water. From the large number of observations, it is difficult to draw certain conclusions that are generally applicable other than those proposed in this chapter. Extremes of pH and alkalinity, which are often but not always associated with very high or low salinity are more likely to affect plants than animals owing to their connections with photosynthesis. The apparently insignificant ecological effects of ionic differences within the freshwater range should not surprise us because, in contrast to the sea, inland waters are chemically unstable and variable and could have been colonised only by organisms with ionic regulating mechanisms that can function under a wide range of chemical conditions. Experimental research on freshwater animals has demonstrated that they are very adaptable in this respect. But within the salinity range with which we are now concerned correlations have been found in Europe between faunas and floras and the calcium content ('hardness') of the water. (Strictly speaking 'hardness' denotes the combined concentration of the divalent ions Mg^{++} and Ca^{++}.) The most critical level of calcium seems to be about $0 \cdot 25$ to $0 \cdot 50$ meq/l ($5 - 10$ mg/l), above and below which species appear or disappear. There is, however, no certain indication that this is a directt effect of calcium level rather than of some other associated factor such as food supply, pH or even the physical effects of the presence or absence of deposited calcium carbonate on the substratum or other surfaces (see discussion in Macan, 1963, Ch. 11).

The very low calcium concentration sufficient for calcium balance in the very few freshwater animals that have so far been investigated experimentally, even in a crab and a mollusc with special calcium requirements, suggest that the concentration of this ion must fall considerably lower than $0 \cdot 25$ meq/l to have a directly limiting effect on most freshwater animals. This is the conclusion from the data in Table 5.2 which sets out the minimum equilibrium concentrations of sodium, potassium and calcium obtained by experiment with a few freshwater invertebrates. They are the lowest concentrations from which the animals can, when deprived of food, take up these ions rapidly enough to balance the loss by diffusion. The figures are the means of several and were obtained from laboratory experiments under standard conditions. In the natural environment, they would be modified by other factors (e.g. temperature). But they demonstrate what these regulatory mechanisms are capable of, and it is probable that all natural waters contain enough sodium and potassium to enable most freshwater animals to maintain equilibrium (see below p. 72). Recent experiments have shown that the minimum ionic concentrations in the water required from normal development to hatching of the embryos of *Lymnaea stagnalis*, *L. natalensis* and *Biomphalaria sudanica* (the last two both tropical African molluscs) are approximately 0.025 meq/l Na^+, 0.001 meq/l K^+ and 0.1 meq/l Ca^{++} (Beadle and Taylor, unpublished). Embryos with their developing shells would be expected to have a higher demand for calcium than the adults.

Symoens (1968, p. 81 ff) has brought together data, admittedly incomplete, on mineral composition and the occurrence or absence of species of algae, fish and molluscs in waters of the Bangweulu and River Luapula region in the southeast section of the Zaire basin. It is possible to group the algae according to their apparent preferences for different ranges of pH, salinity or calcium level, though there are many species that flourish under a wide range of chemical conditions. The level of pH can, of course, be determined by several factors such as alkalinity, organic acids, and the photosynthetic activity of the plants themselves. Without a knowledge of the physiology of the algae concerned we cannot conclude that pH, as such, is responsible for the presence or absence of species.

Moss (1972–73) attempted to relate the distribution of a number of species of freshwater planktonic algae with the mineral composition of the water and with their requirements for growth in artificial media of controlled ionic composition. This and related work reviewed by Moss does not on the whole support a clear relation between type and abundance of the flora and the mineral composition of the water, there must be some other as yet unknown factors at work such as organic nutrients or trace elements. But experiments designed to study the effects of changing pH and bicarbonate might be interpreted as indicating that differences in algal floras between dilute acid and the more alkaline waters are due to the change in the form of carbon available for photosynthesis. In acid waters this is predominately free carbon dioxide and with increasing alkalinity it is progressively replaced by bicarbonate (Fig. 7.1). But direct study of the carbon sources required by the different species is needed.

There are some inland waters in Africa that are exceptionally deficient in all inorganic ions. This is in most cases due to the insoluble nature of the rocks in the region, as along the watershed (Muchinga Mountains) between the southeastern edge of the Zaïre basin and the Luangwa valley which drains into the Zambezi (Symoens, 1968, p. 21), and in other regions, such as the highland catchment of the Upper Niger, where there are considerable outcrops of granitic rocks. Very dilute waters are also found where the catchment area is restricted to the extent that the water comes mostly from direct rainfall, as in some small crater lakes and in the blanket bogs on high plateaux in temperate latitudes. In some of these the concentrations of Na^+, K^+ and Cl^- are lower than in most rainwaters. This is partly due to peat and mosses which act as ion-exchangers, absorbing metallic ions in exchange for H^+ ions.

Though other factors are also responsible in some cases, there is no certain evidence that the faunas of such waters are restricted directly by the scarcity of ions other than calcium. The indications are, however, that it is the low level of calcium that is mainly responsible for the common absence from low salinity waters of animals such as molluscs and the larger Crustacea which, with their calcareous shells and exoskeletons, have an especially great need for this ion. Moreover, low salinity is usually associated with high acidity (low pH) which accelerates the loss of calcium from organisms. Symoens (1968, p. 92) records the calcium content and the occurrence of species of molluscs in fourteen waters in the Bangweulu-Lualaba region covering a range of $0 \cdot 3 - 3 \cdot 0$ meq/l Ca. Even in the three most deficient waters (salinity $0 \cdot 02 - 0 \cdot 03°/_{oo}$, $0 \cdot 03 - 0 \cdot 06$ meq/l Ca. Even in the three most deficient waters (salinity $0 \cdot 02 - 0 \cdot 03°/_{oo}$, $0 \cdot 03 - 0 \cdot 06$ meq/l Ca), there were one or two species. The greatest number of species (13) were found in the River Luapula at Kasenga in water containing little calcium ($0 \cdot 2 - 0 \cdot 6$ meq/l). In the two 'hardest' waters (Ca $2 \cdot 8$ and $3 \cdot 0$ meq/l), which were well within the freshwater salinity range ($0 \cdot 3$ and $0 \cdot 6°/_{oo}$), there were only three and five species respectively. There are clearly influences at work other than the level of calcium that determine the distribution of these species. The same might be said of the field and experimental work done on the relations between mineral composition of the water and the occurrence of the bilharzia-carrying snails of the genera *Bulinus* and *Biomphalaria* (review by Berrie, 1970, p. 53).

A very large number of samples of natural waters have been analysed with a view to explaining the distribution of bilharzia-carrying molluscs (e.g. *Bulinus* and *Biomphalaria* spp.) in terms of the inorganic composition of the water. Unfortunately very little of practical value has emerged from the work. Apart from some comparatively rare examples of very high salinity and very low levels of calcium and most rarely of potassium (discussed in this chapter), chemically limiting conditions (excess or deficiency of a substance) are very seldom encountered in natural waters. On the other hand the level of temperature, the rate of water-flow and, of course, the presence or absence of predators and of suitable food plants certainly determine the distribution of these molluscs in waters that are chemically favourable to them (see review by Appleton, 1978).

There are probably many small lakes in tropical Africa whose salinity is less than $0 \cdot 03°/_{oo}$, but there are three from which we have some relevant chemical and biological information. These are Lakes Lungwe and Tumba in the Zaire basin, and Lake Nabugabo close to the northwest shore of Lake Victoria.

Lake Lungwe, whose water is more deficient in minerals than any other in Africa so far analysed (Table 5.1), is a very small lake with less than 1 km^2 of open water and surrounded by swamps dominated by the moss *Sphagnum* and sedges (Cyperaceae). It is situated in a remote region at an altitude of 2 700 m in the Mitumba Mountains which lie west of the Ruzizi valley between Lakes Kivu and Tanganyika (Fig. 16.1). It was the subject of a short investigation by Marlier *et al.* (1955). The most remarkable biological feature is the complete absence of fish, though the invertebrate fauna is abundant. This is apparently due to earth movements associated with the later stages in the formation of the nearby Rift Valley. There is evidence that the lake was formerly much larger and lay in a basin on the course of the upper Ulindi River which flows into the Zaire system. It was subsequently completely drained by the reversal of its main inflow and through capture by another tributary of the Ulindi. Climatic changes may also have played some part. This, it is suggested, would have destroyed the fish and some at least of the invertebrates. Further small earth movements partially dammed the outflow and formed the present lake which flows over the natural dam, which will perhaps eventually be eroded and the lake will be drained once more. The very low salinity of the water is probably due both to the now very restricted catchment and to the insolubility of the surrounding rocks. The presence of an abundant invertebrate fauna, though composed of relatively few species, is not surprising. Some may have survived the drainage in small pools, but many would be transported into the basin as resistant stages.

Of particular interest is the presence of the molluscs *Gundlachia* sp. (Ancylidae), *Gyraulis* sp. (Planorbidae) and the bivalve *Pisidium* sp. Further searching might well reveal others. The calcium level (0·07 meq/l) must be very near to the lower limit for these animals, if we can judge from the minimum equilibrium concentration (0·12 meq/l) found for *Lymnaea stagnalis* (Table 5.2). Marlier *et al.* (1955) found that the molluscs had exceptionally thin and fragile shells, which must be attributed mainly to the scarcity of calcium.

Lake Tumba is a much larger lake (about 740 km^2) situated in the central Zaïre basin and draining into the Congo near its confluence with the Oubangui (Fig. 9.3). Its chemistry and biology have been studied by Dubois (1959) and Marlier (1955, 1958a) respectively. It is mentioned in Ch. 7 (p. 110) as a lake in which the organic materials and energy are derived almost exclusively from photosynthesis by terrestrial and emergent plants. The salinity is a little higher, but the concentration of calcium is recorded as even lower than that of Lake Lungwe (Table 5.1). It is thought to have been very recently formed by the partial blockage of the mouth of a tributary of the Zaïre through the deposition of sediment brought down by the main river. Most of the common groups of freshwater animals are represented and the number of individuals is large especially around the shores among the emergent vegetation and submerged vegetable debris. In addition, there are over a hundred species of fish and an apparently productive fishery (Matthes, 1964). The molluscs and crabs are significantly absent. We cannot doubt that this is associated with the low calcium content of the water, but the presence of mollusc-eating fish, including the lungfish *Protopterus dolloi*, may have contributed to the situation. From the scanty information so far available concerning the much larger Lake Maindombe in the same area (Fig. 9.3), it would seem that there is a similar situation there, in which an abundant fauna is supported by externally produced organic matter in water of very low mineral content.

Further suggestive evidence concerning the ecology of the Mollusca in relation to the mineral content of the water comes from Lake Nabugabo. This is a small lake (about 20 km^2) off the west coast of Lake Victoria from which it was isolated by a sand bar about 4 000 years ago. The subsequent speciation among the cichlid fishes is discussed in Ch. 14. The low mineral content (Table 5.1) is mainly due to the insoluble rocks of the catchment which drains into the lake through swamp channels. Here again, the invertebrates are well represented, and there are more than twenty species of fish including the mollusc-eaters *Protopterus aethiopicus* and *Astatoreochromis alluaudi*. There are, however no crabs nor molluscs in the lake itself though a few isolated and very small colonies of *Biomphalaria sudanica* (Planorbidae) have been found in the depths of the reed and *Sphagnum* swamp on the east side of the lake (Beadle, unpublished). Most of the waters –

small lakes, streams and swamps – within about twenty kilometres of the north-west coast of Lake Victoria from just south of the Katonga River to south of Bukoba (exclusive of the Kagera River valley) are of this character. They are generally free of molluscs and are not therefore sources of bilharzia. The incidence of this disease among the population of this region is comparatively low (Jordan and Webbe, 1969, p. 26).

From Tables 5.1 and 5.2 it would appear that the levels of calcium in these three lakes are close to the minimum at which freshwater molluscs can maintain their calcium balance. The ecological facts, so far known, are consistent with this conclusion. In Lake Lungwe the molluscs are maintaining a tenuous existence with inadequately calcified shells and are free from the attacks of predatory fish. In Tumba and Nabugabo, molluscs are either absent or are present in very small numbers in dense vegetation where they are, perhaps, protected from the mollusc-eating fish living in these lakes. The low calcium content of these waters may so reduce their fecundity and resistance to other hazards that a foothold can be maintained only in the absence of predators. More ecological and experimental work is needed to confirm this interpretation, but it would seem to provide an interesting demonstration of the combined effects of two limiting factors, chemical and biological. Further research on the ecology of such waters might be relevant to the control of mollusc-borne helminth diseases of Man and domestic animals.

Allochthonous organic matter (e.g. leaves from land plants, especially forest trees), which provide most of the basic food for micro-organisms and animals in many of these very dilute waters, initially contains large quantities of the major ions. Much of these diffuse out into the water and are presumably dispersed into the drainage system, and some are absorbed from the water by the living organisms. There is however evidence that a residue is more firmly bound by the organic matter. Even the very diffusible sodium and potassium ions are not all released from leaf mould after prolonged perfusion with pure distilled water (Beadle, unpublished). Here is a source of ions for detritus-feeding micro-organisms and animals, and thus at second-hand for other animals living in ion-deficient water.

Beauchamp (1953a) suggested that the apparently very low concentration of sulphate in several African lakes, including Lake Victoria, is a sign that sulphur is in short supply and is limiting the rate of primary production. This appeared to be supported by experiments involving the addition of sulphate to cultures of algal cells in lake water (Fish, 1956). But, as pointed out by Talling and Talling (1965, p. 455), these experiments were done with foreign algae and under conditions very different from those in the lake. Moreover, later improvements in the method of estimation have shown that the level of sulphate in the lakes is not as low as previously thought. It should be added that extremely low and even indetectible concentrations of phosphate and nitrate are also very common and may be associated with very high rates of primary production, as in Lake George. There is no reason to suppose that the needs of the algae for sulphur are any greater than for phosphorus.

There is good evidence that seasonal and other changes in the rate of primary production in tropical lakes are induced mainly by the presence or absence in the photozone of available nutrient such as nitrate and phosphate. The stimulating effect on production of wind-induced deep stirring is a clear demonstration (Chs 6 and 77). But this is not necessarily associated with a change in the concentration of the major inorganic ions.

It is, however, sometimes assumed that a water of low mineral content, as indicated by conductivity, is correspondingly unproductive. We are handicapped by the scarcity of measurements of primary production in African waters and by the unequal relevance of some of those that have been made to the productivity of the water system as a whole (Ch. 7). Nevertheless, if we can base an opinion on records of the abundance of organisms and of the prosperity of the fishery, this assumption is not obviously justified. The conductivity of a water is almost entirely due to the major ions listed in Table 5.1, and, as already pointed out, we have no reason to suppose that there exists a water in which any of these are so deficient as to exclude all organisms, though in some waters molluscs and decapod Crustacea may be excluded because of insufficient calcium. The water of the Upper Niger is peculiar in that, though the conductivity is low, there is an adequate amount of calcium, and molluscs are abundant (p. 159). There are some examples of waters of low conductivity (less than 100) that are apparently very productive. Lake Mweru in the southeast Zaïre basin is an example that has been investigated (de Kimpe, 1964). This supports abundant populations of many species of algae, invertebrates and fish, and the fishery is assessed as productive. Even Lake Bangweulu further upstream in the same valley (Fig. 9.3), with a conductivity of less than 40 (Table 5.1) has a rich flora and fauna and an apparently productive fishery. There are some lakes of very low mineral content, such as Lake Tumba in the central Zaire basin, in which the rich invertebrate and fish fauna are largely supported by particulate organic matter washed in from the surrounding forests which has been photosynthesised outside the lake. But a predominantly external (*allochthonous*) source of organic matter is not confined to very dilute and acid waters. It is characteristic also of the dense papyrus swamps of eastern Africa where the waters are generally of normal conductivity and with a near neutral reaction (Ch. 18).

Some of the most mineral-deficient natural waters in the world are to be found in the Amazon basin in South America. The limnological work done by the Brazilian National Research Institute at Manaus in collaboration with the Max-Planck Institute for Limnology at Plön, Germany can be studied with advantage by workers in Africa.[*] Many of these waters have a salinity similar to that of the lowest quoted in this chapter, and a few are even more dilute ($K_{20} < 6$) with a correspondingly low level of calcium. Nevertheless a wide range of animals (fish and invertebrates) inhabit them and are nourished mainly from allochthonous sources of food, as in Lake Tumba in the Zaire Basin. Though some workers seem to assume that a very low level of the major ions is necessarily inhibiting, the published data do not seem to contradict the general conclusions which have been drawn here from the work on Tropical African waters. Of particular interest is the occurrence in some of the most acid and mineral-deficient waters in the

---

[*] From 1969 most of this work has been published in the journal *Amazoniana* (Mühlau, Kiel).

[†] Bayly and Williams (1973, p. 74) report that some coastal dune lakes in southern Queensland, Australia, contain less than 1 mg/l Ca^{++} (0·05 meq/l), yet support large numbers of the crayfish *Charax* and the shrimp *Caridina*.

Amazon basin of molluscs with shells composed of pure conchyolin and devoid of calcium. *Ampullarius papyraceus* has a dark brown shell and that of some small ancylids is as clear as glass (pers. comm. H. Sioli). Some waters harbouring these molluscs are recorded as having a 'hardness' equivalent to 0·05 meq/l of calcium (Sioli, 1955). Since magnesium is included in this, the actual level of calcium was even lower.[†]

Marlier (1967) described a Lake Jari in the Amazon basin, shallow (6 m deep) with clear brown water and no emergent or floating vegetation. The phytoplankton was very scarce and there was a negative net primary production in the surface water, pH 4·7–4·8 and K_{20} 2·7. No molluscs were found but the zooplankton though scarce were varied in species composition – planktonic crustacea, tanypid and chaoborid larvae, corixids, ostracods, oligochaets and hydrachnids. The main source of food was undoubtedly the débris from the surrounding forest whose source of nutrients is the soil rather than the lake water. It may be supposed that this organic matter fallen into the lake would bind a significant amount of inorganic ions thus available to animals in their food (see below). Lake Jari has the lowest conductivity of any lake so far recorded in the world, though Lake Waldo in Oregon, U.S.A., is a close contender for this distinction (Larson and Donaldson, 1970). Some relevant data from Lake Waldo are: K_{20} 5·2, pH 6·1–6·6, Na^+ 0·007, K^+ 0·002, Ca^{++} 0·004, Mg^{++} 0·004 meq/l (compare Fig. 5.1). The phytoplankton comprised a very few diatoms whose rate of primary production per volume of water was extremely low, but the bottom of the shallow water was coated with a mat of aquatic mosses with encrusting diatoms. The zooplankton was very scarce but included at least four species of Crustacea. The animals that live in these two lakes are thus capable of maintaining themselves in water with ionic concentrations lower than those found by experiment to exclude some species of animals (Table 5.2).

Recent work on tropical rainforests suggests that much of the nutrients and inorganic ions are taken back by the trees directly from the fallen litter (leaves, wood, etc.) through the action of some of the soil fungi (Mycorrhiza) that pass them from the litter to the tree roots (Stark, 1970). This could account for the fact that though some large areas of the Amazon forests have soils which, for geological reasons, are deficient in minerals and nutrients, they nevertheless support an exceedingly luxuriant vegetation. There is thus an almost closed circulation between the trees and their ground litter augmented by some uptake of minerals from rainwater. The ecosystem works very economically. In these areas cutting and burning of the forest will provide agricultural land whose fertility is likely to be exhausted within a very few years. Such conditions will surely be found in the mineral-deficient regions of the Zaïre basin.

That the rivers and other surface waters in such forest regions are also poor in minerals is understandable. The phytoplankton is characteristically scarce and the rate of primary production very low. The basic food for the aquatic animals is organic matter fallen into the water (leaf litter, insects, etc). The litter decomposes very slowly, mainly through the action of aquatic fungi (Padgett, 1976).

In addition to the above so-called 'major inorganic constituents', the minor or 'trace' elements must be mentioned ('micronutrients', Wetzel, 1975, Ch. 13). In recent years, with improved methods for concentrating samples by exchange-resins and with spectrographic analysis, it has been shown that all natural waters, inland and sea, contain a number of elements that are usually present in only microgram (μg/l) quantities. For example, zinc (Zn), copper (Cu), cobalt (Co), manganese (Mn), molybdenum (Mo), boron (B) and vanadium (V), have been detected and estimated in lakes (Bradford *et al.*, 1968; Groth, 1971). The presence of all these, as well as iron, in minute quantities have been proved necessary for the growth of many freshwater algae (see Hutchinson, 1967, Ch. 21). Their ability to extract these elements from such very low concentrations is indeed remarkable. Unlike nitrogen, phosphorus and silica, however, evidence that deficiencies of any of these elements are limiting to growth of organisms under natural conditions is rare (see Goldman, 1960, for apparent molybdenum deficiency in a lake). But, though evidence for this is difficult to obtain, there remains the possibility that some of the otherwise inexplicable peculiarities in the flora or fauna or in the level of productivity of certain waters may be due to such deficiencies. Some of the elements, notably Fe and Cu, are certainly required also by animals, but it is not known whether they can satisfy their needs from their food or whether, to maintain equilibrium, they depend also upon uptake from the water, as most of them do for preserving the necessary internal concentrations of the major ions. But there is no doubt of their direct importance to plants, and the possibility of deficiencies in African inland waters could be investigated.

It seems therefore from observations and experiments made so far that in 'normal' freshwaters (salinity range about $0\cdot03-5\cdot0^o/_{oo}$) differences in the faunas and floras are rarely explicable in terms of the total or relative concentrations of the major ions in the water. On the other hand there is growing evidence of a general relation between the species composition of the phytoplankton and the level of dissolved nutrients and organic matter which do not contribute significantly to the conductivity or salinity. But the causal links are little understood and must involve other organisms such as bacteria and protozoa, and the nutrient levels are of course partly controlled by the algae themselves (e.g. Fig. 20.5).

Talling and Talling (1965) published inorganic analyses of sixty-seven eastern African lakes done by them and collected from other sources. In general Na^+ and HCO_3^- are the predominant ions, though, partly owing to the nature of the local rocks, the relative concentrations of the other ions such as K^+, Mg^{++} and Ca^{++} vary considerably. The pH, except in the most dilute waters ($K_{20} <100$) is very high (pH $8-10\cdot5$) varying of course with the intensity of photosynthesis at the time of measurement. As would be expected, the very wide range of salinities presented by these waters (K_{20} $15-162, 500$) is correlated with great difference in the proportions of the ions. The ratios Ca^{++}/Na^+ and Mg^{++}/Na^+, for example, are much reduced in the more saline waters owing to precipitation of Ca^{++} and Mg^{++} as carbonates. The very low level of sulphate in those lakes that have an anoxic lower layer is perhaps due to reduction and escape as H_2S or fixation as insoluble sulphide (e.g. of iron) in the sediments. The general scarcity of nitrate in surface waters is more difficult to explain as is the relatively high level of PO_4^- phosphorus and of silicate, especially in the higher salinity waters, but, as explained on p. 44, in comparison with the major ions the concentrations of the nutrients are very low but are subject to much greater percentage fluctuations.

The chemical changes associated with differences in salinity are thus recognisable. To what extent are these changes associated with differences in the fauna and flora? It should first be emphasised that extensive and repeated collecting from many African lakes has shown that each has its characteristic assemblage of phyto- and zooplankton species, of which certain species tend to be dominant (see also p. 111). There are often some seasonal changes in the biomass and in the relative proportions of the species, but these are not usually sufficient to alter the general pattern of species which distinguishes the fauna and flora of one lake from that of another. A notable example is Lake George (p. 237) whose present diatom flora has persisted for about a thousand years. Attempts have been made to explain these differences and stabilities in terms of the chemical composition of the water and the availability of nutrients. It must however be remembered that certain species, especially of animals, are absent from some waters merely because they have been inaccessible since their formation or the water has recently suffered some annihilating catastrophe (e.g. volcanic, or desiccation) and have not yet been fully recolonised. Some species are confined to certain isolated waters only because they were evolved there and have been unable to spread to other waters in which it is now known they can live as well or even better than in their original habitat.

Talling and Talling (1965), though primarily concerned with the ionic composition of the eastern African lakes, discuss the possible ecological effects of differences in water composition as judged from the recorded presence or absence of phytoplankton species. An example is the apparent negative correlation between the occurrence of the common planktonic diatom *Melosira* and the pH and alkalinity of the water.

In an investigation on the distribution of diatoms in a large number of African lakes Kilham (1971b) found a correlation between the concentration of dissolved silicate and the species composition of the diatom flora. Silicate is of course an essential nutrient only for diatoms, and these findings seem to suggest that each species has its range of silicate concentration from which it can most efficiently absorb it.

The Tallings suggested 'conveniently but arbitrarily' a classification of their sixty-seven lakes on the basis of conductivity.
Three classes were proposed.

1. K_{20} below 600 (*c.* $0 \cdot 5^{o}/_{oo}$) 32 samples;
2. K_{20} 600–6 000 ($0 \cdot 5$–$5 \cdot 00^{o}/_{oo}$) 20 samples;
3. K_{20} above 6 000 (*c.* $5 \cdot 0^{o}/_{oo}$) 15 samples.

As mentioned above, correlations have been found between the distribution of some species of algae and the concentration of a certain nutrient. But conductivity is a measure of the sum total of the major ions (salinity) and there is no evidence that in waters within the 'freshwater' range (K_{20} 40–6 000), which are the majority, salinity as such (or the associated ionic composition) is an important ecological determinant for most animals and plants.

It seems to me therefore that no classification based on conductivity can have a wide ecological application other than one which merely distinguishes between

1. the very dilute waters ($K_{20} < 40$);
2. the 'freshwater' range (K_{20} 40–6 000) which includes the majority;
3. the hypersaline ($K_{20} > 6$ 000).

Though the boundaries between these are rather blurred they do in general correspond with very great differences in the fauna and flora. More detailed classifications will depend upon further knowledge of the distribution of individual groups or species in relation to some special determining condition which will not be the same for all species. Salinity is not the only criterion upon which a classification of inland waters might be based. Kilham (1971a, Part 1) studied the inorganic composition of a very large number of African lakes and rivers from his own and others' analytical data. The resulting rather complicated classification takes into account the type of climate, the nature of the predominant rocks in the catchment, the salinity and most particularly the relative proportion of the ions. This classification would seem to relate mainly to the geological origins and to the climates that have determined the present chemical characteristics, but it emphasises the important fact that there are a large number of different environmental conditions that can determine the presence or absence of living organisms. To these must be added the interrelations of the different organisms (e.g. presence or absence of predators and parasites).

In conclusion it may be said that in most tropical African freshwaters (within the approximate salinity range $0.03-5°/_{oo}$, K_{20} 40–6 000) there is so far no clear evidence of the direct ecological effects of salinity or of the relative concentration of any of the major inorganic ions (Na^+, K^+, Ca^{++}, Mg^+, Cl^-, HCO_3^-, SO_4^{--}), though some planktonic algae, as might be expected, seem to be sensitive to the concentration of bicarbonate (HCO_3^-) and the type of diatom flora to be associated with the level of dissolved silicate. Animals appear to be very little influenced by the inorganic composition of the water within this range of salinity. In waters of salinity below $0.03°/_{oo}$ the low level of calcium can apparently exclude molluscs and decapod crustacea both directly and by reducing their resistance to other unfavourable influences. The most dilute waters, such as Lakes Tumba and Lungwe in the Zaïre basin and some of the Amazonian waters, can support a very large number of invertebrate animals but few species. We do not yet know whether or not other species are absent because they have lesser powers for maintaining ionic equilibrium in extremely dilute water: only experiment can decide. The physical and biological conditions in a laboratory culture will however usually differ from those in the natural situation and the sensitivity of an organism to ionic deficiencies may thereby be altered. Another important consideration is that an organism may be adversely affected only at a particular point in its life-history and not at other times. It will nevertheless be eliminated.

Only relatively few species of plants and animals can live in hypersaline waters (K_{20} over 6 000) and possession of an appropriate osmotic regulatory mechanism is certainly one of the necessary conditions for this. These waters are discussed in Ch. 19.

6

Water circulation and stratification in temperate and tropical lakes

Though lakes and ponds are by definition relatively stationary bodies of water, it has been explained in a previous chapter how internal movements of water, especially in a vertical direction, are biologically very important. The ultimate source of the necessary dissolved substances is external, but the continued production of living matter in a lake depends very much on circulation between the upper and lower layers by which the products of organic decomposition, which have sunk below, are brought back to the surface and used again. The energy for these movements and recycling of materials comes ultimately from solar radiation which causes both temperature changes and winds. Climate has in this way an important influence on organic productivity.*

A summary of the principal ways in which water moves may help the reader to appreciate the discussion that follows.

1. Random or turbulent movement

This is ultimately the most important type of movement and can be the end result of most other types. For example, turbulence can develop at the interface between a horizontal current and the substratum, as in a river, especially if the latter is uneven, or between two contiguous water masses of different densities and moving at different rates or in different directions. They 'rub against' each other causing both travelling waves (p. 77) and turbulent mixing along their common boundary (Figs. 6.1 and 6.2) (see Smith, 1975, for a monograph on turbulence in lakes and rivers.

2. Mass-flow of water in a determined direction

An obvious example is the flow of a river downhill which, on discharging into a lake, may continue for some distance as a horizontal current at a depth deter-

---

* An introduction to the physical hydrology of temperate lakes is that of Ruttner (1963) and a more advanced treatment can be found in Hutchinson (1957), Dussart (1966) and Wetzel (1975, Ch. 7). Only the first two deal with tropical lakes, but much more is now known about them.

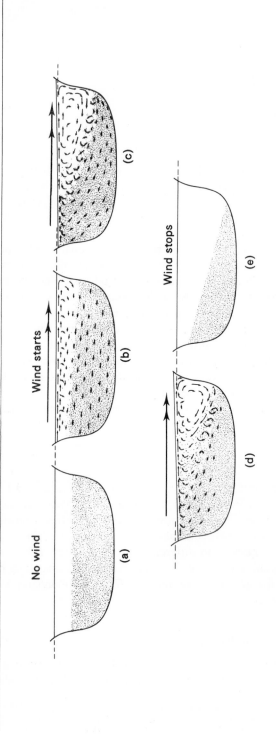

Fig. 6.1 A diagrammatic and simplified representation of the currents caused by a steady wind from one direction in a two-layered stratified lake. The windward end of the lower layer is tipped upwards and its upper surface is eroded by the return current in the upper layer. When the wind stops the currents cease and the volume of the lower layer has been reduced because some has been stirred into the upper layer. There follows a series of internal waves illustrated in Fig. 6.2 (based on Mortimer, 1959, 1969)

mined by its density relative to that of the lake water due to difference in temperature, salinity or turbidity. There must be some turbulence and consequent mixing between such density currents and the overlying lake water. Vertical convection currents arise through loss of heat at the surface resulting in an upper layer which is colder and, thus, denser than the water below it. Conversely, heating at the surface produces a reverse density gradient which resists the stirring effects of winds impinging on the surface.

The deep currents set up by wind stress at the surface of a lake (in addition to the surface waves to be discussed on p. 79) are illustrated in a very simplified manner in Fig. 6.1. The diagram refers to a stratified two-layered system in which there is an upper layer (epilimnion) overlying a denser hypolimnion (see p. 81), but in an unstratified and thus homogeneous lake the currents are similar to those initially set up in the upper layer (epilimnion) of a stratified lake (Fig. 6.1b) and have not therefore been separately illustrated. In this, the surface water forced forward by the wind causes a rise in level at the far end and returns as a deeper current in the opposite direction and there is a turbulent interface between the two currents. The whole water column in an unstratified lake can thus be brought into circulation.

In a lake stratified into two layers of different density the principal water movements induced by a powerful and steady wind are:

(a) a piling-up of surface water at the downwind end of the lake and a compensatory return current in the hypolimnion which brings it nearer to the surface at the windward end;

(b) a rotating circulation in the epilimnion which progressively erodes the upper surface of the hypolimnion and thereby brings some of the lower water into the upper circulation;

(c) finally, if the wind stops before the mixing of the two layers is complete, the volume of the hypolimnion has been reduced and the thermocline has been lowered.

Figure 6.1e represents the moment at which the wind stops. The interface between the two layers then swings back to the horizontal and beyond. In this manner an internal oscillating wave is started, and this will now be described.

3. Standing waves

In an unstratified lake the piling-up of surface water downwind (as described above) produces a disequilibrium as soon as the wind stops. The water then flows back to bring the surface to the horizontal and the original levels are restored. But this position is overshot and the whole oscillates back and forth like a pendulum with decreasing amplitude, and finally, in the absence of more wind, settles down with a horizontal surface. This process can be imitated by momentarily tipping up one end of a bath.

Such oscillations, known as 'seiches', marked by a periodic rise and fall of level at each end, have been known in European lakes for very many years. The period of oscillation around a middle horizontal line (node) depends primarily upon the length and depth of the basin. In European lakes, periods vary from a few minutes to over two hours with maximum vertical amplitudes at each end of a few

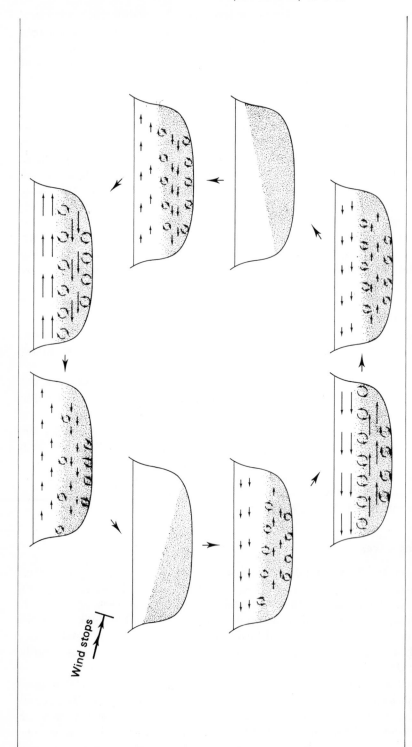

Fig. 6.2 One cycle of a simple internal wave in a two-layered lake following the cessation of a steady wind in one direction (see Fig. 6.1). The two layers swing reciprocally to and fro with decreasing amplitude. The opposing currents in the two layers gather speed to a maximum as they swing through the horizontal position. The turbulence between the layers and between the lower layer and the bottom are then maximal. The two lengths of the arrows and the two diameters of the circles indicate changes in the rate of flow and in the violence of the turbulence respectively

centimetres to over one metre. The situation is often complicated by several simultaneous seiches each with its own node; or even by one with its node line at a different angle to the others; i.e. produced by a wind from another direction.

These oscillations, which are often visible as large changes of water level (surface standing waves), are characteristic of lakes whose water is of more or less uniform density. When, however, there are two or more layers of different density, as during summer stratification (p. 81), the periodic shift of water in the upper layer induces a corresponding movement in the opposite direction in the layer below it. Such 'internal seiches' are detected from periodic changes of temperature or chemical composition at a series of depths measured at a fixed point in the lake, and for this reason are also known as 'temperature seiches'.

Temperature and chemical measurements in some of the large tropical African lakes (e.g. Lakes Tanganyika and Malawi) have disclosed some internal standing waves of considerable magnitude, and the associated chemical and biological changes suggest that they are important in maintaining productivity.

The oscillatory cycle in a two-layered system is represented diagramatically in Fig. 6.2. For further details see Mortimer (1952, 1953, 1959, 1969) and Hutchinson (1957). Biologically the most important features are:
(a) the periodic rise towards the surface at each end of the lower, denser and nutrient-rich water and
(b) the course of the currents in the two layers which flow in opposite directions and thus provide conditions for turbulent mixing within the layers, at the interface between them and along the bottom.

Both the period of an internal seiche, which depends upon the depths and densities of the layers as well as upon the dimensions of the lake and its amplitude, are greater than those of a surface seiche. Periods range from a few days to three or four weeks with maximum amplitudes of 10 to 50 m. There may be more than two layers, in which case the horizontal currents caused by the oscillations, flow in alternate directions in one layer and the next. There are often more than one wave system due to successive winds from different directions, and an additional complication may arise in a large lake from the drag due to the earth's rotation (Coriolis force) which may influence the direction of the tilt and the course of horizontal currents and may involve a rotating wave system (Mortimer, 1969). Such complications will surely be discovered in the large African lakes. In a particular lake however the situation is not likely to be as hopelessly complicated as might be inferred from the above remarks. Since the dimensions of the lake are fixed, and the climate, particularly the prevailing winds, of the region have certain characteristic features, the water movements in any one period are likely to be dominated by a particular wave system.

4. Travelling waves

These are the most familiar of the several types of oscillating systems in water.* Winds that impinge on the water surface not only, as described above, cause

---

* For an introduction to the study of travelling surface waves and their subsurface effects see Smith and Sinclair (1972).

mass movements of surface water and consequent deep current systems. They also set up surface waves travelling in the same direction which persist with decreasing amplitude after the wind has dropped. Travelling surface waves are associated with subsurface oscillations that are of considerable biological importance in shallow water and in the upper regions of deeper lakes. The water particles beneath a wave move as though attached to the edge of a vertical disc rolling forwards in the same direction (trochoidal wave motion, illustrated in Fig. 6.3). Their net forward motion is thus considerably less than that of the surface waves above them. As in the successive falling of a row of upstanding dominoes, it is the energy that travels, not the material. Such a wave may transfer energy over very great distances – even thousands of miles across oceans. But it is the vertical oscillation and the resulting mixing of the water that interests us most. The vertical amplitude clearly determines the depth to which the water is directly stirred by surface wave action. The most important factors that determine the amplitude of the subsurface trochoidal waves (as well as the height and wavelength of the surface waves themselves), and thus the depth to which the water is stirred by them, are the force of the wind and the 'fetch'. The latter is the distance over which the wind impinges on the surface.

The controlling conditions are illustrated in Fig. 6.4. In theory, and approximately in practice, the depth of mixing is about one half of the wavelength (the distance between wave-crests). It is clear from the shape of the curves in Fig. 6.4 that a powerful wind of 19·5 m/sec (45 mph) would cause mixing to a depth of not more than 30 m with a fetch of 100 km and over. The corresponding depth with a wind of 4·5 m/sec would be about 11 m. But, with a normal overall annual regime of winds, surface waves can be effective only in very shallow water as agents in the direct circulation of water from the bottom to the surface (There are presumably some less powerful secondarily induced movements set up below the level immediately stirrred by the surface waves, which may thus affect in some degree deeper water than is deduced from Fig. 6.4). Many lakes in Africa, however, including the very large Lake Chad and the relatively small Lake George are shallow enough over most, if not all, of their extent to be stirred to the bottom by surface wave action during the frequent periods of heavy wind. Moreover, the shallow peripheral regions of all deep lakes must also be so affected, which is one of the several reasons why the inshore waters are in general more productive than the deeper pelagic regions. The back and forth motion along the mud–water interface caused by wind-induced surface waves in shallow water with consequent stirring of the mud are discussed by Viner and Smith (1973) in relation to Lake George. Apart from the distribution of sufficiently shallow water and the length of fetch, the effectiveness of this kind of stirring will obviously depend upon the total annual duration of winds that are powerful enough to cause stirring to the bottom. The controlling factors change from time to time both seasonally and with less predictable storms and calms. In some regions of a lake the depth will be such that surface waves will affect the bottom only at rare intervals. The oscillating rise and fall of the nutrient-rich lower water at both ends of a lake following a powerful prevailing wind (Fig. 6.2) could periodically bring this water within the range of surface waves.

In lakes too deep ever to be stirred to the bottom by surface wind-action, the depth of the superficial region of mixing, marked by the lower limit of oxygen, is therefore partly determined by the strength of the winds and the length of the fetch. For example in Lake Tanganyika, 650 km in length and well exposed to the heavy southeast trade winds, this limit is 150–200 m below the surface (p. 00). In the small and well-sheltered crater-lake Nkugute, 1 km in diameter, it is 7–10 m (p. 96). There are of course many lakes whose indented coastline and islands provide shelter from wind and reduce the available wind-fetch in some parts of the lake. These are therefore less deeply stirred than the unsheltered but maybe deeper water of the same lake. The permanently stratified and deoxygenated lower water in the sheltered branches of some of the volcanic barrier lakes, such as Lake Bunyoni, are examples of this (p. 102).

Surface waves have an important ecological influence on the shoreline. The arching and breaking of the crests of waves are due to the circular motion of the subsurface water illustrated in Fig. 6.3. Shoreline erosion, with deposition and shifting of beaches, is caused mainly by surface wave action which thus determines the type of fauna and flora that can be established along the shore and in the very shallow water. A well-known example of special adaptations to these violent conditions are the prosobranch molluscs living on the wave-battered rocks along the shores of Lake Tanganyika (p. 288).

Surface travelling waves are a special case of waves generated at the interface between two fluids (in this case air and water) of different densities and moving at different speeds. They can also arise between two contiguous water layers in a stratified lake and even between the lowest water layer and a liquid mud over the sediments. Such differential horizontal movements are associated with the internal standing waves already discussed and lead to the turbulence and consequent mixing between layers as represented in Fig. 6.2.

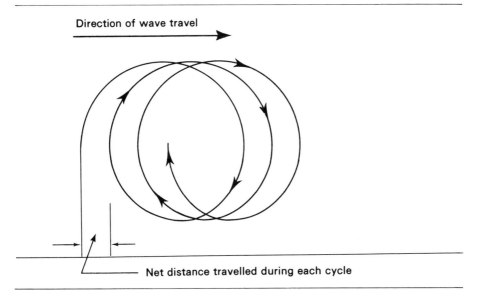

Fig. 6.3 Travelling surface waves. Subsurface motion of water particles (redrawn from Smith and Sinclair, 1972)

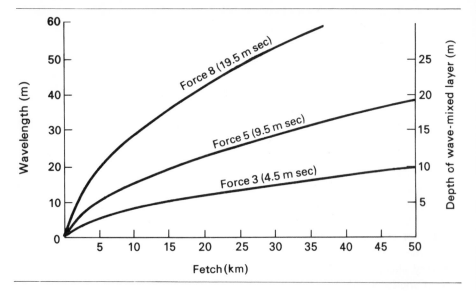

Fig. 6.4 Travelling surface waves. Relation between depth of wave-mixed layer (wavelength/2), fetch and windspeed (redrawn from Smith and Sinclair, 1972)

With the foregoing discussion as background we can now consider the main features of the water circulation in tropical lakes. As a basis for comparison with lakes on the lower latitudes of Africa, we shall first describe briefly the seasonal course of events in a typical lake under northern temperate conditions where winter temperatures are low enough to freeze the surfae for several weeks and the water is deep enough to develop a stable stratification in summer. In the spring and early summer, the intensity and duration of solar radiation increase and most of what is not reflected at the surface is absorbed in the upper 50 m, even in the clearest water in midsummer. Some of this radiant energy is absorbed by the algae (Ch. 7), but most, especially the infrared component, is immediately dissipated as heat which in early summer accumulated more rapidly than it is lost by back-radiation to the air or through evaporation at the surface. The temperature of the upper layer thus increases and the density decreases compared with that of the water below it, which is not reached by the radiation.

In the early stages the density gradient may not be sufficiently steep to resist stirring by winds, but, eventually, except in very shallow lakes, a stable condition of *stratification* is reached, which is usually characterised by two main layers – an *epilimnion* of warmer, less dense water at the surface and a colder and denser *hypolimnion* at the bottom (Fig. 6.5). Between these there is often a *metalimnion*, a region of discontinuity across which the temperature gradient is steepest (the *thermocline*).* Only the epilimnion is subject to diurnal heating and cooling and to

---

* On calm days in the temperate summer and throughout the year in the tropics the water at the very surface may become very much hotter (by seveal degrees) than at a few centimetres depth. There is, then, a thermocline just below the surface and a stratification that is normally dispersed by cooling at night. Such temporary diurnal stratification near the surface can have important biological effects only in shallow lakes in which the water is not deep enough for the development of a stable lower layer because the disturbances due to the daily fluctuations of temperature affect most of the water column.

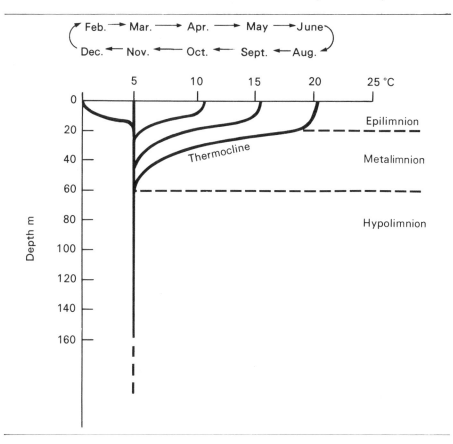

Fig. 6.5 Theoretical temperature changes and stratification in a lake in a temperate climate

direct stirring by winds, and the depth of the thermocline is a reflection of the strength and duration of windy weather since the surface heating began. The hypolimnion, which is thus insulated from disturbance so long as these conditions persist, has a temperature which reflects the severity of the previous winter and the stability of stratification. It also cannot be much less than 4°C at which the density of water is maximal (Fig. 6.6).

In midsummer the stratification is usually very stable and resists deep stirring by the winds. The degree of stability can be estimated quantiavely in terms of the amount of energy required to overcome it and to disperse the thermocline in the manner shown in Fig. 6.5. From the temperature-depth curve the corresponding density curve can be calculated (corrected if the density is also significantly determined by differences of salinity. From this the amount of energy required to stir the column to a homogeneous condition can be calculated on the assumption that there is no gain or loss of heat during the stirring. The figure so obtained represents the *stability* and can be expressed as kilogram-metres per volume of water below one square metre of surface to a specified depth (Kg m m²h). It is the energy needed to lift the weight of the column from its present centre of gravity to that of the same column when completely mixed (see Ruttner,

1963, p. 28 ff. for a simple treatment of Schmidt's formula, with references to further developments). It is obvious that the stability of a temperate lake comes to a maximum during the summer and is zero at the overturn.

The very great stability which can be attained during summer stratification was well demonstrated by Linsley Pond, Connecticut, during a hurricane in September 1938 which lowered the thermocline by only about two metres; that is to say, only a small proportion of the hypolimnion was brought into circulation (Hutchinson, 1957, p. 452). Alternate periods of wind and calm, especially during the early summer when stability is low, can lower the thermocline and produce another above it, resulting in a condition of multiple stratification.

At temperate latitudes stable stratification thus occurs just before and during the period of maximum illumination and rising temperature, when photosynthetic production by the planktonic algae is at a maximum (Ch. 7). The consequent outburst of plant and animal life in the epilimnion soon produces a shower of dead organic matter which falls into the hypolimnion and begins to decompose by autolysis and bacterial action. There is thus a progressive depletion by the algae of nutrient salts at the surface and accumulation below the thermocline through decomposition of the descending organic matter, and the decomposing microorganisms consume oxygen in the hypolimnion. The two layers become chemically differentiated to an extent which depends upon the rate at which organic matter is produced and on the stability of stratification.

Only a portion of the organic matter is decomposed before it reaches the bottom, and decomposition is retarded as soon as the oxygen is exhausted. This, as stated in the previous chapter, is the origin of the organic mud which is deposited at the bottom at a rate depending partly on the rate of production at the surface. Below the surface of the mud oxygen is always absent and the organic matter is preserved indefinitely in the manner of silage (Fig. 4.1). It follows that oligotrophic lakes accumulate little organic mud.

There are, as would be expected, many intermediate stages between extreme oligotrophic and eutrophic conditions. It can be well understood from the above how research on lakes in the north temperate zone led to the conclusion that the degree of depletion of oxygen in the hypolimnion during summer stratification is a measure of productivity (Ruttner, 1963, p. 69). How far this concept can be applied to tropical lakes will be discussed later.

The seasonal changes can be more dramatically illustrated by plotting the depths of the isotherms and oxygen isopleths (lines joining points of equal concentration) throughout the year. As an example, Fig. 6.7 represents the annual thermal changes in Esthwaite Water, a small lake in northwestern England (Mortimer, 1941). Stratification is indicated by a horizontal arrangement of isotherms and mixing by a vertical arrangement. A thermocline is shown by the crowding together of horizontal isotherms. During the period of stratification (late May to early September) the initial thermocline was forced downwards and another appeared temporarily at 6 m in later July. This was the result of changeable and stormy weather. From September onwards, as the air temperature decreased, the isotherms one after the other reached the surface until the whole column had come to a uniform temperature of 11°C, i.e. was 'homothermic'. With the loss of stability the lake was then completely stirred by winds as shown by the uniform

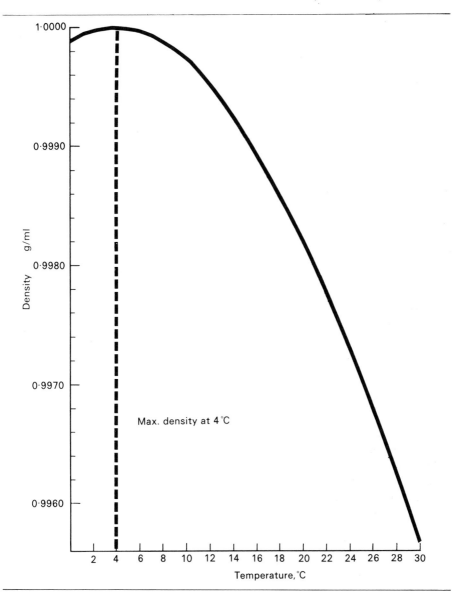

Fig. 6.6 The relation between the temperature and density of water

distribution of oxygen at this time (Fig. 6.7b), and the whole had fallen to 4°C by the end of December. Since the density of water is maximal at 4°C (Fig. 6.6) further cooling at the surface initiated another period of stratification and the floating sheet of ice gave additional protection against stirring (Fig. 6.7a). In the spring, as soon as the surface temperature passed 4°C, there was another over-turn, and the whole remained unstable until summer stratification was firmly re-established.

There was a progressive fall in oxygen in the hypolimnion during summer

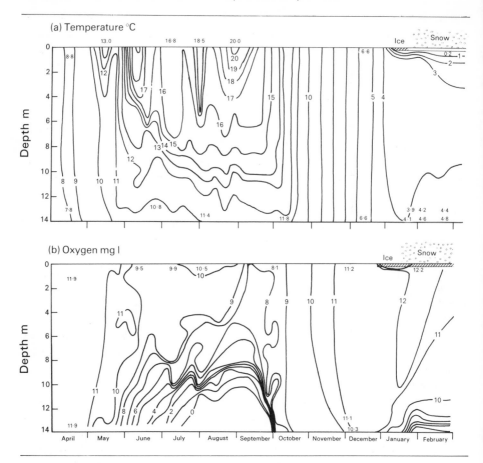

Fig. 6.7 Temperature and oxygen changes in a typical eutrophic temperate lake (Esthwaite Water, English Lake District) (redrawn from Mortimer, 1941)

stratification and at the overturn it was distributed through the whole mass (Fig. 6.7b). The higher concentration at the bottom during the winter stagnation than in summer was due mainly to the much lower temperature, which reduces both the number of organisms and their rate of respiration.

Simultaneous measurements of the concentrations of substances derived from decomposition, such as ammonia, nitrate, nitrite and silicate, demonstrated how these accumulated in the hypolimnion during summer stratification and were dispersed through the whole column at the autumn overturn (Mortimer, 1941).

Lakes at high latitudes are thus subjected to two overturns in the year (in autumn and spring) and are called *dimictic*. At lower latitudes, where the winter is not severe enough to cool the surface below 4°C, mixing continues from autumn to spring as a single period of overturning. In such *monomictic* lakes the bottom temperature during summer stratification is always above 4°C, and reflects the average winter temperature, which, of course, is inversely related to latitude and altitude. For example, the summer temperature at the bottom of Lake Garda in the southern European Alps (lat. 45°30′N, alt. 65 m, max. depth

346 m) is about 7°C, whereas that of Lac d'El Kansera on the coastal plains of Morocco (lat. 33°0'N, alt. 135 m, max. depth 45 m) is 13–14°C. On the other hand, Aguelmane Sidi Ali, a small lake high up in the Moroccan Atlas Mountains (lat. 32°30'N, alt. 2 050 m, max. depth 40 m) is stratified in summer with a bottom temperature of about 7°C (Gayral, 1954).

The development of summer stratification is thus promoted by increasing solar radiation, but can be delayed or even, under some circumstances, prevented by windy weather. It follows from the previous discussion on types of water movement that the effectiveness of the wind as a mixer depends not only on its strength, but also on its direction in relation to the orientation of the lake, on the surrounding topography which may expose it to or protect it from winds, and on the relation between the fetch or distance over which the wind operates and the depth of water to be stirred. In very shallow lakes and ponds there may be insufficient water below the level directly heated by radiation for summer stratification to become stable enough to resist normal winds. Such *polymictic* waters can be, at most, temporarily stratified during periods of warm and calm weather. But with the peculiar temperature conditions in the tropics, as we shall see, the degree to which the surface is exposed to winds is a major factor determining the extent and duration of stratification even in deep lakes.

The stability of stratification is in some cases greatly increased by a relatively high concentration of dissolved salts and thus an increased density of the lower water, and in extreme cases the overturn cannot reach the bottom, leaving a mass of water more or less isolated from the wind-induced circulation in the upper layers. Such a lake is called *meromictic* because it is only partially stirred, a completely circulated lake being *holomictic*. There are several examples in Europe and North America of lakes which are in this condition through an inflow of highly saline water derived from the surrounding or underlying rocks (Hutchinson, 1957, p. 482 f.). A striking example from Africa is Lake Kivu in Zaïre (Ch. 15). Some others, as Lake Lugano on the Swiss-Italian border, have recently become meromictic from industrial and domestic pollution which provides an inflow of dense saline water and of organic matter decomposing to inorganic salts. There is also evidence that meromixis can result from natural biological causes. At the bottom of relatively deep and sheltered lakes, and particularly during the undisturbed period of winter ice-cover, natural organic decomposition at the bottom can release enough dissolved substances to increase significantly the density of the lower water (Hutchinson, 1957, p. 489 f.). In some moderately deep African lakes (e.g. Lake Edward, p. 00) the density gradient due to dissolved substances is much less than in Lake Kivu, but is probably still a factor in maintaining a condition of prolonged stratification which is resistant to stirring by normal winds. But in the very deep meromictic lakes, Lakes Malawi and Tanganyika, the density gradient is apparently very slight. Whatever the cause, meromictic lakes are evidently less productive than they would otherwise be if they were periodically completely overturned. The stagnant bottom layer is continuously accumulating nutrients derived from organic decomposition, of which only a portion is brought to the surface and re-enters the production cycle. It would, however, be a mistake to conclude that the hypolimnion of a temperate stratified lake is a sink in which organic matter and the nutrients derived therefrom are irrevocably locked up and fail to reach the surface for resynthesis until the autumn overturn. The state of temporary or permanent stratification is a *dynamic* equilibrium. It is clear that the position of a discontinuity between layers of different density is determined by a balance between wind-induced stirring and, in the case of the thermocline, the influx of heat from above which tends to maintain the temperature and density gradients. The position of the anoxic boundary is determined by a balance between the rate at which it is being eroded by turbulence and mixing with oxygenated water from above, and the rate at which oxygen is being consumed by decomposition of organic matter below. In calm weather these boundaries tend to rise, and conversely heavy winds cause them to sink. In fact, there is good reason to suppose that both the hypolimnion of a temporarily stratified lake and the permanently anoxic layer of many tropical lake are only relatively 'stagnant'. The possible modes of exchange between hypolimnion and epilimnion in a temperate lake during summer stratification have been discussed by Kraus and Turner (1967) and by Blanton (1973). These are (1) by erosion of the upper surface of the hypolimnion through turbulence in the epilimnion, and (2) by 'leakage' round the periphery of the hypolimnion where its surface touches the bottom. There is evidence that primary production in some lakes is stimulated by the consequent recycling of nutrients from below during the period of stratification. The seasonal rise and fall of the anoxic boundary and the remarkable uniformity in chemical composition throughout the entire 1 400 m depth of water in Lake Tanganyika are surely indications of considerable vertical circulation (p. 97). The abundance of pelagic fish in Lake Tanganyika and of the organisms upon which they depend makes it difficult to believe that the recycling of nutrients is very much retarded. Future research will no doubt disclose the truth.

Our main concern in this chapter is to compare the regimes of circulation in lakes under temperate with those under tropical conditions for which Africa, with its enormous expanses of inland water within the tropics, provides abundant material. Except under the peculiar conditions at very high altitudes (p. 107) waters within the tropics are not subjected to freezing nor to temperatures as low as 4°C. We might at first sight expect that the comparatively small seasonal variation in tropical temperature would be less conductive to stable stratification than in temperate lakes, where the annual stratification and overturn is primarily dependent on a wide temperature swing. A glance, however, at the density-temperature curve of water (Fig. 6.6) will at least provide some food for thought. The increase in density between 20° and 24°C is about eight times that between 4° and 8°C. It follows that, other conditions being the same, a relatively slight temperature gradient in a tropical lake will maintain the same stability as a much steeper gradient in a temperate one. Conversely, a much smaller drop in temperature at the surface is needed to obliterate stratification and, on further cooling, the same reversed gradient would cause greater instability in a tropical than in a temperate lake.

With these general points in mind we must now consider the most relevant information available from the tropics (mostly Africa) and decide how much can be concluded from it. The first important limnological investigations on tropical lakes were made by the German Expedition to Java, Sumatra and Bali in 1929 (Ruttner, 1931). It was clear from this work that the seasonal hydrological regime was very different from what had been found in temperate lakes. Ruttner concluded that regular seasonal temperature changes were not large enough to induce deep stirring but that overturns, which were not actually observed at the time, probably occurred irregularly at least in the shallower lakes as the result of occasional and abnormally violent winds and low temperatures. The observations could not be continued for long enough to confirm this suggestion, but an expedition in 1974 to one of these lakes, Ranu Lamongan in Java, was fortunate to arrive at the time of a complete overturn shown by vertical homothermy, reduction of oxygen in the surface water and a consequent fish-kill. The progressive return to stratification was followed during the subsequent few weeks (Green et al., 1976). Ranu Lamongan, at a latitude of 7°59'S, is a small volcanic lake (diameter 800 m, maximum depth 30 m), but apparently sufficiently sheltered from normal winds to be stratified for long periods. The anoxic boundary is then at a depth of 5–10 m. In dimensions and in state of stratification it is somewhat similar to Lake Nkugute (p. 102) but it has a much richer fauna and appears to be considerably more productive. It is however not as well protected by hills from stirring by winds and the nutrients in the lower water are no doubt more frequently recycled. Interesting comparisons were made between the fauna and flora of Lake Lamongan in 1929 and 1974. They had altered very little.

During the past forty years chemical and physical measurements have been made at intervals in the large African lakes, and the conditions prevailing at the time of investigation have been described. Only very recently have sufficiently continuous measurements been made over a long enough period to justify some general conclusions regarding the regimes of circulation. Lakes that are probably shallow enough to be frequently stirred to the bottom by surface wave action are often highly productive. Such are Lake Chad (Ch. 12) on the southern edge of the central Sahara (max. depth variable c. 5 m), and the much smaller Lake George (Ch. 13) on the Uganda–Zaïre border (max. depth 4·5 m). The four small lakes on the Nyawarongo Plain in Rwanda, investigated by Damas (1954II), are all less than 5 m deep and are subject only to the usual diurnal surface stratification. The same polymictic condition is shown by the small artificial lake Nungua (max. depth 4·5 m) in southern Ghana (Thomas, 1966). Lake Upemba in the upper Lualaba valley in Zaïre (1500 km^2, max. depth 2 m) is the site of intense algal production. The water is almost continuously stirred and can be supersaturated with oxygen to the bottom during the day (van Meel, 1953).

The high gross primary productivity of Lake George will be discussed in the next chapter, but production is maintained at a high level mainly by frequent stirring of the entire water column including the surface of the liquid mud, whereby the euphotic zone is well supplied with nutrients. Nitrate and phosphate are however so rapidly assimilated by the abundant phytoplankton that their actual concentration in the water is normally very low. The typical daily regime involves a calm night and early morning. As the solar radiation increases there develops a steep temperature gradient from the surface. Some time towards midday comes the wind, building up to a maximum and falling away during the later afternoon. This, added by a fall in surface temperature, effectively breaks down the stratification and all the water and some of the mud are thoroughly stirred. The typical daily regime may be augmented by heavy rains, thunderstorms and extra strong winds at other times of day or night. The winds are also responsible for surface seiches detectable by a continuous level recorder which oscillates over a few centimetres within a mean period of 1·7 hours. During the shortlived stratification the surface and lower water oscillate horizontally in different directions as do the more prolonged internal waves in deeper lakes (Viner and Smith, 1973). Dramatic departures from this rather regular regime have been demonstrated in recent years on some rare occasions of prolonged calm weather. In October 1957 there was a windless period of more than two weeks followed by a violent storm. It was estimated that over 1·3 million commercially valuable fish, and a much greater number of smaller ones, died of asphyxiation within a few hours. The main cause of this disaster, though not investigated chemically at the time, seems to be clear.* Owing to the very high concentration of algae and other organisms which obstruct the light, the extremely rapid photosynthetic production of oxygen during the day is limited to a surface layer of much less than a metre in depth (euphotic zone; Table 7.1). The temperature is high (24–28°C) and in the lower water oxygen is rapidly consumed by the abundant organic matter, both alive and decomposing, but frequent and thorough mixing by wind normally maintains sufficient oxygen throughout, even at night when photosynthesis stops. During periods of prolonged and quite calm weather, which are fortunately rare, the daytime stratification is probably not dispersed by wind, the lower water would soon become deoxygenated, and the fish confined to the upper oxygenated layer which at night would become even shallower owing to the cessation of photosynthesis. A sudden storm would then mix the whole and drastically reduce the oxygen concentration at all depths. The fish on this occasion were seen gasping at the surface and the conclusion that death was due to asphyxiation was supported by the fact that the victims were all of the genera *Sarotherodon*, *Haplochromis* and *Bagrus* which are gill-breathers; the lungfish (*Protopterus*) and the mudfish (*Clarias*) survived, and it is these which have aerial respiratory organs and can live without dissolved oxygen.

Here is an example of the phenomenon referred to above, in which surface thermal stratification is prolonged owing to lack of wind-stirring and to insufficient cooling at night. The water is so shallow that the decomposing mud and suspended plankton and detritus remove the oxygen from all of the water except for a thin layer at the surface which is oxygenated by photosynthesis.

The phenomenon is of great practical importance for the pond-culture of fish in the tropics (see Hickling, 1962, pp. 45–8). In Uganda, it has been shown that exposure to wind by suitable siting of ponds and removal of protecting floating vegetation, such as water lilies, are essential to ensure adequate oxygenation. In some cases deoxygenation at the surface at night, augmented after a calm day by an overturn due to surface cooling, can bring the fish gasping to the surface in the morning before photosynthetic oxygenation begins to revive them. The problem of oxygen supply is complicated by the addition of fertilisers to ponds, which through increasing production of oxygen at the surface, reduces the depth of the

---

* Another disturbance of this kind occurred in 1969, and observations supported, in general, the explanation proposed here, and disclosed a striking, though temporary, effect on the nutrient cycles.

euphotic zone through self-shading by the much augmented algal population and increases oxygen consumption throughout.

Damas (1937) made one set of observations on Lake Ndalaga, a small and shallow volcanic-barrier lake in a steep-sided valley in eastern Zaïre (Fig. 15.1). The bottom 3 m of water was anoxic (max. depth 20 m), but the upper water was also low in oxygen and was only about 65% saturated at the surface. This Damas interpreted as evidence of very recent stirring following a period of stratification with deoxygenation of much of the lower water. There appeared to be no fish in the lake at the time (April) during the cool dry season, but the local fishermen assured him that this was a regular seasonal event and that at other times of the year the fish were abundant. Perhaps the fish which survive the overturn take refuge in the outflow or inflows, or the lake may be repopulated from other small lakes in the neighbourbood during the rainy season. This situation would repay further investigation.

The occasional but less disastrous deaths of fish in deeper lakes have been reported from time to time from East Africa and Indonesia (Ruttner, 1931). They are undoubtedly due to sudden upwelling at the surface of stagnant water from below, with consequent deoxygenation and perhaps release of hydrogen sulphide. These situations have not been examined in detail, but they are presumably caused either by internal waves bringing anoxic water at some points to within range of surface mixing, or by a particularly violent storm deepening the wave-mixed layer to reach the lower anoxic water.

It is to be expected that there are some relatively *shallow* tropical lakes that are sufficiently protected from winds to allow stratification to develop but not to be continuously maintained during the calmer season. This is demonstrated by the work of Imevbore (1967) on Lake Eleiyele, the artificial reservoir for the town of Ibadan in Nigeria. It is about 10 m deep. Fortnightly daytime measurements of temperature, oxygen and other chemical constituents during a whole year suggested that stratification was permanently maintained, though less strongly between June and September than from October to May when the temperature gradient was greatest. But some dawn and midday measurements taken over three days in June showed that during one night the water became homothermic from top to bottom and the *status quo* was rapidly restored as the surface water was warmed by the sun next day. Moreover, though oxygen disappeared below 5 m from October to December (i.e. at the beginning of the period of strong stratification), some oxygen subsequently reached the bottom, though the isotherms from daytime measurements continued to show an apparently very definite stratification. This case goes to show that, under certain conditions of temperature, depth and of exposure to winds, the circulatory regime of some shallow tropical lakes is of a character intermediate between that involving a usually daily overturn, as in Lake George and Lake Chad, and that with a normally single seasonal period of deep stirring characteristic of deeper lakes. Such occasional nocturnal overturns could be spotted only if routine measurements of temperature are taken at dawn.

There are two other investigated examples of *shallow* tropical lakes whose dimensions, situation, surrounding topography and climatic regime favour a regular seasonal alternation of overturn and some degree of stratification. These are the artificial lakes Mwandingusha in the southeast Zaïre basin (lat. 9°S, max. depth 14 m: Magis, 1962) and McIlwaine in Zimbabwe (lat. 18°S, max, depth 24 m: Falconer et al. 1970, and Fig. 26.6). In both lakes a temperature gradient develops during the calmer and warmer period from October to April, though the water column is not sufficiently stable to prevent some vertical movement of water as shown by the increase of temperature at the bottom during this period. It is possible, though not recorded, that, as with Lake Eleiyele (above), the temperature gradient may be sufficiently reduced at least on some nights to cause some mixing to the bottom, especially if there happens also to be a heavy wind. During the windy and cooler season from May to September there is a homothermic water column with more or less continuous deep stirring.

Lake Albert at the north end of the Western Rift (Fig. 3.4) is an example of a large tropical lake (5 000 km²) which was not originally thought to develop a deep stratification nor to show marked seasonal changes in circulation, owing to its shallower depth (maximum 55 m) relative to the other large Rift Valley lakes and to the violence of the winds which blow along the deep trenchlike valley (Worthington, 1929; Verbeke, 1957a). Nevertheless Fish (1952) and Talling (1963) have shown from measurements of temperature, oxygen and other chemical constituents, that thermal and chemical gradients develop, but without a marked thermocline, below the diurnally heated surface layer during the calmer parts of the year (e.g. November and March). At these times oxygen in the bottom water is much depleted but not exhausted, and nitrate and phosphate accumulate. In August, after strong southerly winds, the water column becomes thermally and chemically almost homogeneous, showing that it is mixed to the bottom. Talling (1963) suggested from the study of Lake Albert that thermal stratification might be augmented by 'marginal cooling'. He found that the water in the extensive shallows at the south end was sometimes cooled at night to a lower temperature than that of the surface water offshore, and that this denser water apparently flowed down the slope under the warmer surface. This could, in principle, accentuate the temperature gradient and might contribute to the general circulation. The possible importance of such profile-bound density currents for the circulation of other tropical African lakes will be mentioned later. Further investigations are needed on other lakes.

Lake Turkana in the Eastern Rift in northern Kenya is longer and deeper (9 000 km², max. depth 120 m at the south end) but otherwise similar to Lake Albert in its shape and in its surrounding topography. But local heavy winds almost daily through the year ensure that the lower water is continuously well-circulated and oxygenated (p. 181). In neither lake have internal wave systems been demonstrated. More can be said with certainty about Lake Victoria.

Lake Victoria (Fig. 14.1 and Ch. 14) has an enormous surface area (about 69 000 km²) but is relatively shallow (max. depth 79 m, av. 40 m). It lies across the Equator between latitudes 0°30'N and 2°50'S. The climate of the northwest (Uganda) region of the lake is more uniform with a higher rainfall than that in the southeast (Tanzanian coast) where the seasons are more marked with a much longer warm and dry season. These differences and the broken nature of the coastline, with its many shallow gulfs and bays, might well discourage attempts to find a hydrological regime for the lake as a whole. The first fisheries survey of the lake in 1926–27 by Graham (Graham, 1929) provided evidence of both temperature stratification and of homothermal conditions in different regions, but it was generally supposed that there was no regular seasonal regime of alternating stratification and overturn in view of the apparent uniformity of the tropical climate; a conclusion also reached by Ruttner (1931) from his work in Java. The measurements made by Beauchamp (1939, 1953b) on Lakes Tanganyika and Malawi gave an indication of seasonal water movements, but it was in Lake Victoria in 1952–53 that the first clear evidence of a regulr seasonal cycle in a tropical lake was found by Fish (1957). This has been confirmed by others (Newell, 1960; Talling, 1957b, 1964, 1969) and is now recognised as a characteristic feature of several other large lakes in tropical Africa.

There has been some disagreement on the detailed interpretation of the data, especially on the evidence for internal waves (see Newell, 1960; Talling, 1966) that will no doubt be resolved by further research, but the main features of the annual cycles in the open water are now clear. These are illustrated by the isotherms in Fig. 6.8a plotted from Fish's data got in 1952–53 at a station about 25 km off the north coast. An essentially similar picture was found in 1960–61 at the same station by Talling (1966).

It is very striking how small, in comparison with temperate lakes, are the changes in water temperature throughout the year. From September to January there was a small but definite rise in temperature and the heat content of the whole column was increased. This trend led ultimately to stratification with a thermocline at 40 to 60 m between January and May, but apart from the daily heating of the upper few centimetres, there was no more than 2°C difference between surface and bottom. During January there was a temporary disappearance of the thermocline with the warmer surface water (25·0°C) reaching the bottom. This was also found by Talling (1957b, 1964) and was interpreted by both as a recession of thermocline into deeper water due to wind induced displacements or perhaps internal oscillation, rather than to a breakdown of stratification. Signs which might be interpreted as oscillations of lesser amplitude were seen during the period from January to May (Fig. 6.8a) though the nature of these movements cannot be decided without more extensive data. The breakdown of stratification during May to July was shown by the homothermic condition throughout the column at a low temperature of 24·0–24·4°C. This is the period of the heavy southeast trade winds (Fig. 3.8), and it is reasonable to agree with Newell (1960) that, in view of the more or less constant duration and intensity of solar radiation throughout the year, the lowered water temperature at this time is probably due mainly to increased evaporation caused by the winds. Since this is a period of relatively clear skies, there may also be a greater loss of heat by radiation at night. The onset of calmer weather in August inaugurates another period of progressively increasing stratification The major determinant of this cycle therefore is the wind, which regularly rises to a peak of intensity and persistence during May to July and which probably reduces the stability of stratification by surface cooling and certainly provides the energy for mixing.

These water movements, disclosed by measurements of temperature, are accompanied by some expected changes in the distribution of dissolved substances. Simultaneous measurements of oxygen are shown in Fig. 6.8b, and it is

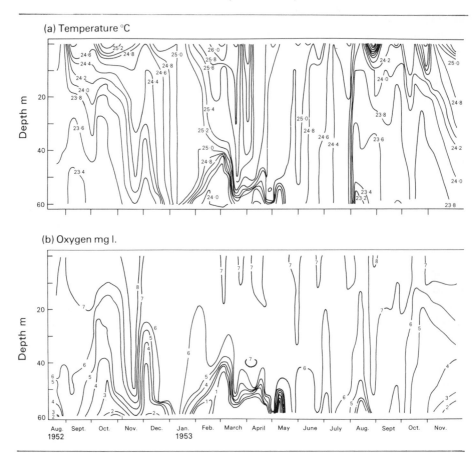

Fig. 6.8 Temperature and oxygen changes in Lake Victoria from measurements by Fish and put together in this form by Talling (1966).

clear that the oxygen isopleths are closely related to the isotherms. During the period when the thermal discontinuity is well marked (February–May), the oxygen concentration at the bottom is much reduced. With the homothermal conditions in January and July, oxygen in high concentrations reaches the bottom. Very similar oxygen curves were obtained in 1960–61 by Talling (1964) who also followed the distribution of other substances which was related to changes in temperature and oxygen. Thus, from February to May, the bottom water, with minor fluctuations, had a low pH (indicating increase of CO_2) and relatively high concentrations of nitrate, phosphate and silicate, all of which can be attributed to decomposition and to a thermal (density) barrier isolating it from the upper layers. Water with a higher pH and lower concentrations of these nutrient salts, resulting from photosynthesis and growth of algae near the surface, reached the bottom with the oxygen-rich water during the homothermal period between June and August.

These data from one station give a rather clear picture of a regular seasonal cycle of vertical water movements and associated chemical changes. In the following chapter, the corresponding fluctuations in the intensity of photosynthetic production will be described. But we are still far from a full understanding of the hydrological and production cycles in the lake as a whole. Much remains to be discovered of the part played by horizontal currents, types of stratification and internal waves. Moreover, the vast expanse of open water is only a portion of the lake, which is bounded by innumerable shallow gulfs and bays of varying sizes and degrees of isolation. The total surface area of these is probably nearly quarter of the whole. Owing to their shallowness and to nutrient-rich inflows, they are more fertile than the open lake, and in some northern bays stratification has been found to be less prolonged than further offshore (Fish, 1957).

Kitaka (1971) has presented temperature and oxygen data from a point near the centre of the lake (depth 60 m) obtained during stormy weather. These suggest that the lower water was then being drawn to the surface by a cyclonic swirl in the centre of a storm depression. In view of the frequency of violent storms on Lake Victoria, he suggests that such upwelling by cyclonic circulation may be common and, by implication, of importance in the production cycle.

It was pointed out in Ch. 3 (p. 37) that the climate of the humid forest region inland from the Guinea Coast of West Africa (Fig. 3.7: Kumasi) differs from that of the other tropical rainforest areas, such as the upper Zaïre basin. The generally high humidity and even temperature is interrupted for a few weeks in December and January by the Harmattan wind from the northeast, which brings warmer, dry air with clear skies from the Sahara. This has a marked effect on the regime of circulation in standing waters. There are no large natural lakes and very few small ones in this area, but the manmade Volta Lake has its northern half in the savanna and its southern in the forest region. There are indications that the water in the northern half has one period of instability, and that in the south two (p. 378).

The latter type of double regime is also shown by the small Lake Bosumtwi in the neighbouring forest region of Ghana, and has been studied in greater detail by Whyte (unpublished, seen in MS). It was formed by meteoric impact about 1·3 million years ago. It is situated 30 km southwest of Kumasi (Fig. 11.3), and is roughly circular with a diameter of about 8 km and a maximum depth of 78 m. Stratification involving deoxygenation below 7–10 m is disturbed for about a month twice annually. The more violent of the two disturbances occurs in August, during the relatively cool period between the two rainfall peaks. At this time the water column is more or less homothermal owing to the dense cloud cover and consequent relatively low daytime atmospheric temperature which fails to warm the water surface. The resulting instability offers little resistance to wind stress from the southwesterlies that are then at maximum strength. The resulting overturn lowers the oxygen boundary, but does not always reach the bottom. It often brings enough anoxic and H_2S-charged water to the surface to kill a number of fish. A period of less violent stirring occurs in December and January when the Harmattan from the northeast is producing a warm, clear and dry atmosphere. Then there is considerable surface heating with consequent stratification during the day, but at night, with the clear skies, there is a greater loss of heat by radiation than at other times. The hot and dry Harmattan wind, though less powerful than the southeasterlies in August, probably accelerates evaporative cooling at the surface and certainly provides the energy for stirring; but this stirring does not penetrate deeply enough to bring toxic water to the surface.

We now come to those very deep lakes of tropical Africa that are apparently always stratified and in which a large amount of the lower water is permanently free of oxygen. The best known of these are Lakes Tanganyika and Malawi which occupy long deep trenches in the floor of the Western Rift (Fig. 3.4). Compared with Lake Victoria their shorelines are relatively free of enclosed gulfs and bays and their hydrology is consequently more uniform.*

Lake Tanganyika (Fig. 16.1) is 650 km long and the maximum depths of its north and south basins are 1 310 m and 1 470 m respectively. The depth at which oxygen disappears, and H_2S is detectable, varies between about 100 m at the north end and 200 m at the south end. About three-quarters of the total volume is therefore permanently anoxic. Lake Malawi is somewhat smaller and much shallower (length 560 km, max. depth about 685 m) and the anoxic water (below 200–250 m) occupies a smaller proportion of the whole.

It would appear from the flourishing state of the fisheries in the deep water of these two lakes that the lower, apparently stagnant, water is not a permanent trap for all the substances derived from the decomposition of the organic particles which fall into it. We are led to suspect that a considerable amount of nutrients is somehow being brought to the surface. It has already been argued (p. 86) that this must occur continuously, though on a scale as yet to be decided, through the erosion of the upper part of the anoxic layer by turbulence at the interface and through movements within the anoxic layer itself associated with internal waves. These would also increase the erosion by periodically bringing the upper edge of the anoxic layer nearer to the surface.

Both Lakes Tanganyika and Malawi lie in deep rifts orientated north–south, up which the powerful southeast trade winds are funnelled during the period from April to September. The seasonal alternation between this period of strong unidirectional winds with low rainfall, and the remainder of the year with variable and generally less violent winds and more rain, is the predominant influence on the hydrological regime. The power, direction and duration of the wind, the orientation of the lake, the surrounding topography and the marked stratification, all favour internal waves along the main axis.

Typical temperature profiles from the deep water of the north basin of Lake Tanganyika in August (end of the dry windy period) and in April (end of the calmer and wetter period) are shown in Fig. 6.9. They are based on data from the Belgian Expedition in 1946–47 (Capart, 1952b). The seasonal changes involve only the upper 200 m, the temperature below this remaining remarkably constant. During the period from October to May there was a well-marked thermocline between about 50 m and 80 m, above which the diurnal fluctuations are apparent. During the windy season the thermocline is disrupted and the temperature profile is affected down to about 200 m.

The more recent investigations of Coulter (1966, 1968) on the upper 200 m in

---

* The development of ideas on the hydrology of these two lakes can be followed from the publications of Beauchamp (1939), Capart (1952a, b), Dubois (1958), and Coulter (1963, 1966, 1968) for Lake Tanganyika, and of Beauchamp (1953b), Eccles (1962, 1974), and Jackson et al. (1963) for Lake Malawi.

the south basin of Lake Tanganyika (Fig. 6.10), with some temperature profiles along the length of the lake, show what seems to be the main features of the circulatory regime. According to Coulter the temperature data can best be interpreted as demonstrating the existence of four layers which are well differentiated during the period of stable stratification (October–April):
1. from the surface down to 50–80 m, subject to daily circulation;
2. below about 80 m to at least 200 m, which is seasonally circulated, shows one or more shifting thermoclines and a changing gradient of oxygen depletion;
3. below 200 to about 700 m, a zone of minimum temperature and absence of oxygen;
4. below 700 m to the bottom, oxygen-free and a very slightly higher temperature.

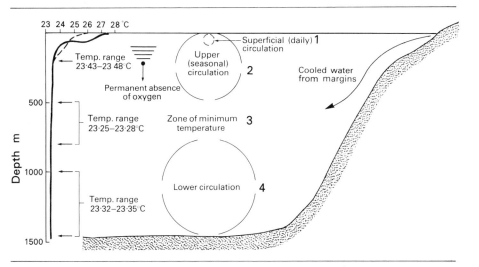

Fig. 6.9 Lake Tanganyika temperature profiles in North Basin (from Capart, 1952b) and regions of circulation suggested by Coulter (1968).

It would be misleading to apply to any of these layers the terms epi-, meta- and hypo-limnion coined for temperate lakes during summer stratification (Fig. 6.5). These could reasonably be applied only to the seasonal stratification in the water *above* the anoxic boundary where thermoclines periodically develop and disperse in response to the seasonal wind regime. In calm periods there are thus epi-, meta- and hypo-limnia above the anoxic boundary which lies at 150–200 m (Fig. 6.10).

There is no good evidence that the stability of the lower oxygen-free layer is dependent on a thermocline above it. Nor is the increase of density, which must be assumed to exist below the boundary, due to an increase in inorganic ions, which is a characteristic feature of the lower 'monimolimnion' of meromictic lakes in temperate climates. Lake Kivu (Fig. 6.14) is exceptional in this respect because of the influx of saline water from below. In Lake Tanganyika the decomposition of falling organisms may perhaps release relatively large organic molecules or even small particles in sufficient quantity to make a significant contribu-

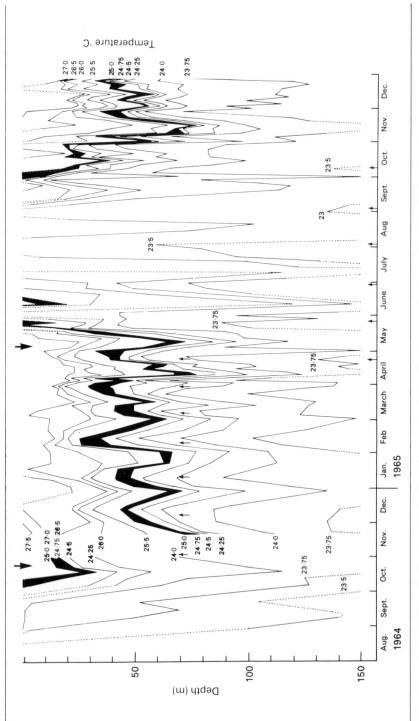

Fig. 6.10 Lake Tanganyika. Isotherms to a depth of 150 m. Dark shading between isotherms 24·75 and 25·00 is added merely to emphasise vertical oscillations which are evidence for internal waves of period approximately one month (redrawn from Coulter, 1963). Curves showing the longitudinal distribution of the thermocline are found in Coulter (1968).

tion to the density of the water without contributing to the conductivity. Direct measurements of density, and organic carbon or detection of suspected organic compounds might support or refute this suggestion.

The lower permanently anoxic layer also found in Lake Malawi and many smaller tropical lakes in Africa and elsewhere should not be confused with the 'hypolimnion'. In some, but certainly not all, respects it might be regarded as equivalent to the bottom mud of a holomictic lake, the surface of which is periodically stirred. Occasional thermal stratification characteristic of all lakes is restricted to the water above the anoxic boundary.

In Lake Tanganyika the southeast trade winds between April and September caused a flow of surface water to the north and thus a downward tilt of the isotherms along the lake. The upper thermocline was thus brought to the surface in the south basin, and was obliterated in June with consequent mixing to well below 100 m (Fig. 6.10). The curves show that, during the windy period, there was a periodic upwelling of the lower water into the upper 150 m at intervals of about one month. These movements were confirmed by the periodic reduction of oxygen at these levels, and the general mixing during these months was accompanied by an increase in the growth of the plankton, which would be expected to follow the upwelling of nutrients from below. The occasional localised mass deaths of fish reported during the windy seasons are presumably due to a particularly violent wind contemporary with a periodic upwelling of oxygen-free water which may, in such circumstances, actually reach the surface.

During the rest of the year (October–March), when stratification in the upper 150 m was most stable, the thermocline at 30–70 m continued to oscillate with about the same frequency (Fig. 6.10). Turbulence at the interfaces between the layers, due to horizontal flow in opposing directions associated with internal waves, is probably an important factor in the mixing. But the question remains whether anything more than a small proportion of the lower anoxic water is involved; whether most of it is really permanently cut off. The importance of this question for an understanding of the production cycle is obvious.

The rather uniform chemical composition of the water between surface and bottom was regarded by Beauchamp (1939) as evidence of occasional deep mixing, and by Kufferath (1952) as due to continuous slow, deep circulation. Analyses by the Belgian Expedition (Fig. 6.11) showed that the concentrations of the inorganic ions Mg^{++}, Cl^- and SO_4^- (Na^+ and K^+ were not measured), were almost identical at the surface and at 1 300 m in the north basin, and Ca^{++} showed only a slight increase with depth. The nutrient salts which are taken up in quantity by the phytoplankton and released on decomposition – nitrates, phosphates, silicates and sulphates – were, as expected, depleted at the surface and increased below the oxygen level (200–300 m), but below this their concentration was very uniform. The electrical conductivity, which is approximately proportional to the sum of the ionic concentrations, showed only a small increase below 300 m (Fig. 6.11).

The analyses by Degens et al. (1971b), though differing in some details from those of Kufferath, give the same overall picture of vertical uniformity in composition (Table 6.1). Most remarkable is the identity in the concentration of sodium at the surface and at 1 460 m.

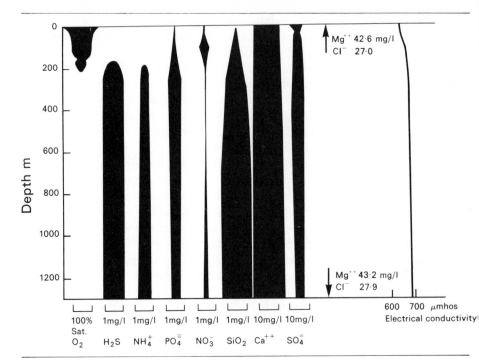

Fig. 6.11 Lake Tanganyika, North Basin. Vertical distribution of some dissolved substances (redrawn from Kufferath, 1952)

Table 6.1 Data from Lakes Tanganyika and Kivu selected from Degens *et al.*, 1971a, b

	Ca	*meq/l* Na	NH₄	*μA★/l* PO₄
Lake Tanganyika				
Surface	0·41	2·9	5·1	1·1
460 m	0·27	3·0	2·5	4·9
1 460 m	0·26	2·9	2·0	5·2
Lake Kivu				
Surface	0·24	5·2	18	0·8
100 m	3·2	8·4	487	18·8
200 m	4·1	10·6	1 314	32·7
300 mm	5·6	20·2	5 460	53·2
440 m	5·7	21·2	7 105	54·8

★ A: Formula weight in g

If we can take the above as evidence of deep circulation within the deep anoxic layer, we are faced with the question as to what kind of circulation this could be. The uniformity in depth of chemical composition would suggest that, unlike Lake Kivu (p. 106), the density differences, at least below 300–400 m, are determined mainly by temperature. Measurements sensitive to 0·01°C made by the Belgian Expedition (Capart, 1952b) disclosed that a progressive small decrease in temperature in the upper part of the anoxic layer down to about 800 m is followed by a rise of up to 0·1°C below 1 000 m. At such depths latter could, in principle, be due to pressure (adiabatic heating) or to flow of heat from the earth (Hutchinson, 1967, pp. 207, 460, 465). It is difficult to account for the level of the minimum temperature between 500 and 800 m, since it is lower than any which has ever been recorded at the surface of the deep water. There may be periods, as yet unrecorded, of exceptional cooling at the surface with mixing to this depth, but there is another factor which might contribute to the heat budget and to circulation in the deepest water. Nocturnal cooling of the shallow inshore waters to a temperature below that of the outer surface water might cause a flow of colder and denser water down the slope to 500–800 m, as suggested by Talling (1963) for Lake Albert (p. 90), and by Eccles (1962) at the shallow south end of Lake Malawi. These might provide a frequent source of colder water below which a reversed temperature gradient is caused by a continuous heat flow from below. It was included by Coulter in his suggested scheme of circulation (Fig. 6.9).

The fact is that we do not yet have enough information upon which to base a convincing model of the circulation in Lake Tanganyika. To get such information from this deep and stormy lake will require more sophisticated oceanographic methods and equipment applied over a longer period than hitherto. We can, however, say with confidence that the major influence is the powerful winds which, during much of the year, are favourably directed down the long axis of the lake. They may therefore be expected to be very effective in causing movements of water through surface waves, by driving forward large masses of surface water and (probably the most important) by the consequent setting up of internal waves with their horizontal oscillating water currents in the layers of different density. These internal waves can be detected during most of the year, and they must cause some turbulent mixing both within and along the interfaces between the layers, as well as along the mud surface, in the manner illustrated in Fig. 6.2. We know that during the windy season the thermoclines in the upper 200 m and the surface of the anoxic layer are driven downwards, presumably by this kind of disturbance. In such a manner, a circulation from top to bottom could, in theory, be accomplished, but much more quantitative data is required on winds, water temperatures, currents and distribution of dissolved substances. It will be important to discover the cause of the temperature inversion in the lower water and to find whether this plays any part in the vertical movements of water in the anoxic region.

Nevertheless it must be noted that in the shallower, but still very deep, water of Lake Malawi there is no such temperature inversion, though here, too, the uniformity of chemical composition suggests some form of circulation throughout the entire anoxic layer. It is becoming increasingly clear that water movements in deep lakes are very much more complicated than was previously thought and that there are other types of internal wave motion than those mentioned in this chapter (Mortimer, 1969). There are indications of circular (Kelvin) waves in Lake Tanganyika (pers. comm., G. W. Coulter). We cannot hope to understand fully the processes of nutrient recycling until the nature of the main water movements have been disclosed.

The effects of climate on the hydrology of deep lakes and thus on the nature and distribution of the fauna cannot be better demonstrated than by comparing Lake Tanganyika with Lake Baikal in Siberia. The two are remarkably similar in mode of origin, size, depth (Baikal is even deeper) and in surrounding topography. But the temperature regime on a latitude of over 50°N ensures seasonal circulation and permanent oxygenation of the deepest water in Lake Baikal. The consequent biological differences between the two lakes are discussed on p. 286.

Lake Malawi, though shallower, resembles Lake Tanganyika in structure and orientation, and a similar seasonal hydrological regime is to be expected. Most of the measurements have been made above 300 m, but several temperature profiles down to 600 m showed no signs of an inversion at the bottom (Jackson *et al.*, 1963, p. 42). The depth (maximum 685 m) is probably not great enough for adiabatic or terrestrial heating at the bottom. It is permanently free of oxygen below 200–250 m. But the chemical gradients are no more marked than in Lake Tanganyika and, in the present state of our knowledge, we can only assume that there is a deep circulation associated with internal waves.

Whatever is going on below 300 m, it is clear from the work of Beauchamp (1953b), Harding (in Jackson *et al.*, 1963) and Eccles (1962) that in the upper 200–300 m of Lake Malawi the regime is similar to Lake Tanganyika (Fig. 6.12). During the relatively calm and rainy period from December to March there is stratification with a thermocline between 50 and 70 m, which shows marked oscillations and in 1955 was progressively forced downwards, and was ultimately dispersed in July by the southeast trades. The importance of this annual process for production was shown by the corresponding changes in the concentrations of silicate and phosphate and in the volume of the plankton (Jackson *et al.*, 1963).

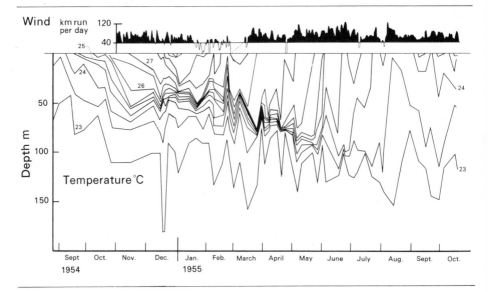

Fig. 6.12 Lake Malawi, Nkata Bay. Seasonal isotherms to depth of 200 m, and wind speed (redrawn from Jackson *et al.*, 1963)

The discussion of these two very deep lakes has so far been centred on the open, deep water. Inshore regions not deep enough to contain water below the level of seasonal circulation (e.g. 200 m) would be expected to be more fertile because the bottom deposits are within range of this circulation. Owing to the steepness of the shorelines, extensive shallow water is however found only at the north and south ends of these lakes. Eccles (1962, 1974) showed how internal oscillations may stimulate the fertility of the large shallow area at the south end of Lake Malawi. These not only bring in nutrient-rich water from lower levels but, in so doing, produce a back and forth horizontal flow along the bottom. Eccles found that the position at which the 26°C isotherm reached the bottom moved shorewards by as much as 10 km in two days. Such movements would presumably disturb the mud surface and accelerate the release of nutrients into the water. He also suggested that marginal cooling may be another source of horizontal movements of the bottom water.

Lake Edward in the Western Rift (Fig. 13.1) can be regarded as intermediate in character between the above two very deep lakes and the other Rift Lakes Albert and Turkana which are fully circulated at least annually and remain oxygenated to the bottom. The latter are apparently exposed to more powerful winds than Lake Edward, though this impression requires confirmation from measurements. The profile of Lake Edward is peculiar in that, though about 40 km wide, the deepest water (112 m) lies within 4 km of the Zaïre shore where the western escarpment falls precipitously into and under the water. Below 50–60 m there is normally no oxygen and the stagnant anoxic layer which is highly reducing with free H_2S, is thus confined to a trench along the western shore. More than two-thirds of the lake is less than 50 m deep and is frequently stirred and is always oxygenated to the bottom. The deep water is therefore of doubtful significance in the total production cycle. There are flourishing fisheries based on both Uganda and Zaïre shores. It is, moreover, uncertain as to what extent the water below 50 m is excluded from the upper circulation and whether there are regular seasonal movements of water. Measurements made in July 1931 (Beadle, 1932b), in May, November and December 1935 (Damas, 1937) and in February 1964 (Beadle, 1966) all showed an oxygen-free layer below 50–60 m. Damas, however, found that the thermocline was lowered by about 20 m between May and December 1935, and from this and chemical evidence he concluded that the lake is completely overturned annually, though the possibility of internal waves was not considered. Verbeke (1957a) reported, without quoting figures, that oxygen reached the bottom in August 1953 following a period of particularly violent weather. Oxygenation of the great mass of anoxic water could hardly be effected by anything less than a long-continued violent disturbance from the surface, and there is no evidence that this is an annual event. Repeat observations over a period of more than a year are now needed to establish the nature and frequency of the circulation. The profile of the basin, with its extensive shallow eastern half sloping gently to the western depths (Fig. 13.1), might provide favourable conditions for marginal cooling causing profile-bound density currents to flow into the deep water.

It is being argued in this chapter that the major determinant of the circulation, and hence of production, in tropical lakes is wind rather than seasonal fluctuations of radiation and atmospheric temperature. As well as causing water movements and consequent circulation of nutrients in the manner already described, winds increase the rate of evaporative cooling at the surface, especially during periods of low atmospheric humidity. This may be sufficient to reduce the thermal stability of the water and favour deep stirring by winds. We should therefore expect that small sheltered lakes, deep relative to their surface area, would be the least productive. Both the force and the fetch of the winds would be comparatively small.

Melack (1978) in a study of sixteen stratified crater lakes in Western Uganda found a correlation between (a) the ratio maximum diameter/minimum height of the crater rim, and (b) the depth of the anoxic boundary. In view of other complications that must be involved, such as the direction and force of the prevailing winds, this seems to demonstrate very well the effects on the one hand of the length of the wind fetch and on the other of the degree of protection from winds on the depth to which the water is stirred from the surface. Four small lakes in Ethiopia and Cameroon and six in Australia fitted reasonably well into the same correlation diagram.

There are two kinds of lake which provide examples of these conditions:
1. Some volcanic crater lakes with high steep-sided crater walls.
2. Lakes formed by blockage of steep-sided river valleys by volcanic lava flows (Ch. 15).

Both of these types are well represented in tropical Africa and a few have been investigated. For example, there is the small but relatively deep crater-lake Nkugute in western Uganda (alt. 1 500 m, diameter 1 km, max. depth 58 m), surrounded on all sides by hills and the wall of the crater (Fig. 13.1), and Lake Bunyoni formed by volcanic blockage of a steep-sided narrow branching valley in the Kigezi Highlands of southwestern Uganda (alt. 1 950 m, max. depth 45 m) (Fig. 15.1). Some relevant hydrological data are set out in Fig. 6.13 (Beadle, 1966; Denny, 1972). The following features are immediately apparent:
1. the temperature profiles are almost identical except that the difference in altitude (450 m) is reflected in a difference of about $2 \cdot 7°C$ between the constant temperatures at the bottom of each lake;
2. oxygen penetrates in both to a depth varying from time to time between 10 and 20 m, below which it is apparently absent;
3. though the temperature gradient in both is much less steep than in a temperate climate lake during summer stratification, the difference in density between the surface and bottom water in the afternoon is about as great because of the temperature–density relation of water (Fig. 6.6). Temperature and density curves from Lake La Ronge in Canada are included as representative of a temperate climate lake during the summer.

Before drawing conclusions from this it must be noted that normal surface cooling at night (temperature curves for Lake Nkugute) and the not infrequent falls in surface temperature during the day (Lake Bunyoni temperature curves) due to rainfall and cloud cover can both reduce the density gradient. The stability of the water column is therefore frequently reduced to a much lower level than that of a typical temperate lake in summer. Nevertheless, repeated measurements have indicated that the water in both lakes below about 20 m has been free of

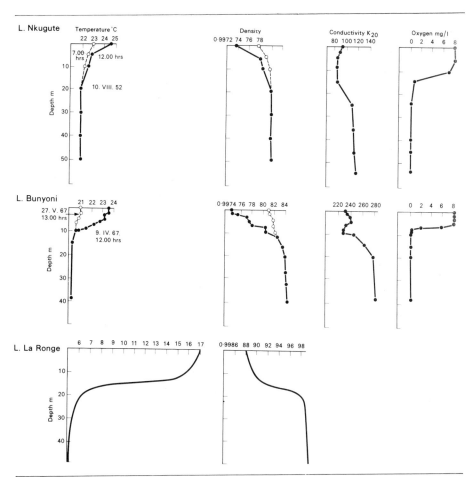

Fig. 6.13 Temperature, oxygen and relative density profiles of Lakes Nkugute and Bunyoni in western Uganda compared with Lake La Ronge in Canada during summer stratification. Curves from the two Uganda lakes selected and redrawn from Beadle (1966) and Denny (1972), for L. La Ronge from Hutchinson (1957, Fig. 135). Relative density derived from temperature according to Hutchinson (1957, Table 7)

oxygen for at least ten years. The apparent absence of a complete overturn during the periods of reduced stability must be attributed to the short wind fetch and consequently insufficient surface disturbance to reach a depth of more than about 10 m. In the case of Lake Nkugute the fetch is about 1 km, and the surface waves induced by a gale of force 8 (20 m sec) would in theory mix no more than the upper 4 m of water (Fig. 6.4). On Lake Bunyoni, owing to its branching shape, the fetch would vary with the direction of the wind but could not be more than 2–3 km. An exceedingly strong gale would be needed to increse the depth of the wave-mixed layer to more than 10 m (Fig. 6.4). As suggested in the discussion on Lake Tanganyika (p. 95) the stability, even at its minimum, may be greater than that calculated from the temperature and conductivity profiles. The density of the lower anoxic and H$_2$S-laden water is perhaps augmented by dissolved and particulate organic substances that do not significantly raise the conductivity.

In both lakes the temperature at the surface has never been found to be less than 1°C higher than that of the bottom water. How then is the lower temperature at the bottom maintained without occasional surface cooling to this temperature and a complete overturn? No evidence could be found in Lake Nkugute of nocturnal marginal cooling with underflow of colder water (Beadle, 1966), and the scarcity of inshore shallow water in Lake Bunyoni would appear to preclude it. This is a characteristic feature of the temperature profile of all permanently stratified tropical lakes, and more detailed and prolonged investigations are needed to provide the explanation.

There is, however, no doubt of a frequent exchange between the upper edge of the anoxic layer and the water above it which is directly stirred from the surface. This is shown by the fluctuation in the depth to which oxygen penetrates and by the fact that stirring from above prevents the anoxic boundary from rising higher than about 10 m from the surface. As in the large stratified lakes, such as Lake Tanganyika, the vertical distribution of the ions Na^+, K^+, Mg^{++}, Cl^- and $SO_4^=$ within the anoxic layer is sufficiently uniform to suggest that there is some circulation between the anoxic boundary and the bottom. Yet in spite of an abundant accumulation of phosphorus and nitrogen below the boundary, both Lakes Bungoni and Nkugute are unproductive. The fisheries are very poor, even after the introduction of supposedly suitable fish such as *Tilapia nilotica* to augment the previous rather sparse populations of *Haplochromis* spp. and *Clarias lazera*. The supply of nutrients from the inflows and the rate of recycling from below are evidently not normally rapid enough to maintain anything but a low rate of production.

On the other hand, like many other tropical lakes, both are known to be subject to occasional abnormal disturbances due to heavy storms which happen to hit them, maybe after a period of relatively low temperature. Sudden mass deaths of fish have been reported in both lakes by the Uganda Fisheries Department (Denny, 1972). These appear to be due to the sudden stirring up of enough of the anoxic lower water to asphyxiate many of the fish (as in Lake George, p. 88). In Lake Bunyoni they have been severe enough to damage the fishery and, after a very serious mortality in 1964, restocking was stopped and the commercial fishery was abandoned.

An interesting example of permanent stratification in a small sheltered lake is the volcanic crater lake Barombi Mbo in West Cameroon which is devoid of oxygen from about 20 m downwards (Green, Corbet and Betney, 1973). This is larger and deeper than Lake Nkugute, being 2·5 km in diameter and 111 m maximum depth, but it is well sheltered from winds by the surrounding crater rim and dense forest. The presence of an endemic cichlid fish especially adapted to low oxygen levels (p. 321) suggests that this condition has persisted for a very long time.

A comparative study was made between 1963 and 1965 of the five Bishoftu crater lakes southeast of Addis Ababa in Ethiopia (Baxter and Wood, 1965, Prosser *et al.*, 1969, Wood *et al.*, 1976). They lie at about 190 m altitude and 8°N latitude (Fig. 10.1). Their surface areas range from 0·5 ad 1·0 km² and their maximum depths 6·4 to 87 m. They demonstrate the dependence of deep circulation both on the depth and degree of exposure to winds and on evaporative cooling at night. The two deepest and most sheltered, Bishoftu, 87 m and Pawlo, 65 m, show the most stable stratification with an anoxic lower layer. Observations by Wood *et al.* (1969) over two and a half years suggest however that Lake Pawlo is probably annually overturned towards the end of the year. The main cause seems to be the breakdown of thermal stability through cooling at the surface due both to increased evaporation during the day and to a greater loss of heat at night into a drier and clearer atmosphere. The shallowest, Kilotes, is almost continuously stirred to the bottom. Lake Aranguadi is of intermediate depth (32 m) and markedly stratified, but its extremely high algal productivity (Table 7.1) is at least partly due to rather frequent stirring into at least the upper part of the anoxic water.

It is very evident that an assessment of the circulation and productivity regimes in small tropical lakes can be made with confidence only after prolonged investigation. Not only may there be a long-term cycle of more than one year, but the climatic trigger setting off such a cycle may be a rare and unpredictable event.

Apart from Lake Bunyoni there are several other examples that have been investigated of steep-sided valley lakes originating from blockage by lava flows. Such are Lakes Luhondo (max. depth 68 m) in Rwanda, and Mutanda (44 m) in western Uganda (Figs. 15.1 and 2) which are sufficiently deep and sheltered to be stratified and deoxygenated for a very long time, if not permanently (Damas, 1954–55; Baxter *et al.*, 1965). Nothing is as yet known of periods of deep circulation, if any, though Damas considered that the rather slight gradient of concentration of dissolved substances within the lower deoxygenated layer of Lake Luhondo was evidence of some kind of continuous but slow circulation of the whole column. This argument could as well be applied to Lakes Bunyoni and Mutanda, though the signs, noted above, of occasional more violent deep stirring in Nkugute and Bunyoni must cause us to refrain from coming to certain conclusions concerning the regime of circulation in this kind of tropical lake until repeated observations have been made over a long period. It is interesting to note that Lake Luhondo receives most of its water from the neighbouring much deeper Lake Bulera (173 m) which is also formed by volcanic blockage and that the latter is oxygenated to the bottom. Damas (1954II) suggests that this curious condition is due to the very low temperature (less than 15°C) of the water from the main inflow which comes down from high country and drops rapidly into the lake over the Rusumo Falls. Owing to its higher density this water may flow under the surface (temperature 20–22°C) and continuously maintain some oxygen at the bottom of the lake. Analyses of suitable samples should confirm or refute this suggestion but, until more observations are made, it cannot be ruled out that, shortly before Damas's investigations, Lake Bulera had been deeply stirred by abnormally violent winds, and that the neighbouring Lake Luhondo, which is more stable owing to the much steeper salinity gradient, was unaffected.

The most firmly stratified of all known African lakes is Lake Kivu in the Western Rift (Damas, 1937). This is an extensive branching river valley, blocked in the Pleistocene by a volcanic barrier (Ch. 15 and Fig. 15.1). It is very deep (max. 480 m) and surrounded by high mountains, but the most remarkable feature is the very high density, due to high salinity, of the lower water (Fig. 6.14). From about 70 m downwards the density increases rapidly, and this in spite of the equally remarkable rise of temperature, which amounts to about 3°C between 70 m and 375 m (the lowest sample taken by Damas). This condition must be due to the deep seepage of warm saline water from volcanic sources. The recent investigations by Degens *et al.* (1971a) have confirmed the very steep salinity gradient and the very high concentrations of nitrogen and phosphorus accumulated in the saline lower water as compared with Lake Tanganyika (Table 6.1). Moreover, their echo-soundings showed 'cones' projecting from the surface of the sediments in the north basin, which they interpret as jets of warm saline water issuing from below. They also found that the salinity gradient downwards through the anoxic region below 100 m is not uniform, but is marked by a succession of more or less discrete layers of increasing salinity. Whatever the origin of these layers, they certainly indicate a lack of vertical circulation within the anoxic region. In this respect Lake Tanganyika is quite different, with an almost uniform vertical composition of the anoxic water between about 200 m and 1 400 m which suggests a vertical circulation of probable importance in the recycling of nutrients.

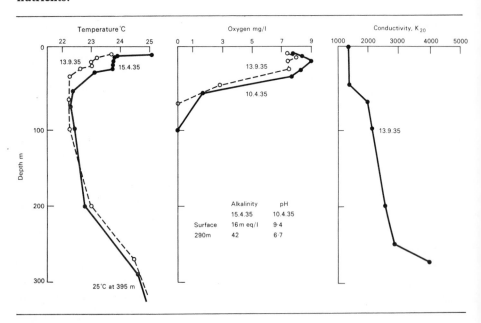

Fig. 6.14 Lake Kivu. Temperature, oxygen and conductivity profiles (redrawn from Damas, 1937)

Though more prolonged observations are needed for confirmation, Damas concluded that all circulation in Lake Kivu is confined to the upper 70 m, where density gradients are mainly determined by temperature differences, and that the lower very saline oxygen-free zone is permanently excluded (see also Ch. 15). Though this and the mud form a sink into which most of the organic matter produced near the surface falls and is presumably lost, there is some evidence of seasonal fluctuations in productivity (p. 129). This increases during the dry and windy season and is presumably due to deeper stirring by the wind, facilitated by surface cooling through increased evaporation. At this time more of the upper region of the stagnant zone is probably brought into circulation than at other seasons. More direct evidence of the high nutritive value of the bottom water of Kivu is given on p. 129.

All of the lakes so far considered are at altitudes less than 2 000 m. They range from Lake Bunyoni (1 973 m, bottom water temperature c. 19°C) to Lake Turkana (375 m, bottom water temperature c. 27°C). There are a number of small lakes at very high altitudes on the mountains of tropical Africa, where the seasonal temperature fluctuations may be as small as at low altitudes in the tropics, but the actual level of temperature can be very low, Löffler (1964) has investigated some small lakes, mostly of glacial origin, at altitudes exceeding 4 500 m on the East African mountains. Lewis Tarn and Curling Pond at the foot of the glaciers at over 5 000 m on Mount Kenya are permanently below 4°C from top to bottom, and show little signs of stratification. Others, at somewhat lower altitudes, have correspondingly higher temperatures and a greater tendency to temporary stratification. There is no doubt, however, that they are all frequently mixed except Curling Pond, which is frozen over for long periods.

There is no large and deep lake in the African tropics at an altitude of over 2 000 m comparable with Lake Titicaca in the Peruvian Andes (lat. c. 16°S, alt. 3 800 m, max. depth 280 m). It would therefore be of interest to conclude with a brief comparison between Titicaca and the lakes of comparable dimensions at the lower altitudes of the African tropics. The limnology of Lake Titicaca is discussed by Gilson (1964), Richerson et al. (1975), Widmer et al. (1975) and Carmouze et al. (1977). Its depth is intermediate between Lakes Edward and Kivu, and, if it were placed at an altitude around 1 000 m in one of the African Rifts, we would confidently expect a well-marked and prolonged stratification with an oxygen-free lower layer. There would be some seasonal disturbance of the upper part of the stagnant layer, and a slow circulation might be operating within it. In fact, both the temperature profiles and the continuous presence of a high concentration of oxygen (rarely below 4 mg/l) at the bottom suggest that full circulation occurs frequently. This state of affairs is probably due mainly to the continuously low level of temperature (c. 11°C at the bottom and 11·5–14°C at the surface). A low rate of decomposition, unrelieved by a 'summer' rise in temperature, would both reduce oxygen consumption and release less soluble material to augment the density of the lower water. But the most important factor may be the effect of the low temperature level on stability. With the same temperature gradient, the change in density (Fig. 6.6) between 11° and 13°C is about one-third that between 25° and 27°C – the temperature level of Lake Edward which it most closely resembles in shape and profile, though Titicaca is much deeper. A continuously low temperature, without seasonal fluctuations, would result in low stability and frequent mixing might be further favoured by the cooling effect of dry winds across the surface. It has been calculated by Monheim (quoted by Richerson et al., 1975) that evaporation accounts for about 95% of the water lost from the lake.

From all the information now available from tropical lakes, other than those at very high altitudes, it would be reasonable to draw the following general conclusions. (See also discussion by Talling, 1969.) The tropical temperature regime is, in general, conducive to prolonged stratification and deoxygenation, a tendency which is opposed by circumstances favouring deep stirring by winds. Even very shallow waters, which are normally well stirred, if protected from wind or during exceptionally calm periods, may remain stratified and deoxygenated below the surface, the diurnal changes of temperature being usually insufficient to induce convectional mixing at night.

The terminology relating to stratification (e.g. epi-, meta- and hypo-limnion) was originally proposed to apply to lakes in temperate climates where density and chemical layering are normally associated with marked thermal discontinuities. So often, however, in deep or well-sheltered tropical lakes the most clearly marked chemical discontinuity, the lower limit of oxygen, is not associated with a thermocline. In Lake Tanganyika, for example, there are moving and fluctuating thermoclines and periodic overturns only in the upper oxygenated water, the temperature of the lower water being unaffected. The anoxic lower layer is not therefore the equivalent of the hypolimnion of temperate lakes in summer. The above terms cannot be used without some qualification, but there is so much still in doubt about stratification in tropical lakes that it would be premature to propose a special terminology.

Several lakes of medium and great depth show evidence of regular seasonal increases of algal production associated with periods of deep stirring. But whereas in *temperate* lakes an outburst of production is triggered by rising temperature and illumination in spring (the surface water having previously been enriched with nutrients by the winter mixing), in the *tropics* temperature and illumination are always adequate and production is stimulated by the onset of seasonal heavy winds which bring to the surface nutrients, which had been depleted above and had accumulated below during the previous period of stratification. The recycling of nutrients in deep stratified lakes with a permanently anoxic lower layer is probably brought about mainly by water movements associated with internal waves, which are ultimately induced by winds. The wind not only provides the energy for stirring but also, when atmospheric humidity is low, may reduce the stability of stratification by evaporative cooling at the surface. Some small, relatively deep and well-sheltered lakes are protected from the seasonal heavy winds and their prolonged state of stratification is disturbed only at rare and irregular intervals by abnormally violent weather. To what extent these conclusions are supported by direct observations and measurements of photosynthetic production will be discussed in the next chapter*.

Any conclusions regarding seasonal regimes of circulation must be qualified by the warning that the tropics are subject to irregular violent disturbances, examples of which have been given. These cannot be predicted from the seasonal meteorological data available for the area, and such storms are usually very local and may hit or miss a lake by a narrow margin. But their effects can be catas-

* Lake Lanao in the Philippines is a tropical lake which is seasonally stratified, often developing several successive thermoclines, but is sufficiently exposed to winds to be completely overturned in the cool season. It is one of the few tropical lakes outside Africa that have been kept under observation for a year or more (Lewis, 1973).

trophic and their immediate influence on productivity may be considerable. Whether there are any lakes whose general level of productivity is maintained by sufficiently frequent random disturbances of this kind, is not known. It may ultimately transpire that 'fish kills' are just extreme cases of many irregular but less violent disturbances, most of which do not reach the threshold of catastrophe, but may, nevertheless, in some lakes, play a significant part in the circulation of nutrients.

7

Primary production in temperate and tropical lakes

The energy of solar radiation is responsible both for the state and for the circulation of water (Ch. 6) and, indeed, for many features of the environment that affect aquatic organisms. But above all it drives the machinery that manufactures living matter and sets up the stores of energy required by the organisms. In order to compare one water with another and to identify the factors that control the rate of primary (photosynthetic) production, it is clearly necessary to measure it. It is the object of this chapter to consider methods of measurement and the results that have been obtained from tropical lakes. Comparisons are made between different types of lake and, finally, by the association of information on both water circulation and primary production, some conclusions are drawn concerning regimes of production in tropical and temperate lakes.

It should first be mentioned that, apart from dense swamps (Ch. 18), there are some lakes that are highly productive of fish and invertebrates, but whose supply of organic matter is mostly photosynthesised by emergent or terrestrial plants. These are common in the forests of the Zaïre and Amazon basins (p. 64). Lake Tumba, on a tributary of the River Zaïre, is a shallow lake about 765 km² in area, surrounded by dense forest. The salinity of the water is very low (Ch. 5 and Table 5.1), and it is coloured deep brown by humic substances. Planktonic algae are very scarce, but there is nevertheless a rich fish fauna of more than seventy species. These are nourished directly or indirectly via invertebrates from forest litter and debris from other land vegetation (Marlier, 1958a). This is therefore a productive lake in which sources of organic matter and energy are wholly allocthonous. It is probable that there is a similar ecological situation in Maindombe, a very much larger lake (2 300 km²) in the same region of the Middle Zaïre basin and surrounded by dense forest. It is swampy and shallow (3–6 m deep). Very little research has been done, but its water is extremely acid (pH less than 4) and dark in colour. A very low salinity is to be expected. The phytoplankton is very scarce and photosynthetic production in the water is probably negligible. Nevertheless it is reported that the fish fauna is rather rich, apparently confirmed by the presence of many crocodiles, and it seems that the invertebrates are abundant, though there is no report on the molluscs, which are likely to be absent (de

Bont, 1969). Further investigations on Lakes Tumba and Maindombe would surely provide some interesting information on the sources of energy and materials under these curious conditions.

There are in fact no waters, even the large lakes, whose energy resources are entirely dependent on photosynthesis within the ecosystem itself. In this chapter, however, we are primarily concerned with underwater photosynthesis which is the principal source of production in the great majority of large lakes.

The rate of primary production by the submerged and emergent vegetation in rivers and the shallow littoral regions of lakes is more difficult to assess, though not necessarily less important to the ecosystem than production by the planktonic algae. But the latter are more evenly distributed and more easily sampled and good methods for estimating production by aquatic macrophytes are only recently being developed (see Mann, 1969, and Westlake in Vollenweider, 1969).

Estimates of primary production by the submerged beds of *Potamogeton* in shallow sheltered bays of Lake Turkana have however recently been made (p. 182). Otherwise there have been no systematic measurements, or at least none published, on other sites of production in tropical African lakes, and the work discussed in this chapter relates only to primary production by the phytoplankton.

It is to be expected that the rate of production is highest in waters with the densest populations of planktonic algae. Extreme examples are those usually rather shallow waters in which the growth of algae is intense enough to give a green colour visible from a distance. There are a few 'green' lakes in which the dominant alga is a green chlorophycean, such as *Oocystis sp.* in Lake Abiata in Ethiopia, but in most cases it is the blue-greens (Cyanophyceae) that are responsible. The conditions that favour dense populations of blue-greens cannot easily be defined; different species are favoured by different conditions. However, vigorous growth of several common blue-green species occurs in water of high alkalinity and pH (over 7·5), with an abundance of nutrients in circulation and of organic matter in solution as well as a moderately high temperature (Fogg *et al.*, 1973, Ch. 12). Such conditions are found in Lake George where 70–80% of the phytoplankton is composed of the blue-green algae *Microcystis* and *Anabaenopsis* which are so abundant that up to 600 mg/m^2 of chlorophyll *a* have been recorded and the gross rate of photosynthetic production is extremely high (Table 7.1) (Ganf, 1975). 'Green' waters are common in all continents and climates. Artificial eutrophication from organic wastes usually produces conditions favourable to blue-green algae, dense growths of which are a characteristic symptom of organic pollution. Several blue-green algae such as *Spirulina* and *Chroococcus sp.* as well as some diatoms such as *Nitzchia spp.* are adapted to extremely saline water, both neutral and highly alkaline, which is often coloured bright green, and the gross rate of photosynthetic production is enormous. Such are Lake Aranguadi in Ethiopia (Table 7.1), the saline lakes in Kanem northeast of Lake Chad, many small lakes in the Rift Valleys (Hecky and Kilham, 1973; Melack and Kilham, 1974; Melack, 1976) and numerous small patches of saline water in the Sahara (pp. 339–341). It is to be noted that extremely dense populations of algae with very high rates of primary production are found both in freshwaters, e.g. Lake George, and in hypersaline lakes. The latter may be of very high alkalinity and pH, e.g. Lake Arangwadi, or of relatively low alkalinity, and near neutral pH, as in Lake Mariut (Table 7.1). The only feature that may be common to them all, and would favour a high rate of production, is an abundant supply of nutrients and especially efficiently recycling by frequent vertical circulation.

The functional versatility of the blue-green algae as a group is remarkable. Apart from the ability of some species to photosynthesise, to fix atmospheric nitrogen, and to assimilate heterotrophically dissolved organic matter in the water (pp. 46–47), some produce intracellular gas vacuoles whose volume is controlled. All these features are found in some bacteria which they also resemble in the absence of an intracellular membrane enclosing the nuclear material. It appears that the function of the gas vacuoles is to adjust the buoyancy of the cells so as to retain them in the region of optimum illumination by compensating for the differences in density and movements of water at different levels. Among other external influences, low light intensity favours expansion of the vacuoles and *vice versa*. Consequently during calm weather at night masses of algae may rise to the surface to form a green floating mass known as a 'water bloom' (Reynolds and Walsby, 1975). Such a floating green scum is a common feature of many ponds, lakes and polluted waters in all climates, and is not necessarily associated with an excessively high density of algae under the surface. Blooms seem to have no functional significance but are merely an incidental result of gas vacuolation. The algae are 'caught out' on the surface in calm weather. Of the African lakes described in this book Lake George produces frequent water blooms of the blue-green *Microcystis* and *Anabaena* whose occurrence and characteristics are discussed by Ganf (1974, 1975).

Primary production is thus effected by four types of micro-organism – photosynthetic bacteria, diatoms, green algae and blue-green algae.*

Even for one well-defined section of the ecosystem, such as the phytoplankton, there are several difficulties in the way of estimating the total gross primary production over the whole lake during a long period of, say, several years. Even if measurements are made at number of points and at frequent intervals, this may not adequately allow for the discontinuous horizontal distribution of the algae nor for the vagaries of the climate which, particularly in the tropics, are likely to vary from year to year and to alter the overall rate of production through changes in the supply of nutrients. We are tempted to conclude that it is more profitable, and certainly easier, to study *differences* in the rate of production, and to discover the factors which cause them, than to attempt to arrive at an overall integrated figure for the whole system.

We are concerned in this chapter mainly with *gross* primary production, that is, with the total photosynthesised material before it is consumed and dissipated by both the plants themselves and by other organisms. The primary producers may in some respects be compared with a power station from which not all the energy generated (*gross* production) is actually sent out. Some is consumed in the functioning, maintenance and expansion of the station itself. What remains for transmission is the *net* production. Similarly, a proportion, often rather high, of the

---

* For lists, illustrations and distribution of tropical African freshwater algae see Van Meel (1952) and Compère (1970, 1977).

gross production of plants is consumed in internal maintenance and growth of the plants themselves. Plants, unlike a continuously generating power station, produce only during the day, but the energy-consuming processes of metabolism continue through the twenty-four hours. During the summer in a temperate climate, with long days of high light intensity and short nights, net primary production in a lake can be high, with the consequent rapid increase in biomass of all organisms. Conversely, in winter, with short dull days and long nights, the net primary production is often insufficient to maintain the rest of the population of organisms whose biomass decreases accordingly. In the tropics we might at first expect a more even rate of primary production, both gross and net, throughout the year. We shall see, however, that even in the tropics seasonal changes in the rate of production can be well marked and are mainly due to fluctuations in the nutrients available to the plants that are correlated with the movements of water discussed in the preceding chapter.

In a deep, open lake photosynthesis by the phytoplankton is, in practice, the most satisfactory point in the production cycle for a quantitative attack. The source of the energy, solar radiation, can be measured at different depths, and the rate of synthesis can be estimated from measurements of the resulting chemical changes. In a lake, whose water is mainly 'open' and not cut up into basins or bays which differ much from each other, the algae may appear sufficiently uniformly distributed to justify an overall assessment of primary productivity per unit of surface area of the open water from experiments done at a few points. But it must be emphasised that each lake differs in the homogeneity of its algal population. The small Lake Mariut near the Mediterranean coast of Egypt has three regions which differ very greatly in the density of the phytoplankton and the rate of primary production – a very low rate in the *Potamogeton* zone, higher, in the open water and very rapid indeed in the polluted region (Aleem and Samaan, 1969, and Table 7.1). The value which can be attached to measurements of photosynthesis at points in the open water in an assessment of overall productivity will vary greatly. Even in Lakes Malawi and Tanganyika, with their mainly rocky and sandy unindented shorelines, the density of planktonic algae is in general greater in the inshore waters, and in Lake Tanganyika there are three main regions in the deep water which differ significantly in the density and species composition of the phytoplankton and in the rate of primary production (Hecky *et al.*, 1977). Regional differences have been found also in Lake Turkana. The most meaningful measurements relating to overall productivity are likely to come from small shallow lakes, such as Lake George in Uganda, which are frequently mixed vertically and horizontally and have a rather uniform population of planktonic algae. But for most lakes the value of calculated figures for total production is questionable. More justifiable deductions from the data are *changes* in the rate of production with a view to investigating their causes. It is, in fact, of great practical value to know of seasonal and other fluctuations in the rate of gross primary production and to understand the factors that control them. There is here the possibility of experiment and even, in some cases, of artificial control.

The mechanism by which solar energy is trapped and stored by green plants is basically the same whether the environment is air or water (see Fogg, 1968). Chlorophyll absorbs radiation within the wavelength range 400–700 mμ, and the energy so gained is applied to splitting molecules of water into molecular oxygen, which is evolved, and active hydrogen which is immediately attached to a carrier molecule (a pyridine nucleotide). This sets off a train of enzyme-controlled reducing reactions whereby carbon dioxide (taken up from the environment) is ultimately built into molecules of carbohydrate, protein and, in some cases, fat. The overall photosynthetic and respiratory processes may be represented as

$$CO_2 + H_2O + Energy \underset{\text{Respiration}}{\overset{\text{Photosynthesis}}{\rightleftarrows}} (CH_2O) + O_2$$

(CH_2O) represents the basic of the carbohydrate molecule, whose energy can be released in respiration by the reverse (oxidative) process involving another long chain of enzyme-regulated reactions. In the present connection the above formula represents the fact that CO_2 is removed and oxygen is evolved in *equivalent* quantities which are related to the energy content of the carbohydrate synthesised. Measurements of either of these two gases can therefore be used as a means of estimating the rate of photosynthesis.*

There are, however, some important differences between air and water which affect the supply both of solar energy and of carbon dioxide. Of the radiation which reaches the surface of water some is reflected. This will depend on the angle of incidence and thus on season and latitude as well as on cloudiness, but throughout the year 5–15% of the total incident radiation may be so reflected, and a similar proportion may be back-scattered into the atmosphere by suspended particles in the water. Of the radiation absorbed below the surface most is dissipated as heat raising the temperature of the water, and only a small proportion actually activates chlorophyll molecules in green plants and initiates the chain of ractions leading to photosynthesis.

Owing to the absorption by the water and by dissolved and suspended matter, the depth to which enough light for photosynthesis can penetrate (euphotic zone) is limited to less than 0·5 m in the most turbid water to 20–30 m in the clearest. In the shallow but relatively unproductive Lake Nabugabo off the west shore of Lake Victoria the euphotic zone has been shown to be confined to about the upper 0·6 m mainly due to the dense suspension of mud stirred from the bottom (pers. comm. from T. R. Milburn). In the highly productive Lake George the euphotic zone is of similar depth, but is limited mainly by the high concentration of algae in the surface water (self-shading).

Carbon dioxide, which in air is necessarily a gas (about 0·03% by volume), in natural waters can exist in three forms – free dissolved CO_2, bicarbonate ion (HCO_3^-) and carbonate ion $(CO^=)$. The proportions of these three are reflected in the pH of the water (Fig. 7.1) which rises progressively as CO_2 is removed by photosynthesis during the day, and falls at night as CO_2 is released by respiration. The combined concentration of bicarbonate and carbonate is not altered by

---

* The above equation implies that the photosynthetic quotient O_2/CO_2 for carbohydrate synthesis is 1·0. For protein it is about 1·1 and for fat it may exceed 3·0. But during experiments on algal photosynthesis in water over short periods of daylight, such as are described below, carbohydrate synthesis predominates and the oxygen production may, for practical purposes, be taken as equivalent to the amount of carbon fixed.

this shift in CO_2 and is known as the 'alkalinity'.* Most natural waters have a pH of between 6 and 9 and thus contain both free CO_2 and HCO_3^- (Fig. 7.1). Aquatic mosses appear to be confined to relatively acid (low pH) waters because they can fix only free CO_2, whereas many aquatic flowering plants and algae can also obtain CO_2 from bicarbonate to give carbonate $2HCO_3^- = CO_2 + CO_3^- + H_2O$ (Steeman Nielsen, 1947; Ruttner, 1953). At a pH higher than 8 the proportion of free CO_2 becomes very low indeed (Fig. 7.1), though the absolute concentration is of course greater the higher the total alkalinity. There are many algae which thrive in water of higher pH than 9·5 and thus must obtain their inorganic carbon entirely from HCO_3^-. At pH 10·3 about 50% of the inorganic carbon is still in the form of HCO_3^-, but above pH 11·0 the HCO_3^- is largely replaced by CO_3^- and OH^-. In principle, therefore, under otherwise optimal conditions, vigorous photosynthesis would finally be limited by depletion of HCO_3^- unles, as has been shown with some plants, there is a direct inhibiting effect on other cellular processes at a high pH but lower than 11.

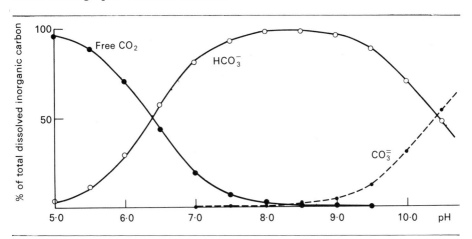

Fig. 7.1 Relationship between free dissolved CO_2, bicarbonate ion ($HOCO_3^-$), and carbonate ion (CO_3^-)

This matter is of particular interest because there are some tropical waters of not abnormally high alkalinity such as Lake George, in which photosynthesis is so intense as to raise the pH to 10 during the afternoon (Table 5.1). Moreover, in the volcanic regions of the African Rift Valleys are many very saline waters of very high alkalinity ($HCO_3^- + CO_3^-$), whose pH well exceeds 10, and in some cases has been reported as high as 11, but which nevertheless support massive growths of blue-green algae (Jenkin, 1936; Beadle, 1932b). Whether the algae, such as *Spirulina*, which characteristically inhabit extremely alkaline waters, can actually assimilate CO_2 from carbonate is a question which invites investigation. Since, however, the alkalinity of some of these waters exceeds 250 meq/l (Table 19.1), the actual concentration of HCO_3^- at pH 10·5 would be vastly greater than that

* More precisely, alkalinity = $HCO^- + CO^- + OH^- - H^-$, but the last two ions are at a very low level in most natural waters, except in highly alkaline (soda) lakes above pH 10, in which the OH^- ions can have important direct effects on organisms.

of, say, Lake Victoria with an alkalinity of about 1 meq/l at pH 8. It is more surprising that any aquatic organism can maintain itself at a pH of over 10·5.

The rate of primary or photosynthetic production in a body of water is usually expressed as the weight of oxygen released or of carbon fixed per unit surface area of water in unit time (gC/m^2t). There are three kinds of method by which the necessary data may be obtained:

1. The quantity of living plant material produced can be estimated by counting or weighing algal cells or measuring spectrophotometrically the concentration of chlorophyll a in samples of known volume. Though an indeterminate number of cells is continuously lost through decomposition, grazing by animals and falling to the bottom, there is a correlation between the population density (cells per unit volume) and net productivity.

2. The rate of photosynthetic fixation of carbon can be estimated by measuring either the removal of CO_2 (directly or from the associated change of pH) or the equivalent increase of dissolved oxygen. These changes can be studied either in the lake itself or in bottles of lake water submerged at different depths. It can also be measured from the rate of uptake of radio-carbon (^{14}C) added as bicarbonate in submerged bottles of lake water. From simultaneous measurements of oxygen consumption in shaded samples the rate of breakdown can be estimated and hence the gross productivity can be calculated (Fig. 7.2). Another variant of this method, which allows measurement in a short time of a large number of samples, taken from a ship at many widespread points, involves 'incubation' of the samples. They are exposed in a closed chamber on board ship to controlled temperature and to a range of artificial light intensities based on photoelectric measurements in the lake.

3. The rate of other chemical processes which are linked to photosynthesis, such as removal of inorganic nutrients from the water, can also be followed. Nitrate and phosphate are in principle the most suitable, but silicate is the easiest to estimate accurately, though its depletion is a measure only of the production of the diatoms which have silicious skeletons.

Method 2 also provides a means for determining the depth of the euphotic zone. Since light intensity and hence the rate of photosynthesis tend to fall exponentially with depth, the euphotic zone is often regarded, for convenience, as extending to the depth at which the light intensity is reduced to 1% of that just under the surface. The light is measured by progressively lowering a photoelectric cell with appropriate filters to enable integration of the total radiation between wavelengths 400 and 700 mμ, which is the region of the spectrum from which most of the energy for photosynthesis is obtained by green plants.

The principles involved in making these measurements can be illustrated by describing an imaginary experiment with paired bottles in a lake with euphotic zone of about 10 m (Fig. 7.2). Measurements of concentration of chlorophyll a and of % light extinction at different depths are made at the same time. The water in each pair of bottles (usually of capacity 30–200 ml) has previously been taken by sampler from the depth at which they are afterwards set and the members of each pair have the same initial concentration of algae and of oxygen. If we wish also to know the rate of respiration (i.e., destruction of organic matter) the initial oxygen must be measured at the start in a third bottle. One of the paired

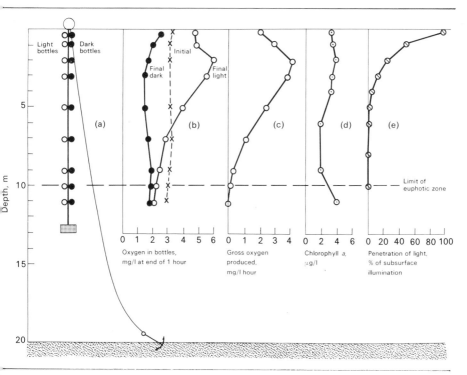

Fig. 7.2 Measurement of primary production. An imaginary paired bottle experiment conducted in a lake with a euphotic zone of about 10 m.

(a) Paired light ○ and dark ● bottles exposed at different depths for 1–3 hours around midday.

(b) The initial and final oxygen concentrations in the bottles.

(c) Gross production of oxygen in the light bottles during the period of the experiment. This is the difference between the final oxygen in light and dark bottles and can be converted to the equivalent of carbon fixed.

(d) Concentration of extracted chlorophyll *a* measured by spectrophotometer at wavelength 665 mμ.

(e) Per cent absorption of radiation (400–700 mμ) which penetrates the surface, measured with photoelectric cell and appropriate filters

bottles is shaded from light (dark bottle) to prevent photosynthesis, the other is fully exposed (light bottle). The bottles are left submerged for a measured length of time, such as one to three hours around midday, after which they are withdrawn and the final oxygen concentration in each is measured. We now have the data for calculating for each depth:

1. The rate of respiration of the whole plankton community (plant, animal and microbial), from the difference between initial and final oxygen in the dark bottle.
2. *Net* production by the whole community of organisms in the euphotic zone, from the difference between initial and final oxygen in the light bottle.
3. *Gross* production by the phytoplankton, from the difference between final oxygen in the light and dark bottles. This does not involve measurement of the initial oxygen.

Though the overall rate of photosynthesis must obviously be determined partly by the concentration of chlorophyll (i.e. of algae), this is often fairly evenly distributed in the euphotic zone, as represented in Fig. 7.2d, and the shape of the gross production curve is more closely related to light extinction. There is, however, a maximum rate, shown in Fig. 7.2c at 2 m, achieved with an optimum range of light intensity above and below which the rate falls off.

To obtain a figure for primary productivity for comparison with other lakes it is first necessary to calculate the average rate of photosynthesis per unit surface area throughout the euphotic zone during the period of the experiment. This is represented by the area circumscribed by the gross production curve (Fig. 7.2c). To obtain an integrated figure expressing the mean rate of photosynthesis over the whole day, either a series of experiments must be conducted during the period of sunlight or an equation must be applied which takes into account the shape of the curve, the density of the algal population and its photosynthetic capacity per unit quantity of chlorophyll and the mean intensity of radiation throughout the day. An integrated figure for longer periods, such as a year, requires a sufficient number of such daily estimates suitably spaced over the seasons to arrive at an average daily figure for the year. For information and discussion of methods of measurement and computation see Talling (1965a) and Vollenweider (1969). In the latter are some critical discussions on the advantages and limitations of the various methods.

In comparing the rates of primary production in different lakes we should bear in mind the main factors which can influence the rate of photosynthesis. They are:
1. the quantity of incident radiation and the depth to which it penetrates at sufficient intensity (depth of euphotic zone);
2. the quantity of photosynthesising plant material in the euphotic zone;
3. the availability of carbon dioxide, which is related to the amount of dissolved bicarbonate (Fig. 7.1);
4. the quantity of nutrient salts, such as phosphate, nitrate and silicate, available in the euphotic zone, which depends on their circulation after release through organic decomposition;
5. the temperature, though its general influence on the biochemical reactions involved in both decomposition and photosynthesis.

In attempting to make a general comparison between productivity of waters in temperate and tropical regions, it must be remembered that, though a rather greater total amount of solar radiation per year reaches the outer atmosphere above a unit area of land surface at the equator, the amount available for photosynthesis in a lake is not related to latitude in any simple manner. It is dependent, in practice, also an such local conditions as altitude, humidity and cloud cover. The proximity of the sea or a large lake, owing to the high specific heat of water, may reduce the effects of diurnal and seasonal fluctuations in intensity of radiation, giving a more even atmospheric temperature. At first sight we might expert a higher productivity in the tropics in view of the absence of a winter season when solar radiation and temperature are at a very low level. We might suppose that conditions in the tropics are more continuously favourable. This expectation is, however, not justified until other essential factors have been considered and in particular the availability of nutrient salts whose production and circulation are indirectly dependent upon solar radiation. On the whole, as was shown in Ch. 6, the tropical temperature regime is less favourable for deep-water circulation; moreover, the winter overturn in a temperate lake provides a maximum concentration of nutrients in the euphotic zone in the spring just when both temperature and illumination are increasing (Lund *et al.*, 1963). It is questionable, therefore, whether the very high rate of photosynthesis during the summer gives an annual production greatly different from that of a tropical lake which, we might presume, produces continuously at a moderately high level. It must also be remembered that a high temperature inducing more rapid photosynthesis is also likely to stimulate all other processes in the metabolic cycle – grazing and assimilation by herbivores, predation by carnivores, bacterial decomposition with release of nutrients and reassimilation by the algae, all reflected in an increased oxygen consumption. Though *gross* primary production and the circulation of energy and materials may be more rapid at a higher temperature, the *net* production is not necessarily increased (see Ryther, 1963, for discussion of this subject). These questions can be answered only by direct observations and measurements.

In view of the number of variables other than those arising directly from differences of latitude, and because data from the tropics are still relatively scarce, comparisons must be rather tentative. Nevertheless the data are sufficient to provide interesting comparisons between a few tropical African lakes and between these and some typical temperate lakes (Table 7.1). The table includes data from a high altitude lake in Ethiopia, from one at low altitude in Egypt (both extremely productive) and from some continental European lakes of different productivities.

From only three tropical lakes (Victoria, George and Turkana) have sufficient measurements between made over a year to justify calculation of an integrated annual figure for photosynthetic production and to demonstrate quantitatively the occurrence or absence of seasonal changes. The first two of these lakes, both of which are on the Equator and at not greatly different altitudes, show two quite different regimes of production. Some at least of the reasons for the differences can be given.

Fig. 7.3 shows curves representing the depth distribution of photosynthetic rate (during an hour or two around midday) in four East African lakes. These are chosen from Talling (1965a) to illustrate a wide range of conditions. Some basic data on these lakes are included in Table 7.1. The following points should be noted from Fig. 7.3:

1. The average depth of the euphotic zone ranged from over 13 m in the open water of Lake Victoria to about 0·7 m in Lake George. This is an inverse measure of the density of the light-absorbing particles of which the algae themselves are an important component. A test tube of Lake George water is perceptibly coloured by blue-green algae.

2. The maximum rate of photosynthesis was in no case found actually at the surface, but at a depth which is related to that of the euphotic zone, i.e. to the degree of light absorption. The intensity of light just under the surface at midday was thus above optimum and photosynthesis was thereby partially inhibited. Curves of this shape implying a maximum rate well below the sur-

Table 7.1 Estimated figures for gross planktonic photosynthetic production in tropical and temperate lakes and some related data

Lake	Mean latitude	Altitude m	Depth at sites m	Euphotic zone approximate depth m	Approximate temperature in euphotic zone °C
Victoria (offshore) Uganda	1°S	1 230	79	13–14	24–26
Windermere (N. Basin) England	54°N	<100	64	10	5–18
Tanganyika (open water)	7°S	773	500	20–25	25–27
Bunyoni Uganda	1°16′S	1 970	40	4	20
Kivu Zaïre	2°S	1 500	480		22–24
Mulehe Uganda	1°13′S	1 750	7·5		20
George Uganda	Equator	913	4·5	0·7	24–35
Chad*	13°N	283	12		23–29
Turkana Kenya	3°N	375			26–28
Mariut Egypt	31°N	<100	1·5	0·1–0·5	
Nakuru Kenya	0·2°S	1 758	3·3		
Reshitani Crater Tanzania	3·2°S	1 448	29		
Ponta Negra Bay Rio Negro Brazil	3°S	44	10	2	27–31
Aranguadi Ethiopia	9°N	1 910	28·3	0·14	19·4–21·0
Lyngby Sø Denmark	55°N	<100	2·8	1·0	5·8–22·8
Esrom Sø Denmark	55°	<100	22	8–10	2·8–19·9
Lunzer Untersee Austria	48°N	608	34	16–20	

* Measured at stations both in the open lake and near Bol among the sand-dune archipelago. The latter showed the highest rate of production. There were regular seasonal changes in rate of production (p. 227).

† The very wide range is due to three ecologically distinct regions in this small lake, one of which was polluted.

Notes:

Euphotic zone: From the surface to the depth at which 1% of the light (of wavelength 400–700 mμ) which penetrates the surface remains unabsorbed.

pb O_2: Paired bottle technique for estimating oxygen production described on p. 117.

pb ^{14}C: From estimation of uptake of radio-carbon from bicarbonate in paired bottles.

Alkalinity surface HCO^- + $CO^=$ meq/l	Chlorophyll a in euphotic zone mg/m²	Photosynthetic (gross) g C/m². day	production g C/m². year	Conditions	Reference
0·92	35–100	1·08–4·20	950	pb O₂ 14 expts. over 10 months	Talling, 1965b
0·17	5–100		20	from crop growth and nutrient depletion 1947	Lund et al., 1963; Talling, 1965b
6·7–7·0		0·8–1·1		pb ¹⁴C	Melack, 1976
1·8		1·80		pb O₂ 2 expts. December 1960, June 1965	Talling, 1965a
12–18		1·44		pb ¹⁴C	Degens et al., 1971b
2·2		0·96		pb O₂ 1 expt. June 1961	Talling, 1965a
1·25–2·0	70–280	5·4	1 980	pb O₂ over one year 1967/68	Granf, 1969, 1975
		0·7–2·7		pb O₂	Lemoalle, 1969, 1975
20–25		0·25–6·2		pb O₂	Harbott, 19??
4·2–9·2		0·01–10·75†		pB ¹⁴C	Aleem and Samaan, 1969
122 (K_{20} 10 000)		2·0–3·2		pb and O₂ electrode	Melack and Kilham, 1974
164 (K_{20} 13 500)		7·5		pb O₂	Melack and Kilham, 1974
0·01–0·1 (very low salinity)	3–40	0·03–0·43		pb ¹⁴C	Schmidt, 1976
54	221–325	13–22		pb O₂	Baxter et al., 1965; Talling et al., 1973
3·9		1·2–6·0 (May–October)	660	pb ¹⁴C 1958–59 heavily polluted	Jonassen and Mathiesen, 1959
2·1–2·4		0·24–1·2 (May–October)	144–204	pb ¹⁴C 1956–58 a natural eutrophic temperature lake	Jónassen and Mathieson, 1959
		0·012–0·24 (May–October)	30	pb ¹⁴C Alpine oligotrophic lake	Steeman Nielsen, 1959

The temperatures of the equatorial lakes show very little seasonal variation, though very near the surface, to which the euphotic zone in Lake George is confined, the diurnal change can be very great.

It must be emphasised that most of the figures, and especially those representing rates of production, have been calculated from measurements made by different methods, over varying lengths of time and in restricted positions in each lake. The individual figures quoted must therefore be regarded only as approximations to the overall rates of primary production during longer periods. Their main interest lies in the comparison between them. Owing to the uniformity of conditions in Lake George and because the measurements were made at many points over more than a year, the figure is likely to be the most reliable indication of annual production for a lake as a whole.

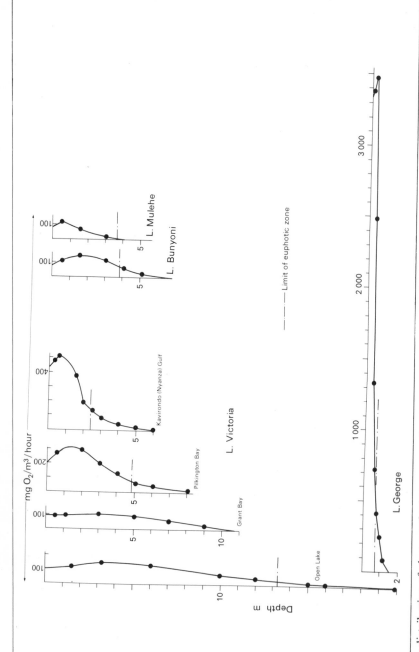

Fig. 7.3 Depth distribution of photosynthetic rate per unit volume of water (mg $O_2/m^3 \cdot$ hour) during the period of each experiment (1–3 hours around midday). The depth to which 1% of the subsurface illumination was measured is shown, lower limit of the euphotic zone (redrawn from Talling, 1965a)

face are characteristic also of temperate lakes at midday in summer. With lower light intensity towards either end of the day, the maximum, as expected, rises to the surface.

3. The maximum hourly rate of photosynthesis (expressed as oxygen production) varied enormously from about 100 mg O_2/m^3 hr at 3 m in Grant Bay, Lake Victoria, to about 3 500 mg at 0·2 m in Lake George.

Primary production in Lake George has been the subject of a continuous investigation over more than a year (1967–68) by Ganf (1969, 1975), and its part in the metabolism of the lake as a whole was discussed by Ganf and Viner (1973). It has been established that an extremely high rate of gross production is maintained at a surprising constant rate throughout the year. There were, however, some indications of a small rise in the mean chlorophyll *a* level during the wet and more windy season (September and October) suggesting a minor seasonal increase in gross production at a time of greater turbulence and, perhaps, increased inflow of nutrients from the land (Viner, 1970c). But, compared with other lakes investigated, the rate of gross production is very constant, and we cannot confidently infer, from observations over one or two years, the regular occurrence of seasonal changes. The nutrients derived from organic decomposition in the mud, as far as can be judged from the nitrogen cycle, are so rapidly taken up by the algae that they are normally hardly detectable in the water. The very high rate of turnover is maintained by the very frequent, usually daily, disturbance by winds which, owing to the shallowness of the water, seems to affect the mud to a depth of about 20 cm. The rate of recycling of nutrients is probably increased by the diurnal movement of much of the phytoplankton between the surface and bottom.

The temperature of the water in Lake George is continuously favourable and there is always an abundant supply of inorganic carbon as HCO_3^-. The factors which actually determine the upper limit of production rate appear to be:

1. the rate at which nutrients are released from the mud and from organisms throughout the water column;
2. the increase in density of the phytoplankton through growth, which puts a brake on photosynthesis by reducing the depth of the euphotic zone;
3. as the result of photosynthesis the dissolved oxygen rises during the day to a high level of supersaturation and this is known to be a partial inhibitor of photosynthesis.

Thus, in the normally very uniform climatic conditions of temperature and daily winds the rate of gross primary production is high and rather constant.

The biomass of algae in Lake George has been estimated at more than 95% of the total biomass, and it was found from paired-bottle experiments that the photosynthetic oxygen production during daylight is almost balanced by the consumption of oxygen during the twenty-four hours by all the planktonic organisms, plant, animal and bacterial in the water and mud. This at first suggests a negligible net production inadequate for the support of the fish and other non-planktonic organisms. It must, however, be remembered that the gross photosynthetic production is so very great that a relatively small proportion, as net, might suffice. There are also the external (allochthonous) sources of organic matter in the form of particles which are consumed directly by many of the animals.

Though most of the shore of Lake George is devoid of emergent vegetation, the extensive swamps which choke the inflows from the north must contribute some organic particles especially in times of flood. Thre is also a very large population of *Hippopotamus* which feed on land at night and rest and defaecate in the water during the day. These allochthonous sources of nutriment may be important, but reliable quantitative estimates are rare.

Calculations based on quantitative studies of populations and feeding habits of the hippopotamus by C. R. Field and from analyses of their faeces suggest that these animals make a significant contribution to the nutrient budget of Lake George. It was reckoned that they add about 2% of the NO_3-N and about 10% of the PO_4-P lost annually via the outflow (Viner and Smith, 1973; Viner, 1975).

Lake George is thus an example of a tropical lake that, being shallow and well exposed to winds, reflects the constancy of the equatorial climate with its very uniform rate of production throughout the year. It is therefore of great interest that in August 1968 the lake was subjected for a few hours to one of the rare and extra violent wind storms, causing the death of a very large number of fish (see also p. 88). Within twenty-four hours the mean cholorophyll *a* content of the phytoplankton had more than trebled and nitrate and ammonia were detectable in the water. The normal balance was restored within about two weeks (Viner, 1970c; Ganf and Viner, 1973). Though some of the extra chlorophyll *a* must have derived from disintegrating algae stirred up from the mud, production was undoubtedly much increased. This was presumably due to the release of a greater amount of nutrients from deeper levels in the mud than normal.

The figure $13-22$ g C/m^2 day for the crater lake Aranguadi in Ethiopia (Table 7.1) is one of the highest yet estimated for any natural water. With the relatively uniform climate at that latitude this rate is probably not very far from the daily mean, and the annual production must be several kg C/m^2. It is likely, though it remains to be proved, that a productivity as high as this and as that of Lake Mariut in Egypt is possible only because of the very high concentration of HCO_3^- ions in these alkaline waters (Table 7.1) from which CO_2 can be extracted by the algae without raising the pH to a toxic level. On the other hand the pH of the surface water of Lake George, with its alkalinity ($1\cdot8$ meq/l) low compared with other Rift Valley lakes, occasionally approaches $10\cdot0$ during the afternoon. This, we may suppose, is near the lethal limit. Many of the extremely alkaline and shallow lakes of East Africa, such as Lakes Elmenteita and Nakuru with their enormous concentrations of blue-green algae, which provide the main diet for large flocks of flamingo (p. 342), are fixing carbon at a prodigious rate (Melack and Kilham, 1974; and Table 7.1).

Given an optimum temperature with unlimited nutrients and carbon dioxide, the theoretical limit to the rate of photosynthesis would be set mainly by the number of algal cells that can be exposed to the most light (below the inhibiting intensity) for the longest time. When, however, a certain density of cells is reached, they themselves begin significantly to obstruct the light and reduce the photosynthetic zone (self-shading). There is thus an amount of chlorophyll per unit surface area above which there can be no increase in photosynthetic rate per unit area. According to Steeman Nielsen (1962) this limiting level of chlorophyll is likely to be about $200-300$ mg/m^2. The photosynthetic productivity of Lake

Aranguadi with 220–325 mg chlorophyll/m^2 must therefore be close to the possible maximum, at least under natural conditions (Talling et al., 1973).* This phenomenon can be of practical importance for fish culture in ponds. It follows that the progressive addition of fertiliser, while increasing the growth of algae and the rate of production in the upper layers, will also decrease the depth of the euphotic zone through self-shading to a point at which overall productivity begins to fall. Hepher (1962) has demonstrated this quantitatively on fish ponds in Israel, and Robarts (1979) on Lake McIlwaine, Zimbabwe.

Comparable figures for gross photosynthetic production in temperate lakes are mostly less than 3 g C/m^2. day during the summer. Lyngby Sø in Denmark (Table 7.1) is exceptionally poductive through heavy artificial pollution which provides abundant nutrients, but its overall annual productivity measured by Jónassen and Mathiesen (1959) was 660 g C/m^2 and thus considerably less than that of Lake Victoria. Natural eutrophic lakes in temperate regions, of which Esrom Sø is typical (Table 7.1), provide figures for annual productivity between 100 and 300 g C/m^2. Oligotrophic lakes are mostly below this range as shown by the Lunzer Untersee in the European Alps and the lower Rio Negro in the Amazon basin whose water is extremely deficient in salt and nutrients (Table 7.1).

It appears therefore that the overall primary productivity of eutrophic natural freshwaters may be somewhat higher in the tropics. In order to be quite certain that this is a consequence of the climatic difference and not due to fortuitous non-climatic factors in the lakes investigated, it would be necessary to have data from many more lakes. In principle, we might compare two lakes, tropical and temperate, for which conditions other than climate are identical or at least close enough to ensure that climatic differences predominate. This, in practice, well-nigh impossible to ensure, but Talling (1965b) has attempted such a comparison between Lakes Victoria and Windermere, which provides some interesting information on relative productivity. In addition, it demonstrates the seasonal changes in productivity which are dependent on very different climatic factors in the two lakes.

Lake Victoria lies in a tectonic basin (Ch. 14), Lake Windermere in a glacial excavated valley. Their maximum depths are approximately equal (Table 7.1), but the area of Victoria (75 000 km^2) is much greater than that of Windermere (17 km^2). Other differences shown in Table 7.1 which are not obviously connected with climate are:

1. The alkalinity ($HCO^- + CO^=$) of Victoria which, though the lowest of the large East African lakes, is about six times that of Windermere.
2. The depth of the euphotic zone which is greater in Victoria. This difference may be due to the inflows into Windermere which introduce some coloured organic matter, whereas about 90% of the water income of Victoria is from rainfall direct onto the surface.
3. The amounts of the main nutrient salts are different in that Victoria has a higher concentration of silicate and inorganic phosphate and less nitrate. These chemical characteristics seem to be common, but not universal, in tropical freshwaters. Measurements of incident radiation (as 10-day means)

* A similarly exceedingly high rate of photosynthesis was recorded by Melack (1979) in the small alkaline saline Lake Simbi, Kenya, by a dense population of the blue-green alga *Spirulina platensis*.

showed a higher mean but smaller range for Victoria (300–500 cal/cm^2 day) than for Windermere (30–400 cal/cm^2 day) and the range of temperature in the surface water showed the same relation (23·8–25·4°C and 5–18°C respectively: Fig. 7.4).

In view of the considerable seasonal changes which, as will be shown, occur even in Lake Victoria, a comparison of gross photosynthetic productivity was made from measurements during the seasonal maximum in each lake and from a rough estimate of overall annual productivity. The rate of photosynthesis per hour around midday during the seasonal maximum in Victoria was about five times that in Windermere, though the concentration of chlorophyll a in the

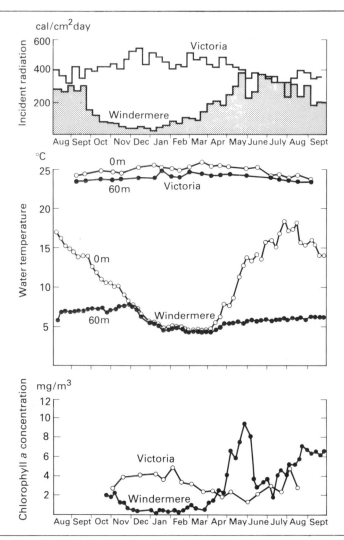

Fig. 7.4 Lakes Victoria 1960–61 (equatorial) and Windermere, England, 1963–64 (54°N). Monthly fluctuations in incident radiation, water temperature and total phytoplankton as indicated by concentration of chlorophyll a (rearranged from Talling, 1965b).

euphotic zones was about the same ($75-100$ mg/m^2). The photosynthetic capacity or the rate of carbon fixation per unit weight of chlorophyll under optimum illumination, was thus greater in Victoria. Talling suggests that the main factor involved in this is the temperature difference between the two lakes, which amounts to $10-12°C$. If this is indeed the explanation we are led to the rather simple conclusion that one reason why the tropical climatic is conducive to higher primary productivity is the direct effect of higher temperature on the photosynthetic mechanism.

Comparison of annual production was more difficult because, though a daily mean of fourteen measurements spread over ten months could be taken as a reasonable basis for calculation in the case of Victoria, only spring measurements were available for Windermere. The annual productivity of the latter could only be indirectly calculated from estimates of crop production and nutrient depletion made during 1947 (Lund *et al.*, 1963). Lake Windermere gave a *net* figure of $20·4$ g C/m^2 year, which, if it could be corrected for respiration, would probably be multiplied by two or three, but would certainly give a gross figure far smaller than the 950 g C/m^2 year for Lake Victoria (Table 7.1). The difference is so great that we must regard it as evidence supporting the conclusions from direct comparison of photosynthetic rate during the algal maxima.

There is, however, a need for many more measurements using identical methods and covering the seasons before we can be sure of disentangling the effect of a tropical climate as such from other possible influential factors such as alkalinity and the availability of nutrient salts, some of which may be dependent indirectly on climatic changes. In particular, the effect of temperature on the photosynthetic capacity of temperate and tropical algae should be further examined. Though the rates of growth and of photosynthesis of the temperature lake algae *Asterionella formosa* are, as might be expected, temperature dependent (Q_{10}* $2·1-2·3$ between $6°$ and $16°C$), it cannot survive at tropical temperatures (Talling, 1957b, 1965a). The genetically distinct features of those species that are confined to the tropics include adaptation to a higher temperature range. This, however, does not necessarily involve a photosynthetic mechanism working at a higher rate. Like respiration, it is dependent upon enzymes the concentration of which, it has been suggested, may be adjusted to maintain a more or less constant rate in the face of limited changes of temperature (Steeman Nielsen and Hansen, 1959). More relevant information might come from observations and experiments on species of alga such as *Microcystis aeruginosa*, which are found both in temperate and tropical waters, but even here there may be genetically determined enzymatic differences.

The measurements on Lake Victoria suggest that the tropical climate favours a high rate of gross primary production. This, however, as already stated, does not imply a high net production available to heterotrophic organisms and particularly to fish. The respiratory rate of the algae themselves and of all other organisms will also be raised by the higher temperature and a greater proportion of the material and energy will be dissipated before reaching the fish. This has been demonstrated by the work on Lake George (p. 123) and more detailed investiga-

---

* Q_{10}, the factor by which the rate of a process is multiplied per $10°C$ rise in temperature.

tion is needed before it can be stated with confidence that tropical conditions as such favour a higher net production available to the fisheries.

It might have been expected that the large number of radio-carbon measurements of primary productivity which has now been made in many parts of the oceans would provide a clear demonstration of the comparative productivities of temperate and tropical seas. But the major limiting factor is the supply of nutrients and, owing to thermal stratification, the primary productivity of tropical seas is characteristically low. The most productive regions are those where deep currents rise to the surface, and these are often at high latitudes. Some of the highest figures, probably equivalent to several hundreds of g C/m^2 year, have been got from around South Georgia and the Antarctic (Ryther, 1963).

In Lake Victoria the alternation between stratification in the calm season of February to June and complete mixing by the heavy southeast trade winds from June to August has already been discussed (Ch. 6). That this annual mixing initiates an outburst of algal growth has been known for some time (Fish, 1956, 1957). Talling (1957b, 1965a) has investigated the seasonal changes in chlorophyll concentration and in the species of algae in the euphotic zone. The former is demonstrated in Fig. 7.4 and contrasted with Lake Windermere. It is striking how little the July to August outburst in Lake Victoria is associated with any change in incident radiation or water temperature, whereas the spring and summer growths in temperate lakes are contemporary with a great increase in radiation and in temperature of the surface water. With continuously adequate radiation and temperature, phytoplankton production in Lake Victoria as shown by chlorophyll concentration never falls very low and rises to a peak in July and August as a consequence of stirring by heavy winds; this is aided perhaps by increased evaporation causing a lowering of the surface temperature and a reduction of the density gradient, thus reducing the amount of energy required to stir the water to the bottom (Newell, 1960). This is likely to provide the euphotic zone with nutrients bought up from below which were in short supply during the period of stratification. The subsequent lowering of production was possibly due to the exhaustion of nitrate (Talling, 1965b). A less pronounced rise in photosynthesis was recorded in January and February, associated with a partial breakdown of stratification.

The course of events in a temperate lake, such as Windermere, is very different. During the winter period of complete mixing the limiting factor is usually not nutrient supply, which is then at its maximum, but inadequate illumination and low temperature. They both rise rapidly in spring, and the consequent outburst of production is later limited by exhaustion of nutrient, which in Windermere is mainly silicate. The second (autumn) outburst is probably stimulated by a further supply of nutrients brought into the epilimnion and thus into the euphotic zone with inflowing rivers (Lund et al., 1963). This subsides with decreasing illumination as winter approaches, though the consequent overturn replenishes the nutrients at the surface.

The timing of the production regime in Lake Windermere is thus mainly determined by the seasonal changes in solar radiation. This is typical of moderately deep lakes in temperate regions, but how far can we generalise about tropical lakes with so much less information at our disposal? No systematic measurements have been made on any of these lakes except those already discussed, and in none other than Lakes Victoria and George have seasonal changes in primary production been studied quantitatively. There is, however, little doubt from the reports of competent observers on several other lakes over the past thirty years that there is a seasonal increased growth of algae associated with heavy winds. At one extreme are the very shallow lakes such as Lake George and Lake Abiata in which the enormous algal production, giving a permanent green colour to the water, is largely maintained by frequent stirring of the bottom deposits. At the other extreme are the deep but well-sheltered lakes, such as Lake Nkugute, whose usual algal production is probably very low, but which may on rare occasions of exceptionally violent weather be deeply stirred causing a temporary rise in productivity. Apart from in Lake Victoria sudden increases in the algal population have been noted and associated with windy periods in parts of Lakes Tanganyika and Malawi. In Lake Albert outbursts of algae have also been recorded at various points and times (Evans, 1962).

Even in Lake Kivu, which of all the large lakes seems to have the most stable stratification (p. 106), there is apparently an annual outburst of production as judged by the decrease in the transparency of the water and by the increased quantity of zooplankton (Fig. 15.3) during the dry and windy season (Damas, 1937; Verbeke, 1957a). This can involve, at most, a very small fraction of the very dense lower water and, as already explained (p. 106), Lake Kivu can be regarded as a lake about 70 m in depth floating upon a mass of dense and more saline water. This upper, floating lake shows seasonal alternations between stratification and complete mixing. But compared with the other permanently stratified lakes (e.g. Tanganyika and Malawi) the lower stagnant water is probably more completely out of circulation and most of the nutrients trapped within it can never reach the surface. The general impression of a seasonal alternation between low and rather higher primary productivity (with some seasonal fluctuations) would probably be supported by quantitative measurements of photosynthesis. The great amount of nutrients trapped within the oxygen-free deep water of Kivu was strikingly demonstrated some years ago by a Belgian experiment at the north end of the lake for the commercial production of methane by pumping bottom water to the surface when the gas escapes by effervescence due to reduced pressure (p. 48). This stimulated an abundant growth of phytoplankton and attracted a great many fish (pers. comm. from A. Capart). There is no doubt that the productivity of the deep and permanently stratified tropical lakes could be very greatly increased at least temporarily if means could be found for circulating a large amount of the lower water.

The higher density of phytoplankton in the shallow water at the north and south ends of Lake Tanganyika, as of Lake Malawi (Coulter, 1963; Eccles, 1962, 1974), indicates that the rate of production per unit volume of water is, as would be expected, higher than in the bulk of the lake where the deep water is very clear. The latter necessarily contains a low concentration of solid particles including algae and other planktonic organisms. The clarity of the offshore water of Lake Tanganyika was previously regarded by some biologists as evidence of low productivity (oligotrophy) though this was difficult to reconcile with the flourishing fishery for 'sardines' and their predators (pp. 292–293). Measurements in 1971 by Melack (1976) by ^{14}C in paired bottles in 500 m water towards the north end indicated that the rate of gross primary production is as high as in the open water of Lake Victoria (Table 7.1). Since they were made in April 1971 before the onset of the vigorous southerly winds the annual rate of production would probably be higher than this. The depth of the euphotic zone was found to be 20–25 m and thus deeper than in the pelagic water of Lake Victoria (13–14 m, Fig. 7.3). Subsequently in 1975 Hecky, Fee and Kling (unpublished, pers. comm.) made many measurements along the length of the lake by means of an incubation method used on board ship (p. 116). Their results surprisingly seem to suggest that production by the planktonic algae is insufficient to provide for the energy needs of all the plankton – still less of the fish. This conclusion was based on the high rate of oxygen consumption by water samples in the dark. It was suggested that organic matter from the sediment must be brought to the surface and recycled through direct consumption by bacteria and protozoa, which are abundant in the plankton. But this raises some important problems, especially (1) the origin of the organic matter which must have been photosynthesized at some time and place, and (2) the means by which it could now be circulated sufficiently rapidly. Our present knowledge of the lake's hydrology would hardly support such a powerful vertical circulation from the bottom. Nevertheless the sources and the means of organic production in Lake Tanganyika now appear even more interesting and stimulating to the researcher. We can at least be certain that considerable quantities of nutrients and organic matter are being raised in some manner from the abyssal region and that primary production is spread through a depth of over 20 m.

In drawing the general conclusions concerning seasonal changes in rate of production, to which the foregoing examples seem to be leading, we must be prepared for exceptional cases in which there are other controlling influences. The three sets of measurements by Talling (1965a) on Lake Albert in 1960–61 showed that the rate of gross primary production was actually higher during the season of stratification in March than when the whole water-column was stirred in August. Speculation on possible causes of this might suggest (1) the water from the Semliki inflow in the south may provide a major supply of nutrients to the surface water during stratification and thus stimulate production, (2) the nature of the phytoplankton – in Lake Albert (Talling, 1963) small green flagellates and the blue-green *Anabaena* are important constituents and both tend to accumulate in the upper layers if not dispersed by stirring, (3) increased turbidity of the stirred water-column might significantly reduce the penetration of light. Some increase in turbidity during the stirred period in March was in fact shown by Talling (1965a, Table 1) from measurements of the extinction coefficients of light of wavelengths 380–685. There is evidence from Lake Mariut in Egypt (Aleem and Samaan, 1969) that in very shallow water heavy winds may increase the turbidity of the water and reduce light-penetration enough to retard photosynthesis, whereas moderate winds will increase production by circulating nutrients without raising the mud.

There is clearly a need for more prolonged investigations on a greater range of tropical African lakes before we can properly understand the regimes of production and the factors that control them. We must nevertheless attempt to summarise the main conclusions that seem to be justified from the studies of hydrology and primary production discussed in this and the previous chapter.

In the equatorial region, between about 10°N and S lat., except at very high altitudes, water temperature and solar radiation are always adequate and relatively constant over the year. Gross primary production, unlike in temperate climates, is not seasonally limited by a low temperature or low intensity of surface incident radiation. The main controlling factor is the supply of nutrients. In very shallow lakes so situated that the lower water and mud are stirred by winds at frequent intervals at all seasons, gross primary production may be maintained at a continuously high and rather constant level. This has been shown to be true of Lake George and is likely to be true of the many other shallow and well-exposed waters near the Equator. It is difficult to assess all the controls, but there are indications that the gross primary production of equatorial lakes under otherwise similar conditions is probably higher than their counterparts in temperate regions. Since, however, the water temperature is high throughout the twenty-four-hour day, respiration and decomposition processes are also maintained at a high level. It is therefore more doubtful whether equatorial lakes have a higher *net* primary production available to the heterotrophic animals, including fish, than temperate lakes under otherwise similar conditions. Evidence on this matter comes from recent work on Lake George (Ch. 13).

In the open water of somewhat deeper lakes, such as Lake Victoria, deep stirring is reduced during the calmer season and there is a tendency for stratification and accumulation of nutrients at the bottom. These are dispersed by the seasonal stirring of the southeast trade winds, and there is a detectable seasonal increase in primary production.

In Lake Tanganyika wind stress at the surface has a direct stirring effect down to 100–200 m, which is only a fraction of the total depth and an enormous volume of water is not affected in this manner. We have, however, reason to suppose that there is some significant circulation of water and nutrients throughout the whole column associated with internal waves. The latter are ultimately caused by wind stress at the surface and are thus subject to seasonal changes. The biological effects of these events in both Lake Tanganyika and Malawi have yet to be fully understood though this periodic upwelling of nutrients with a visible increase in the algal population in shallow water has been observed in both lakes.

A number of relatively small but deep lakes, such as some craters in the bays of some volcanic barrier lakes, have a short wind fetch, are sometimes well sheltered from wind and are permanently stratified and generally rather unproductive. There are, however, indications in some of rare deep stirring with a temporary rise in production caused by an unusually violent storm.

Towards the northern and southern margins of the tropics (beyond 10°N and S) the range of seasonal temperature change increases (Fig. 3.7). In Lake Kariba the marked seasonal overturn is associated with a considerable drop in temperature approaching that in a lake in the warm temperate zone. Lake Chad in the northern tropics is shallow enough to be more or less continuously stirred to the bottom, but the water temperature in winter seems to be low enough to retard the growth and development of most organisms and thus to impose a seasonal regime on production (Ch. 12).

The changes in the hydrology and productivity during the early stages of the evolution of manmade lakes are of exceptional interest. They are discussed in Ch. 20.

It cannot be too strongly emphasised that the data so far available (as in Table 7.1) relate mainly to the open water of lakes where it is easier to make measurements that would seem to relate to the majority of the water of the lake. But it must be remembered that the inshore waters are shallow and are subject to very different hydrological and production regimes. They are usually much more productive owing to frequent stirring to the bottom, to the proximity of inflows of nutrient-rich water and to the presence of large submerged and emergent plants (macrophytes) which are often the major primary producers inshore. In Lake Victoria shallow inshore waters occupy a large proportion of the whole, are clearly very productive and have so far provided the most profitable fishing grounds. Their productivity probably depends more upon the supply of nutrients from the land and their recycling in shallow water than upon conditions out in the open lake.

The degree of seclusion of a bay and the direction faced by the opening in relation to winds and currents will presumably determine the extent to which the production cycles in bay and lake are interconnected. Fish (1957) found evidence of occasional mass movement of nutrient-rich lower water from the open lake into some of the shallow bays near Jinja on the northern shore of Lake Victoria.

It would therefore be wrong to assume without question that the figures set out in Table 7.1 represent the primary productivity of each lake as a whole. How nearly a figure may represent a reliable overall estimate can be judged only after studying the records of the original measurements. One must decide whether they have been spread sufficiently in space and time to justify integration for the whole lake over a long period, having in mind the variety and relative extents of the different habitats in the lake and the vagaries of the climate that have been recorded from the past.

The maintenance of the organic production cycle even in the open water of a lake is, in the long run, dependent upon external sources of nutrients and decomposable organic matter from inflows, from seasonal flooded land, from rainfall and in some cases from nitrogen fixation (p. 43). It is more difficult to assess these contributions and to measure production in the inshore regions than to study the cycle in the open offshore water. The situation is very different in different lakes and, indeed, at different times and places in the same lake, as will be apparent from the chapters devoted to the various lakes and water systems. The proportion of shallow inshore water supporting macrophytes varies very greatly. In some such as Lakes Chad and Kioga, it is a major habitat. But in none is it entirely negligible and it should be studied as an important site of both primary and secondary production. Very little quantitive work has so far been done.

As in other aspects of an aquatic ecosystem, the development of manmade lakes provides an interesting demonstration of the stimulating effects on primary productivity of a sudden increase in plant nutrients, followed by a slowing down a production to a lower level as the nutrients are consumed and become limiting. (See Ch. 20, and particularly the seasonal changes in the Jebel Aulyia Reservoir, Fig. 20.5).

8

The evolution and distribution of the African inland water fauna (with a note on the tilapias)

Inland water systems are more or less isolated from each other by land and sea barriers, and we might think that there is little opportunity for interchange of faunas and floras. It is therefore at first sight surprising that, on the whole, the world's freshwater faunas and floras are rather uniformly distributed and many species are cosmopolitan and restricted only by suitable water and climate.

Compared with the oceans and major land masses most inland waters are temporary and, on the geological time-scale, ephemeral. Apart from those that exist only in wet seasons there are very few which have not been radically changed in form and interconnections during the past million years. With a few notable exceptions, some of which are in Africa, there are no inland water systems which have been isolated for long geological periods. Nevertheless, as we shall see, even comparatively short periods of isolation have had a profound effect on the organisms in several of the African water systems.

The majority of plant and animal species in freshwaters are of very small size, such as the small plants and invertebrate animals of the floating plankton, on the bottom and among the larger vegetation in shallow water, e.g. unicellular algae, Crustacea, Rotifers and Protozoa. A small and temporary connection between normally separate water systems, e.g. a flooded watershed, could allow many of these organisms to pass from one area to another. Moreover, very many of the small animals are, at one stage or another of their life history, resistant to drought and many of the algae produce resistant spores. This seems to be a response to the generally impermanent nature of inland waters and enables the organisms to survive desiccation (Ch. 19). It is certain that these very small drought-resisting forms are carried considerable distances by winds, and it is probable that they are often transported on the bodies of birds and land animals. The aquatic insects, which are an important component of the freshwater fauna, are a special case. The young stages are usually confined to water, but the adults either live entirely out of water, e.g. mosquitoes, lake flies, mayflies, dragonflies, etc., or, like many water-bugs (Hemiptera) and beetles (Coleoptera), occasionally leave the water and move considerable distances overland. The African bivalve mollusc *Aspatharia* and the amphibious prosobranch snails of the genera *Pila* and *Lanistes* (Ampullariidae) can survive drought for at least a year. The air-breathing pulmonate snails can also exist for some time out of water and some can aestivate in dried mud during a dry season, e.g. *Bulinus* spp. On the other hand, most of the prosobranch snails are purely aquatic and in Africa it is these molluscs which demonstrate the effects of isolation in a particularly striking manner.

To the biogeographer, however, the fish, of all freshwater organisms, are the most significant. To quote from Darlington (1957):

> Although they live in water, those that are confined to freshwater are as closely bound to land masses as are any animals. On land they are almost inescapably confined to their own drainage systems and can pass from one isolated stream basin to the next only by slow changes of the land itself or by rare accident, so that they disperse over the world slowly and are likely to preserve old patterns of distribution.

On the other hand, many of the watersheds between the African river basins are rather flat areas which, in flood seasons, provide a more or less continuous sheet of water between the opposing tributaries. Worthington (1933) suggested that such watersheds, and in particular the Luapula-Kafue section of the Zaïre-Zambezi divide (Figs. 9.3 and 9.4), are usually blocked by swamps in which the water is low in oxygen and therefore impassable to most fish. But only one such watershed has actually been examined from this point of view. This is the Muhinga plain between the Kamawafura (River Zaïre) and the Kanjita (River Zambezi) streams (Bell-Cross, 1965). Here during much of the wet season (November–March) there is a continuous link of adequately oxygenated water in channels and open pools. Of the fourteen species of fish in the Zaïre stream nine were found also in the Zambezi stream and six were found crossing the flooded watershed. There is thus some exchange of fish, but it is, of course, confined to those few species which can live under the peculiar conditions of low temperature and rapid flow in the upper reaches. The fact remains that the great majority of species in the Nile, Zaïre and Zambezi basins which have common watersheds are now quite isolated and are inhibited by physical and behavioural factors from approaching the upper reaches of the streams. Nevertheless, the composition of the present fauna of the Zambezi River indicates some exchange of fish with the Zaïre during the Pleistocene (Ch. 9). This may have been due to river-capture or perhaps to earth movements diverting sections of rivers from one basin to the other. Periods of much higher rainfall may also have flooded the watersheds more heavily than at present. There is, however, both faunistic and geological evidence for very recent and much more substantial connections between the Nile, Niger and Lake Chad which has allowed an exchange of species on a large scale (Ch. 11). The Nile and Niger basins are now separated from each other by large stretches of quite arid country.

The effectiveness of a barrier is thus subject to change from time to time and those that exist at present between contiguous basins are only relatively effective. As examples, the Murchison Falls on the Victoria Nile and the rapids on the Semliki River appear at present to be complete barriers, the rapids on the Ruzizi near Bukavu have been surmounted by only one fish (*Barilius moori*) from Lake Tanganyika since the Pleistocene volcanics, which formed the present Lake Kivu with its overflow down the Ruzizi. There are many smaller falls and rapids on the major rivers (e.g. Karuma Falls on the Victoria Nile) that are surmountable by a large number of species. It is interesting to note that the distribution of many of the species of fish in some of the rivers flowing into the Gulf of Guinea is more related to the many rapids and falls on these rivers than to the watersheds between them (Daget, 1962)

It is well established that in the past 'river capture' has been an important process in determining the course of many river systems in all parts of the world (Holmes, 1965, pp. 559–63). The upper reaches of a tributary of river A may erode backwards through the watershed separating it from the neighbouring tributary of river B. The upper reaches of the latter may thus be provided with an easier outlet and will become a tributary of river A. The point of capture is normally marked by an abrupt change of direction, as between the Middle and Lower Niger at Tosaye (Fig. 9.2), and there is sometimes a sudden drop in level with rapids or a waterfall that may be insurmountable to fish. Such 'captures' have probably occurred at several points between the river basins of West Africa during the late Pleistocene.

It should be added that there are a few fish in tropical Africa which can migrate over damp ground at night. These include the lungfish *Protopterus* and some species of the genus *Clarias*, the airbreathing catfish, the former well known for its resistance to complete drought by encysting in a cocoon in dry mud. Some very small species and the young of larger species can move over long distances in very shallow floodwater. The remarkable small cyprinodont fishes of the genus *Notobranchius* actually produce eggs that can survive in the mud when the swamps are dry during which time the embryo's development is arrested (dispause) (p. 349).

Nevertheless, it is generally true that nearly all fish are incapable of life in anything but water and are relatively easily and effectively isolated. More often than other freshwater animals, they find themselves in situations favouring the production of new species in restricted regions. It is therefore not surprising that the distribution of the species of fish in the African inland waters is often clearly and dramatically related to the rather violent geological and climatic events in the recent past (Ch. 3). In discussing the different African water systems it will consequently be easier to characterise the faunas by reference to the fish than to other groups of animals. The distribution of the species of Crustacea that are external parasites on fish has provided an interesting addition to the evidence to be got from the fish themselves (Fryer, 1968b).

The faunas and floras of many of the waters under discussion have been derived by immigration from elsewhere and through the production of new species *in situ* from some of the immigrants. The occurrence and distribution of species cannot therefore be discussed without reference to the events which have led to the formation of new species (speciation). It is not the function of this book to deal more than superficially with genetical ecology, but enough must be said here and later to indicate the significance of genetic factors and to emphasise the wealth of material provided by the African inland waters for studies in this field. The mechanism of speciation is one of the central problems of biology and stimulating discussions can be found in the works of Dobzhansky (1959), Cain (1969), Mayr (1963, 1970) and White (1978). Brooks (1950) reviewed and discussed the problem of speciation in the ancient lakes of the world, and included some of the large African lakes. More information is now available from

Africa which will be presented when the different water systems are discussed. A brief review of some salient features of speciation in East African lakes is to be found in Beadle (1962) and a more wide-ranging discussion in Lowe-McConnell (1969, 1975). Fryer and Iles (1972) have discussed at length problems of speciation in the African cichlid fishes.

The genetic machinery in sexually reproducing organisms essentially involves the shuffling of genes into different patterns at each generation and the occasional appearance of new gene mutations of various kinds and (mainly in certain plants) the duplication of chromosomes. This ensures that in a given environment the individuals of an interbreeding population of a species will vary one from the other, but that the variation is limited and the average characteristics of the population remain rather constant. This is one of the main criteria by which we can define a species. This constancy, however, depends upon two conditions:

1. The environment, which involves chemical and physical factors (e.g. climate, etc.), food and relations with other competing organisms, that must not alter beyond the range to which the population is at present capable of adapting through selection of appropriate combinations of the genes in its possession.
2. The population must be reproductively isolated from (i.e. must not interbreed with any other population of the same species having different characteristics.

The population possesses a set or 'pool' of genes which are shuffled around between the individuals in a limited range of patterns consistent with successful adaptation of the population as a whole to the particular environment. Unfavourable variations beyond these limits will automatically be eliminated because they are not so adapted and their bearers, if they survive, produce fewer or no offspring.

Each individual is therefore endowed with a set of genes (genotype) which is somewhat different from that of other individuals in the population. The expression of the genotype in the characters of the developed organism (phenotype) may be altered by the normal changes in the environment. This sometimes confuses the situation which may require careful investigation in order to decide whether certain variations are a reflection of genetic or of environmental factors.

A good example of such a situation is provided by certain species of the freshwater planktonic crustacean *Daphnia* (Brooks, 1957, 1965). Under favourable conditions they reproduce parthenogenetically from unfertilised eggs giving a range of forms from those with a large dorsal ridge or 'helmet' on the head to those with none. These variations have been shown to be determined during development by the temperature and turbulence of the water and are thus related to climatic changes. The phenomenon is known as cyclomorphosis, and is discussed at length by Hutchinson (1967, Ch. 26). That they are not irreversible variations in genetic factors can be demonstrated by rearing clones from single parthenogenetic females of widely different head-form. The progeny of those with large helmets show the same range of variation as those of non-helmeted individuals. These characters have not therefore been fixed by selection and are not steps towards the production of new species. They do, however, provide additional means for successful adaptation to a variable environment. The presence of what appears to be two forms of the same species of *Daphnia lumholtzi* in Lake Albert has been described by Green (1967a). A large-helmeted form is commoner in the coastal lagoons and a smooth-headed form predominates in the open water of the lake. Green discusses the possible selective value of these in the face of predation by the fish *Alestes baremose* and *Engraulicypris bredoi*. This kind of variation may be more widespread among freshwater animals than is generally realised. It has been found among species of the whitefish *Coregonus* which show different varieties in different Swedish lakes, and these can be shown not to be genetically determined but are phenotype variations due to environmental differences (Brooks, 1957). The variations have made it difficult to separate the closely related species of these genera, and it is obviously important to be aware of this phenomenon when investigating supposed cases of speciation.

It is, however, genetic variability due to the recombination of genes at each generation, that makes survival possible in the face of an inconstant environment, and is the basis of evolutionary change. A population is plastic and capable of adapting to a certain range of conditions, it can also survive as a whole in the face of limited environmental changes, though ill-adapted individuals are sacrificed. Some species habitually produce the same range of well-marked variants, known as polymorphs, which can be recognised as types. They are common in many species of animals – e.g. insects, molluscs and birds – and certain groups of fish, such as the African cichlids, are particularly variable and polymorphic. It is possible that in some situations polymorphism is an adaptation to different subdivisions of the niche occupied by the population as a whole. No environment is quite uniform. There is no doubt that some species are genetically more variable than others, and it is possible that loss of variability can result from prolonged residence in a relatively constant environment and was perhaps a cause of extinction in the past. The genotype has been so selected that, when the environment changes, the organism has lost the plasticity needed for successful adaptation.

When there is a considerable change in the environment beyond the range of its previous normal fluctuations, but not so drastic as to exterminate the population, different patterns of genetically determined characters will be required for adaptation and survival, and these will be naturally selected and the average character of the population will change. Such an event can result from geological or climatic causes, e.g. the appearance of a new lake or the reduction of an old one to a swamp, or from the appearance or disappearance of other organisms which normally react with the species either as predators or as food. The same situation may arise from the spread of part of the population into other areas where different conditions prevail.

Genetical research has shown that considerable adaptive alterations are effected without new mutations by selection of more favourable patterns of already existing genes. Mutations will be selected and will persist only if they can be fitted successfully into the pattern of the genotype, and there may be several generations after the mutation has appeared before the pattern is rearranged to produce a well-adapted phenotype.

In theory, the environment might change simultaneously in the same manner over the entire range of the population, whose genetic patterns would become uniformly rearranged by selection to produce eventually a single new species in place of the old one. In most cases, however, it seems that populations of a species become differentiated into 'geographical races', with mean characteristics which diverge from each other in adaptation to different conditions in different parts of the range of the species. The less the opportunity for interbreeding between these sections of the population, depending on distance and the barriers between them, the more will they tend to diverge. The divergences in genetic patterns result in progressive incompatibility and decreasing fertility of crosses between them, until finally the races are reproductively isolated from each other and are, by definition, separate species. These now maintain their identity, and may continue to diverge, even if circumstances should bring them together again.

We have now hinted at a definition of 'species' – an actually or potentially interbreeding group of organisms which thus share a common gene pool and are incapable of successful crossing with other groups from which they may have diverged. This is the ultimate stage of speciation, and it is not necessarily accompanied by well-marked changes in visible appearance of the phenotype. It must be emphasised that observations in the field and experiments have demonstrated every degree of reproductive divergence between two incipient species from impaired fertility of otherwise healthy hybrids to complete sexual incompatibility. In investigating cases of speciation which are indicated by the divergence of visible characters, the extent to which the process has actually gone can, strictly speaking, only be decided from experiments in interbreeding or from examination of the results of natural hybridisation between divergent populations in regions where they are in contact.

There is abundant evidence from all parts of the world that speciation is usually 'allopatric', that is, it involves some degree of spatial isolation at the start. The distribution of the faunas and floras of continents, isolated islands, river systems, mountain ranges, etc., can only be explained on this basis. The African inland waters provide many good examples of the origin of new species in isolated places, some of them, e.g. the Zaïre basin, Lakes Tanganyika and Malawi, showing a remarkable array of species which are endemic and found nowhere else.

It is, however, doubtful whether actual spatial separation of sections of a population is invariably a necessary condition for divergence leading to speciation, and there are cases which have seemed difficult to explain on this basis (discussion in White, 1978). The so-called 'species flocks' of cichlid fishes in Lake Victoria are composed of a number of closely related species which are clearly now, or have been in the recent past, in the process of divergent speciation. But they apparently occupy the same habitat and share the same food, and there does not seem to be any spatial isolation or external physical barriers to interbreeding. Similar apparent sharing of identical habitats is shown by several related species of rock-browsing cichlid fishes in Lake Malawi (Fryer, 1959a, and see p. 302). Because of the difficulty of conceiving a mechanism by which sympatric speciation (i.e. without some degree of separation) could occur, there have been several different theories to account for the evolution of species flocks of fish in lakes. For example, some species may have diverged when isolated in inflowing rivers or in bodies of water detached from the main lake during a period of low rainfall and subsequently reunited with a rise in water level (see discussion in Mayr, 1970, Chs 15 and 16; Fryer and Iles, 1972, Chs 15–17). The amount of speculation involved in discussions on the evolution of the African cichlids is beginning to overrun the established information on the adaptations in structure and habits of these fascinating fishes and on the conditions under which speciation may have occurred. It must be remembered that the situation in which most feeding is done is usually the most accessible to observers and collectors, but there are other situations, such as breeding sites and refuges from predators, where competition and selection may be intense enough both to cause spatial separation and to control the numbers sufficiently to prevent competition on the feeding grounds (Lowe-McConnell, 1969). In other words, natural selection may operate in situations other than those in which the animals are most frequently found.* It is rarely certain in any particular case that spatial isolation has played no part in the initial stages of divergence.

The present situation is not necessarily, and in certain cases definitely is not, the one in which the species originated. Earth movements or changes of climate in the past may have divided the habitat into separate portions. It has been suggested for instance, that Lake Victoria was reduced in the past to several separate bodies of water and that this may have been responsible for the divergence of the main species flocks of the cichlid *Haplochromis* which later came together in the present lake (Ch. 14). That Lake Tanganyika was probably two lakes during part of the Pleistocene has been suggested as one of the factors responsible for the abundant speciation in that lake (Ch. 12).

The case of the fishes of Lake Barombi Mbo in western Cameroon, described by Trewavas *et al.* (1972) and Green *et al.* (1973), is of particular interest. This small volcanic crater lake, 2·5 km in diameter and 111 m maximum depth, is surrounded by dense forest. It is one of a group of craters whose age is considered to be late Pleistocene. The water below 20 m is devoid of oxygen. The volume of water and the area of bottom available to aerobic organisms, particularly fish, is therefore much restricted. Nevertheless, the lake and its inflowing stream support seventeen species of fish of which twelve are endemic and include all of the eleven species of cichlids found in the lake. There are even two endemic genera. The cichlids are now ecologically well separated in respect both of their feeding habits and depth of water in which they commonly feed. One of the cichlids, *Konia dikume*, feeds on the larvae of the lake-fly *Chaoborus* which migrate daily up and down across the anoxic boundary for which it is physiologically adapted (p. 321). It is suggested that two or three species may originally have invaded the lake, though by a route which is difficult to imagine, and have subsequently evolved into the present group of species. But the greater mystery is the nature of the stimulus for the initial stages of divergence. The lake is small and, apart from the anoxic boundary, appears to be free of physical barriers to the movement of fish, nor is there reason to suppose that such barriers have ever existed since the lake was formed.

Compare this fauna with that of Lake Bosumtwi in Ghana, also an isolated crater lake of similar age, though larger (7–8 km diameter, 78 m deep). For

---

* See Lowe-McConnell (1959) for a discussion of the evolution of the species of *Tilapia* (Cichlidae) in the African lakes based on a study of breeding behaviour.

climatic reasons discussed on p. 93 the deoxygenated lower water is partially dispersed every year. It seems to offer a greater variety of habitats than Lake Marombi Mbo but, of the eleven species of fish, only one of the four cichlids, *Tilapia discolor*, is endemic (Whyte, 1975).

It must be emphasised that, though speciation is progressive, the final consummation is reached only with reproductive isolation which, in itself, is a structural and behavioural feature and does not require spatial separation to maintain it. All those authors cited in this book who have discussed the speciation of organisms in African waters, seem to have assumed the impossibility of any but allopatric speciation. It is however the opinion of some prominent biologists that the initial and even all the changes leading to speciation could occur, and in some cases have occurred, within a single population (see White, 1978, Ch. 7; and Endler, 1977). There are many other difficult cases besides those mentioned here, and an exact definition of the term 'sympatric' is not so easy to formulate. We must however concede that spatial separation has played a major part in the evolution of most species in tropical inland waters, but the reader should adopt a critical attitude. Clearly, we cannot yet be confident that we understand all the possible causes and mechanisms of speciation, and the present allopatric–sympatric speciation controversy may well be superseded.

The rate at which new species have appeared in the past was certainly not uniform and great changes in the environment, such as the appearance of a new lake, were followed by an outburst of speciation in adaptation to the new conditions. Just how rapid this can be is apparently demonstrated by the cichlid fish of Lake Nabugabo (p. 254), which was cut off from Lake Victoria about 4 000 years ago. Greenwood (1965b) has described five endemic species of *Haplochromis*, though the two *Tilapia* spp. of the main lake have remained unchanged. It would, however, be interesting to discover by experiment just how far these endemic *Haplochromis* spp. are now reproductively isolated from their relatives in Lake Victoria from which they are structually distinguishable. The estimated length of time involved would have more significance if the extent of the genetic divergence were known.

The fauna and flora of a water system are thus determined both by its accessibility now and in the past and to the circumstances which may favour the evolution of new local species. The latter involve both the physical features of the environment and the genetic variability of those organisms that have managed to colonisee it. There are, of course, the obvious physical factors which make a water system a favourable or unfavourable (or impossible) environment for certain organisms. Temperature as determined by latitude and altitude clearly has a very potent influence. For example, the four species of the African lungfish *Protopterus* are confined to the tropics from the southern edge of the Sahara and middle Nile to the Zambezi. These are very mobile animals and there is no obvious barrier other than temperature that could limit them to the tropics. Though there are many other organisms which are confined to a relatively small temperature range, there are others which are remarkably tolerant. Several species of planktonic algae (e.g. *Microcystis aeruginosa*), of cladoceran Crustacea (e.g. *Moina dubia* and *Alona affinis*), of copepod Crustacea (e.g. *Mesocyclops leuckarti*), and of Rotifera (e.g. *Keratella tropica, Brachionus angularis*) are found both in tropical Africa and in tropical, temperate and even subpolar regions of other continents.

The amount of rain and its distribution in time and space is another important limiting factor. In the most arid regions, as in the Central Sahara, rain falls at very rare and irregular intervals producing standing water which may last for only a few weeks. In these appear in a miraculously short time a great variety of organisms such as algae, Protozoa, Crustacea and Rotifera, all of which have developed from dried resistant eggs or spores. At the other extreme are the large lakes and watercourses (e.g. Rift Valley lakes, Rivers Zaïre, Niger, Nile and Zambezi) which are fed from regions of high and frequent rainfall. These have developed a varied fauna and flora much of which is entirely dependent on permanent water.

There are also many examples of habitats in which conditions in certain respects are so extreme that only a few organisms have managed to adapt to them. There are the hot springs in volcanic regions, and the saline waters due to excessive evaporation in arid regions and to salt springs (Ch. 19). In East Africa there is an abundance of very alkaline saline water mainly derived from the carbonatites of the igneous rocks and volcanic lavas that abound there. We should also include the subterranean waters found in West Africa whose fauna is specially adapted to total darkness. But the most extensive of the physiologically 'difficult' aquatic habitats are those deficient in oxygen. There are vast areas of swamps in tropical Africa where a scarcity or complete absence of oxygen is the predominant limiting factor for animal life (Ch. 18), and a large proportion of the water of the deepest lakes is permanently anoxic and often charged with hydrogen sulphide (Ch. 6).

It is generally true, and not unexpected, that the few species that have managed to adapt to extreme conditions are little hampered by competition and are often extremely abundant. The green soupy colour of some alkaline saline ponds due to a dense growth of the alga *Spirulina platensis* and the swarms of the cichlid fish *Sarotherodon grahami* in the hot alkaline springs of Magadi in Kenya (p. 344) are good examples.

It must not be forgotten that the environment to which organisms must be adapted has a biological component. Other organisms present as food, as competitors for food and as parasites and predators are as important as the physical and chemical features of the environment. They may be present or absent from the ecosystem for fortuitous geographical reasons. It can well be imagined that the natural or artificial introduction or elimination of a species could have profound effects on other species with which they have close ecological relations. It could alter the course of evolution of some of the resident species, or even cause their extinction. Many of the features of animals and plants are clearly adaptations to the presence of other organisms. Among others are the special structures, mechanisms and behavioural reactions for predation, protection and escape. There can be no doubt that these have evolved in response to relations with other organisms.

It is a striking fact that the cichlid fishes, especially of the genus *Haplochromis* and related genera, have speciated on a very much larger scale in Lakes Victoria, Tanganyika and Malawi, where there are great numbers of endemic species, than in other large lakes such as Albert and Turkana. Worthington (1954) suggested, and was later supported by Jackson (1961b), that the presence of the voracious soudanian predator fish *Lates* (Nile Perch) and *Hydrocynus* (Tigerfish) in Turkana, Albert and Chad have inhibited the speciation of the cichlids upon which they feed. Conversely, the absence of these predators from Lakes Victoria and Malawi has left them free to do so on a grand scale. This suggestion has been contested by others (e.g. Fryer and Iles, 1955) on the grounds that there are plenty of other fish (species of *Bagrus*, *Clarias* and *Barbus*) that prey on the cichlids in Lakes Victoria and Malawi. Indeed, there are many predators on cichlids, including *Lates*, in Lake Tanganyika in which a very large number of endemic species of cichlids are found. It is not immediately obvious how predation could, in principle, retard speciation of the prey. It is possible to imagine circumstances in which it might even act as a stimulus. To quote Lowe-Mc-Connell (1969, p. 69).

> the predators' presence may be both *restricting* speciation by preventing colonization by cichlids of areas without shelter and *promoting* speciation by enhancing the isolation of sheltered areas, such as rock patches, predators thus exerting their influence in both ways simultaneously.

The problem is also briefly discussed by Mayr (1963, pp. 274–5).

This very speculative controversy has continued, with some heat, for several years.* In my opinion, the evidence is insufficient for making a decision. It is not easy to see how conclusive evidence could, in fact, be got: the circumstances vary so much from one lake to another. A similar problem is presented by the very large numbers of endemic species of prosobranch molluscs in Lake Tanganyika and their scarcity in Lake Malawi.

We can now discuss in brief outline the main distribution patterns of freshwater fish in Africa, which for reasons already stated, more clearly reflect the present interconnections and past history of the water drainage systems than other groups of organisms. The main biological features of the inland waters in the north and south temperate regions of the continent will also be mentioned here, since the interchanges which have taken place between them and tropical Africa in the past are of very great interest. In following chapters the tropical African fauna will be discussed in more detail in the setting of the different inland water systems of the tropical belt and of the peripheral regions (e.g. the Sahara) into which this fauna has spread.

If all the species of freshwater fishes were marked on the map of Africa we should discern at once the general pattern which is summarised in Table 8.1.* The vast majority of families and species are to be found in the tropical belt which includes the West African rivers Senegal, Gambia, Niger, Zaïre and those draining the eastern highlands towards the north and east between the Nile and the Zambezi. In addition are included Lakes Chad and Victoria and all the lakes of the two Rift Valleys from Ethiopia to Malawi and from Lake Albert to Lake Tanganyika. This enormous area of warm and more or less permanent water is not only an admirable environment for fish to live and breed in, but provides a

---

* Those interested in following it further should refer to the authors mentioned above and to several other publications to be found from Fryer and Iles (1972), who discuss the matter at length.

* For summaries of the distribution of the families and genera of Africa freshwater fish see Blanc (1954) and Poll (1957).

great variety of conditions and degrees of isolation which has stimulated a spectacular outburst of evolutionary change. Of the twenty-seven families which are recorded from the tropical belt (Table 8.1), eleven are endemic, i.e. have been evolved in this region. A very large number of the species of the non-endemic families, especially the Cichlidae, are also endemic.

In spite of great regional differences, which will be discussed later, a general survey reveals a certain basic uniformity of fauna within the tropical belt. Apart from such fish as the airbreathing catfish *Clarias lazera* (Clariidae) and of the lungfish *Protopterus* spp. (Lepidosirenidae), which are able to move easily

Table 8.1 Tropical and extra-tropical distribution of freshwater fishes in Africa.*

Family	Common English names of some representatives	Tropical Africa Genera	Tropical Africa Species	N. Africa (N. Maghreb)† Genera	N. Africa (N. Maghreb)† Species	S.W. Cape Genera	S.W. Cape Species
Polypteridae	Bichirs	2	6	—	—	—	—
Lepidosirenidae	Lungfish	1	4	—	—	—	—
Cromeriidae		1	1	—	—	—	—
Clupeidae	Herrings	8	+ +	—	—	—	—
Kneriidae		1	+	—	—	—	—
Phractolaemidae		1	1	—	—	—	—
Galaxiidae		—	—	—	—	1	1
Osteoglossidae		1	1	—	—	—	—
Pantodontidae	Butterfly fish	1	1	—	—	—	—
Notopteridae	Knife fish	2	2	—	—	—	—
Mormyridae	Elephant snout fish	11	+ +	—	—	—	—
Gymnarchidae		1	1	—	—	—	—
Salmonidae	Salmon, trout	—	—	1	1	—	—
Characidae	Tigerfish	9	+	—	—	—	—
Citharinidae		19	+	—	—	—	—
Cyprinidae	Barbels, minnows, carp, tench	14	+ + +	3	17	2	15
Bagridae	Catfish	12	+	—	—	1	1
Clariidae	Airbreathing catfish	9	+ +	—	—	—	—
Schilbeidae	Butterfish	8	+ +	—	—	—	—
Mochokidae		7	+	—	—	—	—
Amphiliidae		8	+	—	—	—	—
Malapteruridae	Electric catfish	1	1	—	—	—	—
Cyprinodontidae	Mosquito fish	10	+ +	2	3	—	—
Centropomidae	Nile perch	2	7	- -	–	—	—
Cichlidae		77	+ + +	—	—	—	—
Anabantidae	Labyrinth fish	2	2	—	—	1	2
Tetraodontidae	Puffer fish	1	5	—	—	—	—
Ophiocephalidae	Snake heads	1	2	—	—	—	—
Mastacembelidae	Spiny eels	1	+	—	—	—	—
Gasterosteidae	Sticklebacks	—	—	1	1	—	—

* Marine families some of whose members are very recent or temporary inhabitants of coastal fresh-waters have been omitted, e.g. Auguillidae, Synbranchidae, Mugilidae. It is difficult to discover the total number of recorded species of many tropical African genera, but a very rough idea of the relative abundance is indicated by +. Of the Cichlidae there are certainly several hundred species, and of the Cyprinidae more than a hundred.

† North of the Atlas watershed.

Table 8.2

Lake	All families			Cichlidae only		
	Number of species	Endemic genera	Endemic species %	Number of species	Endemic genera	Endemic species %
Albert	46	0	13	9	0	44
Turkana	48	0	21	7	0	43
Victoria	180+	5	83	170+	4	98
Malawi	245+	20	93	200+	20	97
Tanganyika	214+	38	80	134+	30	98

Approximate figures to illustrate the extent to which new (endemic) species and genera of fish have evolved in five of the Great Lakes. They are arranged in order of increasing degree and time of isolation, as suggested by the geological evidence. The figures are approximate since new finds and reappraisals of the status of species already recorded continue to be made. + indicates that these figures probably be will augmented, but the main conclusions will certainly not be altered.

between water systems, there are a number of genera and even some species of 'waterbound' fish which are common to many of the main basins. Such are the Nile perch *Lates niloticus* (Centropomidae), *Sarotherodon galilaeus* and *Citharinus citharus* (Citharinidae). This, together with a limited amount of fossil evidence, suggests that, at least as far back as the Miocene, there was a rather uniform fish fauna over much of tropical Africa. The formation of the Rift Valleys and associated lakes and other large alterations in drainage systems were due to major earth movements after the Miocene (Ch. 3). These fish were then presented with a great variety of new habitats to which they became adapted and thereby altered. The descendants of the ancient fauna have changed relatively little and are now most clearly represented by fish in the Gambia, Senegal, Niger, Volta, Chad and Nile basins which have been connected with each other in more recent times (Fig. B1). This fauna has usually been called *nilotic* by English writers. But the term is unnecessarily restrictive, since the area concerned embraces most of northern tropical Africa. More appropriate is the word *soudanien*, often used by French zoologists (see Daget, 1968), which is derived from the Arabic *bilad es sudan*, 'the country of the blacks'. This originally denoted that part of tropical Africa south of the desert which was known to the Arab traders before the nineteenth century; that is, most of the tropical belt north of the rain forests and Great Lakes, which is now mainly savanna. Since *Sudan* is now the name of a country occupying a small part of this area the word *soudanian* is adopted in this book as the appropriate English equivalent for this ichthyogeographical region. The now commonly quoted name *Sahel* (borderland) is applied to the belt of semi-arid country between the Sahara and Equatoria which suffers disastrous periodic droughts. The *Soudanian* region, which relates only to the distribution of fish, includes the Sahel but spreads far into the humid tropics and into the Sahara.

The greatest multiplication of species and widest divergence from the original Miocene fish fauna have occurred in the Zaïre, Lake Malawi and Lake Tanganyika basins which have had the longest history of isolation from the soudanian river basins. Enough divergent speciation has occurred in the rivers flowing into the Gulf of Guinea to warrant a western subdivision of the soudanian region and a separate Guinean region along the coast (Daget and Iltis, 1965; Daget, 1962, 1968). It is into these rivers and into those flowing into the Atlantic along the Cameroon-Zaïre coast that marine clupeid fish have migrated and produced a number of species adapted to freshwater. Some have penetrated the Zaïre basin even as far as Lake Tanganyika (p. 291). The present complicated distribution of species in the upper reaches of the rivers along the Guinea Coast can only be explained by vertical earth movements in the recent past, though on a much smaller scale than in eastern Africa. These have left many rapids and waterfalls that are barriers to the movement of fish. It has been suggested by Daget (1962) that river capture has also played an important part in isolating populations of species in this region.

The long-continued isolation of the Zaïre basin and the formation since the Miocene of the Great Lakes of eastern Africa were the principal external influences in the evolution of the tropical African fish. The change from river to lake conditions not only provided a stimulus to speciation, but also, as might be expected, altered the pattern and composition of the fauna as a whole. This is made clear by the diagrams in Fig. 8.1. In terms of numbers of species the faunas of the older lakes Tanganyika, Malawi and Victoria are dominated by cichlids, which are particularly adaptable to lake conditions. At the other extreme, the

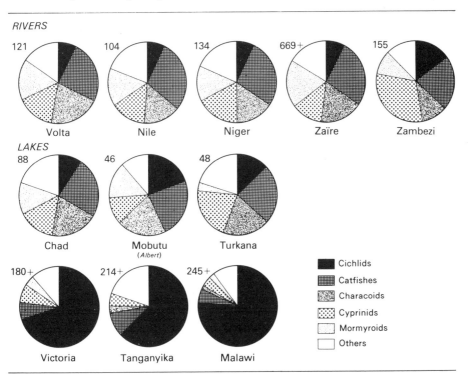

Fig. 8.1 The proportions of fish species of the main groups in various lakes and rivers. The number adjacent to each diagram indicates the total number of species so far found. The + sign indicates that the present list of species will certainly be increased: new species have been found but not yet described (based on Lowe-McConnell, 1969, with additions)

rivers Nile, Niger, Zaïre and Zambezi have relatively few cichlid species, but a much greater proportion of catfishes (Siluroids). The more or less intermediate character of the faunas of Lakes Albert and Turkana may be a reflection of the present (Lake Albert) and the very recent (Lake Turkana) connections with the Nile, from which they have diverged very little. The case of Lake Chad is of particular interest. Not only does its large soudanian fish fauna include only one endemic species, but with its very small proportion of cichlids, its composition is that of a river (Fig. 8.1). Here there have also been very recent connections with the soundanian rivers Nile and Niger (Ch. 12).

It is a striking fact that the number of species of fish in any one of the large tropical African lakes or rivers (Fig. 8.1) is very much greater than in comparable lakes and rivers in temperate regions. This is true of all tropical inland water systems; there are about 670 species of fish already known from the Zaïre basin and perhaps, when exploration is complete, the number will rise to a thousand, which is not far short of the number in the Amazon basin. To compare with these figures are the 172 species in all of the Great Lakes of North America and the 92 species in all inland waters of Africa south of the tropic of Capricorn (Lowe-McConnell, 1969).

It is generally true that the number of species of animals and plants in most habitats in the humid tropics is very much greater than at higher latitudes. The explanation of this phenomenon is not as clear as might at first be thought, and it is generally agreed that the direct effect of continuously higher temperature on the rate of mutation and on the number of generations per year is unlikely to be significant. Published discussions on the subject have led to the general conclusion that the climatic differences may result in differences in the operation of natural selection. At high latitudes the seasonal swing in climate, especially in temperature, and the much greater long-term fluctuations (glacials and interglacials) during the Pleistocene have put a premium on adaptation to a changing physical environment. Wholesale destruction of species and their habitats, as during the ice ages, have occurred from time to time, and the seasonal changes have demanded adaptation to at least two kinds of climate or periodic migration to avoid the change. This lack of environmental stability, it is thought, has prevented the development of the complex ecosystems to be found in the tropics in which selection seems to be concerned mainly with relations between organisms. With the relatively stable equatorial climate, physical environmental factors are of less importance, and there has been plenty of time for the evolution by natural selection of very complicated ecosystems in which a great number of species are adapted to a large number of varied component habitats (see Dobzhansky, 1959; Fischer, 1960; Cain, 1969; and for a critical review of ideas on this subject see Pianka, 1966).

It could reasonably be objected that Lake Baikal, at over 50°N latitude in Siberia, is an exception. It has a very much greater number of species of animals inhabiting its abyssal region, most of the endemic, (including 240 species of amphipod Crustacea) than any other lake in the world. Supporters of the views expressed above might reply that this is perhaps the exception which proves the rule. The physical conditions in the abyssal region (below 250 m) are extremely constant and have probably remained so throughout the Pleistocene. It is the prolonged stability rather than the level of temperature which is important. On the other hand, the number of species of fish in Lake Baikal, which live mainly in the surface waters, is small (50) compared with the number in Lake Tanganyika (214+). These two lakes are remarkably similar in size and structure and are both of great age (p. 286).

Another interesting example is Lake Ohrid on the Albanian-Yugoslav border (Stanković, 1960). Unlike most of the European and North American lakes, which are post-glacial in origin, Lake Ohrid dates back at least to the early Pliocene, and moreover lies beyond the maximum extension of the Pleistocene ice-sheets. It is rather small (250 km^2) but has a maximum depth of 286 m. The temperature below 50 m is permanently below 7°C. Of the 256 recorded species of invertebrates more than 150 are endemic, and of the 17 fish species 10 are endemic. The extent of speciation is vastly greater than in any other lake, except Baikal, in the north temperate zone. Like Lake Tanganyika they have both experienced a very long period of stability and isolation, which seems to be a more important condition for speciation than actual level of temperature.

The biologically exuberant tropical belt is bounded on the north and south by the subtropical arid zones across which the aquatic fauna for obvious reasons becomes progressively attenuated and impoverished. But the Sahara is particularly interesting in that there are a number of tropical African fish and a few endemic species derived from them, which are surviving under conditions of extreme isolation (Ch. 11). The arid zones abut on small but biologically distinct regions at the northern and southern extremities of the continent – the northern Maghreb and the southwest Cape region.

The Maghreb embraces those parts of present Morocco, Algeria and Tunisia which are north of the Sahara Desert – the north and northwestern coastal plains, the Atlas Mountains with high plateaux and the more arid southern foothills which project into the desert. These are the 'Lands of the West', 'al Maghreb', which were overwhelmed by the Arab invasions from the east at the end of the seventh century A.D. But the most effective barrier to the movement of freshwater fish in this region is the Atlas watershed. The country north of this could be called the 'northern Maghreb' comprising the only part of North Africa not dominated by the Sahara, which approaches the Mediterranean coast from Tripoli to Port Said. Conditions are, however, rather less arid in the coastal hills of Cyrenaica, and the desert is interrupted by the Delta of the Nile which has enabled some of the tropical African fish to approach the Mediterranean. The northern Maghreb is more effectively isolated than the Southern Cape region from the rest of Africa because of the formidable barrier of the Atlas range and the greater aridity of the Sahara which extends northwards with undiminished severity to the southern foothills of the mountains.

The freshwater fish of the northern Maghreb are, like its terrestial flora, almost exclusively of Eurasian (Palaearctic) origin (Pellegrin, 1921). The birds, however, have a significant proportion of Ethiopian (i.e. African) species and mammalian fauna is at least 40% African. The explanation of these differences is not altogether clear but must lie in the complicated history of this region which has more than once during the Pleistocene been connected with tropical Africa by a belt of relatively well-watered country across what is now the western Sahara

(Ch. 11, and see Moreau, 1966, Chs 3 and 4). During such periods the Atlas would have presented less of an obstruction to the passage of birds and mammals than to freshwater fish, but some tropical African species of fish managed to reach the southern foothills of the Algerian Atlas. *Tilapia zillii*, *Sarotherodon galilaeus* and *Clarias lazera* are to be found living in small permanent pools near Biskra (Pellegrin, 1921, and Fig. 11.3).

There are no large lakes in the Maghreb and the inland waters are mostly rapid mountain streams and rivers across the coastal plains. These are favourable environments for the cyprinid fish which are represented throughout Eurasia as well as Africa. There are, for instance, thirteen species of the barbel (*Barbus*) endemic to this region, some having close affinities with species in southern Europe. Five other species recorded by Pellegrin (1921) are found south of the Atlas and belong properly to the Sahara. There are eight species of the genus *Phoxinella* (minnow) which is not represented elsewhere in Africa but is widespread in Eurasia. Four of the species are also found in the Middle East, the other four are endemic to North Africa.

A variety of the common trout (*Salmo trutta*) is found in the rocky streams of the Atlas. This and other freshwater species of the family Salmonidae are otherwise absent from Africa (except where recently imported), but are common throughout Eurasia and North America. The same can be said of the stickleback *Gasterosteus aculeatus* (Gasterosteidae) which, in Africa, is found only in the coastal plains of the Maghreb.

Of particular biological interest is the family Cyprinodontidae. There are two species of which one, *Aphanius fasciatus*, is a small and highly adaptable fish found in hot springs, brackish and very saline water, as well as in fresh water in southern Europe, the Middle East, the Maghreb and the northern Sahara. It is one of the few inland water fish from the Mediterranean region which has penetrated into northern edges of the Sahara where the other species (e.g. of the families Cichlidae, Cyprinidae and Clariidae) are mainly derived from tropical Africa. The single species *apoda* of the related genus *Tellia*, which has lost its pelvic fins, is a remarkable example of isolated speciation. It is confined to a few warm-spring pools on the high plateau of the Algerian Atlas.*

As in other parts of the world the fresh waters near the coast are frequently invaded by species which also inhabit the sea, e.g. members of the families Serranidae (sea-perches), Anguillidae (eels), Gobiidae (gobies), Atherinidae (silversides), Mugilidae (grey mullets), Blennidae (blennies) and Syngnathidae (pipe-fishes).

The tropical African aquatic fauna has been more successful in approaching the southern than the northern extremities of the continent. The deserts of the southern subtropical belt are, even during the present comparatively arid period, confined to the mid and central regions (Kalahari and South West Africa). There are many perennial rivers traversing the eastern and southern coastal regions from the Limpopo to the Cape. These, which, together with the Orange and Vaal Rivers, drain the whole of the Republic of South Africa have been subjected to

---

* For a limnological study of four small lakes in the Moroccan Atlas Mountains see Dumont *et al.*, 1973.

geological and climatic changes in the past and have provided the means by which a few of the tropical fish adaptable to lower temperatures have managed to reach the Cape Region. South of the Zambezi the number of species of freshwater fish is markedly reduced. Owing to the warm east coastal (Agulhas) current, subtropical conditions extend to the southeast tip of the continent, and some of the tropical fish are found further down the east coast than would otherwise be expected. *Sarotherodon mossambicus*, for example, inhabits rivers almost as far south as Port Elizabeth (Lat. 34°S). This species, like some other species of *Sarotherodon*, has obviously been assisted by its ability to withstand a relatively high salinity. In the southern part of its range it is confined to the lower reaches of the rivers, since at higher altitudes away from the coast the water is not warm enough at these latitudes. But this involves exposure to the brackish waters of the estuaries. It is even possible that it has occasionally extended its range southward by migrating from one river mouth to the next, perhaps during heavy-floods which might dilute to some extent the coastal seawater (Jubb, 1967).

Only four of the tropical African families of fish have reached the Southwest Cape Region (Barnard, 1943; Jubb, 1967). This is the coastal strip round the southwest corner of the continent where the climate is in general cooler than elsewhere at the same altitude in southern Africa. All the species are endemic to this region. The Cyprinidae are represented by seventeen species and thirteen are of the genus *Barbus* for which the rivers provide particularly suitable conditions. The other families are Bagridae (two spp.), Clariidae (one sp.) and Anabantidae (two spp.). The tropical African fish fauna has thus been much attenuated in its passage to the Cape. This, we may suppose, was due both to barriers to dispersal and to lowering of temperature. Those few that have become established there have flourished and have produced isolated endemic faunas. In addition, there is the anomalous family Galaxiidae, represented by two species of the genus *Galaxias* found nowhere else on the continent. The habitats of *G. zebratus* are typically mountain streams and those of *G. punctifer* small lakes (Barnard, 1943). The members of this remarkable family all live in fresh water though several, but not the South African species, enter, and some even breed, in the sea. They are otherwise found in Australasia, some of the subantarctic islands and southern South America (Darlington, 1957, p. 107, for further references).

The invertebrate freshwater fauna of Cape Province is particularly interesting (Barnard, 1927, 1940; Harrison, 1964, 1965, 1978). There appear to be two main elements. The first is mostly confined to the upper reaches of streams and includes some genera of aquatic insects and the crustacean families Phraeatocidae (one sp.) and Jaeridae (one sp.) both of the order Isopoda, and Gammaridae (ten spp.) of the order Amphipoda. These two orders, which are common in the fresh waters of North Africa, including the northern Sahara as well as the whole of Eurasia, are not otherwise represented in Africa south of the Sahara except for those species of Isopod parasitic on fish in Lake Tanganyika and the Zaïre basin, the rare cave-dwelling Isopod *Stenasellus* in Côte d'Ivoire (Monod, 1945b) and the amphipod genus *Eucrangonyx* which has twice been found in subterranean waters in the Transvaal and Rhodesia (Methuen, 1911, and pers. comm. from A. D. Harrison). It would seem that the free-living isopods and amphipods, and at least some of the insects, in the hill-streams of Cape Province have reached South

Africa by a route other than through the tropics from North Africa. The second element in the fauna mainly occupies the plains, both highland and lowland, and is more closely related to the tropical African fauna to the north.

So far as we can see from the evidence, it appears that those members of the tropical African freshwater fauna that have managed to reach the Cape region have partially replaced a more ancient fauna which has taken refuge mainly in waters at the higher altitudes (Harrison, 1964, 1965). The origin of this old fauna, in which we may perhaps include the fish *Galaxias*, provides some matter for speculation. The amphipod and isopod Crustacea, and the fish *Galaxias*, might have been independently derived from marine forms but this could hardly apply to the aquatic insects. On the other hand, the overall distribution of the Galaxiidae and of the close relatives of these Crustacea is not inconsistent with Wegener's concept of continental drift. This involved a southern land-mass, Gondwanaland, during the Mesozoic which later split up and drifted apart to separate the previously connected continents of South America, Africa and Australia. Many other facts of biogeography have in the past been presented as evidence for continental drift, but in the absence of adequate geological and geophysical evidence the whole concept was generally regarded as insecurely based. In recent years, however, the physical evidence for continental drift has been dramatically augmented particularly from the study of palaeomagnetism and of the structure of the ocean floor (Holmes, 1965; Tarling and Tarling, 1971 Scientific American, 1977). The origin of the ancient freshwater fauna of the Cape region has to be reconsidered in the light of these discoveries.

A Note on the Tilapias

Those cichlid fishes, that until recently have all been included in the genus *Tilapia*, figure prominently in this book as the most important of the commercial fish in tropical Africa and are of great evolutionary interest. To those readers who are not familiar with the taxonomy of these fishes and may wish to consult some of the original literature, it should be pointed out that they have now been separated into two distinct genera:

1. *Tilapia*, of which there are very few species and only two, *T. zillii* and *T. rendalli* (formerly *melanopleura*) are mentioned in this book. They feed mainly on submerged vegetation, such as waterweeds, they have few gill rakers, and they lay and guard their eggs in excavated pits in the sand under shallow water.

2. *Sarotherodon*. The majority of the species belong to this genus and about twenty appear in this book. They have many gill rakers covered with mucus glands which are part of the mechanism for filtering out the phytoplankton and small organic particles that are their food. Soon after laying, the eggs are taken into the mouth of one or both parents (according to the species) and are kept there until the yolk is absorbed. Both genera continue collectively to be called 'tilapias' (no capital T). For more details and discussion on the justification for this division see Trewavas (1966b, 1973, 1978) and Lowe-McConnell (1975, p. 25).

9

The great rivers of Africa

The courses of the permanent African rivers are shown in Fig. 3.5. This well illustrates the general distribution of high rainfall and humidity over the continent. Of the four great rivers only the Zaïre (Congo) is 'complete' in the sense that all of its tributaries are permanent, the entire basin being within the humid tropical belt. The left bank of the Upper and Middle Niger, all of the Lower Nile, the left bank of the Middle Nile, and the southwest edge of the Zambezi basin, are bounded by very arid country that contributes no permanent surface drainage water. The catchments of these rivers are basically the ancient basins that existed long before the Miocene when the major earth movements began to transform the drainage systems of eastern tropical Africa (Ch. 3). They are therefore considerably older than all of the large lakes except Chad, but they have been affected by earth movements which have altered their courses and produced many abrupt changes of level. Consequently, there are more rapids and waterfalls than on any other of the great rivers of the world and, unlike the Amazon in tropical South America, none of the African rivers has provided unimpeded access by non-portable boats into the interior of the continent. This has had some influence on history and development, and has restricted subsequent water transport to internal stretches of rivers. On the other hand, for the same reason, the potential hydroelectric power is immense. It is difficult to imagine that more than a small fraction will be needed in the foreseeable future. The present schemes are mobilising an infinitesimal proportion of the available potential.*
Until now however the African peoples have benefited from the rivers mainly as a source of biological water for themselves, their domestic and wild animals (including fish) and, in the more arid regions, for their crops. Through the ages they have extended and improved the methods of exploitation, and recent large-scale innovations, especially in irrigation, have transformed the life of many peoples.

The importance of the river fisheries is however not generally realised. It has been estimated that the rivers and their floodplains contribute about 40% of the

* The distribution of hydroelectric power in Africa was plotted in the *Oxford Economic Atlas of Africa* (1965, p. 79) and reproduced with subsequent additions in Grove (1970a, Map 16), though the latter includes the Murchison Falls Scheme in Uganda which has been abandoned.

total (over a million tonnes) of fish caught annually in the African inland waters (Welcome, 1976).

The falls and rapids and the fast-flowing upper reaches of rivers are inhabited only by those organisms that can hold onto solid objects and are shaped to reduce the impact of the rushing water, or can find places, as under stones or in rock crevices, where conditions are relatively calm. A few special adaptations will be mentioned later, but it is appropriate at this point to refer to the Simuliidae (black flies), a cosmopolitan family of minute biting flies whose aquatic immature stages are confined to fast-flowing water (Fig. 9.1). The larva is attached to solid objects by a posterior sucker and feeds on organic particles filtered from the water by its mouth brushes. The pupa is secured in a case glued to the rock or other submerged object from which filamentous 'gills' project into the water, the rest of the body being enclosed in the case. In the rapid sections of the tropical

Simulium sp

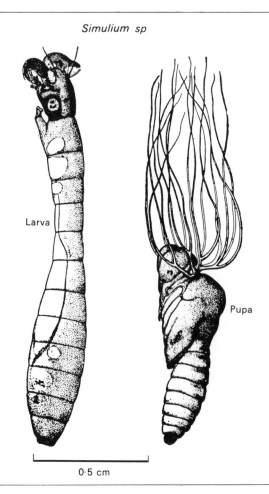

Larva

Pupa

0·5 cm

Fig. 9.1 *Simulium* sp. larva showing mouth brushes for capturing food and the posterior sucker by which it is attached to a solid object in fast-flowing rivers. The pupa is enclosed in a case from which the filamentous gills project (from Smart, 1943)

African rivers there are species of *Simulium*, particularly *S. damnosum*, that are vectors of the human parasitic nematode worm *Onchocerca volvulis*. Onchocerciasis, due to the several varieties of this species of worm, is one of the most important and distressing, though relatively little known, tropical disease which, during the past twenty years, has been found to be more widespread than was previously thought, and is well known also in Central and tropical South America. Its various symptoms include greatly swollen lymph glands, severe skin eruptions and damage to the eyes ('river blindness'). Its distribution is naturally determined by the presence of fast-flowing waters both in the steep sections of large rivers and in the smallest of hill streams. But in humid forest regions the flies can travel a considerable distance, and *S. damnosum* has been recorded as much as 100 km from its nearest suitable breeding water. The larvae of *S. neavei*, a species common in the smaller hill streams of tropical Africa, has the curious and obligatory habit of attaching itself to the surface of crabs, *Potamonautes* spp., and the mobility of its habitat has some influence on the distribution of the disease. A recent review (with references) of onchocerciasis and of the interesting biological and ecological problems associated with the disease and its control is that of Nelson (1970).

Most of the tropical African lakes drain into one or other of the four great rivers, and some of their features are therefore discussed in other chapters. Moreover, Ch. 20 is devoted to manmade lakes which are, after all, the most important developments in river exploitation. More has been written about the Nile than about any other river in the world and general information on all four is easily found in geographical textbooks. This chapter is therefore restricted to summarising some basic facts as a necessary back-ground for discussing limnological research on the Nile, Niger, Zaïre and Zambezi Rivers that is not included in other chapters dealing with the lakes or reservoirs. So far, the lakes have received much more attention from limnologists than the rivers.

The Nile

The origin, history and ecology of the lakes of the Upper Nile basin – Victoria, Kioga, Albert, Edward, George and Turkana (recently isolated) – are discussed in the appropriate chapters. We are here concerned with the river itself, north of the Great Lakes (see Fig. 11.3).

A short introduction to the history and ecology of the Nile is to be found in Hammerton (1972)* and an account and discussion of its physical history in Berry and Whiteman (1968). These provide starting-points for the study of the literature on these subjects. For obvious practical reasons the hydrology of the river was studied by the ancient Egyptians before 3000 B.C., but the basis of our present extensive knowledge of the hydrology of the whole basin we owe mainly to the work between the World Wars of the Physical Department of the Egyptian Ministry of Public Works under Dr Hurst, which was published in a series of monographs (Hurst and Phillips, 1931 onwards) and summarised in a book

---

* The special ecological problems associated with the manmade reservoirs on the Nile and with the Sudd swamps are discussed in Chs 18 and 20.

(Hurst, 1957). The Jonglei investigation Team (1954) also contributed much to the hydrology of the Upper Nile (Howell, 1953). A notable early biological publication is that of Boulenger (1907) on the fish, and for a more recent guide to the fishes of the Upper Nile see Sandon (1950). Many aspects of the history and present biology of the Nile and its tributaries are discussed in Rzóska (1976a). This includes articles by Talling on the physical and chemical properties of the water and on the phytoplankton down the course of the river from the Great Lakes.

Recently established centres for ecological and fisheries research on the Nile are the Institute of Freshwater Biology, Cairo; the Lake Nasser Development Centre, Aswan; the Inland Fisheries Research Institute and the University Hydrobiological Research Unit at Khartoum.

There has been some controversy concerning the geological age of the river, but, contrary to previous opinion, it is now apparently agreed that the valley of the Blue Nile descending from Ethiopia and its subsequent main course to the Mediterranean are very old, dating back at least to the early Tertiary (Eocene) (Berry and Whiteman, 1968, Berry, 1976). There have been several changes of course and river captures associated with climatic changes and earth movements. The most important were the great tectonic upheavals which gave rise to the Rift Valley Lakes and Lake Victoria. Lakes Albert, Edward and Victoria are now the natural reservoirs which collect and store great quantities of water from the high rainfall regions of eastern equatorial Africa and maintain a permanent flow down the White Nile with relatively small seasonal fluctuations. The upwarping during the late Pleistocene of the western side of the Victoria basin reversed the previous west-flowing rivers and formed the present Lake Victoria with its outflow to the north through Lake Kioga, Lake Albert and into the Albert Nile (Ch. 14). Apart, however, from providing storage, this diversion, according to present evidence, would have made little difference to the total annual discharge into the White Nile. Presumably the west-flowing rivers across the site of the future Victoria basin had previously been draining into the Albert basin and thus into the White Nile. During the late Pleistocene (before 12 500 B.P.) a dry period lowered the levels of both Lake Albert (for a short period) and of Lake Victoria (for a longer period) below their outlets and thus the flow of the White Nile must have been reduced (Williams and Adamson, 1974). The volume of the Blue Nile was probably likewise affected (Livingstone, 1976).

There is as yet disappointingly little geological and palaeontological information from the Upper Nile basin north of the Great Lakes to disclose its previous hydrological relations with neighbouring water sytems. The distribution of the aquatic faunas, especially the fish, can only be explained in these terms. For example, the nearly identical soudanian faunas of Lake Chad and the Nile must imply a recent (probably late Pleistocene) connection between the two basins, but our knowlege of the geology of the intervening desert regions is very slight.

The amount and distribution of surface water in the Upper Nile basin during a high rainfall period in the late Pleistocene (between 12 000 and 8000 B.P.) is an important matter in this connection. Several geologists in the past have suggested a previous very large lake in the region between Juba and Khartoum of which the Sudd swamps were supposedly a remnant. The lastest exponent of the theory of

'Lake Sudd' was Ball (1939) who suggested that it was fed by both Blue and White Niles, was held in a closed basin and eventually broke out to the north through the Sabaloka Gorge down the Sixth Cataract about 150 km north of Khartoum. It then joined the main Nile whose principal affluent had previously been the Atbara (Fig. 11.3). The theory was based primarily on the conformation and relative elevations of the land surfaces and on the presence of widespread flood sediments. It was severely criticised by Berry and Whiteman (1968) because of the absence of direct evidence of lake shorelines, beach deposits, etc. The surveys of Williams (1966) and of Berry and Whiteman (1968) produced evidence from both topographical features and beach terraces (some containing mollusc shells) for the existence during the late Pleistocene of two long but narrow lakes on the course of the river between Malakal and Khartoum. Each was 500–600 km long and 30–40 km wide. Unlike the hypothetical Lake Sudd, these lakes were not necessarily associated with an especially high rainfall, since it was suggested that they may have been caused by a plug of clay deposited by the Blue Nile. A curious and noteworthy feature is the gradient of the river through the Sudd which is greater than that on its subsequent course from Malakal to Khartoum. The Sudd is not a shallow lake, comparable, for example, with Lake Kioga, but is a large area of country flooded by overspill from the river which flows through rather rapidly on a well-defined course and augmented by local inflows from the surrounding hills during the rains.

Nevertheless, it can hardly be doubted that during the Lake Pleistocene, when the Chad basin and the western Sahara had a higher rainfall and much standing water, the Upper Nile basin must also have contained considerably more exposed water than at present in extensive flooded plains and swamps, if not in a large Lake. As already stated, it does not seem possible to explain the present distribution of the soudanian fish fauna other than by the recent existence of an aquatic connection between the Chad and Nile basins, though this was not necessarily continuous at any one time. Some of these fish are very large, such as the Nile perch *Lates niloticus* which can survive only in a considerable volume of permanent and well-oxygenated, but not necessarily deep, water. *Lates* is flourishing in the shallow Lake Kioga to which it has recently been introduced (p. 364).

During part of the late Pleistocene the Nile Basin included the southern section of the Ethiopian Rift Valley, which drained into Lake Turkana whose outflow to the north then descended the Sobat River to join the White Nile. The Nile perch still survives in Lake Chamo (p. 191).

The well-known hydrological regime of the Nile can be summarised briefly. It has two components:

1. The relatively steady flow down the White Nile derived from the high rainfall regions of eastern equatorial Africa and buffered and freed of sediment by the natural reservoirs of Lakes Victoria and Albert. This basic flow is reduced to about half by evaporation from the Sudd swamps, but is augmented by the seasonal floods down the Sobat River so that the flow between Malakal and Khartoum is about doubled around October or November. The main function of the Jebel Aulyia reservoir above Khartoum is to regulate and stabilise this flow (p. 386).

2. The great seasonal sediment-laden floods in northern Sudan and Egypt from

September to November are due to the heavy rains in the highlands of Ethiopia in July and August, which descend into the Nile valley in the Blue Nile and Atbara Rivers. For about two months the flow in the main river below the Atbara confluence is thereby increased about sevenfold (Fig. 11.3). Some buffering of the downflow of these two rivers has been accomplished by the Sennar and Roseires Dams on the Blue Nile and by the Kashm-el-Girba Dam on the Atbara, all of which provide water for local irrigation schemes (Fig. 11.3). It is therefore the Ethiopian highlands that provide the silty floodwater upon which the irrigated agriculture of Egypt has always depended. Recent work on the Atbara River and on the Blue Nile gorge, which has been cut back by the river on its descent from the highlands, has shown that nearly all of the sediments in the Mediterranean Delta came from Ethiopia via these two rivers. The rocks from which they were derived can be recognised in the Delta sediments (McDougall et al.. 1975).

Lake Nasser-Nubia, recently impounded by the heightening of the Aswan Dam, has greatly altered the character of the Egyptian Nile. The ecological, medical, economic and social effects of this development are very great. They are introduced in Ch. 20.

The Niger

It is now generally agreed that the present River Niger was born of two parents (Ch. 11). In the late Pleistocene the lower southeast-flowing section was fed from the southern slopes of the Ahaggar Mountains by affluents such as the Tilemsi and Azaouak, which are now practically extinct (Fig. 11.3). It was augmented, as now, by tributaries such as the Sokoto and the Benue. On the other hand, what is now the Upper Niger flowing northeastwards from the mountains on the Guinea-Sierra Leone border, flowed during the late Pliocene and early Pleistocene westwards into the Gulf of Senegal. This accounts for the close affinities between the fish faunas of the middle Niger and Senegal Rivers (Daget, 1954). The subsequent dry period produced a barrier of sand-dunes (Erg Ouagadou, Fig. 9.2) which subsequently, when the last wet phase arrived (10 000–15 000 years ago), blocked the previous westward flow of the upper river. It was thus diverted into a large closed basin west and northwest of Timbuktu, where there was much standing water with at least one big lake, Lake Araouane (Fig. 11.3). The flooded basin, whose previous existence is proved by the great expanse of recent sediments in the present Middle Niger Basin (Fig. 9.2), later began to drain away (either by a breakthrough or a capture) and, with a turn almost at a right-angle, joined the Lower Niger. This final episode in the river's history may have occurred no more than 5 000–6 000 years ago.

The middle and lower sections of the river are thus distinguished not only by their different histories, but also by their very different characters. The upper river debouches onto an immense seasonally inundated flood plain on the very edge of the desert where the extreme fluctuations in hydrological and ecological conditions are of great scientific interest and human importance. The lower river, though spreading outwards in some places during the flood season, especially in the Sokoto valley, is generally restricted to a relatively well-defined course

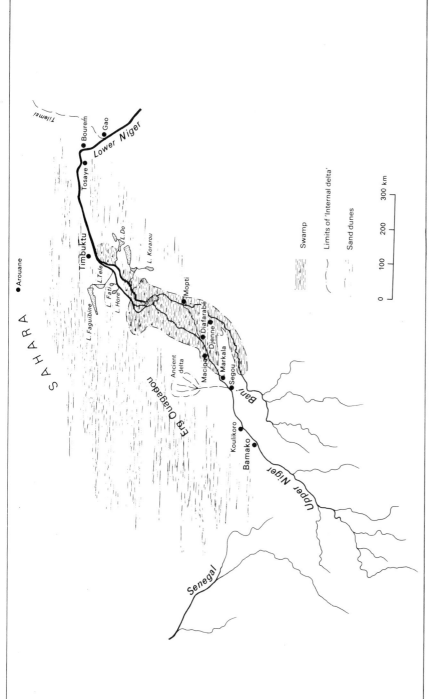

Fig. 9.2 The Upper and Middle Niger River with its lakes and the 'Internal Delta' which is annually flooded

through savanna and forest, and has provided a suitable situation and conditions for making a very large artificial lake, Kainji (Ch. 20. Fig. 11.3).

We owe our knowledge of the recent geological history and geomorphology of the Upper and Middle Niger basins to Urvoy (1942). Recent French limnological work has been based on the I.F.A.N. Laboratory founded in 1949 at Diafarabe on the flood plain, and later transferred to the independent government of Mali and moved to Mopti.

The source of the river is at about 800 m altitude in the Fouta Jallon Mountains within 250 km of the Atlantic coast. It is joined on its downward course to the plains by several tributaries draining the same high-rainfall forested mountain range. It leaves the forest a little above Bamako and from there descends about 60 km of rapids to Koulikoro whence, after a further 300 km with a more gentle gradient, it spreads out onto the flood plain. This extends over about 17 000 km^2 and reaches almost to Timbuktu. The main tributary, the River Bani bringing water from the northern highlands of the Côte d'Ivoire, discharges separately onto the plain and joins the Niger at Mopti (Fig. 9.2).

The sediments carried down by the Upper Niger and Bani Rivers are deposited over the plain to form what has been called the 'internal delta' of the Niger (Gallais, 1967). The water courses have greatly fluctuated in the past and there are at present several channels which diverge and reunite on their way across the plain. An account of this section of the river and its hydrological regime is given by Blanc et al. ((1955a) and Daget (1957). The local rainfall is insignificant compared with that in the highland catchment region where the wet season is from June to October with a peak in August. As a result a flood wave passes down the river and is marked by a maximum water level at Diafarabé in October and at Gao in the following January (Fig. 9.2). There are a large number of small lakes scattered over the plain that are connected with the many branches of the river. Only a few of them are represented in Fig. 9.2, and these are the principal ones that survive during the dry season in regions where the substratum is least permeable. Some of these, such as Lakes Hore and Fati are fed by inflow from the rising river and drain back into the latter as its level falls. Others, such as Lake Faguibine, are prevented from draining back because they lie behind banks that are higher than the river water level when not in full flood. A large amount of water is, however, lost from the lakes and swamps by infiltration into the soil and by evaporation, particulary during the warm dry season on the plain which is driest between April and June.

The vegetation and the predominant features of the fish and molluscan faunas of the various kinds of lakes, pools and water courses on the flood plain are summarised by Daget (1954) and by Blanc et al. (1955a). The aquatic vegetation is of a kind that would be expected in such a situation – aquatic grasses, *Polygonum*, water lilies (*Nymphaea*), floating Nile cabbage (*Pistia*), submerged water weeds, etc. The most remarkable feature is the absence of papyrus, the scarcity of which in West Africa generally (though it is abundant in Lake Chad) is one of the major mysteries of African plant geography (p. 312).

The low mineral content of the water of the Middle Niger is presumably due to the low solubility of the pre-Cambrian granites and schists in the upper catchment area. Table 9.1 gives an analysis of a sample taken from the river at Diafar-

Table 9.1 Some chemical features of the water of the Middle Niger. Sample taken at Diafarabé on 14 July 1954. Converted to meq/l from figures given by Daget (1957)

pH	7·2	Ca^{2+}	0·20
Conductivity (K_{20})	32·4 μmhos/cm	Mg^{2+}	0*
		Cl^-	0·02
Alkalinity	0·4	SO_4^{2-}	0*
($HCO_3^- + CO_3^{2-}$			
Na^+	0·13		
K^+	0·05		

* Not detectable by the methods used. Grove (1972) reports Mg^{2+} 0·08 meq/l in a sample taken at Segou (Fig. 9.2) in October 1969.

abé in July 1954, that is during the dry season when the level is low. It was recorded that in relict pools during the dry season evaporation did not increase the alkalinity beyond 1 15 meq/l. It has been suggested that this comparatively low mineral content would be expected to have a restraining influence on organic productivity and might be a factor limiting the potential of the fisheries (Blanc *et al.*, 1955a, b; Daget, 1957). A comparison with the figures in Table 5.1 (p. 61) shows, however, that the conductivity and ionic levels in the Niger sample are close to those in Lake Bangweulu and that the level of calcium is actually not far short of that recorded for Lake Victoria. It is argued in Ch. 5 that there is probably no natural water containing quantities of the major ions, other than calcium, too low for most freshwater animals. The amount of calcium in the Niger sample is evidently adequate. In fact, Daget (1954, p. 28) records that in the muddy bottom of some sections of the river the molluscs are abundant enough to be exploited as a source of lime. Information on the availability and circulation of nitrogen, phosphorus and silicate has not yet been published.

All the information so far available concerning the flora, fauna and the fisheries of both the Niger flood plain and Lake Bangweulu (p. 316) suggest a productivity as high as many lakes with a higher minera content. The two situations are similar in that they are both fed by rivers of low salinity which spread over an extensive flood plain in the wet season leaving some contracted swampy lakes when the level falls. The apparently flourishing fishery in Lake Tumba in the Zaïre basin, with its equally low salinity, extremely low alkalinity and much lower pH (Table 5.1), must be considered in this context. The basic source of nourishment in Lake Tumba here seems to be organic particles washed in from the surrounding forest land. The sediment brought down from the Upper Niger catchment may make an important contribution to the supply of nutrients.

The primarily soudanian character (p. 144) of the fish fauna of the Niger is very obvious. Of the 112 species recorded by Daget (1954) in the Upper and Middle Niger, 66 are found also in the Nile and 52 in the Gambia River. Of these 33 are common to all three rivers and include the well-known species that are distributed over the entire soudanian region, such as *Lates niloticus* (the Nile perch), *Hydrocynus forskali* (tigerfish), *Citharinus citharus*, *Sarotherodon niloticus* and many others. There is little doubt that more detailed examination of some of the widespread soudanian species will reveal small subspecific differences between populations from widely separated regions. Such have already been described by Daget (1954, 1957) in *Alestes dentex* from the Niger and Nile. It is very surprising that there is not more obvious evidence of incipient speciation over the whole soudanian region where some of the present water systems are isolated from each other by immense stretches of arid country. The comparatively recent more humid climate with surface waters across much of the southern Sahara is certainly part of the explanation (see Chs. 8 and 11).

Of the remaining twenty-seven species seven are found in the Volta, Senegal or Lake Chad. Twenty have not yet been found outside the Niger basin to which some at least are certainly endemic. A few are apparently confined to the Middle Niger. For example, the small clupeid *Microthrissa miri*, which feeds on zooplankton, is common in the river, channels and lakes on the flood plain. Another clupeid, *Pellonula afziuli*, is recorded only from the lower river. These are two of the many species of a mainly marine family that have penetrated the West African rivers and are now confined to fresh water. Two species have established themselves in Lake Tanganyika (Ch. 16). One marine species, *Pellonula vorax*, regularly enters the mouth of the Niger and ascends some distance to breed. Another interesting fish which is endemic to the Middle Niger is the cichlid *Gobiocichla wonderi*. This is a small, elongated fish, up to about seven centimetres long, common in the fast-flowing section of the river between Bamako and Koulikoro. It lives mainly in rock crevices and feeds by browsing on the algae and other organisms encrusted on the rock surfaces for which its mouth and teeth are specially adapted. There are ten species of cichlid fishes in the Upper and Middle Niger (Daget, 1954), five of which are tilapias, but only *G. wonderi* is endemic. In view of the remarkable adaptive changes undergone by this fish, it is all the more surprising that there are so few examples of speciation on the part of the cichlids in the West Africa rivers, small lakes and Lake Chad compared with the Great Lakes of eastern Africa. But there has been enough divergence between the fish faunas as a whole in the West African river basins to warrant a subdivision of the western soudanian region and the designation of a separate 'guinean subregion' within 200–400 km of the coast (discussion in Daget, 1962, 1968; and Daget and Iltis, 1965)*.

The annual rhythm of flooding, with expansion and contraction of the water over the internal delta, is clearly the predominant feature of the environment affecting the life of the aquatic organisms. Its effects on the fish are particularly striking. It happens that, like Lake Chad, the delta is situated in a climatic region on the edge of the Sahara (lat. 15°N) where the seasonal change of temperature is comparatively large. For example, the mean monthly temperature of the water in the river at Diafarabé in 1952 was about 20°C in December and January and about 28°C from May to October which is the period during which the water rises and the flood reaches its peak (Daget, 1957). There is therefore a progressive rise of temperature between January and May and a more rapid fall between October and December.

The great majority of the fish breed during the period of rising and warming water (July–October). The inflowing rivers are then bringing in fresh nutrients and the newly flooded land provides more food and better chances of escape from

---

* McGregor Reid and Sydenham (1979) review the ichthyogeography of the Benue River, the most important tributary of the lower Niger.

predators for the young fish. On the other hand during the low water period growth of the fish is arrested, the West African lungfish, *Protopterus annectens*, encysts and aestivates in the mud of dried pools, and the cyprinodont *Notobranchius (Aphyosemion) walkeri* lays eggs in the mud that survive until the next flood (p. 349). A few species, such as *Microthrissa miri* (Clupeidae) and *Barilius niloticus* (Cyprinidae) are known to show the reverse rhythm, breeding in the river itself when the water is falling (November–January) and the temperature is lower. But these are apparently exceptional and the breeding seasons are generally geared to the floods.

There are some striking mass migrations of fish that have been observed to be directed up the course of the main river. These, however, do not appear to be connected with breeding cycles and their significance is so far unknown. The extensive lateral migrations from the river over the flooded land are, however, more obviously related to breeding and feeding.

The economic importance of the delta region is threefold:

1. For the growing of crops, especially rice, on the flooded plain. A dam was constructed at Markala and completed in 1946 for controlling irrigation of rice and grain crops.
2. For dry season grazing of cattle. Like the Sudd region of the Upper Nile (p. 327), the delta provides good grass at a time when much of the surrounding country is too desiccated for grazing.
3. For the fisheries, which are in operation mainly during the flood period (Blanc et al., 1955b).

Few people know that fishes are sometimes an agricultural pest, but this is in fact the situation in the rice fields of the internal delta (Matthes, 1978). Several herbivorous species, especially *Tilapia zillii*, *Distochodus brevipinnis* and *Alestes baremose*, graze on the plants during the early stages of flooding before the water exceeds more than 1 m in depth. The main culprit is *T. zillii* which chews the leaves and stems. *Alestes spp.* have been seen jumping out of the water to reach the panicles containing the seeds. Under the most favourable conditions for the fish entire rice fields have been laid waste.

Below the internal delta, the Niger flows through very arid country and receives no significant permanent affluent until it reaches the confluence with the River Sokoto about 1 200 km downstrream from Timbuktu. The late Pleistocene tributaries Tilemsi and Azaouak (Fig. 11.3) are normally dry. The River Sokoto and its main branches rise in the highlands of northwestern Nigeria and the upper reaches flow only during the rainy season between May and October. Below the town of Sokoto, where the River Sokoto is joined by its principal tributary the Rima, there is some flowing water in the river bed even during the dry season. The valley then flattens out into a plain which is extensively flooded every year. Though covering a smaller area than the internal delta, the floods provide for extensive cultivation of deep-water rice as well as for fishing, and for the grazing of cattle in the dry season. But the most intensive fishing is done in the dry period when the fish are concentrated in great numbers in the very many ponds and pools left on the flood plain.

A report on the seasonal changes in the hydrology and plankton of this section of the Sokoto river and flood plain was published by Holden and Green (1960). As in the delta, the ecological situation is dominated by the regular seasonal alternation between flooding and contraction of the water to relict ponds and a slow-flowing stream. The flooding dilutes the mineral constituents of the water, though according to the analyses the composition does not depart from the normal freshwater range (alkalinity $0·47-1·7$, Ca^{++} $0·5-2·1$ meq/l, pH $7-8$). Assessment of the abundance of the phyto- and zooplankton at different times is complicated by the flooding which spreads the organisms through a much greater volume of water, and both are most concentrated towards the end of the dry season. It is not clear how far recorded fluctuations in numbers of organisms per unit volume are due to changes in water volume and how far to changes in the rate of production. It is to be expected that as a similar situations elsewhere, the annual flooding of the land will release plant nutrients and stimulate production of the phytoplankton, though the actual concentration of these nutrients may not increase until the volume of water begins to contract when the rains have stopped.

It is suggested that the rate of primary production is limited by the low level of NO_3, PO_4, and, more especially, of SO_4 which was undetectable. But the arguments against low concentrations of these nutrients being regarded as a necessary indication of limiting action have already been discussed (Chs 4 and 5). It is not possible from the available information to decide what is the main factor controlling the production and abundance of organisms.

The assessment of the biomass and production of the fish is likewise made difficult by the great fluctuations in volume of water. Holden (1963) describes the fishing methods and the research done on the species composition and abundance of the fish in the dry season relict pools, in which they are sometimes concentrated in enormous numbers.

The River Zaïre (Congo)

Lakes Kivu and Tanganyika are hydrologically part of the Zaïre basin, but are only tenuously connected by the Lukuga River. They are discussed separately in Chs 15 and 16. Much information concerning the river and its fauna and flora (mainly from the middle and lower sections) was collected by explorers and naturalists well before the beginning of this century, and Boulenger's large work on the African freshwater fishes (1909–16) describes very many species from the Zaïre River. Organised limnological work from local bases established in the previous Belgian Congo date mainly from after the Second World War. Except for the Hydrobiological Station at Uvira on Lake Tanganyika under the Institut pour la Recherche Scientifique en Afrique Centrale (I.R.S.A.C.), they have been continued under the Republic of Zaïre. Some of the previous history of research in this field has been reviewed by Symoens (1963). The present centres for limnological research are:

1. Station de Recherche Piscicole at Kipopo near Lubumbashi.
2. Institut Nationale pour l'Étude Agronomique du Congo at Yangambi near Kisangani.
3. The Universities of Kinshasa and Lubumbashi. (Fig. 9.3).

    A limnological survey of the Lake Bangweulu–River Luapula basin continued for several years from 1956 under Dr J. J. Symoens of the University of Lubumbashi. Some of the results have appeared in a series published by the Cercle Hydrobiologique de Bruxelles. These are concerned with mineral composition of waters (Symoens, 1968) and with algae and several groups of invertebrates. The main repositories of basic material and information concerning the limnology of the Zaïre basin are the Institut Royal des Sciences Naturelles, Brussels, and the Musée Royal de l'Afrique Centrale, Tervuren, Belgium.

The geological history of the Zaïre basin is discussed by Robert (1946) and Cahen (1954). It covers an area of over four million square kilomeres across the Equator from about 7°N to 12°S, and drains the largest expanse of lowland tropical forest in Africa. The relatively flat central region lies in one of the ancient continental basins (Figs. 3.1, 3.2) which was more than once invaded by the sea during the Mesozoic. The subsequent uplifting of the peripheral land accentuated the basin and obstructed its drainage to the coast. As a result a large lake formed during the Pliocene which later (before the beginning of the Pleistocene) was drained away to the Atlantic through cutting back and capture by a coastal stream, which then became the lower Zaïre River. The great swampy area with the shallow Lakes Tumba and Maindombe in the western half of the central basin can be regarded as remnants of this lake. The point of capture was probably just below Stanley Pool between Brazzaville and Kinshasa where the river begins its plunge down thirty-two rapids in the 350 km to Matadi, with a vertical drop of 275 m. Here is probably the greatest concentration of potential hydroelectric power in the world. It can be seen from the map (Fig. 9.3) that many of the rivers descending from the surrounding high land are interrupted by falls and rapids – a legacy from previous earth movements. These are barriers that have contributed to the relative isolation of the aquatic fauna, especially the fish, both in the basin as a whole and in its different sections.

Earth movements and erosion have made some changes in the watersheds separating the Zaïre from the neighbouring basins, some of which will be mentioned later, but the river system as a whole seems to have had a continuous existence since well before the Pleistocene. Even the late Pleistocene dry periods in the central basin, the latest 10 000–15 000 years ago (p. 25), though completely transforming the terrestrial vegetation, must have left the river itself intact. It continued to drain the surrounding higher land where dense forest and a high rainfall persisted. It seems that the Zaïre river system survived in its present state without great change at least throughout the Pleistocene, when the water systems of eastern tropical Africa were being drastically altered by violent and disruptive earth movements and fluctuations of climate.

The main middle section of the Zaïre River along its 2 000 km course from Kisangani (formerly Stanleyville) to Kinshasa (Leopoldville) falls only about 100 m and is thus a wide and slow flowing river, much of it 3–15 km broad. The same is true of the lower reaches of the two main tributaries Oubangui and Kasai. There is therefore a great expanse of slow moving water in which conditions during most of the year are almost 'lacustrine' and probably not far removed from those in the former lake which occupied the Central Basin.

An important present ecological feature is that, though the rainy seasons cause an annual increase in the flow of the main river and tributaries in the central basin, the depth and rate of flow are maintained at a rather high level throughout the dry season. This is because it drains humid tropical forests on both sides of the Equator, where rainfall is never very low. Owing to the seasonal migration of the tropical rainbelt the heavier rainfall period in the north is partly balanced by a lower rainfall in the south, and vice versa (Fig. 3.8). Nevertheless, there are net fluctuations in the rate of flow in the water courses in the flat central basin which reach maximum levels in December and January and flood over large areas of swampland, but the fall in July and August is not sufficient to contract the main rivers in their beds to an ecologically significant degree.

Environmental stability over a very long period, a wide range of habitats, and prolonged isolation are thus characteristic features of the waters of the Zaïre Basin. We have reason to suppose that these are conditions that favour the evolution of a large endemic fish fauna (Ch. 8). Some of the species were probably evolved in adaptation to life in the large Pliocene lake which might well have provided a great variety of environmental conditions.

In the entire basin, exclusive of Lake Tanganyika, 669 valid species of fish have been recorded, of which 558 are endemic, belonging to 25 families and 168 genera (Poll, 1959 and pers. comm.). The ichthyology of this vast region is only partly known and the total number of species may well approach one thousand.* A few tropics of special interest will be briefly discussed here.

In the main slow-flowing river of the central basin, the Mormyridae (elephant snout fish) is the dominant family both in number of (75 of a total of 408 in the central basin) and of individuals (Poll, 1963). Some species swim in large shoals just above the mud surface, persumably feeding, like most of the mormyrids in the Great Lakes, on insect larvae and worms in the mud. The shoals are thought to be kept together by the signals emitted from the electric organs on either side of the tail. These organs are modified muscles and have attracted much interest among physiologists (Fessard. 1958; Blunnet, 1971, and p. 186).

The swamps bordering the middle Zaïre River provide an environment to which the lungfish *Protopterus dolloi* is particularly adapted. Its breeding habits are compared with those of *P. aethiopicus* in eastern tropical Africa in Ch. 18 (p. 326). Associated with the swamps of the central basin there are several shallow lakes. Some of these are known to have water of very low mineral content. Such is Lake Tumba whose ecology is discussed in this connection in Ch. 5.

The 350 km of rapids on the lower river between Stanley Pool and Matadi provide what must be the most extensive example of this kind of environment in the tropical world. The difficulties and dangers in collecting fish from such a place can well be imagined, but some remarkable endemic adaptations have now been discovered (Poll, 1959, giving references to some of the original publications). Adaptation' of fish to torrential waters, has occurred in many parts of the Zaïre basin other than in the Central region, and species from seven families have taken part – Kneriidae, Cyprinidae, Amphiliidae, Mochokidae, Clariidae, Cichlidae and Mastacembalidae. Apart from the rapid flow, the special characteristics

---

* There are records of many other species which in Poll's opinion are not yet properly substantiated ('valid'). Another example of speciation within the Zaïre basin is shown by the freshwater crabs (*Potamonidae*). According to Bott (1955) there are twenty-two recorded species of which twelve are endemic to the basin and include one endemic genus *Longipotamonautes*.

of this environment are the abundance of oxygen and the low intensity of light in the situations (under large rocks, etc.) in which the fish have to spend much of their time. Consequently, the characteristic structural modifications include suctorial mouths, flattening of the body, a horizontal position of the often enlarged paired fins, loss of accessory air-breathing organs (in the Clariidae) and (in some species) reduction of the eyes. For example the clariid catfish *Cymnallabes tihoni* in the Lower Zaïre rapids has a much flattened head, a narrow eel-like body, and large horizontal paired fins, has lost the aerial respiratory organs in the upper branchial cavity characteristic of this family, and its eyes are very much reduced. Also living in the same stretch of the river is the spiny eel *Caecomastacembalus brichardi* (Mastacembelidae) whose eyes are buried under muscle and are reduced in size but are anatomically complete. Another five species have been found in these rapids with all the above adaptations other than reduced eyes. They include the cichlids *Leptotilapia tinanti* and *L. rouxi*. If it were not for the difficulties of collection there is little doubt that several more would be known. Loss of sight is a feature of cave-dwelling animals and it is noteworthy that in a cave at Thysville near to the lower rapids and about midway down their course, there lives one of the three species of blind cave fish so far recorded from tropical West Africa – *Caecobarbus geertsi* (Cyprinidae). The other two species, *Channallabes apus* and *Dolichallabes microphtalmus* (Clariidae), are found in the Katanga province of Zaïre. There are two blind fish in subterranean springs in Somalia – *Phreatichthys andruzzii* (Cyprinidae) and *Uegitglanis zammaranoi* (Clariidae) (Thines, 1955).

The compostion of the fish fauna of the upper Lualaba River poses an intriguing problem in zoogeography (Poll, 1963). The Lualaba is the upper section of the main Zaïre River above Kisangani (Fig. 9.3). It rises about 1 500 km to the south of Kisangani in the highlands on the borders between the Zaïran province of Katanga, Zambia and Tanzania, and thus shares a watershed with some tributaries of the River Zambezi. It has two main branches:
1. The main Lualaba in western Katanga flowing north to the swampy Upemba Lakes where it is joined by the Lufira.
2. Further north is the confluence with the Luvua River coming from Lake Mweru on the border of north-eastern Zambia. The main affluent to Lake Mweru is the River Luapula flowing on a U-shaped course from the swamps of the Zambian Lake Bangweulu.

The fish faunas of the above two river systems together with that of the upper Kasai River further west (Fig. 9.3) were reviewed by Poll (1963). Though these three are tributaries of the River Zaïre they all have falls and rapids along their courses which have restricted the movements of fish. Consequently, though Zaïrean species form the basis of their faunas, there are a considerable number of species endemic to each river system, e.g. 33/120 in the Luapula between Lakes Bangweulu and Mweru, and 19/60 in the Upper Kasai. Fifty-two species are endemic to the two rivers together. There are also a number of Zambezian species in all three rivers, and these, too, are not found in the Central Zaïre basin. In the Luapula, between Lakes Bangweulu and Mweru, there are as many as thirty Zambezian species. These facts, briefly summarised, can be explained only by the existence of barriers to the free movement of fish between the Middle Zaïre

Fig. 9.3 The Zaïre basin and adjacent water systems. The 500 m contour is shown in order to emphasise the approximate boundaries of the relatively flat central basin

River and these upper tributaries, and by the occasional water connections in the past with the Zambezi basin, and between the three tributaries themselves across their dividing watersheds. The recent, and now intermittent overflow from Lake Tanganyika via the Lukuga River into the Lualaba has not yet significantly affected the fauna of the latter.

Poll (1963) reported the presence of several Soudanian species of fish that were apparently confined to the upper Lualaba River above the rapids at the Portes d'Enfer and thus not found in the main Zaïre River (Fig. 9.3). These included *Protopterus aethiopius congicus*, *Polypterus bichir katangae*, *Ctenopoma muriei*, *Ichthyborus besse* and *Sarotherodon* (formerly *Tilapia*) *niloticus*. To explain this remarkable phenomenon he suggested that the upper Lualaba may formerly have been a tributary of the Nile system and was later blocked by the upwarping and volcanic action associated with the formation of the Western Rift. It may then have been captured by the Zaïre River at the Portes d'Enfer where it takes a turn to the northwest and whose rapids have prevented further downstream movement of these fish (see also Poll, 1976, pp. 22–3). According to Cahen (1954) these events are not inconceivable on geological grounds and, in fact, were suggested prior to Poll's work on the fish.

The validity of the theory of course hangs upon the soundness of the geological evidence, the correctness of the species identifications and upon the certainty that the soudanian species are absent from the main Zaïre River.

There was never any direct and positive geological evidence. In fact, investigations on the geological history of Lakes Victoria, Albert, Edward and the Nile basin indicate that from well before the Miocene the drainage into the Zaïre basin from the east was via rivers rising east of the future Victoria basin, the then Nile basin being confined to the north. These rivers were ultimately disrupted by the rifting and consequent formation of the Rift Valley lakes and the later associated appearance of Lake Victoria. Lakes Victoria and Albert at present flow into the Nile, though their outflows have been interrupted by low levels more than once during the later Pleistocene (Chs 10 and 14). There is however no geological or biological evidence to support an outflow from Lake Tanganyika to the north into the Nile basin at any period during the history of the lake (Ch. 16). There was, therefore, never the possiblity of a Lualaba–Nile connection via Lake Tanganyika.

The evidence from the known distribution of fish species has recently been augmented and reassessed by Banister and Bailey (1979). The distribution of the archaic general *Protopterus* and *Polypterus* is likely to be related to conditions in the more remote past and anyhow those in the upper Lualaba are subspecies distinguishable from those in the Nile basin. They are ancient widespread fish and less dependent upon intact river systems because of their aerial respiratory organs and adaptability to swampy conditions. The *Ctenopoma* found in the upper Lualaba is now reidentified as *C. ctenotis* widespread in the Zaïre basin. The problem centres mainly on *Sarotherodon niloticus* and *Ichthyborus besse*. Subsequent work by Thys (1968) has in fact established that *S. niloticus* is distinguishable from the Nile species, and it was recognised as a subspecies *upembae*. It has now been designated to a separate species *upembae*. Moreoer it has now been found in the main Zaïre River at least as far down as Yangambi, and recently *Ichthyborus besse* too has been discovered in the main river.

The fauna of the upper Lualaba is thus not as distinct from that of the central Zaïre basin as was previously thought. Owing, however, to the much greater variety of habitats and the upper reaches of many of its tributaries isolated by falls and rapids, the number of species (many endemic) is vastly greater in the main basin.

It seems then that we must reluctantly abandon this attractive theory, but the distribution of the *Sarotherodon upembae* is of great interest because it includes the Malagarasi River but not Lake Tanganyika, and thus adds to the evidence that the Malagarasi flowed directly into the Zaïre basin before it was interrupted by the Lake Tanganyika Rift (p. 281).

The southwest region of the Zaïre Basin is situated in Angola and includes the upper reaches of the tributaries of the Kasai River from the Cuango to the main

Kasai. These share watersheds with the Angolan rivers flowing into the Atlantic, such as the River Cuanza and Cunene. The latter in turn share watersheds with the Kubango and Cuando Rivers which drain into the Okavango Basin and River Zambezi respectively (Figs. 9.3, 9.4). Some species of fish have moved across these watersheds, and there is geological evidence that some of these exchanges have been effected by river capture. Some may have moved across flat watersheds flooded in very wet seasons as was suggested by Bell-Cross (1965 and see p. 134). In his monograph Poll (1967) describes the fish of the Angolan Rivers and in another publication (1966) discusses the ichthyogeographical problems of this interesting region. The exchange of fish species across the watershed between the northern Zaïre basin via the upper reaches of the Oubangui River and the Chad Basin is mentioned in Ch. 12 (p. 216).

A discussion of the ecological and social effects of the Highway Construction Programme in the Amazon Basin is that of Goodland and Irwin (1975). Under its sensational title this is a short but authoritative account of the ecosystem of the world's largest humid tropical forest of which the river and lakes are components. There is an expert assessment of the environmental disasters that are likely to follow, and indeed can already be witnessed, from the large-scale and indiscriminate destruction of the trees (see also p. 69). The research and experience gained in Amazonia could surely help to guide future developments in the Zaïre Basin. The physical and hydrological features of the Amazon and Zaïre Basins were compared by Marlier (1973). For an introduction to limnological research in Amazonia see Sioli (1964).

The basic source of food for the aquatic animals in the very extensive mineral-deficient waters of the Amazon Basin is the forest litter (p. 69). Examination of the stomach contents of the fish has shown that most species are omnivorous, but some feed almost exclusively on ants and other insects that fall into the water (pers. comm. H. Sioli). These are therefore an important source of food in this peculiar ecosystem and there is little doubt the same will be true of many of the waters that traverse the Zaïre forests.

I hope that this short discussion of some aspects of the limnology of the Zaïre basin will impress the reader with the biological wealth of this great river, which is as yet only half explored. In view of the immense resources of power, minerals and exploitable terrestrial and aquatic organisms future developments in the basin will surely raise some serious problems of pollution and conservation. We must hope that these will be foreseen and solved in a more enlightened manner than is at present being shown by the developers in the Amazon Basin.

The Zambezi

The 3 000-kilometre course of the Zambezi River from its common watershed with the Zaïre to the Indian Ocean crosses a number of rather flat plains subjected to seasonal flooding and interrupted by steep sections of falls and rapids the most notable of which are the Victoria Falls. In this respect it is a typical African river. The sum of the length of those sections along the entire course of the river and its tributaries that are navigable and exceed 150 km is about 6 000 km. From the mouth of the river, however, it is navigable for only shallow

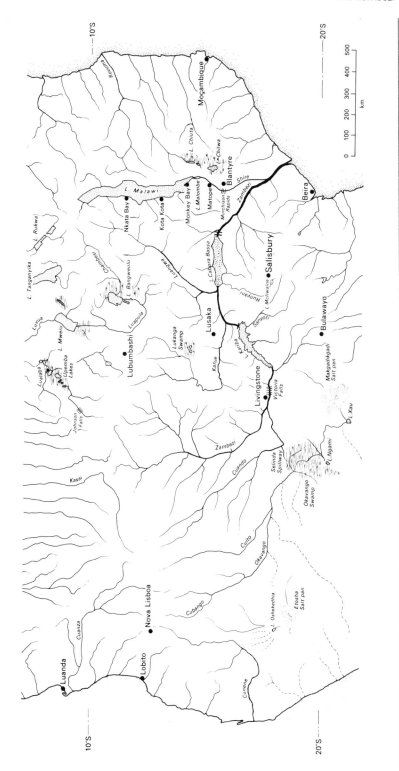

Fig. 9.4 The Zambezi and its relations with the neighbouring basins

draft vessels for 550 km to the Cabora Bassa Rapids in Mocambique, where a large hydroelectric dam and lake have now been constructed (p. 372). When the dam is completed at the Gorge on the lower Kafue River, a major tributary of the Zambezi (Fig. 9.4), together with the second power-station on the Zambian shore at Kariba (Ch. 20), the Zambezi will produce more hydroelectric power than any of the other African rivers.

Apart from investigations on the fish, little limnologicl work has been published on the Zambezi except in relation to the manmade Lake Kariba (Ch. 20), and Lake Malawi (Ch. 17) which is hydrologically part of the Zambezi Basin, but is biogeographically isolated from it by the Murchison Rapids on the Shire River. The main features of the fish fauna and the distribution of the recorded species in the river itself are well known (Jackson, 1961a; Jubb, 1961; Bell-Cross, 1968, 1972). The total number of indigenous species in the entire basin, excluding Lake Malawi and neglecting the estuarine fish at the mouth, is, according to Bell-Cross, 122. This is only a small fraction of the number in the Zaïre basin (excluding Lake Tanganyika), but is similar to that in the Nile basin north of Lake Albert. The falls and rapids have in varying degrees been barriers to the movements of fish whose distribution in the basin they have to some extent determined. As would be expected, some species have been more successful than others in traversing them, or have done so when they were more passable. The most effective of the barriers on the main river is the Victoria Falls which marks the lower limit of the Upper Zambezi. The fish fauna of the Middle Zambezi (58 species) is poorer than that of the upper river basin (84 species). This was atttributed by Jackson (1961a) to the generally less favourable ecological environment in the middle section which is mostly faster flowing water broken by frequent rapids. Such conditions would preclude species dependent upon relatively calm water with submerged and emergent vegetation providing food and protection against predator fish such as *Hydrocynus vittatus*. The lungfish *Protopterus annectens* has a rather spasmotic distribution in the Zambezi Basin and is confined to the lower and middle sections below the Victoria Falls. Its appearance in the swampy margins of Lake Kariba would not be surprising. Otherwise it is found in the upper Zaïre River, and in Luapula and Lualaba. Most surprisingly, it inhabits Lake Chad and the West Africa Rivers west of the lower Niger (Poll, 1954).

An interesting consequence of the construction of the Kariba Dam has been the recent appearance in the new lake of several species of fish not previously recorded in the Middle Zambezi but well known in the river above the Victoria Falls (p. 000).

The Okavango River which rises from watersheds shared with the west-flowing Angolan Rivers Cuanza and Cunene, flows into the mainly arid Okavango basin and collects in the Okavango Swamps (Fig. 9.4 and p. 328 ff.). There is no doubt that during the higher rainfall period in the late Pleistocene, the Okavango River was a tributary of the Zambezi. Even now, in seasons of heavy rain, the swamps overflow into the Zambezi via the Selinda Spillway, and the Okavango basin is thus not quite 'closed'. Consequently, the composition of its fish fauna is very close to that of the Upper Zambezi and to that of the Kafue River which, though independently flowing into the Middle Zambezi, is partially isolated from it by the rapids in the Kafue Gorge.

As far as the distribution of fish is concerned the Okavango, Upper Zambezi and Kafue Rivers form a single zoogeographic region. It has a number of species common with the southern tributaries of the River Zaïre, especially the River Kasai and with the west-flowing rivers of Angola. Exchange of species in this area has already been discussed. Bell-Cross (1972) has suggested that he distribution of the Zambezi fishes can best be explained as the result of a previous separation of the Okavango/Upper Zambezi/Kafue river system from the Middle-Lower Zambezi, with consequent different origins of much of their faunas. Prior to the Pleistocene, it is supposed that the former may have flowed westwards to join the Zaïre basin and was later reversed into the Zambezi. There are also species, especially in the Middle and Lower Zambezi, that have apparently come from Lake Malawi and even some nilotic (soudanian) species, as well as species endemic to the river and its tributaries. The past history of the Zambezian fish fauna is exceptionally difficult to unravel, and there have been some necessarily speculative suggestions for routes by which the various elements of the fauna may have reached it (Bell-Cross, 1972).

10

Lakes Albert (Mobutu) and Turkana (Rudolf), the Ethiopian Rift Lakes and Lake Tsana

Though these two lakes are more than 500 km apart, Lake Albert lying within the Western Rift, Lake Turkana in a depression associated with the Eastern Rift, a comparison between them is particularly interesting (Fig. 3.4). Both are of similar size and shape and are probably subjected to a similar seasonal hydrological regime (p. 90). The present fish faunas are soudanian and very closely related to each other and to that of the Nile. The total number of species and the proportion of endemics in both lakes are similar and very much smaller than in the other three large and long-isolated lakes shown in Table 8.2 and Fig. 8.1. Of the 46–48 species in each lake 22 are common to both and to the Nile. This is not surprising for Lake Albert is in open communication with the Nile, and Lake Turkana, though in a closed basin and having no surface outlet was probably connected with the Nile as recently as 7 500 years B.P.

The Ethiopian Rift Valley lakes have been included in this chapter because the southern half of the valley recently drained into Lake Turkana. Lake Tsana, though within the catchment of the Blue Nile, is biogeographically more isolated from the main Nile by barriers than is Lake Turkana, and has a biota characteristic of the Ethiopian Highlands.

Lake Albert (Mobutu)

The first fisheries survey of the lake was that of E. B. Worthington in 1928 (Worthington 1929, 1932). Subsequently a more comprehensive ecological survey of Lake Albert was made by the Belgian Hydrobiological Expedition to Lakes Kivu, Edward and Albert (1952–54) the results of which were published by the Institut Royal des Sciences Naturelles de Belgique (Brussells) between 1957 and 1962.

When Samuel Baker arrived in 1864 on the shores of Lake Albert, traced the course of the Victoria Nile down the Karuma and Murchison Falls into the lake, and found its outlet by the Albert Nile, the final confirmation was given for Speke's assertion that Lake Victoria was the great reservoir of the White Nile (Baker, 1866).

Lake Albert is a typical Rift Valley lake lying at an altitude of 615 m between two parallel escarpments, that on the western (Zaïre) side rising abruptly to nearly 2 000 m above the water surface. The lake is about 150 km long, with an average width of about 35 km, and a maximum depth of 56 m within 7 km of the mid-western shore. The main inflow is at the south end via the River Semliki which comes from Lake Edward through the western edge of the great Ituri rain forest in Zaïre, augmented by streams from the northern slopes of the Rwenzoris. On its course through the forest are several kilometres of rapids which are an effective barrier to faunal interchange between the two lakes (Fig. 13.3). Most of the lateral inflows into the lake from the escarpments are seasonal and contribute very little, since their catchments are small. Owing to an accident of geological history, the overflow from Lakes Victoria and Kioga, known as the Victoria Nile, originated by uptilting of the Victoria basin in the late Pleistocene (Ch. 14) and made its way via a previous river valley to a low point along the Rift wall to plunge over the Murchison Falls and to reach Lake Albert at its very northernmost end almost directly into the outflowing Albert Nile.

The water of the Victoria Nile is much less saline than that of Lake Albert (approx. 0.1 and $0.6^o/_{oo}$ respectively). It has therefore been possible to demonstrate by conductivity measurements that even in times of flood the river water does not affect the lake beyond about 10 km from the north end (Beauchamp, 1956). The Victoria Nile thus serves to maintain the level but has no other influence on the water of the lake except at its northern end though its rate of flow is considerably greater than that of the Semliki. The hydrology and ecology of the lake would have been different if the Victoria Nile had flowed into it near the south end, or alternatively, had joined the Albert Nile further north (e.g. by the valley of the River Aswa to beyond the Sudan border) where it could have had no controlling influence on the level of the lake. In the latter event a small reduction in the rainfall in the Albert and Edward basins and thus a smaller inflow from the Semliki, could result in a excess of evaporation over inflow with a consequent fall in level and stoppage of the outflow. The level of Lake Albert is now maintained above the exit partly by the Victoria Nile which functions in the manner of the inflow to a constant level water-still.

Important information concerning the very recent history of the lake has been provided by Harvey (1976) from a 10·6 m sediment core in the nothern section of the lake under 46 m of water. This covers a period of 25 000 years. Estimations of moisture, bicarbonate and organic content as well as the abundance and species composition of the diatoms indicate that there were two periods (12 500–14 000 and 18 000–25 000 years ago) during which the water level was lower, and the salinity and alkalinity higher than at present. There is little doubt that for some time during these two periods there was no outlet from the lake and the then scarcity of diatoms would suggest a very high alkalinity causing the ultimate dissolution of the diatom silica shells in the sediment. There is however no evidence to show how these conditions affected the fish and other fauna.

Though these reductions in level were no doubt caused mainly by a general decrease of rainfall in the catchment, it is noteworthy that the more recent of the two dry periods was contemporary with stoppage of the outflow from Lake Victoria (14 500–12 000 years B.C.), as deduced from similar evidence from sediment cores (p. 252). The great reduction in the flow down the Victoria Nile in the absence of a contribution from Lake Victoria would surely have prevented the river from maintaining the level of Lake Albert.

These facts, together with any further geological and palaeontological discoveries, must be taken into account in reconstructing the biological history of Lake Albert and the White Nile.

The present drainage system in and on either side of the Albert basin demonstrates very clearly how uplifting and associated subsidence along the crest to form the Western Rift has cut across a previous east–west drainage line (Fig. 14.1 and 2). This was previously fed from the highlands northeast of Lake Victoria and flowed in valleys which now hold the flooded branches of Lake Kioga, and then via the Kafu River into the Zaïre basin. The uptilting of the Victoria basin towards the eastern wall of the Albert Rift caused the Kafu to flow back into Lake Kioga and started the overflow from the north end of Lake Victoria which flooded Lake Kioga and finally escaped via the Victoria Nile. These dramatic events occurred as recently as the late Pleistocene and account for the very extraordinary disposition of the present water systems in this area (see Ch. 14 on the origins of Lake Victoria). The many rapids and falls on the Victoria Nile before and after its passage through Lake Kioga and the form of its bed show that this powerful stream is a comparatively recent one. The final plunge was originally over the main fault scarp at Fajao, but it has now cut back nearly two kilometres to the present Murchison Falls, leaving the gorge behind it. These Falls have always been an effective barrier to fish and have prevented the soudanian fauna of Lake Albert and the Nile from reaching Lakes Kioga and Victoria.

The earth movements that formed the Albert basin probably started in the Miocene. From then on there were repeated sinkings of the valley floor, more pronounced at the south end and associated with the uprising of the Rwenzori Mountains. At the same time there have been enormous depositions of sediments brought down from the neighbouring high land. At the south end of the basin the sediments are more than 2 400 m deep and the true valley floor has sunk to about 1 800 m *below* sea level. The present maximum depth of the lake (56 m) is thus only a fraction of the total depth of the trench in which it lies over a great mass of sediment. For discussion and further references see Harris *et al.* (1956). Bishop (1965), Bishop and Trendall (1967), Bishop (1969), Livingstone (1976).

It is fortunate that the progressive sinking of the valley floor did not include all of the area previously covered by water. Some of the sediments were left at a higher level around the basin and are conveniently exposed by subsequent uplifting, faulting and erosion. From these we have a record, tough incomplete, of the Pleistocene history of the lake and its fauna. Three main periods are distinguishable, involving a succession of three lakes:

1. The 'Kaiso' lake, early to mid Pleistocene.
2. The 'Semliki series' lake, mid to late Pleistocene.
3. The present lake.

The most illuminating for biological history are the Kaiso beds dating from early to mid Pleistocene, named after a village on the eastern shore near which they were originally discovered by Wayland (1925). More than 600 m of sediment were deposited during this period. Compared with contemporary sediments in the Lake Edward basin (p. 234) the number of fossil fish so far found in the Albert Kaiso beds is small, but there are three soudanian genera now living in the lake (*Lates, Bagrus* and *Heterobranchus*), and one (*Clarotes*) which is not now in the lake but is found in the Nile. The most abundant fossils are, however, the molluscs of which about twenty-five species have been described, all of which except the Nile Oyster *Etheria elliptica* are now extinct (Adam, 1959). A similar 'Kaiso' molluscan fauna existed at the same time in the Lake Edward basin (Fig. 13.2) and the discovery of Kaiso species in the early Pleistocene beds in the Lake Turkana basin suggests that this fauna was at that time very widespread (Fuchs, 1937, 1939). This is also supported by the remarkable fact that some of the otherwise extinct genera in the Kaiso beds (e.g. *Neothauma*) have survived to this day in Lake Tanganyika. The conditions that may have existed in the Albert and Edward Kaiso lakes and in the subsequent Semliki series lakes, also in both basins, and the final transition to the present lakes, are discussed further in Ch. 13.

Later Pleistocene (post Semliki) beds in the Albert basin show a molluscan fauna composed of species closely related to or identical with those now in the lake. Though much remains obscure, it is difficult to avoid the conclusion that something drastic happened in the mid Pleistocene, such as severe desiccation due to tectonic and/or climatic causes, which destroyed the Kaiso fauna, and that the lake was later rejuvenated and repopulated from the Nile. Extinction of the early Pleistocene mollusc fauna seems to have occurred also in the Edward and Turkana basins, the present fauna of the latter also being derived from the Nile.

The geological and palaeontological evidence from the Albert and Edward basins is such as to invite endless speculation on their Pleistocene and recent history, but is not enough to present a convincing picture of the changing distribution of lake water surfaces during the period. It must be remembered that, prior to the late Pleistocene, it is supposd from the evidence that neither the River Semliki nor the Victoria Nile flowed into Lake Albert. The main inflows were presumably from the rivers flowing westwards across what later became the Victoria Basin by the upwarping of its western edge (p. 248 ff).

It is evident from the general ecological survey of the invertebrate animals by Verbeke (1957a), and from the work of Green (1967a, b, 1971) on the Cladocera and Rotifera of the zooplankton, and from the studies of the feeding habits of the fish by Worthington (1929), Verbeke (1959b), Hamblyn (1966) and Holden (1967, 1970) that most of the available ecological niches are occupied and that the resources of the lake are now in general well exploited to provide a basis for the productive fishery (see Cadwalladr and Stoneman, 1968). The main species of fish involved at the different trophic levels are the same as those in Lake Turkana (p. 184) and the most commercially valuable species are also identical, e.g. *Sarotherodon niloticus, Lates niloticus, Alestes baremose* and *Citharinus citharus*.

The windstorms that sweep the lake are notoriously violent, as Samuel Baker learnt in 1864 to his discomfort and near destruction. This is especially so from May to August when the water column is well stirred to the bottom. Some degree of stratification appears in the calmer period around the turn of the year, shown by a reduction, though not very great, of oxygen in the lower water. But on the whole it appears that vertical water movements are sufficiently prolonged to en-

sure a good circulation of nutrients between the bottom sediments and the euphotic zone. The possibility that profile-bound density currents, due to nocturnal
cooling of the shallow inshore water, may contribute to vertical movements of
water-masses in this lake has already been discussed (p. 90).

Measurements of primary production in the open water are discussed on
p. 130. There are indications of periodic fluctuations in the planktonic diatoms
Stephanodiscus and *Nitzschia*, though it is not very clear how these are related to
hydrological events (Talling, 1963; pers. comm. J. H. Evans). Occasional outbursts of the blue-green alga *Anabaena* have been recorded in various parts of the
lake. These and the local sudden death of large numbers of fish, especially of
Lates, the Nile perch – a phenomenon occasionally seen in many tropical lakes
(Ch. 6) – is evidence of stratification with occasional violent stirring into the bottom sediments, though the importance of such to overall productivity is doubtful.

Lake Turkana (Rudolf)

This was the last of the large African Rift Valley lakes to be found by Europeans,
Teleki and von Höhnel, in 1888 (von Höhnel, 1894, 1938). The first specifically
limnological and geological expedition, led by E. B. Worthington, worked there
for a few months in 1930–31 (Cambridge Expedition, 1932–33). In 1934 an expedition under V. E. Fuchs (1939) contributed further to the geology of the lake
basin. From 1932 onwards French and American scientists have been investigating at intervals the geology, palaeontology and archaeology of the lower Omo
River and delta region (Arambourg, 1935–48; Butzer, 1971a). This and the work
of R. Leakey's team and associated workers on the remains of early man in the
old exposed sediments east of the lake (Coppens *et al.*, 1976) have produced
important information on its past history. The Lake Turkana Fisheries Research
Project led by A. J. Hopson has done some extensive limnological work over a
continuous five-year period (1970–75). The report (Hopson, 1978a), which also
reviews previous work, is now the most authoritative and comprehensive source
of information on the hydrology and biology of the lake itself.

Lake Turkana is situated in a very hot and arid region in the extreme northwest of Kenya, surrounded by vast stretches of very dry country, Turkana to the
west and the Northern Frontier Province to the east, suitable only for the existence of a sparse and partly nomadic population (Fig. 3.4). At the northern tip of
the lake is a point at which Sudan, Kenya and Ethiopia meet, still one of the most
remote regions of sub-Saharan Africa. The potentiality of the lake fishery has
been considered to be very great, and to establish this was one of the objectives of
the Fisheries Research Project.

The lake lies in a branch of the lowest section of the Eastern Rift at an altitude
of about 375 m (Fig. 10.1). Its length (265 km) is greater than that of Lake
Albert, but its average width (30 km) is a little less. The maximum depth
(120 m) is in a small depression near the south end. Otherwise the greatest depth
is about 80 m near the centre of the lake.

There is only one permanent inlet, the River Omo, which drains a large area of
the Ethiopian highlands and rises some 500 km from the lake. It supplies more
than 98% of the inflow, though its catchment area is only about 38% of that of

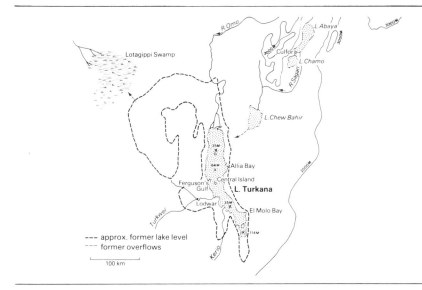

Fig. 10.1 Lake Turkana and the lower Ethiopian Rift Lakes

the whole lake. It is therefore the rainfall in the southwestern highlands of Ethiopia that determines the seasonal and longer-term changes in level of the lake. The seasonal fluctuations have at present an amplitude of 0·5–1 m (maximum levels in September–October, minimum June–July).

From the records and observations of visitors to the region there is clear evidence of greater changes of level during the past ninety years. The peak mean levels (relative to that in 1968) were in 1895 (+15 m), in 1917 (+5 m). A low level (about –5 m) was held between 1950 and 1960. 'The level of this lake fluctuated over a range of 20 m within the past 75 years, an amplitude exceeding that of any other world lake of natural origin' (Butzer, 1971b, pp. 123, 146). The flat plains in the delta region and west of the lake are littered with the remains of contemporary species of fish and molluscs up to several kilometers from the present lake shore. Measurements of salinity taken during the past fifty years (Table 10.1) show clearly the concentrating effect of the dry period between 1950 and 1960, the dilution following the wet period in the 1960s and the final dry spell starting in the early 1970s. In general, the timing of these recent changes in level was correlated with those of the major Rift Valley lakes and Lake Victoria.

Von Höhnel (1894) in 1888 noticed a large area of submerged dead forest trees in the Omo delta region, indicating a fairly recent rise in level preceded by a recession long enough to establish a forest. It is obvious that such fluctuations would have had disturbing effects on the pastoral and agricultural practices of the inhabitants of this very flat region. Between 1888 and 1895 about 200 km² of land were flooded, between 1898 and 1955 800 km² emerged from the water, 350 km² of which were again inundated between 1962 and 1965 (Butzer, 1971b, p. 175). It might be added that inundation of such large areas of land would have had a stimulating effect on the fertility of the lake.

The next most important inflows are the Kerio and Turkwel Rivers

Table 10.1 Lake Turkana. Open water. Samples taken in mid lake offshore from Ferguson's Gulf

	Conductivity (K_{20})	Approx. salinity $^o/_{oo}$ (g/l)	Alkalinity (meq/l)	
1931	2 860	2·2	20·0	Beadle, 1932b
1934	3 220	2·4	22·9	Beadle, in Fuchs, 1939
1953	—	—	21·6	Fish (unpublished)
1961	3 300	2·5	24·5	Talling and Talling, 1965
1970	2 200	1·7	—	Hopson (unpublished)
1975	3 600	2·7	21·3	Ferguson and Harbott, 1979

Total analysis of the 1961 sample gave 2·5% from the sum of the ions (Table 5.1). The other salinity figures were calculated on the assumption that the relation between conductivity and salinity is 3 300 : 2·5 (as in 1961).

(Fig. 10.1). These are seasonal in their lower reaches, but flow into the lake for several months in the year. The streams that come from the surrounding hills and escarpments are dependent on the local weather, and some may flow as little as a few hours in the year. There are however several permanent springs that flow into the lake from below the escarpments, and holes dug into the dry river beds can fortunately provide a good source of drinking water in place of the almost undrinkable saline-alkaline water of the lake. Travellers on the semi-desert plain west of the lake, seeing no signs of rain in the neighbourhood apart perhaps from a storm cloud over the escarpment some 50 km distant, may suddenly be confronted with a torrent of water rushing down a river bed carrying trees, drowned animals and other debris. It will subside within a few hours leaving nothing but pools and ripples on the wet sand. The pools will serve as temporary drinking holes for the mammals and birds of the region, but within a few days the scorching sun will have removed all traces of surface moisture.

The earth movements that formed the Turkana basin began at least as far back as the early Miocene and the earliest known exposed lacustrine deposits date from that period (Savage and Williamson, 1978). Fuchs (1937, 1939) found three species of molluscs in the early Pleistocene beds at the Gaza waterhole east of the lake that are identical with species found in the Kaiso beds of the Lake Albert basin (p. 175). No record has yet been published of fossil fish from sediments around Lake Turkana, but in the early Pleistocene sediments of the lower Omo valley Arambourg (1948, Vol. 3) identified seven species that also occur in the Kaiso beds.

More recently a large number of fossil molluscs have been found in Plio-Pleistocene deposits east of the lake, which can be compared with those from the Albert-Edward basin (Gautier, 1970; Williamson, 1978; van Damme, 1979, unpublished manuscript). It is clear that before the mid-Pleistocene the two lakes had a very similar history. The evidence suggests that during the late Pliocene there was an outburst of speciation in both lakes from a previously widespread Miocene molluscan fauna. These endemic faunas were extinguished during the first half of the Pleistocene presumably because of low water level and consequent high salinity or even complete drying up. The subsequent rise in level culminating at about 9 000 years B.P. allowed the influx of species from the Nile. Since the fall in level below the outlet after about 7 500 years B.P. the molluscan fauna of Lake Turkana has been impoverished owing, it is suggested, to the increase in salinity (p. 183). Analyses of late Pleistocene sediment cores (Fig. 3.3) indicate that there was a high lake level for some time prior to 20 000 B.P. followed by a very dry period up to at least 12 000 B.P. during which the fauna of the previous lake may well have been destroyed by very high salinity or even by dessication. Subsequently the lake rose once more to about 80 m above present level to establish a connection with the Nile and its fauna, which lasted until about 7 500 B.P. Since then there has been no surface outlet, which distinguishes the present Lake Turkana from all the other African Great Lakes except Chad. Its predominantly soudanian fish fauna is further good evidence for a very recent connection with the Nile.

The lake would at that time have flooded the depression now occupied by the Lotagippi Swamps northwest of the lake, which drain into the Pibor River and via the Sobat into the Nile. The contour round the lake in Fig. 10.1 thus represents approximately the shoreline at the time, with the supposed position of the outlet. The Omo River would then have flowed into the lake close to the outlet to the Nile – an unusual situation curiously similar to that of the present Lake Albert. The higher rainfall would presumably have made the Kerio and Turkwel into more powerful and permanent rivers entering the lake near the south end, and thus supplying most of the water to the lake, except at its northern end, as does the Semliki to Lake Albert at the present time (p. 173). There is no evidence that the two more recent increases in level (Fig. 3.3) were large enough to cause an overflow.

The salinity of the water is higher than that of any of the other very large African lakes (Table 5.1). This is due mainly to the absence of an outlet and great contraction in volume during the past 7 500 years. Nevertheless when the water was first analysed (Beadle, 1932b) and the beaches of the old outflowing lake had been located, it was clear that the present salinity of the water was much lower than would result from simple concentration of the salts by evaporation. We now have more information to apply to this problem, though some very uncertain assumptions have to be made in estimating the quantity of salts apparently removed from the water during the past 7 500 years.

In the first edition of this book (p. 143) the possibility was raised of a subterranean seepage of water, which is known to be an important cause of the unexpectedly low salinity of Lake Chad (p. 215). Against this it may be objected that Lake Turkana lies at the lowest point in this part of the Rift, but in such an unstable and faulted region the possibility of a subterranean outflow cannot certainly be excluded on these grounds.

There is no doubt that much calcium and magnesium are removed by precipitation from the inflowing water as it enters the very alkaline lake. This is reflected in the low levels of these ions relative to that of the other major cations (Table 5.1). It has however recently been suggested that Na^+ and K^+ may also have been removed from the water in combination with some of the minerals known to be deposited in the sediments (Yuretich, 1976; Ferguson and Harbott, 1978). In favour of this suggestion and against a subterranean seepage they point to the fact that the concentration of chloride in the present lake is about 310 times the mean in the Omo River and the same order as the theoretical concentration factory (x280) due to the estimated decrease in volume and evaporation of the lake since 7 500 B.P. Chloride is not known to be 'bound' in any way to minerals nor is it precipitated in combination with the common cations in the water. But the assumptions upon which this thesis is based – the initial volume of the isolated lake and the rate of the inflows and their composition during the whole period – are very doubtful. Moreover the very recent volcanic activity in the basin must have contributed a significant amount of salts to the lake during the period. The nearly saturated salt water in the crater Lake Katwe (p. 000) is almost entirely derived from below and most of the salt in Lake Magadi now comes from numerous saline springs along the shore (p. 332). The high salinity of Lake Kivu, which has a copious outlet, is due mainly to seepage of saline water through the floor of the lake (p. 106).

Calculations based on the figures assumed by Ferguson and Harbott (1978) would suggest that the amount of salt lost from the water during the period was of the order of 900 x 10^{10} kg (10^{10} tons). The depth of sediment deposited during the same period was according to Yuretich (1976) 3–7 m. This would imply that more than 1 000 kg of salt have been deposited in the sediments per m^2 of lake surface. It would surely be possible to discover from sediment cores whether a m^2 column of sediment to a depth of 7 m contains something approaching this quantity of the major ions.

Simple calculation of the present water balance based on the mean annual rate of inflow by the Omo (18·6 km^3), the surface area of the lake (7 560 km^2) and the measured rate of evaporation with the Piche evaporimeter on the shore (2 m or 15·1 km^3/year) shows a loss of 3·4 km^3 of water not accounted for by evaporation. But here again the assumption that the Piche evaporimeter truly represents the rate of evaporation from the surface of a heavily stirred lake is doubtful.

Another question might be raised at this point – why have many lakes in closed basins become excessively saline after a comparable contraction in volume? Such are Lakes Nakuru and Elmenteita in the Kenya Rift, whereas Lake Naivasha, also devoid of an outlet and derived by contraction of the same large late Pleistocene lake, has remained fresh (Fig. 3.6, and p. 333). In this case a subterranean seepage would seem the most likely explanation. Whatever the solution to this intriguing problem, Lake Turkana has been saved from excessive salinification and harbours a rich fauna and flora, though the absence of some invertebrate species may be attributable to high salinity (p. 183).

Though subject to violent diurnal changes of wind, the mean monthly climate is on the whole surprisingly uniform over the year and in fact less variable than that around any of the other Great Lakes of Africa. The reasons for this relative stability are not fully understood (Butzer, 1971b; Griffiths, 1972). The mean monthly maximum air temperature at present ranges only between 31° and 33°C, but the period October–January is the warmest and driest and is subject to strong south, and southeast winds. In contrast April–August is somewhat cooler and more humid and the winds are less violent and more variable in direction. The small amount of rain that falls on the lake and its immediate surroundings may occur at any time, though it is least likely around the turn of the year.

The most important influences on the productivity of the lake are the winds and the Omo floods. Though subject to seasonal variation the winds are always sufficiently frequent and powerful to maintain an adequate circulation of water at all depths in the open lake. This circulation is augmented by horizontal currents due to the inflow and to winds. Almost every day the wind builds up during the morning and subsides around midday for a calmer afternoon and evening. The temperature stratification normally developed in the afternoon is dispersed during the morning. Consequently the oxygen content of the deepest water is never less than about 70% saturation, and the recycling of nutrients from the bottom must be subject to few restraints. Conditions in the inshore very shallow waters, especially in sheltered bays, are different (see below).

In spite however of the regularity of the local climate there is a definite seasonal increase in the rate of primary production by the phytoplankton in the open water (Harbott, 1978), mainly stimulated by the inflow of nutrient-rich flood water from the Omo River. Owing to the much lower conductivity and higher turbidity of the river water its course down the lake can be traced. The effects of the flood can be detected some weeks later at the south end. The southerly currents are pushed back along the west coast by the prevailing winds (Ferguson and Harbott, 1978).

Most of the phytoplankton in the open water, though varying seasonally and from place to place, is composed mainly of eight species and is dominated by the blue-green *Microcystis aeruginosa*. Measurements of rates of gross primary production at four stations down the length of the lake showed a maximum in the north and a considerable gradient to a much lower level at the south end. Moreover the rate at the north end was accelerated about five times by the Omo flood, but the delayed affect of this at the south end, though detectable, was very small (Harbott, 1978). These spatial differences and seasonal changes are not unexpected, but they serve as a warning that measurements made at only one station, as has been done in several of the African Great Lakes, may give misleading impressions of the primary production of all the open water of the lake, and of course give no indication of production in the inshore regions. The computed figures for daily gross production in Table 7.1 (0·25–6·2 g c/m^2 day) represent the entire range from the minimum in the south to the maximum in the north, where, incidentally, the water is least saline. The latter is surpassed only by the two alkaline crater lakes in the Table.

Measurements were also made at several points in the shallow inshore water particularly in Ferguson's Gulf on the West Coast which is nearly isolated from the lake by a sandspit formed by the prevailing winds and currents from the south (Fig. 10.1). It supports a very productive fishery. The gulf is about 8 km long and 3 km maximum width and is very shallow, of maximum depth 4 m but mostly less than 2 m. It is therefore thoroughly stirred by winds. The phytoplankton varies in different regions but is much richer than in the open lake in numbers both of individuals and of species. The dominant species are blue-greens especially *Anabaena circinalis* and *Botryococcus braunii*, and diatoms of the genus *Nitzschia* are very common. The overall estimate of primary production was about 4 g c/m^2. day which, owing to the apparently stable conditions in the Gulf is probably maintained throughout the year. This rate exceeds that anywhere in the open lake except the seasonal maximum at the north end.

Until recently the lake has been approached by most biologists from the west, particularly from Ferguson's Gulf, and impressions of the littoral have been based mainly on the midwestern shore where the scarcity of large aquatic plants (submerged and emergent macrophytes) is a striking feature. This may be due to the scouring water currents flowing northwards up this coast with consequent very unstable sandy substratum. There are many stretches too of the eastern shore, both sandy and rocky, that are similarly denuded, and in spite of subsequent finds mentioned below it remains true that aquatic macrophytes are much less common and are represented by fewer species than in the other large lakes, and this must be attributed at least in part to the comparative scarcity of sheltered shallow water.

The commonest emergent plants in Lake Turkana are the grasses, *Sporobolus spicatus* and *Paspalidium geminatum* confined to sheltered and shallow water, but able to maintain themselves in the damp soil during seasonal low water levels. In spite of their relative scarcity the macrophytes are important for the inshore fisheries which are mostly based on sheltered bays (Hopson et al., 1978). Not only do they provide the smaller fish with shelter from wave battering and from predators but are themselves food for some species. *Alestes dentex*, for example, consume quantities of *Potamogeton*, and the algae encrusting the leaves and stems are grazed by the herbivours tilapias. In dry periods cattle feed on the emergent grasses and *Potamogeton* is collected for fodder.

Estimates of the rate of gross primary production in *Potamogeton* beds were made by measuring dissolved oxygen changes in enclosed samples as well as the gaseous oxygen emitted by the plants, and the figures represent the sum of the production by the macrophytes, their encrusting algae, the phytoplankton in the water and those lying on the bottom. They ranged from 0.93 g c/m^2. day in El Molo Bay to 4.6 g. c/m^2. day in Allia Bay (Harbott, 1978). The latter figure is close to that got from the phytoplankton in Ferguson's Gulf. It is clear that submerged plants in shallow sheltered bays contribute much to the productivity of the inshore fisheries.

A most remarkable feature is the complete absence from the lake of papyrus which is the dominant macrophyte in much of the Nile and especially in the Sudd swamps into which Lake Turkana previously flowed. Though some other emergent macrophytes inhabit the lake one might have guessed, subject to experimental confirmation, that the high salinity or alkalinity is in some way responsible, were it not that the lower Omo River is apparently also devoid of this plant (pers. comm. A. J. Hopson). The Botanic Gardens at Kew have no records of papyrus from the Omo Valley nor from the Rift Valley lakes Abaya and Chamo (pers. comm. F. N. Hepper). All of these were recently connected with the Nile through a Lake Turkana of much lower salinity. It is of course possible that plant collectors in this region have not been interested in such a well known plant. But if present, it must be very scarce.

As with the phytoplankton there is considerable variation in numbers and species composition of the zooplankton in different parts of the lake, the number of species being greatest at the north end near the inflow from the Omo. There is also a noticeable population increase during the flood period (Ferguson, 1978a).

In numbers of individuals the Protozoa are the most abundant of the three main groups of planktonic animals in the open water. The commonest species are *Raphidiophrys pallida* (Heliozoa) and *Vaginicola sp.* (Ciliata). Their food is probably detritus particles and bacteria. Their predators are the carnivorous planktonic crustaceans which in turn are an important food for most fish at some stage in their lives. The Protozoa therefore play an important part in the ecosystem.

The Rotifera are fewer in species and individuals than would be expected from reports on other African lakes, and their contribution to the circulation of organic matter is probably small. They are however interesting because the species composition appears to be influenced by high salinity and/or alkalinity as judged by their distribution in lakes of different salinities in Africa and elsewhere, e.g. Lake Chad (Pourriot *et al.*, 1967; Robinson and Robinson, 1971). Many species common in most freshwaters are absent from Lake Turkana or, as *Keratella tropica*, are confined to the most dilute water near the mouth of the Omo. On the other hand *Brachionus plicatilis*, *B. dimidiatus* and *Hexarthra jenkini*, common in Lake Turkana, are typical of relatively saline waters elsewhere.

The biomass of the Crustacea in all regions of the open water is greater than that of the other two groups. Of the twelve species recorded the most abundant are *Trophodiaptomus banforanus* (Calanoida) which is predominantly a detritus feeder, and the carnivorous *Mesocyclops leuckarti* (Cyclopoida) whose diet includes the immature stages of other crustaceans. The Cladocera are relatively scarce.

The study by Labarbera and Kilham (1974) on the distribution of copepods in forty-eight African lakes suggests that, as with the rotifera, some species absent from Lake Turkana are confined to lakes of lower salinity, whereas some of the Lake Turkana species, notably *Mesocyclops leuckarti*, are typical of waters more saline than Lake Turkana. Though we must not jump to the conclusion that salinity or alkalinity as such is the direct limiting factor in all these cases, it certainly appears that the salinity of Lake Turkana has reached a range in which some of the fauna is beginning to show features characteristic of 'saline' water.

Of the decapod Crustacea two species of prawns have been recorded, *Caridina nilotica* and *Macrobrachium niloticum*. These are important items in the diet of several fish. There are apparently no potomonid crabs which are common in most of the other large African lakes including Lake Albert and the Nile. Another peculiarity is the small number (five) of species of gastropod molluscs that have been found, all of which are prosobranchs, namely *Melanoides tuberculata*, *Cleopatra pirothi*, *Gabbiella rosea*, *Ceratophallus natalensis* and *Bellamya unicolor*. The first two are benthic and live over the deep water sediments; the others are restricted to the shallow inshore water. It might be suggested, though without other supporting evidence, that the crabs and the many molluscs that are missing but common elsewhere, are excluded in some way by the high salinity and alkalinity of the water. They certainly need more calcium than the other invertebrates for their shells and exoskeletons, but the level of Ca^{++}, though low relative to the other ions, is actually similar to that in Lake Victoria (Table 5.1). We are faced with the fact that the prawns and the above gastropods are not excluded. There are other environmental conditions that might account for this anomaly such as the scarcity of the appropriate plant food, the absence of suitable types of substratum and the generally very turbulent waters.

A number of snail shells have been found on the plain a few kilometers from the west shore (J. Landye, unpublished, quoted by Ferguson, 1978b). These included all the species at present alive in the lake and five others – four pulmonates: *Biomphalaria pfeiferi*, *Gyraulus costulatus*, *Segmentorbis augustus* and *Bulinus sp.*, and *Pila speciosa*. These species, also found in Lake Albert, must have lived in the lake when its level was several metres higher than any recorded during the past eighty years.

The soudanian character of the fish fauna of Lake Turkana has already been emphasised. Hopson and Hopson (1978) recorded 48 species in the lake of which 30 are spread over the entire soudanian region. Of these 22 are also in Lake Albert. Eight are found in the Nile. The remaining 10 species are endemic to the lake. The association of a small number of endemic species with a relatively short period of isolation is well demonstrated by Table 8.2. Both the total number of species and the proportion of endemics are similar in Lakes Albert and Turkana, but are much less than in the three long-isolated lakes which also have a number of endemic genera. This comparison is reinforced by the cichlids alone which in the other three lakes have speciated on a very grand scale. The remarkable closeness of all the figures in Table 8.2 from Lakes Albert and Turkana is surely a reflection of the common origin of the two faunas and of their very similar recent history which has included a few thousand years of isolation from the Nile to which Lake Albert has been reconnected.

In both lakes, most of the main ecological niches appear to be well exploited by the same genera and often by the same species of fish. For example the inshore phytoplankton is consumed by the cichlids *Sarotherodon niloticus* and *S. galilaeus*, *Potamogeton* with its encrusting algae by *Alestes dentex* and *Tilapia zillii*, the zooplankton by many species especially of the genera *Alestes*, *Engraulicypris* and *Aplocheilichthys* some of which also feed on insects and the prawns *Caridina nilotica* and *Macrobrachium niloticum* (McLeod, 1978). The benthic fauna and flora are food for fish such as *Citharinus citharus* and *Barbus bynni*. The major predators on other fish in both lakes are *Lates niloticus* (Nile perch), *Hydrocynus forskali* (Tiger fish) and *Bagrus bayad* (Catfish).

The habitats of the different species of fish – inshore, offshore and at different depths in the open water are discussed by Hopson and Hopson (1978). During the day many species congregate in characteristic positions in the lake and disperse at night. These movements can often be followed by echo-tracing (Fig. 10.2). Most species breed during the time of the Omo floods and some, such as *Alesters baremose* and *Citharinus citharus* are anadromous and ascend the Omo to breed.

Fish speciation has followed a similar pattern in both lakes Turkana and Albert, and the endemic species are mostly found in the mid depths of the open water and lower regions of the offshore waters. For example the huge Nile Perch *Lates niloticus*, common to both lakes, has in Lake Turkana given rise to the much smaller endemic *L. longispinis*. The former can take fish up to half their own length; the latter feed on small fish and prawns. Their ranges overlap, but in general *L. niloticus* is more concentrated in the offshore waters and *L. longispinis* is more widespread in the pelagic region. In lake Albert there is a similar rela-

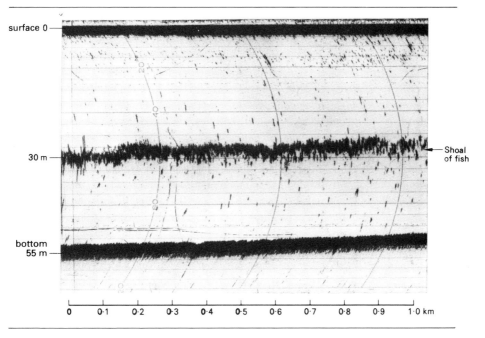

Fig. 10.2 Part of an echo-transect taken during the daytime between Ferguson Spit and Central Island, Lake Rudolf, showing a very large shoal of fish at about 30 m (A. J. Hopson, unpublished)

tionship between *L. niloticus* and the smaller endemic *L. macrophthalmus* (Holden, 1967).

A very interesting problem is raised by the absence, or extreme scarcity, of certain otherwise widespread soudanian fish. Until recently no mormyrids (elephant snout fishes, Fig. A4.1c) nor the electric catfish *Malapterurus electricus* had been found in the lake and *Gymnarchus niloticus* (Gymnarchidae) seemed to be restricted to the Omo delta. These are all common members of the soudanian fauna including the Nile and Lake Albert and the mormyrids especially have evolved a number of species in tropical Africa.

Chappuis (1939) discussed this problem and suggested that *Mormyrus spp.* may have evolved in the Nile system after the lake was isolated. This theory however cannot possibly be supported in view of our present knowledge of the very recent outlet to the Nile (Fig. 3.3).

More recent investigations have now shown that, though *Gymnarchus* is indeed confined in the lake to the Omo delta, *Mormyrus kannume* and *Hyperopisus bebe* are in fact present too in the Omo River and in other inflows, and have been found in the lake near the mouths of these, especially during floods. Moreover *Malapterurus* (though only two specimens) was taken in the open lake well away from any inflow (Hopson and Hopson, 1978). There still therefore appears to be a restriction to the free colonisation of Lake Turkana by these particular fish, which are abundant in Lake Albert.

It happens that these species are distinguished from all others in Lake Turkana by the possession of organs which produce an electric field in the water around them. The organs have been evolved apparently independently in seven families of fish, marine and freshwater. They are usually composed of batteries of modified muscle fibres each of which generates a potential across its bounding membrane, and the battery discharges under the control of the nervous system. There are two kinds of electric organ: (1) as in the mormyrids and *Gymnarchus*, whose electric charge is weak (mostly less than 10 volts) and probably functions as a means of locating objects that have an electrical conductivity different from that of the water and of communicating with the members of the species which in fact possess special receptor organs developed from the lateral line system, (2) as in *Malapterurus*, which discharges strongly at more than 300 volts, whose function is mainly for defence and for stunning prey (Lissmann, 1958; Bennet, 1971).

It will by now be clear where the argument is leading – a large change in the electrical conductivity of the water might be expected to have important effects on the electric fields generated by these fish. Lake Turkana has a conductivity (at present over 3 000 μmhos, Table 5.1) much greater than that of any other African lake possessing a normal freshwater fauna. But evidence from experiments that changes of salinity actually have important effects on the functioning of the electric organs and that the distribution of the fish can be determined by these effects, is so far inadequate. Few ecologically minded physiologists have considered the problem.

Electric fish are in fact found in a great range of salinities and there are both freshwater and marine species. Few observations seem to have been published on the distribution of any one species in the salinity gradient of, for example, an estuary, though freshwater electric gymnotids have been reported to be abundant in the tidal brackish water regions of the Amazon estuary in South America (pers. comm. H. W. Lissmann, to whom I am grateful for a discussion on this problem). As an example of the rare experimental work Harder *et al.* (1964) found that changes in the salinity of the water altered the amplitude of the discharge from electric organs of several mormyrids.

We have to admit that we do not even know whether the proper functioning of the electric organs is essential for the existence and reproduction of these fish or at least whether some species can manage without them for considerable periods. If the high salinity of Lake Turkana is the limiting factor, the relatively recent increase has almost, but not quite, banished the electric fish from the lake. The salinity may continue to increase too rapidly for adaptation (if adaptation is possible) or the onset of another high rainfall period may bring the lake back to its previous condition with the electric fish well distributed within the lake. In Lake Chad mormyrids have been found only in water of conductivity less than 400 μmhos (Benech *et al.*, 1976). The basic theory is however not yet fully supported though there must surely be more information available on the distribution and habits of freshwater electric fish that is relevant to this problem. Some collaboration between ecologists and physiologists would surely be profitable.

Apart from the electric fish there are no signs that the salinity or alkalinity of the water has affected the fish in Lake Turkana. Moreover one of the crater lakes on Central Island, probably connected with the main lake less than a hundred years ago, has a salinity about three times that of the lake. In this are populations of *Clarias lazera, Synodontis schall, Sarotherodon niloticus* and *Haplochromis rudolfianus* (Hopson and Hopson, 1978). One cannot however assume that other species would survive if the salinity of the lake ever rises to that level.

Central Island is one of the important remaining refuges for the Nile crocodile which, being safe from hunters, is found in very large numbers. The less saline crater lakes on the island provide suitable beaches for nesting and nurseries for the young. Their biology has been studied by Modha (1967).

The present state of the fisheries of Lake Turkana, future possibilities and recommendations for development and control, are presented by Hopson (1978b). So far most of the fishing has been inshore, especially on the western side and in sheltered bays along both shores that have rich growths of emergent and submerged macrophytes. The most productive fishery is in Ferguson's Gulf based chiefly on the vegetarian fish *Sarotheroden niloticus* which directly exploits the high level of primary production by the algae (p. 181). The Lake Turkana Fisheries Survey has discovered much about the location and movement of many of the species, and it is clear that, with improved methods, the fishery could be expanded further into the open water and that the level of cropping could be raised considerably without depleting the stocks.

The Ethiopian Rift Lakes

Northwards from Lake Turkana (Rudolf) the Eastern Rift Valley extends more than 600 km into the heart of Ethiopia (Fig. 10.1). The first 100 km crosses steadily rising country in which lies the shallow remnant of Lake Chew Bahir (formerly L. Stephanie) at an altitude of 520 m and thus considerably higher than L. Turkana (375 m). It then cuts through the highlands with a progressive rise in the height of the valley floor to reach a maximum of over about 1 800 m about 100 km south of Addis Ababa. Thence the rift descends to the northeast and the lateral escarpments diverge from each other across the Afar desert to the coast. Down this valley flows the Awash River which ends in some small terminal saline lakes near Djibouti. These show clear evidence of past climatic changes similar to and perhaps contemporary with those in the southern section of the valley discussed below (Gasse and Street, 1978). This beautiful valley striking across the high mountainous country of central Ethiopia with its escarpments, extinct and partially active volcanoes, and its chain of lakes, resembles, though on a less impressive scale, the same valley across the highlands of Kenya. As a demonstration however of the effects of recent climatic changes and geological events on all levels, connections and faunas of the lakes it is equally informative. Most of the research has been done during the past twenty years and further relevant information will certainly be discovered. For a short review of ideas on the recent geological history of the lakes see Grove *et al.* (1975).

The lakes of the main Rift lie in three at present separate closed basins (with no surface outlets, Figs. 10.1 and 10.3, and Table 10.2): (1) the four partially interconnected Galla Lakes Ziway (Zwei), Langano, Abiyata and Shala at the

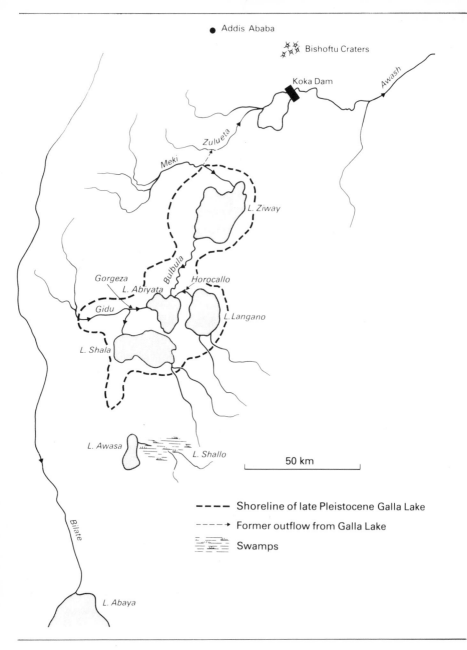

● Addis Ababa

Bishoftu Craters

Koka Dam

Awash

Zulueta

Meki

L. Ziway

Gorgeza   Horocallo

L. Abiyata   Bulbula

Gidu

L. Langano

L. Shala

L. Awasa

L. Shallo

50 km

Bilate

– – – – Shoreline of late Pleistocene Galla Lake

– – – → Former outflow from Galla Lake

Swamps

L. Abaya

Fig. 10.3 The Ethiopian Rift Lakes

northern (upper) end of the valley; (2) Lake Awasa, higher but isolated from Lake Shala by a ridge at about 1 800 m. Its main affluents come from the east and drain through a swampy Lake Shallo. Lake Awasa has no surface outlet and its comparatively low salinity (Table 10.2) is probably due, as with Lake Naivasha (p. 335), to subterranean outflow by seepage through the bed of the

Table 10.2

Lakes	Altitude (m)	Max. Depth (m)	Conductivity (K_{20})
Ziway	1 636	7	370–427
Langano	1 582	46	1 900–2 291
Abiyata	1 580	14	10 700–3 000
Shala	1 558	266	20 400–33 360
Awasa	1 680	22	790–1 080
Abaya	1 285	13	670–900
Chamo (Margherita)	1 233	13	927
Chew Bahir (Stephanie)	520	ephemeral	
Turkana (Rudolf)	375	120	2 200–3 300

The conductivity figures are from measurements made by different people at different times over the past forty-five years. Except for Lake Chamo, they demonstrate the great fluctuations that occurred (in both directions) due to seasonal and longer-term changes of climate (data collected by Talling and Talling, 1965; Grove et al., 1975, and in Table 10.1).

lake. Lake Abaya, the largest of the Ethiopian Rift Lakes, is fed principally by the Bilate River which rises near Addis Ababa and bypasses the Galla Lakes (Fig. 10.3). It overflows into Lake Chamo via the lower Culfor River. In periods of flood Lake Chamo discharges into the Sagan River sufficient water to reach Lake Chew Bahir (formally Stephanie) more than 100 km to the south. The latter is thus an ephemeral lake lying at the southern end of the third closed basin.

The relations between the northern (Galla) lakes can be seen in Fig. 10.3. Their altitudes, depths and some recent measurements of conductivity (representing salinity) are listed in Table 10.2. Lakes Ziway and Langano drain into Lake Abiyata. The Gidu River from the west normally flows into Lake Shala, but in heavy flood reaches Lake Abiyata via the Gorgeza Channel. A rise of about 15 m in the level of Lake Abiyata would be needed for a direct overflow into Lake Shala via the Gorgeza and lower Gidu Rivers (Fig. 10.3). There is evidence that this overflow in fact existed during the course of the regression but was finally severed about 2 500 years ago (pers. comm. F. A. Street).

Evidence for past history has come from (a) a series of old lake sediments on the sides of the valley at different heights above the present levels of the lakes, which contain fossil remains of fish and molluscs, some 14C dated; (b) from study of the faunas of the existing lakes; and (c) from traces of past water connections that no longer exist today. Both Nilsson (1940) and Mohr (1966) studied some of these sediments and concluded that there was probably a very high water level before the major faulting in the Pleistocene which gave the present form to the valley. At that time, it is suggested, there would have been a large lake or group of lakes from the present position of Lake Ziway in the north to that of Lake Abaya in the south and draining northwards into the Awash valley. The subsequent faulting and volcanics ultimately produced the conditions during the late Pleistocene whose effects are more clearly shown by the surviving evidence. A series of sediments at lower levels on the sides of the valley northwest of Lake Ziway and southeast of Lake Shala have been investigated by Grove et al. (1975). These contain beach pebbles, fossil remains of ostracods, crabs, molluscs and fish as well as algal limestones, and several levels have been 14C dated. It is clear that from about 6 000 to 5 000 B.P. the four Galla Lakes were united into one lake whose surface was at least 30 m above that of the present Lake Ziway and thus probably flowed northwards into the River Awash. The remains of an outflow channel was in fact found at about 1 670 m, and down this the lake must have drained into the R. Dubeta and so into the Awash (Fig. 10.3).

The fish at present living in Lake Ziway are clearly derived from those of the Awash and the Ethiopian highland waters generally (including Lake Tsana, p. 193). The presence of an endemic species, *Barbus ethiopicus* (Banister, 1973), is interesting in view of the evidence already mentioned that the final closure of the outflow into the Awash basin occurred less than 3 000 years ago. Other species such as *Sarotherodon niloticus* and *Barbus intermedius* are found over most of this region including Lake Tsana, and are represented in the fossil beds above Lake Shala. They therefore lived in the previous Galla Lake which discharged into the Awash, and have survived in the present Lake Ziway. The Galla Lake basin must therefore previously have drained to the north. There is in fact evidence from exposed sediments that this happened twice during the past ten thousand years (9 400–8 500 and 6 500–4 800 years B.P.). The final regression therefore started about five thousand years ago (Street, pers. comm.). The drier climate during the past few thousand years has resulted in four Galla Lakes whose interconnections are shown by arrows in Fig. 10.3. The flow is now southwards but is blocked by a volcanic ridge south of Lake Shala. The general increase in salinity (conductivity) from Ziway to Shala must be due to evaporation as well as to the influx of saline water from volcanic sources (Table 10.2). There is an abundant fauna, including fish and molluscs, in Lakes Ziway and Langano, but from the figures in Table 10.2 one would conclude that the salinities of the waters of Lakes Abiyata and Shala are much too high for fish or molluscs (Ch. 5). It is known, however, that the salinity of Lake Abiyata varies very much with the season, and at flood times the water within a considerable distance of the inlets is very much less saline. Even during low level periods there must be fresher water for fish and molluscs in and around the mouths of the Bulbula and Horacallo Rivers (Fig. 103). In fact, there have been a number of recent reports of fish in Lake Abiyata, and the *tilapia* fishery is said to be very productive (pers. comm. B. Wood and F. A. Street). Even Lake Shala is reputed to produce *Sarotherodon*, but these must surely be confined to the neighbourhood of the inflow from the Gidu River. Not enough regular and repeated work has been done on these two little-known lakes. They clearly present some interesting problems arising from the great fluctuations in levels and salinities, and neither lake appears to be chemically and biologically homogeneous.

Unlike the Galla basin there is no reason to suppose that, since the rifting started long before the Pleistocene, the more southerly section of the valley now occupied by Lakes Abaya, Chamo and Chew Bahir ever drained northwards into the Awash. On the contrary there is very convincing evidence that the outflow from this region was to the south, particularly during the late Pleistocene. The present rainfall in the basin, and apparently for the past two or three thousand years, has been insufficient to maintain permanent water in Chew Bahir. From the reports of the first Europeans to see the lake in the early 1890s and of all subsequent travellers in the region up to 1975 it is certain that during the past hundred years at least Chew Bahir has been a salt-pan intermittently flooded by the Sagan River to give a shallow expanse of very saline and alkaline water (von Höhnel, 1938; Grove et al., 1975). The extent and depth of the water has varied greatly from season to season, but it has remained during this time a closed basin with no surface outlet.

It is therefore a matter of very great interest that Lakes Abaya and Chamo now have a soudanian fish fauna, including the Nile perch Lates niloticus,* obviously derived via Lake Chew Bahir from Lake Turkana and the Nile. A former over-flow route from Chew Bahir towards Lake Turkana can be traced (Fig. 10.1). Near the south end of the lake-bed there are some small volcanic hills which are islands in times of flooding. On these to a height of about 20 m above the bed there are extensive algal limestone lake deposits containing shells of aquatic mol-luscs, especially of the Nile oyster Etheria elliptica, 14C dated to 5 000–6 000 years B.P. (Grove et al., 1975). At that time a greater rainfall in the Ethiopian highlands, descending perhaps by the Bilate River, must have maintained a high enough water-level in Chew Bahir for a copious overflow into Lake Turkana, which itself was then at a much higher level and overflowed via the Sobat valley into the Nile (p. 179), giving a free passage for the Nile fauna up into the Ethio-pian Rift as far as Lake Abaya.

Little can be said at present about the biological history of the small Lake Awasa between the Galla Lakes and Lake Abaya (Fig. 10.3). Its drainage is sepa-rated from both by volcanic watersheds and its main inflow comes through the swampy Lake Shallo. Parenzan (1939) reported the fish Clarias mossambicus and Barbus gregorii in an inflow, but none in the lake itself. The salinity (conductiv-ity) of Lake Awasa is not unduly high (Table 10.2) – less than half that of Lake Turkana. We need to know more about the fauna of Lake Awasa and its past history.

Research on water circulation and primary production in some of these lakes and in the Bishoftu crater lakes (Fig. 10.3) are discussed on Chs 6 and 7.

Lake Tsana†

Lake Tsana was the first of the lakes in the Nile Basin known to Europeans. It was in fact visited by Portuguese travellers in the seventeenth century. It has the general appearance of a large shallow mountain 'tarn' (about 3 000 km² in area, mean depth 9 m) lying on a platform between mountains with an outflow to the south east which, within 30 km of the lake, drops over the spectacular Tissisat Falls (Fig. 11.3). The Great Abbai River, or Blue Nile, then makes a great U bend round to the northwest and descends from the highlands down a magni-ficent gorge, about 350 km in length, onto the Sudan plains (Morris et al., 1976), and so to the Nile at Khartoum – a total drop of about 1 450 m.

* Amongst more than twenty species are also the well known soudanian fish Bagrus docmac, Synodontis schall, Schilbe mystus, Mormyrus longirostris and Hydrocynus forskali (Parenzan, 1939). All of these are absent from the Galla Lakes.
† Though now commonly called Tana, 'Tsana' is used here to distinguish it from the River Tana in Kenya. The correct Amharic version is in fact 'Tsana' by which it was generally known before this century (Garstin, 1911).

Lying at the head of the Blue Nile whose floods (with those of the Atbara) are the main cause of the seasonal rise and fall of the lower Nile (p. 155), it might be thought that Lake Tsana plays an important part in the hydrology of the Nile. This however is not so. The outflow, even during the flood, makes a relatively small contribution to the Blue Nile, which gets most of its flood water from the subsequent tributaries from the southern Ethiopian Highlands (Hurst, 1957, p. 95). It is in fact a shallow swelling on the course of one of the many tributaries of the Blue Nile. Many years ago there was a plan for a dam by which the lake would become a more effective controlling reservoir for the river, especially during periods of low rainfalls (Hurst, 1957, p. 309).

The Blue Nile is an older river than the White Nile (p. 154) and the Tsana basin is thought to have originated during the Pliocene (Mohr, 1962) though Morandini (1940, p. 55), stated that the volcanic blockage which formed the lake itself was very recent.

The first limnological work was done in the 1930s by Italian scientists and published by the Italian Royal Academy (Missione di Studio, 1940). This and the small amount of more recent work were briefly reviewed by Rzóska and Talling (Rzóska, 1976a, Chs 14 and 27).

At an altitude of 1 829 m and latitude 12°N the water temperature (15–20°C) is lower than in Lake Chad (23–29°C) at the same latitude but much lower altitude (283 m). Though thermal stratification is frequent during the day, the whole water column, as in Lake Chad, is otherwise kept in circulation and the lower water is well oxygenated (Morandini, 1940; Bini, 1940a).

Chemical analyses of the water were done in the 1920s and 1930s and were set out by Talling and Talling (1965, Table 1). There were no measurements of conductivity, sodium or potassium. Assuming no significant amount of dissolved organic matter, the figures given by Bini (1940a) for total dissolved solids (150–170 mg/l) would represent a salinity range of 0.15–0.17‰. When compared with other lakes (e.g. Table 5.1) these figures give no support to Bini's contention that the water is particularly poor in salts, which he suggested would support the geological evidence for the very recent origin of the lake as well as account for its apparently low-productivity as judged by the relatively small number of plankton species and low biomasses. No quantitative measurements have however been published on the latter. The salinity is in fact about that of the very productive Lake George. The general level of pH (7·8–8·0) is much lower than in Lake George (8·5–9·8), which must, in part, be a reflection of the greater activity of the phytoplankton in the latter. Bini was unable to detect nitrate in the water, though this is not in itself a certain indication of the rate of circulation of nitrogen. If however further work confirms the low rate of production, this is likely to be caused by a scarcity of one or more nutrients from external sources. The low temperature must also be considered. A study of production rates at all levels to compare with that made on Lake George would surely increase our understanding of the controlling influences in shallow lakes.

The first general account of the aquatic fauna and flora was that of Brunelli and Canicci (1940), and some further points were added by Rzóska and Talling (Rzóska, 1976a). *Cyperus papyrus* and other *Cyperus spp.* are abundant along the shores, especially at the south end. Most however of the lake surface is open and devoid of emergent vegetation.

As usual the organisms of the greatest biogeographical interest are the molluscs and fish. Of the 15 species of molluscs collected by Bacci (quoted by Rzóska, 1976a, p. 229), 5 are of palaeartic (Eurasian) origin. Fossils of the latter, which include the bivalve *Unio abbysinicus* show that they were widespread in the White Nile until the very late Pleistocene. They were perhaps eliminated by a rise in water temperature as the recent dry conditions developed, but survived in the cooler climate of the Ethiopian highlands. One subspecies only, *Bellamyia unicolor abyssinica*, is endemic to the lake. A peculiar feature is the complete absence of the genera *Cleopatra*, *Pila* and *Lanistes* which are common in most of tropical Africa. Bini (1940b) recorded 31 species and subspecies of fish in the lake, but of the 25 families in tropical Africa only 3 are represented (Cyprinidae, Clariidae and Cichlidae). There is only one cichlid, *Sarotherodon niloticus*, and three clariids of which one, *Clarias tsanensis*, is endemic to the lake.

He reported 25 species of *Barbus* of which 2 species and 12 subspecies were thought to be endemic. Recent examination of this and other collections by Banister (1973, pp. 47–75) has shown that, though there is much individual variation, all of them can reasonably be regarded as belonging to one species *Barbus intermedius*, which is widely distributed in the Ethiopian highlands and as far south as northern Kenya. The number of species of fish at present recognised in the lake is therefore 7.

For many years it has been accepted (e.g. by Pellegrin, 1911b, and Boulenger, 1916, II, p. 217) that the loach *Nemachilus abbysinicus* (*Cobitidae*) lives in Lake Tsana. This family has not been found elsewhere in Africa. If true, it would have to be included on p. 2 among the zoogeographical wonders of Africa. The record originated from the reported finding of one specimen among a large number of fish collected in 1902 and deposited in the British Museum of Natural History (Boulenger, 1902). It is remarkable that this record has been generally accepted without further confirmation. Bini (1940b, p. 142) made a great effort to find it, using all possible methods of capture including dynamite. He failed. We cannot therefore include *Nemachilus abbysinicus* in the list of Lake Tsana fish without more persuasive evidence.

The restricted mollusca and fish faunas suggest a history of at least partial isolation. Rapids and falls on the course of the river between the lake and the Sudan plains have probably existed for a very long time. But much remains to be discovered about the existing fauna of the lake, and we hope that some fossil evidence will be found in the lake basin. Sediment cores from the lake itself would certainly provide some interesting information.

11

The Sahara

The soudanian region (p. 144), based on the distribution of fish, can be roughly defined as the whole of Africa north of the 5th north parallel of latitude or approximately a line 200 km north of the Guinea Coast prolonged eastward, exclusive of Mediterranean North Africa (Maghreb) and the Eritrean and Somali plains in the east. This can be regarded as a single zoogeographical region with respect to the inland water fish fauna.

There are several species of fish which are common to all six of the major basins of this region: Senegal, Gambia, Volta, Niger, Chad and the Nile. Examples are *Polypterus senegalus* (Polypteridae), *Hydrocynus forskali* and *Alestes nurse* (Characidae), *Citharinus citharus* (Citharinidae), *Malapterurus electricus* (Malapteruridae), *Sarotherodon galilaeus* (Cichlidae) (Fig. A4.1).

Several other species, though widely distributed over the region, are curiously missing from certain basins. *Lates niloticus* (the Nile perch) has not yet been reported from the Gambia River, though it is found in the other five basins. There are a few species which are common to this region and to other parts of Africa such as the catfish *Bagrus docmac*, *Malapterurus electricus* and *Sarotherodon niloticus* in Lake Tanganyika and *Schilbe mystus* and *B. docmac* in Lake Victoria. There are also a number of species which are endemic to some of the basins, especially those of the West African rivers flowing into the Gulf of Guinea.

Nevertheless, the fish fauna over this enormous area is remarkably uniform, and this is all the more surprising when it is realised that there are now more than 1 600 km of desert separating the Nile from Lake Chad which is the nearest of the other soudanian basins.

The key to this situation lies in the past history of the Sahara Desert whose climate has more than once during the Pleistocene been more humid than at present. The last major occasion was during the neolithic period not more than 8–10 00 years ago. These climatic fluctuations have already been mentioned in Ch. 3, but more evidence will now be presented to explain the present distribution of the faunas of the water systems bounding this region and of the few remaining standing waters in the Sahara itself.

The reader who wishes to dig more deeply into the fascinating prehistory of the

Sahara may refer to Gautier (1928), Capot-Rey (1953), Mauny (1956, 1978), Lhote (1958, 1959), McBurney (1960), Monod (1963), Moreau (1966, Ch. 3), Faure (1969) and Schiffers (1950, 1971–73). In these he will find discussions of the major problems, geological, biological and archaeological, and references to the most important original works. Much of the research in the western and central Sahara has been done by French scientists, originally based on North Africa. More recently the most important centre has been the Institut Française d'Afrique Noire at Dakar (I.F.A.N., since 1966 renamed Institut Fondamental d'Afrique Noire), founded in 1938 under the direction of Théodore Monod. It was the first of the Centres for African Studies now established in several of the independent countries. Unlike most of them which are devoted mainly to archaeology, history, anthropology and sociology, I.F.A.N. has also strongly promoted biological studies as can be seen from the annual Bulletin and occasional Mémoires which include publications on tropical Africa other than the Sahara. The Institute has also set up substations including one devoted to limnology at Diafarabé on the Middle Niger (p. 158) and at N'Djamena for Lake Chad (Ch. 12).

There is much purely geological evidence pointing to recent higher rainfall at least in the western and central Sahara. The most dramatic is provided by the ancient watercourses which could certainly never have been formed under present climatic conditions. They are very obvious in the southern slopes of the Atlas and in the central mountains, but can be traced across vast expanses of open desert in aerial and spacecraft photographs (Fig. 11.1). Some of these normally dry watercourses ('oueds' in the western and 'wadis' in the eastern Sahara) are flooded annually in their upper reaches for a short period by rains in the hills above, but many can be quite dry for several years, especially those draining the central mountains, and then can be suddenly inundated by a few days of torrential rain. It has sometimes been said that death in the Sahara may result either from thirst or from drowning, and there was a strict regulation for the French Saharan troops forbidding them to camp in dry canyon-like oueds between rocky cliffs which are common in the mountainous regions. They may be flooded without warning by sudden and violent torrents.

One of the most important of the Saharan water courses is the Oued Saoura which rises in the Moroccan Atlas and flows at least once a year some 500 km across the desert to end in a salt-pan (sebkha') (Fig. 11.3). This thread of well-watered land has been for centuries the northern section of one of the great trans-Saharan routes. The Oueds Irharhar and Mya take occasional flood water northwards across the desert from the Tassili-n-Ajjer, the Ahaggar Mountains and the Plateau of Tadmait. But their common lower valley, is always dry, the water continuing underground and eventually seeping to the surface in the large salt-pans ('chotts') at the foot of the Saharan Atlas. This provides, *en route*, artesian water to the oases in the area of Ouargla, Touggouert and Biskra (Fig. 11.3).

These water courses were clearly formed in times of much heavier and more continuous rainfall than at present, a conclusion which is supported by extensive ancient fluvial deposits. A good series have been examined in the valley of the Oued Saoura (Monod, 1963, pp. 133–40) which can be approximately dated from the human stone tools found in them. There is here evidence of three main

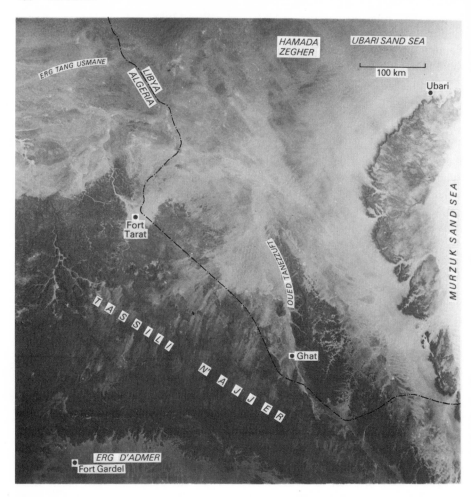

Fig. 11.1 Old Pleistocene water courses traversing Palaeozoic sandstone rocks of Tassili n'Ajjers. Gemini XI pilots' photograph No. S-66-54773 Alt. 326 km

wet periods during the late Pleistocene, the last contemporary with a neolithic culture some 5 000–8 000 years ago. Archaeology has, in fact, amply corroborated geological evidence for more humid periods in the Sahara during the Pleistocene, the most striking evidence coming from the astonishing number of rock paintings and engravings scattered over the whole area but concentrated especially in the central mountains (Ahaggar, Tassili-n-Ajjers, Tibesti and Ennedi). In these are depicted elephant, buffalo, rhinoceros, antelopes, hippopotamus and domestic cattle showing how different was the climate and vegetation of this now desert area no more than a few thousand years ago (Lhote, 1958, 1959; Monod, 1963; Lajoux, 1963; Hugot, 1974). In Fig. 11.2 is a rock painting from the Tassili of a man in a boat and a hippopotamus, dating from probably less than 5 000 B.P. The neolithic cultures of the Sahara seem to have reached a high level before the beginnings of the great agricultural civilisations of the Middle East and Egypt (Sutton, 1974). The desiccation of Northern Africa during the last few thousand

Fig. 11.2 Part of a painting in red ochre in a rock shelter at Aouanrhet about 50 km southeast of Djanet in the Tassil n'Ajjer Mountains, Central Sahara. The whole painting represents three canoes and three hippopotamus – presumably a hippopotamus hunt. From a late period of the Tassili frescoes, probably later than 5 000 B.C. (From Lhote, 1958)

years B.C. was punctuated by advances and recessions both in the Central Sahara and in the Sahel belt south of the present desert (Schiffers, 1971; Dumont, 1978). Eventually however the Saharan neolithic cultures were extinguished without having reached the stage of true agriculture in spite of their technical skills, powers of observation and aesthetic sensibility. It has even been surmised that, but for the late Pleistocene drought in northern Africa, the course of human history might have been different (Hugot, 1974).

In addition to flowing water courses there were large lakes in existence at different times up to a few thousand years ago. Ancient shorelines and sediments with fossil fish, molluscs and diatoms, some of which have been carbon-dated, testify to the previous existence during the late Pleistocene of lakes across the present desert between 14° and 22°N latitude (Faure, 1967, 1969). Several old lake sites in now utterly desert regions, such as Araouane, have yielded bone fishing harpoons and hooks as well as middens (prehistoric refuse heaps) with remains of Nile perch (*Lates*) and other fish (Monod and Mauny, 1957) (Fig. 11.3).

To quote from Faure (1969):

> Large lakes were present 22 000 years ago in the now dry Niger desert. They were still extensive between 10 000 and 8000 years B.P., but towards 7000 B.P. drought set in, and rapid evaporation resulted in salt deposition in the remaining basins (Bilma, Fachi), so that limestones and soluble salts were laid down on diatomites. The deposits were subsequently eroded by sporadic rains. These rains were still sufficient about 5 000 years ago to enable a fairly large neolithic population to live in the Ténéré area, and even in 3350 B.P. hippopotamus were still present west of Fachi.* Ponds persisted even longer in low-lying areas where the reworked material deposited locally was enriched with sodium sulphates, chlorides, and carbonates. The evaporation of the lake surfaces was much more rapid than that of underground water from surrounding areas, so that the watertable was depressed beneath these basins, and pure water springs were formed in oases close to salt deposits at their lowest points.

The area referred to is not in the River Niger basin, but is the extreme desert country of Niger north west of Lake Chad (Fig. 11.3).

Abundant evidence has come from Lake Chad of former much higher water levels (Ch. 12). In the desert northeast of the lake (Borkou), and particularly around Ounianga just southeast of the Tibesti Mountains, fossil evidence has been found of extensive surface water during the Pleistocene (Monod, 1963, pp. 150–3). Widespread alluvial deposits in the now complete desert in this region have yielded pollen of a previous Mediterranean tree flora and the remains of several tropical African animals such as the elephant *Loxodonta africana*, *Hippopotamus amphibius*, *Crocodilus niloticus*, the aquatic tortoise *Trionyx* and a number of soudanian fish such as *Bagrus (Porcus) docmac*, *Lates niloticus* and *Synodontis* sp. Some of the *Lates* (Nile perch), judging from their vertebrae, must have exceeded one metre in length and could have lived only in fairly large and well-established lakes (Daget, 1958, 1959b, 1961). The very significant find of a small live specimen of the Nile crocodile (*Crocodilus niloticus*) was made in 1924 by a French officer Lt Beauval in a stream pool in the Tassili-n-Ajjer Mountains in the central Sahara. The French expedition in 1928 (Mission du Hoggar) found the Eurasian green frog (*Rana ridibunda*) in streams in the Ahaggar Mountains, which are, of course, now separated from the nearest possible habitats for these animals by great expanses of extreme desert (Seurat, 1934, pp. 39–40; Gauthier in Seurat, 1934, pp. 73–4).

From the distribution of the present soudanian aquatic fauna it is impossible to doubt the reality of water connections from time to time during the late Pleistocene across the whole region from the Nile to Senegal. As we shall see (Ch. 12), the composition of the fauna of Lake Chad can only be explained on the assumption of recent connections, not only with the Niger and Zaïre basins whose

---

* The late Pleistocene distribution of the hippopotamus, shown by fossils, rock drawings and historical records, provides an elegant demonstration of the previous extent of standing water in the Sahara (Fig. 11.4).

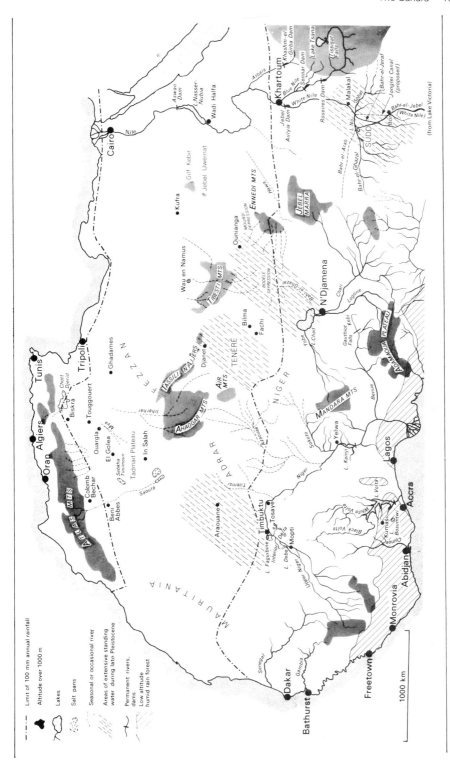

Fig. 11.3 Africa north of latitude 5°N

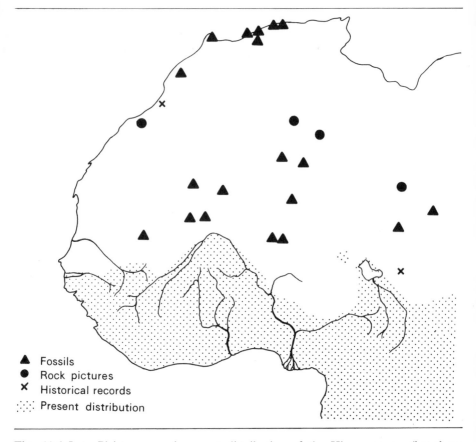

Fig. 11.4 Late Pleistocene and present distribution of the Hippopotamus (based on Mauny, 1956)

upper waters share watersheds at the present time with the inflows to Chad, but also and especially with the Nile. There are now more than a thousand kilometres of complete desert between Lake Chad and the nearest even occasional inflows into the Nile. We have so far no geological or palaeontological information to indicate where the connection might have been. There are two apparent possibilities:

1. to the northeast of Lake Chad and through the gap between the Erdi Plateau and Ennedi via the Mourdi Depression;
2. by the Upper Chari valley the present main inflow to the lake, and either the Wadi Hawa or Bahr-el-Arab (Fig. 11.3).

For a discussion of this problem see Monod (1968a).

Of possible significance in this connection are the reports by Sandford (1933, 1936) of sub-fossil mollusc shells at widely separated points in the southern Libyan Desert. These are mainly of tropical African origin and include the terrestrial snail *Limicolaria* sp. and the aquatic snails *Planorbis* (*Biomphalaria*) *pfeifferi* widespread in tropical Africa, *Bulinus vaneyi* now living in Lake Chad, *Melania tuberculata*, *Lymnaea caillaudi* closely related to the tropical species *natalensis*, the tropical amphibious snail *Pila ovata* and the bivalve *Aspatharia caillaudi*. *Pila*, which was actually found alive in the upper Wadi Hawa, and *Aspatharia* spp. are well known for their ability to survive long periods in dry mud, and species of both *Biomphalaria* and *Bulinus* have been reported to do so (p. 351). The significance of these finds and of the vegetation on the sites in various stages of regression is not that they relate directly to any major humid period in the Pleistocene – they are much more recent – but that they demonstrate how a relatively very small increase of rainfall can transform this country from extreme desert to an environment capable of supporting at least a temporary aquatic fauna in scattered pools and small lakes. This has happened several times in the very recent past and the aquatic molluscs have apparently been recruited from reservoirs in the higher and better-watered country to the south and southwest. Humans have taken advantage of these temporary ameliorations of climate and have migrated back and forth across the area. It has been suggested that they could have transported in water-containers some of the smaller snails (e.g. *Bulinus* and *Biomphalaria*) to the more remote oases.

The climatic change required to produce a water connection to enable fish to move between the Nile and Chad basins, which probably existed in the neolithic period when the upper Nile was expanded into a much larger area of open water or swamp than at present (Fig. 11.3), was thus not as drastic as might have been supposed. The climate of the Sahara, at least since the early Pleistocene, seems to have been rather insecurely poised between the biologically favourable and the almost impossible. There is clearly much of great interest to be discovered from further and more detailed investigations in the remote region of southwestern Sudan from which it may be hoped that the position and nature of the ancient Nile–Chad connection will emerge.

The eastern Sahara north of Khartoum (North Sudan, Egypt and Libya) is more thoroughly eroded to desert, and ancient water courses have mostly been obliterated. Nevertheless, there is much archaeological evidence of higher rainfall at times during the Pleistocene (McBurney, 1960). The abundance of stone artifacts widely scattered, even in the central area between Kufra Oasis and the Nile which is now uninhabitable, the neolithic rock paintings of domestic cattle in the Gilf Kebir and Uweinat hills, and the late Pleistocene fluvial deposits from previous much higher water levels in the Nile valley; all these point to the same conclusion.

Another region with a very interesting past history is the Upper Niger and the desert northwest of Timbuktu. In the early Pleistocene the upper river (above Timbuktu) may have flowed westward to the Gulf of Senegal, but in the late Pleistocene (neolithic) it flowed in its present direction continuing northwards to terminate in the lakes whose alluvial deposits are scattered over a wide area of Araouane that is now extreme desert (Fig. 11.3). The group of small lakes and swamps, including Lake Faguibine and Debo, south and west of Timbuktu are presumably the remains of this flooded area which has been kept in existence by the upper Niger (Ch. 9). The Araouane deposits contain the remains of a varied freshwater fauna and flora, many species of which are now living in tropical African inland waters. Among the molluscs there were *Lymnaea natalensis* and *Biomphalaria gaudi* (Pulmonata); *Melania tuberculata*, *Vivipara multicolor* and

Cleopatra cyclostomoides (Prosobranchiata), *Mutela rostrata* (Lamellibranchia). The aquatic reptiles were represented by *Trionyx triunguis* and *Crocodilus niloticus*, and the mammals by *Hippopotamus amphibius*. There were several soudanian species of fish including *Lates niloticus*, some of whose vertebral centra were up to 5 cm in diameter. There were very many species of diatoms most of which are now living in the existing African lakes, and the stems and rhizomes of the reed *Phragmites* testify to a well-established aquatic flora (Monod, 1958, 1963). In the late Pleistocene the middle and lower Niger formed a separate system draining the southern slopes of Adrar and Aïr which then had a higher rainfall than now (see Fig. 11.3). It is thought that Lake Araouane eventually rose high enough to break through to the southeast to join the Middle Niger, thus deflecting by this route the waters of the upper river. Alternatively the lower river may have cut back through the barrier (Seuil de Tosaye) which retained the upper lakes, whose waters were in this way captured and diverted through the gap at Tosaye into the Lower Niger. The subsequent desiccation destroyed the Araouane Lakes and reduced the original headwaters of the Middle Niger (Oued Azaouak and others) to occasionally inundated river beds (Voute, 1962).

Apart from a minor recovery about 3 000 years ago (evidence from the Chad basin, Servant and Servant, 1970) the past five thousand years have thus seen a progressive overall desiccation. The Great Saharan Lakes have gone, many of the water courses are permanently dry or very rarely inundated. With the exception of the Nile which carries eastern equatorial water to the Mediterranean, and the Niger taking water from the western humid tropics along the southern edge of the desert, there are now no rivers in the region that flow continuously and find an outlet. The remains of permanent streams in the mountains and their occasionally flooded lower courses all drain into closed basins. Nevertheless, though the rainfall in the desert itself is very low and spasmodic, a very large quantity of water from the southern slopes of the Atlas, from the central mountain masses and from the highlands of west and central tropical Africa enters the desert and escapes evaporation by sinking underground. There are a number of immerse subterranean basins under the desert filled with alternate layers of porous and impermeable rocks that date back as far as the Palaeozoic. The water in all but the uppermost (most recent) strata is under high pressure and squirts up to a considerable height from artesian borings. Some of the water in these ancient layers is 'fossil', i.e. it comes from the period of original deposition, but it has subsequently been, and continues to be, augmented by rainwater percolating into the exposed strata round the edges of the basins, that comes mainly from the northern and southern highlands and the central mountains. Much rainwater now flows under the old Pleistocene river beds such as the Saoura and Irharhar which are marked by a chain of oases fed by subterranean water within easy reach. But many water courses have been eroded out of recognition or submerged under enormous masses of sand transported by winds from former alluvial deposits or wind eroded rocks. These immense 'sand seas' or 'ergs' cover thousands of square kilometres, are frequently blown into serried ranks of dunes and are practically devoid of vegetation and natural surface water. But very large quantities of water are in places buried beneath them and can be reached by digging or boring at suitable points. In many old river beds water can be got by digging to

less than 5 m, but under sand dunes it is often necessary to descend more than
50 m. Both the depth and the chemical composition of the water depend upon
the position and nature of the underlying rocks*.

At certain points the slope of the lower rocks brings the water to the surface as
a spring, and a pond or rockpool ('guelta') may be formed. These are particularly
abundant in the Fezzan in western Libya and in the adjacent region of the Algerian Sahara where the many oases are fed with subterranean water from the Tassili-n-Ajjers in the south. Beyond Ghadames, where it nourished an ancient Roman trading post, this water flows under the Great Eastern Erg and joins the
subterranean Oued Rhir on the far side and provides irrigation for the most productive of the Algerian date plantations (palmeraies). The largest remaining
permanent standing waters in the Sahara are at Ounianga Serir. (4·2 km²) and at

Fig. 11.5 Aerial photograph of Ounianga Serir, a salt lake; the largest (over 4 km²) existing lake in the Sahara. See Fig. 11.3 (with acknowledgement to Dr A. Pesce)

* For accounts of Saharan hydrology and water resources see Capot-Rey, 1953; Ambroggi, 1966;
  Klitsch, 1967, 1971; Schiffers, 1967.

Ounianga Kebir (3.7 km^2) southeast of the Tibesti Mountains (Fig. 11.5). The water is strongly saline and alkaline. Wau en Namus ('That Place of Mosquitoes') in the southern Libyan desert about 400 km north of the Tibesti, is perhaps the most isolated of standing waters in the Sahara (Fig. 11.6). There is a ring of large saline pools surrounding the central cone of a volcano. On all sides are vast expanses of extreme desert.

Many of the submerged drainage courses terminate in a salt-pan ('sebka' or 'chott') where there is a slow upward seepage and evaporation at the surface. The amount of water reaching the pan often varies in response to spasmodic or seasonal fluctuations of rainfall in the highlands above, so that a film of water may actually appear at the surface or, as the Sebkha de Timmimoun at the southern edge of the Great Western Erg, a fresh sparkling layer of crystals is periodically deposited as the delayed result of increased rainfall in the Saharan Atlas and the

Fig. 11.6 Aerial photograph of Wau en Namus, 'That place of Mosquitoes'. Small lakes surrounding a volcanic ash cone in the centre of the southern Libyan desert (with acknowlegement to Dr A. Pesce)

consequent greater flow of water down the old river beds which are submerged beneath the sands of the Erg.

During the more humid neolithic period the human population of the Sahara, especially the western half, was obviously much larger than now. The manner in which *Homo sapiens* has managed to maintain a foothold over the past five thousand years and, indeed, to establish empires and to conduct a flourishing trans-Saharan trade between the Mediterranean and the Soudan, under conditions of progressive desiccation and diminution of the means of sustenance, is a remarkable demonstration of human adaptability (see Gautier, 1928; Bovill, 1968a; McBurney, 1960, and Ch. 2).

Survival has depended primarily upon efficient exploitation of the ever-dwindling water resources for the irrigation of crops such as dates and cereals. The techniques that were evolved show great ingenuity from primitive mechanical devices for raising water from holes and shallow wells to deep artesian borings releasing high pressure water at the surface. Most remarkable is the system known as the 'foggara'. This is designed to tap the water descending through alluvial deposits from relatively high ground, as, for instance, the Tadmait Plateau (Fig. 11.3). Though differing in detail, the essential feature is a near-horizontal tunnel, sometimes several kilometres long and dug from a line of vertical shafts, with a descending gradient less than that of the water table and thus emerging at the surface. The water flows out continuously, though rather slowly, to irrigate and maintain an oasis. The idea of the foggara is thought to have originated in Iran where it is still commonly in use. Though most of the Saharan foggaras have now fallen into disuse, there are still a number in action especially in the northwestern Sahara.

During the past five thousand years of gradual desiccation men (negro, berber and arab) have inhabited the Sahara and have, so to say, followed the water as it diminished and disappeared underground.

It was at one time thought that the present reserves of underground water were largely derived from the last (neolithic) period of heavier rainfall and that the inhabitants of the Sahara have subsequently been living on a dwindling capital resource doomed to ultimate exhaustion. However, modern methods of prospecting have revealed a much less gloomy future. There is, for instance, a very large subterranean basin in the El Golea region of the western Algerian Desert. The volume of water now annually entering this basin from contemporary rainfall in the Saharan Atlas and on the Western Erg is much greater than the annual consumption in the oases (making due allowance for evaporation *en route*). The same conclusion has been drawn regarding the water budget of the large subterrnean basin south of the eastern end of the Saharan Atlas in the region of Touggouert from which the most important of the Algerian date plantations are irrigated. This is fed from the rains on the southern slopes of the Atlas and by underground drainage from the Ahaggar Mountains in the far south and from the Tadmait Plateau in the southwest.

The rate of loss by evaporation in the desert is much lower than might be expected. Owing to the high porosity of the sands, especially of the dunes, a large proportion of the water from a rare fall of rain rapidly sinks below the surface and is thus protected from evaporation. It appears, in fact, that the present rainfall in and around the Sahara is more than sufficient to balance evaporation before the water sinks underground, and to provide for present needs, and that there is no need for invoking 'fossil water' to account for the volume of the reserves. Modern techniques for deep boring have however greatly increased the capacity for human occupation of some parts of the Sahara. From now on we may expect an ever-increasing drain on the subterranean reservoirs, especially in the region of the Libyan oilfields.

It is clear therefore that the world's greatest expanse of arid desert is of considerable interest to the limnologist. The hydrology and biology of the present major river basins of northern tropical Africa can only be understood in the light of the recent hydrological history of the Sahara, and the many small and scattered patches of water to be found in the desert itself are of scientific and human importance out of all proportion to their size.

It was a surprise to zoologists when French expeditions during the early years of this century discovered that the Sahara supports a considerable, though widely scattered, freshwater vertebrate fauna (see Pellegrin, 1911a). Later expeditions to the central Sahara have shown very clearly that the fish in the widely scattered and isolated small patches of water are the remains of the former tropical African soudanian fauna that had penetrated as far north as the foot of the Atlas Mountains during the late Pleistocene. Remnants of this fauna are now living in isolated but permant pools from Biskra in the north to Mauritania in the southwest and to the mountains of Ahaggar, Tassili-n-Ajjer and Tibesti in the south. Pellegrin (1921, 1931), Estève (1949), Monod (1951, 1954, 1963) and Daget (1959a) are representative of the many publications on this subject. In pools near Biskra within sight of the Algerian Atlas there are *Tilapia zillii*, *Astotilapia (Haplochromis) desfontainesii* and *Clarias lazera* – species that are very widespread in tropical Africa. *Sarotherodon galilaeus*, whose distribution includes much of the soudanian region from the Nile to Senegal, has been found in the central Sahara as far north as Adrar near the Ahaggar Mountains. The cyprinid fish *Barbus apleurogramma*, known from Lakes Victoria and Tanganyika and other waters south of the Equator, has been found in the Ennedi (Fig. 11.3), where the tropical African clawed toad *Xenopus mülleri* was also discovered (Monod, 1968a). The freshwater prawns of the family Atyidae, which are more water-bound than most freshwater invertebrates, have entered the desert from both directions. The northern species *Atyaëphyra desmaresti* is found isolated in the Saoura Valley (northwestern Sahara), and the tropical species *Caridina africana* now lives in springs in the Ennedi (Monod, 1968b).

In view of the present extreme isolation of the habitats, it is rather surprising that the Saharan fish should have diverged so little from their soudanian stock. Most species, such as those mentioned above, are apparently identical with their relatives outside the desert. There are, however, some that have become detectably different and a few that have diverged enough to be given specific rank. For example, *Sarotherodon borkuanus*, endemic to the desert region south and southeast of the Tibesti Mountains, seems to have evolved from *S. galilaeus* to which it is very closely related*. *Labeo tibestii*, endemic to the Tibesti-Ennedi region,

---

* *E. Trewaras* (pers. comm.) considers this fish to be a subspecies (*borkuanas*) of *S. galilaevs*.

appears to be another example of Saharan speciation. But on the whole the evolutionary effects of isolation are much less than are shown by the soudanian fish in some other parts of tropical Africa, especially in the Great Lakes other than Chad. This is presumably a reflection of the very recent desiccation of at least the western half of the Sahara for which there is much other evidence already summarised in this chapter. Perhaps, also, the habitats present a range of conditions too limited to stimulate many new adaptations.

Throughout the Pleistocene at least, as at the present time, the Atlas watershed is a barrier between the soudanian fish in the desert and those of Eurasian origin in the Maghreb. But *Barbus biscarensis*, 'Barbeau de Biskra', which appears to be most closely related to *B. callensis* of coastal Algeria, is found as far south as the Ahaggar in central Sahara (Pellegrin, 1931; Estève, 1949). The small cyprinodont *Aphanius fasciatus* is another anomalous case. Well known in fresh, saline and thermal waters in southern Europe and coastal North Africa, it has penetrated only the northern fringe of the Sahara as far as the region of Touggouert. Why should this most highly resistant and adaptable fish have failed to move further south? Two Eurasian species of Amphibia, the frog *Rana esculenta* and the toad *Bufo viridis* have been recorded from the Ahaggar. But no tropical species of fish or amphibia have been found north of the Atlas Mountains.

The Sahara is no barrier to the north–south–north migration of birds. Even palaearctic species of aquatic birds (from Eurasia) such as ducks, herons, tern and waders, have been recorded wintering south of the desert in West Africa, and there is good circumstantial evidence that more than fifty species cross annually, and that many do so without resting *en route* (Moreau, 1967). This extraordinary phenomenon is noted here because these birds normally live in and around water and because they provide a great contrast to the restricted and time-consuming movements of the more water-bound fish.

The distribution of the aquatic molluscs both now and during the Pleistocene is rather similar to that of the fish. Some palaearctic species have reached more than half way across the western Sahara and in the region just north of the Ahaggar there was a mixture of the two faunas in the Pleistocene. For example, shells of *Bulinus truncatus* (tropical African) are found with those of *Lymnaea truncatula* (Palaearctic). The palaearctic molluscs along the coast east of Tripoli have failed to move into the Sahara. The Pleistocene and living species in the Fezzan and the Kufra Oasis are all tropical African (Germain, 1932, Fischer-Piette, 1948; Sparks and Grove, 1961).

The fish, amphibia and molluscs of the Sahara have thus come mainly from tropical Africa. The northern (palaearctic) species are relatively few and have penetrated a comparatively short distance into the western Sahara only. These groups of animals are incapable of moving great distances away from water. The distribution of the other aquatic invertebrates that can be transported vast distances without water cannot be expected to provide such a clear picture. But insufficient evidence is available except for the Crustacea, which were studied over a wide area of the western Sahara by Gauthier (1928, 1929, 1938). This work led to the conclusion that the occurrence and distribution of the different groups and species of Crustacea cannot be related to north–south movements of tropical and Mediterranean faunas. Many have a wide range outside the desert, and it is well known that some are spread as eggs over great distances by birds. No species endemic to the Sahara had been found. The kind of crustacean fauna seemed to be determined mostly by the nature of the environment, particularly by the degree of permanence of the water rather than by history and geography. In the temporary pools fed by intermittent rains, often at very rare intervals, there are certain cladoceran and copepod species as well as euphyllopods (e.g. *Triops*, *Estheria* and *Streptocephalus spp.*) whose life-history cannot be completed without occasional desiccation of the eggs. In permanent pools, such as the 'gueltas' which are usually protected from evaporation in rock clefts on the course of dry river beds, there are no euphyllopods and the species of Cladocera and Copepoda are characteristic of more permanent waters.

The above general conclusion concerning the distribution of the aquatic Crustacea in the Sahara must now, however, be revised in the light of recent expeditions led by H. J. Dumont to several regions of the Western and Central Sahara (Dumont and Van de Velde, 1975; Dumont and Decraemer, 1974, 1977; and *pers. comm.* from H. J. Dumont concerning unpublished results). To quote a few examples – several tropical African copepods have reached the northern edge of the desert, notably *Afrocyclops gibsoni* and *Thermocyclops neglectus*. Some are derived from the Mediterranean region such as the cladoceran *Alona elegans lebes*, and in southern Morocco was found a cave-dwelling harpacticoid copepod *Nitocrella ioneli n. sp.*, a genus previously known only from southern Europe. There is even a copepod genus *Metadiaptomus* (with two species *chevreuxi* and *mauritanicus*) that is endemic to the Sahara though widespread in the western and central region. It must be supposed that these north and south movements occurred during the late Pleistocene period of high rainfall. In veiw of the vast area of the Sahara and of the great number of very isolated and small patches of water we can confidently expect that more species of equatic invertebrates will be discovered and further light thrown on the history of the Saharan aquatic faunas.

In conclusion it can be said that, even under the present conditions, the Sahara supports, in aggregate, a considerable aquatic fauna and flora in very small and isolated patches of water.* Some of these, being fed from underground sources with fresh water, have escaped salinification and desiccation for several thousands of years and still harbour fish that were previously living in open lakes and rivers. Most, however, were subject to great increases in salinity both from direct evaporation and from leaching of salt deposits derived by evaporation of the ancient lakes. Periodic desiccation is the fate of most though, according to the situation, the life of a body of water may be anything between a continuous period of several years to a few days once in ten or more years. To live under such conditions requires special adaptations to high and changing salinity and to prolonged desiccation. Acceleration of development can be an advantage in the most transient of these waters (Ch. 19). To the limnologist, as to the archaeologist and historian, it is the hydrological changes in the Sahara since the mid-Pleistocene that are of particular interest. Not only are they fascinating in themselves but they provide an explanation for several curious biological features of the present water systems in northern tropical Africa.

---

* As a further and striking example, 324 species of algae were found in one pond in the Ennedi Mountains (Compère, 1970).

12

Lake Chad

Apart from its geographical position near the centre of the continent and on the southern edge of the Sahara Desert, Lake Chad has many peculiar and interesting features very different from those of the other great lakes of tropical Africa. It is traversed by the common boundaries between four countries – Niger, Chad, Cameroon and Nigeria. Most of the present research is based on the research stations of The Office de la Recherche Scientifique et Technique Outre-Mer, Centre O.R.S.T.O.M. at N'Djamena (Fort Lamy) south of the lake, and of the Federal Nigerian Fisheries Services at Malamfatori on the west coast of the lake. It is not possible to review adequately in one chapter all the recent and present work on Lake Chad. There is a general summary by Carmouze *et al.* (1972).

Unlike the geologically recently formed Lake Victoria and the lakes of the Rift Valleys, it lies in one of the primordial continental basins (Figs. 3.1, 3.2), and has been a focus for drainage from the surrounding highlands at least since the Cretaceous. It seems to have undergone relatively little deformation by earth movements since the Miocene, during which time eastern tropical Africa was distorted beyond recognition by the elevation and rifting that produced the lake basins. But it has been much affected by the great fluctuations of climate that have caused the continent at this latitude from Senegal to the Nile to swing more than once during the Pleistocene between absolute desert with blown sand-dunes and well-watered savannah (Grove and Warren, 1968; and Ch. 11). Since the basin is shallow, the consequent changes in area of the lake were very great indeed, and connections must have been established more than once with the contiguous basins of the Niger, Nile and Zaïre.

There are two other major peculiarities that will be discussed. First, though lying in an apparently closed drainage basin and subject to intense evaporation and reduced to a fragment of its former volume in the late Pleistocene, the water is fresh. Second, in spite of its present hydrological isolation from other water systems, almost all of the fish species are identical with those now living in the neighbouring 'soudanian' basins.

The Chad basin is some two and a half million square kilometres in area (Fig. 12.1). It is bounded by the Aïr and Tibesti Mountains in the north, by the

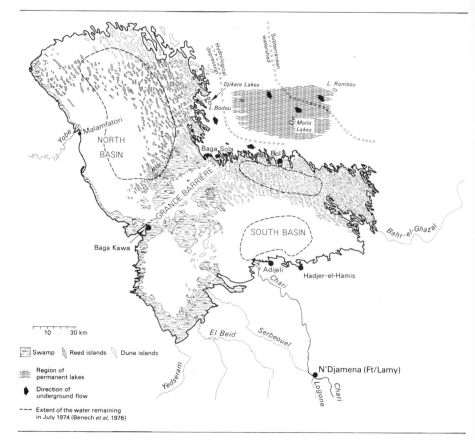

Fig. 12.1 Sketch map of Lake Chad in mid 1960s and the neighbouring salt-lake region (the latter redrawn from Maglione, 1969)

Ennedi and Jebel Marra in the east, and by the northern slopes of the Jos Plateau in Nigeria in the west. The main drainage comes from the Cameroon and Adamaoua highlands in the southwest and most abundantly from the Bongo Massif and the highlands of the Central African Republic in the southeast (Fig. 11.3).

The lake lies towards the western edge of the basin and, during this century (to 1971), it has fluctuated in area between 10 000 and 25 000 km². It is curious that the altitude of the lake surface (about 280 m) is greater than that of the lowest part of the basin, the Bodélé Depression (150 m) 500 km to the east. During the annual peaks between 1962 and 1964, when lake levels were the highest in the present century, water flowed down the Bahr-el-Ghazal leading to the Bodele Depression for a distance of over 50 km. Its further progress was stopped by the very gentle gradient of the water course and by the porosity of the substratum (pers. comm. from A. J. Hopson; Fig. 11.3).

The study of fossiliferous sediments and ancient shorelines around the lake has shown clearly that it has several times expanded and contracted during the Pleistocene (Pias, 1958; Grove and Pullan, 1963; see also Fig. 12.2). According to

Fig. 12.2 'Normal' Lake Chad. Note ancient dunes oriented northwest-southeast, forming promontories, islands and submerged banks (Gemini Space photograph, courtesy of National Aeronautics and Space Administration)

Servant and Servant (1970, 1973) there have been three major expansions of the lake during the past 12 000 years. The last occurred no more than 6 000 years ago. At its greatest extent 'Mega-Tchad' covered 300 000 to 400 000 km^2 and was about five times the size of the present Lake Victoria and thus vastly greater than any lake now in existence except the Caspian Sea. Obvious evidence of a recent period drier than at present is shown by the old sand-dunes north and east of the lake, the edge of which have been partially flooded to form the archipelago of elongated islands and submerged banks all orientated in northwest–southeast parallel lines. This curious feature is unique to Lake Chad and, of all the evidence of past changes of climate to be seen in tropical Africa, this is perhaps the most obvious and striking (Figs. 12.1 and 12.2).

The existence of Lake Chad was known in Europe from hearsay before the end of the eighteenth century and it was thought by some authorities to be a closed basin into which the water of the River Niger was finally discharged (Ch. 2 and Fig. 2.1). It was first found in 1823 by the Europeans Denham, Oudney and Clapperton, after crossing the Sahara from Tripoli. Since then there have been some large changes of level which were recorded by explorers during the last century. The information was put together by Tilho (1911, 1928) and is briefly summarised by Grove and Pullan (1963). There were at least three periods of high level between 1823 and 1870 at the end of which the lake was higher than at any later time up to the mid twentieth century. The early years of this century were marked by a series of very dry seasons, and in 1908, 1914 and 1940–45 the lake had shrunk so much that it no longer extended north of the Yobe River (Figs. 12.6, 12.7a). This progressive extension of the desert climate to the south was at one time thought to be evidence of a long-term expansion of the Sahara into the Soudan, which we now know to have occurred more than once during the Pleistocene. But after 1950 the level and area of the lake again increased and during 1962–64 became as great as ever formerly recorded. Then followed a period of continuous recession to a very low level after 1974.

'Normal' Lake Chad in the 1960s

Most of the research discussed in the folowing pages was done during the 1960s when the area of interconnected water, though falling slowly, did not fluctuate very widely from that shown in Figs. 12.1 and 12.2. In this condition it has become known as the 'normal' Lake Chad. The very drastic recession following a dry period in the early 1970s ('Little Chad') and the ecological consequences thereof will be considered at the end of the chapter. The following statements (up to p. 225), that are made in the present tense, refer to conditions during the 1960s.

As would be expected the rainfall on the lake and its immediate surroundings is at present low – about 200 mm/annum at the north and 500 mm/annum at the south end, with an overall average of 300 mm/annum (Gras et al., 1967, p. 30). This is only a fraction of the amount of water estimated to be lost annually by evaporation from the lake surface – over 2 000 mm (Dussart, pers. comm.). It should be noted for comparison that the climate of Lake Victoria is such that the direct rainfall approximately balances evaporation.

There are no significant inflows from the arid regions in the east and north, and the Yobe in the northwest with the El Beid in the southwest together supply no more than about 5% of the total inflow. The maintenance of the lake level therefore depends upon the water coming via the Chari and its tributary the Logone from the humid highlands of the Cameroons and Central African Republic in the south. The mean annual rainfall on the Adamaoua Plateau is more than 1 200mm (Grove, 1970b). It is the seasonal changes in rainfall in these remote regions that determine the fluctuations in water level that are so important to the ecology of the lake (see Fig. 12.4).*

---

* For a hydrological survey of the country immediately southwest of the lake see Barber (1965).

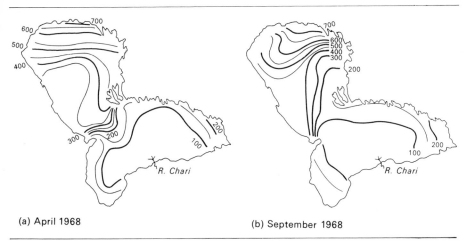

(a) April 1968   (b) September 1968

Fig. 12.3 Seasonal movements and changes in conductivity (K_{20} in μmhos) of water in Lake Chad. (*a*) at the end of the dry season (April 1968), and (*b*) in the middle of the wet season (September 1968). Note the northward shift of water across the central 'grande barrière' under pressure from the floods down the Chari River. The conductivity range 100–200 is approximately equivalent to a salinity range of $0\cdot15-0\cdot8\%$ (redrawn from Carmouze, 1969)

Much can be learnt of the hydrological regime by following the distribution and changes in the salinity of the water. This can most conveniently be measured as electrical conductivity. The original data were collected by Bouchardeau (1958), and have been augmented by Hopson (1969a) and Roche (1970). Changes in salinity an ionic composition are discussed by Carmouze (1969, 1976).

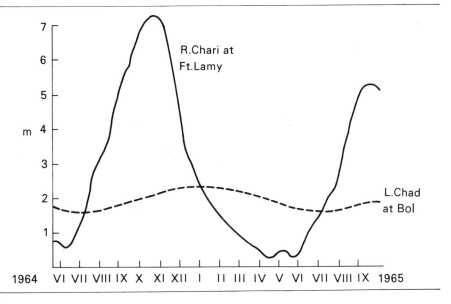

Fig. 12.4 Seasonal changes in level of River Chari and Lake Chad (Gras *et al.*, 1967)

The salinity of the water is of course lowest in the region of the main inflow from the Chari River (Fig. 12.3). However, it should be noted that, though there is a progressive increase from the Chari estuary to the eastern and northern extremities, the maximum salinity does not exceed $1^o/_{oo}$ (K_{20} 1 000 μmhos), and thus remains well within the freshwater range (Ch. 5). The salinites of Lakes Tanganyika, Albert Edward and Turkana are all higher than this (Table 5.1), and they all support a rich and typical freshwater fauna and flora.

The hydrological balance of the lake has been studied by O.R.S.T.O.M., particularly by Bouchardeau (1958) and Carmouze (1971). The volume of the inflows varies from year to year but in some years it has almost equalled the total volume of the lake. This accounts for the very great influence of the seasonal inflow of water on the reproductive and growth cycles of the fauna (see below).

As would be expected, there is a general shift northwards of the salinity contours (isohalines) with the inflow of flood water from the Chari between July and January, followed by a general increase in salinity and a return to the dry season condition (see also Hopson, 1969a). The horizontal movements of water masses are also determined by the prevailing winds and by the distribution of open water and barriers to flow, such as swamps and islands. The pattern of mass horizontal movements throughout the year has been studied in detail by Carmouze (1971) who measured sodium content as the most reliable means of identifying water masses.

Recent research to discover the processes that maintain the relatively low mean salinity, in spite of the continuous inflow of salts into an apparently closed basin, is reviewed by Roche (1970). Some of the factors in the situation are difficult to evaluate quantitatively, but from measurements over several years of the volume and salinities of the inflows, rainfall on the lake and of the rate of evaporation it has been estimated that more than 90% of the incoming water is evaporated. Nine to ten per cent is lost in other ways. There are apparently three routes by which salts may be removed from solution in the lake:

1. Through chemical and biological precipitation in the lake itself. The increasing salinity and alkalinity from south to north cause progressive precipitation of some ions which settle on the bottom and on other surfaces such as the stems of plants. It has been estimated by Carmouze (1970, 1976 and pers. comm.) that the following percentage fractions of the total amount of each ion in the inflow are on the average removed by chemical precipitation and by uptake and desposition by living organisms – calcium 60%,[*] magnesium 54%, potassium 36%, carbonate 44% and silicate 84%. Nevertheless, this does not account for all of the total salt lost and could not in principle remove in quantity such ions as sodium and chloride, the relative proportions of which increase towards the north (Carmouze, 1969, 1973). There are many small lakes elsewhere in closed basins where great quantities of calcium, magnesium and carbonate have been deposited, but whose water is hypersaline through evaporation.

2. The most remarkable superficial feature of Lake Chad is the much serrated

---

[*] Great masses of mollusc shells are deposited in the sediments of the north basin where, presumably owing to the higher alkalinity, their calcium carbonate remains intact. In the south basin the deposited shells decompose rather rapidly and the calcium returns to the water (Lévêque, 1972b).

north and east shores which are broken up into innumerable small bays and inlets between the old sand-dunes. When the level of the lake rises seasonally and when the mean level increases during a period of years owing to higher annual rainfall in the catchment, these inlets are extended inland by flooding with lake water, which along the northern and northeastern periphery is relatively saline. These flooded arms are later abandoned by the falling lake and become isolated pools concentrated or desiccated by evaporation. Subsequent floods deposit more saline water which becomes isolated from the lake.

3. Geological and hydrological† investigations of the recent lacustrine deposits around the lake have shown that the disposition and slope of the impermeable clay and water-bearing sandy layers are in general such as to favour subterranean seepage of water away from the lake. The underground water movements are complicated but it is now generally thought that such 'marginal infiltration' is probably the main route by which salt is lost from the lake (Roche, 1970).

Dieleman and de Ridder (1964) in their special study of the ground waters under the artificial polders between dune promontories near Bol, demonstrated that the water is actually moving outward from the lake. In general, the salinity of the soil water progressively increases in the direction of the flow due to evaporation of the surface of water rising by capillarity. Consequently salinification of the soil is causing some problems for the further expansion of irrigation agriculture in the polders and on land further away from the lake. Further evidence was found by Maglione (1969) for the group of small lakes inland from the main lake north of Bol in the Kanem region (Fig. 12.1). Some of the hollows between the ancient dunes reach the water table, and the lakes so formed have water of varying degrees of salinity. Many are permanent and those nearest to the main lake respond to the seasonal changes in water level in the lake, but those further inland (some 50 km) are fed from a subterranean water table flowing from the north. The biology of these saline ponds is discussed in Ch. 19.

Though we have no direct knowledge of the annual amount of water lost by this route, we cannot easily avoid the conclusio that the freshness of the lake water is maintained primarily by a continuous seepage into the desert sands to the north and east, and that the salt accumulating in that region originally came from the lake. The salinity of the water now seeping into the sands is about ten times that which comes in by the rivers. Much salt is thus removed from the lake in a relatively small volume of water.

The problem of recent connections with the neighbouring basins is raised by the present composition of the fauna of the lake, especially the fish. Blache (1964) has listed the species of fish in the Chad basin and those in the basins that surround it, and has included the Volta basin for a comparison. A summary extracted from his table is given in Table 12.1.

There is no doubt from the figures in the table not only that the Chad basin is part of the soudanian ichthyogeographical region,* but also that it has been the

† Including measurement of the salinity and isotopic composition of the underground waters.

* The following species are found apparently unchanged in Lakes Chad, Albert, Turkana, the Niger, Volta and Nile: *Polypterus senegalus, Heterotis niloticus, Hydrocynus forskali, Citharinus citharus, Malapterurus electricus, Lates niloticus, Alestes baremose, A. dentex, Syndontis schall, Clarias lazera, Tilapia zillii, Sarotherodon niloticus, s. galilaeus.*

Table 12.1

	Total number of species	Number of species confined to the two basins mentioned
Chad basin	179	
Lake Chad	84	
Common to Chad and Niger basins	106	8
Common to Chad and Nile basins	85	7
Common to Chad and Zaïre basins	47	13
Common to Chad and Volta basins	58	0

recent meeting-place of species that have been evolved in the Niger, Nile and Zaïre basins. The absence of species confined to the Chad and the Volta basins, which are not contiguous, gives strong support to this interpretation of the data. The existence of species confined to Chad and to one or other of the surrounding basins suggests that the connections were very recent. Otherwise we should expect some genetic changes to have affected these species through isolation in the Chad basin.

There is good geological evidence of an overflow, less than 10 000 years ago, from the old Mega-Chad into the Kebi and so via the Benue into the Niger (Pias, 1958; Grove and Warren, 1968). Even now, in times of exceptional floods, some of the water in the Logone overflows from its bed to the west and escapes into the Kebi and over the Gauthiot Falls to join the Benue. In fact, there are 105 species of fish that are common to the Kebi and Chad basins and, of these, 21 are not found in the Benue or Niger (Blache, 1964). So, from the distribution of the fish, we are led to the conclusion that the Kebi was recently the connecting link between Lake Chad and the Niger. That this connection is still made at intervals via the Logone presumably accounts for the Chadian species in the Kebi that have not yet extended further down into the Niger basin (Fig. 11.3).

The upper courses of the Chad and Zaïre basins share a common watershed in the humid highlands of the Central African Republic. An exchange of species across this region during periods of heavy rain is not unlikely. Of the forty-seven species of fish that are common to the Chad and Zaïre basins (Table 12.1), as many as eighteen descend to further than the upper reaches of the Chari and Logone and are therefore probably recent immigrants from the Zaïre.

We cannot escape from supposing a recent connection between Chad and the Nile, but the nature and location of the link across what is now a vast expanse of extreme desert will not be decided without further geological investigation. The possible connecting routes have already been discussed (p. 200).

It is interesting also that, though twenty-five to thirty endemic species have been reported from the basin, only one, *Alestes dageti*, is found in the lake itself. The remainder live in the Chari-Logone river system (Blache, 1964). This is perhaps a consequence of the instability and fluctuating conditions in the lake during the past which may even have involved a period of desiccation or very saline water. The affluents from the south, though fluctuating greatly in volume, have presumably persisted as flowing rivers from far back in the Pleistocene. It is possible also that, owing to its relative ecological uniformity, the lake has failed to provide a stimulus to divergent speciation. The only extensive habitats are shallow open water, fine-grained bottom deposits, swamp, and submerged water-weeds between some of the islands. It should also be noted that no endemic species of any other group of aquatic animals has so far been found living in Lake Chad. This applies even to the Mollusca (Lêveque, 1967a), a group of which, after the fish, is prone to speciation with prolonged isolation. The time and mode of origin of the fauna of Lake Chad is a major unsolved problem.

The recent changes of level have altered not only the area, but also the ecological structure of the lake. The present arrangement of ecological zones is therefore very temporary. For the south basin they are described by Blache (1964) and Gras et al. (1967). From the north and east shores, as we have seen, there project into the lake row upon row of partly submerged sand-dunes forming elongated promontories and islands (Fr. archipel) all orientated in a northwest–southeast direction, each partly fringed by a band of papyrus and reed swamp, like so many of the islands in the East African Great Lakes. The prevailing winds that accumulated these dunes during a recent arid period were obviously those in operation at the present time that blow at right-angles to the line of the dunes. In the dry season (November–April) the Harmattan blows from the northeast, and during the rains (May–October) the predominant winds are more variable from the east and southwest.

Further into the southern basin of the lake, the dunes are totally submerged under a belt of swamp (Fr. ilots bancs), composed mainly of papyrus (Cyperus papyrus), a reed (Phragmites mauritianus) and the grass Vossia cuspidata, with many other associated plants (see Hepper, 1970, for a brief description and illustrations). In the north basin the vegetation of many of the swamp islands is predominantly the reed Phragmites australis, which with Typha australis also dominates the shore-fringing swamps in the north, thus taking the place of papyrus in the south basin (Hopson, 1969a). The reasons for this difference are as yet unknown, but the higher salinity of the water and the more powerful and drier winds in the north might, on investigation , be found to inhibit the establishment of papyrus. The plants grow from a floating mat of entangled roots and rhizomes embedded in decomposing plant remains and some silt. Above the ridges of the submerged dunes the water is shallow enough for the major swamp plants to be rooted to the substratum. Elsewhere they are unattached and swamp islands break away into the open water, particularly at times of rising water level, and are moved about by winds and currents. The open channels through the swamp mark the position of valleys between the submerged dunes. The structure of the swamp is very similar to that of the fringing swamps around open lakes and water courses all over tropical Africa wherever the situation is favourable (Ch. 18). It seems that the conditions in the Lake Chad swamps resemble those in the outer (open water) edges of swamps, as in Lakes Victoria and Kioga, where attachment to the substratum is insecure and islands break away in windy weather. The underlying water is mostly well stirred every day and mixed with the open lake water so that, except in the floating mat itself and perhaps in the shallow swampy region in the southeast corner of the lake, no drastic diminution of dissolved oxygen is likely. The great expanses of oxygen-deficient water typical of the stagnant swamps of eastern tropical Africa (Ch. 18) are probably not to be found in

Lake Chad. This, however, remains to be confirmed. Even in the open water the submerged dunes can be detected by echo-sounder (Dejoux *et al.*, 1969). The north and south basins are separated by a belt of swamp attached to an underwater ridge (*grande barrière*) crossing the lake between Baga Sola and Baga Kawa, through which there are many fluctuating channels and spaces (Fig. 12.1). The open water of the south basin is shallower (2·5–4·5 m) than that of the north basin (4–7 m), presumably because of the annual discharge into the former of sediments from the Chari River. Consequently, in the south basin the area of open water is less and the extent of swamp islands and coastal swamps is greater than in the north basin.

Another important habitat is the extensive underwater beds of macrophytes that block the passages between many of the islands. These are composed mainly of *Potamogeton schweinfurthii, Vallisneria spiralis* and *Ceratophyllum demersum*. They shelter several species of fish not normally found in the open lake including some Cichlids and the young stages of *Lates* and *Distichodus*.

Along the southern coast to the east of the Chari estuary is a very flat plain relieved only by a group of small volcanic rocky hills, Hadjer el Hamis, near the lake shore, from which has been found some of the evidence of former high lake levels. The plain is inundated to a varying extent according to the season, giving a great expanse of shallow water (less than 2 m deep) choked with submerged aquatic vegetation (Fr. *zone des herbiers*) including *Potamogeton, Ceratophyllum* and *Vallisneria* with floating leaves of *Nymphaea* (water lilies).

A good general idea of the ecology and productivity* of the River Chari and the southern region of the lake can be got from the work of Dussart and Gras (1966) and Gras *et al.* (1967), Pourriot (1968) and Rey and Saint-Jean (1968, 1969), on the phyto- and zooplankton in the south basin. A survey of the zooplankton in the north basin has been made by Robinson and Robinson (1971). The quantitative data of Gras *et al.* are well summarised by the three-dimensional diagram in Fig. 12.5. The phytoplankton is in general dominated by blue-green algae (Cyanophyceae) especially by *Microcystic* and *Aphanocapsa. Diatoms, mainly Melosira*, are relatively increased when the Chari floods into the lake. The study of the zooplankton was restricted to the Crustacea (Cladocera and Copepoda) and Rotifera. Among the commonest species found were:

1. Cladocera: *Bosmina longirostris, Ceriodaphnia cornuta, Moina dubia* and *Diaphanosoma excisum*.
2. Copepoda: *Thermocyclops neglectus, Mesocyclops leuckarti* and *Tropodiaptomus incognitus*.
3. Rotifera: *Keratella valga, K. tropica* and *Brachionus falcatus*.

Though the presence of young stages throughout the year showed that reproduction of zooplankton is continuous, there are very well-marked seasonal variations in the rate of reproduction similar to those of the phytoplankton (Fig. 12.5). This diagram sets out the changes in density (total number of plant cells per litre) over one year starting in July 1964. The samples were taken from a series of stations from N'Djamena down the Chari and northwards across the

---

* A summary of measurements of primary productivity in the lake by Lemoalle (1969, 1975) is presented in Table 7.1 to compare with those from other lakes, and the changes following the major recession of the water after 1972 are discussed on p. 000.

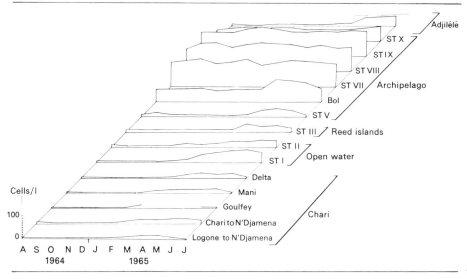

Cells/l

100

0

A S O N D J F M A M J J
    1964   1965

ST X
ST IX
ST VIII
ST VII   Archipelago
Bol
ST V
ST III   Reed islands
ST II
ST I   Open water
Delta
Mani
Goulfey
Chari to N'Djamena
Logone to N'Djamena
Adjïlélé
Chari

Fig. 12.5 River Chari and Lake Chad. Annual variation of phytoplankton in cells/l at different stations (Gras *et al.*, 1967)

open water, through the swamp belt and into the dune-island zone, with a final station (Adjilele) in the marshy zone east of the Chari estuary. Some interesting conclusions can be drawn from them. When the river is in highest flood (October–December) there is a low plankton density in the open water increasing a little across the swamp belt and a relatively high density in the dune-island zone. In the dry season with the river at its lowest level (February–July), the density of both phyto- and zooplankton rises in the open water and swamp belt so that the zooplankton, at least, is more or less uniform in density across the entire transect.

It is clear that in the south basin the dune-island zone is the most continuously productive region and is apparently unaffected by the seasonal regime of flooding from the Chari inflow. The plankton of the open water, however, is much influenced by the flooding and recession of the Chari which flows directly into it. The dune-island zone is protected from these seasonal fluctuations by the 'buffering' effect of the swamp belt. It is suggested by Gras *et al.* (1967) that the diminution of the plankton in the open water during the flood period is mainly due to the great increase in turbidity of the water from suspended matter brought down by the river and from violent stirring of the bottom sediments by the southwest winds at this time. Owing to the relatively high tropical latitude (13°N) and the proximity of the desert, there is a difference of around 10°C between summer and winter water temperatures, and the periods of high and low plankton density are marked by rising and falling temperature respectively. The very dry northeast (Harmattan) winds during the winter must also contribute to the lowering of water temperature by increasing evaporation at the surface. But the rather uniform density of plankton in the dune-island zone throughout the year suggests that continuously favourable conditions (e.g. food supplies) may override the effects of seasonal temperature changes.

The changes that we know to occur in the water during the flood period are: rise in level, increase in temperature (due to seasonal rise in air temperature), decrease in salinity, increase in turbidity, and influx of nutrients and inorganic ions (especially Fe) from the rivers. Which of these factors, and perhaps others, are responsible for the observed seasonal changes in plankton density in the open water, is an open question. It seems likely, however, that the most important are seasonal temperature changes and the periodic great influx of water containing organic matter and plant nutrients. The stimulating effects of the latter might well be delayed and plankton production is in fact at a peak during the low-water but high temperature period.

The zooplankton densities and the seasonal changes estimated by Robinson and Robinson (1969, 1971) in the open water of the north basin, when allowance is made for the large variations and errors inevitably associated with plankton measurement, are consistent with those of Gras et al. (1967) in the south basin. The Robinsons estimated the range of densities in terms of numbers of individuals throughout the year to be about 450 000 to 2 000 000 per m^2 of surface. These figures are much greater than any obtained from other large, but deeper, tropical African lakes, but they are of the same order as those from two other shallow, but much smaller, lakes, Mulehe (Fig. 15.1) and George (Ch. 13). Lake Chad has the greatest expanse of very shallow water in Africa, and the total standing crop of zooplankton is therefore gigantic. It is directly exploited by some of the fish such as *Alestes dageti* and *A. baremose* (Lauzanne, 1969; Robinson and Robinson, 1969).

Before discussing the benthic fauna something much be said about the nature of the bottom sediments which have been studied in detail. As the result of previous desert conditions in the Chad basin there is a large amount of sand under the lake. This, however, is exposed only where turbulence or currents prevent the deposition of sediments above it, as on the ridges of submerged dûnes and in some of the trenches between them where there are frequent currents. The sediments are of five main kinds (see Dupont, 1970; Hopson, 1969b; Dejoux et al., 1971):

1. Recently deposited mud (Fr. *vase*) composed of finely divided brown to black flocculent material with varying quantities of organic matter.
2. Clay (Fr. *argile*). This is a mud sediment that has in varying degrees been compacted and mineralised, from soft muddly to stiff blue-grey clay.
3. Sand (Fr. *sable*) occurs in patches that are common on the west side of the north basin. They are probably the remains of submerged dunes.
4. 'Pseudo-sand' (Fr. *pseudo-sable*). This is a curious sediment of the colour and consistency of coffee grounds, composed of separate rounded particles (about 0·2 mm diameter) each with a centre of clay covered with a ferruginous skin. It is confined to the south basin (Dupont, 1970).
5. 'Peat' (Fr. *tourbe*), mostly organic with partially decomposed plant remains. It appears to be most frequent in the south basin. This, like the similar deposit under the slow-moving waters in eastern Africa, is derived from the swamps, but some has drifted and accumulated in regions of relatively calm open water. There are also layers of peat marking the position of previous fringing swamps during periods of lower water level (Dupont and Lévêque, 1968).

Most of the sediments are mixtures with one or other of the above predominating. A study of their layering has provided evidence of three cycles of expansion and contraction of the lake during the past few hundred years.

Work on the bottom fauna has so far been confined to the molluscs, oligochaet worms and insect larvae. The distribution and seasonal changes in the south basin have been studied by Dupont and Lévêque (1968), Dejoux, et al. (1969), Lévêque (1972a, b), and Lévêque and Gaborit (1972), and a preliminary survey of the north basin has been made by Hopson (1969b). For a systematic study of the molluscs see Lévêque (1967a, 1972a), of the oligochaets Lauzanne (1968) and of the insects Dejoux (1969). Except for the insects the number of species involved is rather small. In the south basin Dejouz et al. (1969) recorded six species of molluscs, six of oligochaets, twelve of chironomid larvae and eight of Ephemeroptera and Trichoptera. Each main type of substratum was found to have its characteristic fauna. For example, on sandy bottoms the molluscs are represented mainly by the bivalves Corbicula africana and Pisidium pirothi with chironomid larvae of the genus Tanytarsus. In mud the prosobranch snails Melania tuberculata and Cleopatra cyclostomoides predominate with the chironomid Clinotanypus clavipennis. Characteristic of the granular clay sediments are the bivalve Byssanodonta parasitica and oligochaets of the family Alluroididae.

But the factors that determine the distribution of the benthic fauna are complicated and difficult to assess. There are possibly many important influences including particle size and the amount and condition of organic matter, as well as currents and rates of deposition of sediment. This is emphasised by the preliminary survey in the north basin by Hopson (1969b). The most striking difference between the benthic faunas of the two basins lies in the molluscs of the muddly sediments. Instead of the Cleopatra, which is dominant in the south basin, the prosobranch snails Bellamya unicolor and Melania tuberculata are by far the commonest species on the muddy bottom off Malamfatori. They provide a part of the diet of the Nile perch (Lates niloticus). With a view to eliminating other factors than quality of the sediments Dejoux et al (1971) studied the benthic fauna of a restricted area among the dune islands near Bol, where it could reasonably be assumed that other factors, such as water composition, were uniform. They found a marked correlation between type of sediment and the molluscan and oligochaet faunas, but the insects appeared to be independent of the nature of the substratum.

In a wide-ranging survey of the overall distribution of the benthic molluscs in the whole lake Lévêque (1972a) came to the conclusion that neither the oxygen concentration, which is always adequate at the bottom, nor the nature of the bottom sediments was important in determining distribution. There was, however, an apparent correlation with salinity, deduced from conductivity. This, as already indicated, shows a gradient increasing from south to north. The absence of benthic molluscs at the extreme north of the lake is very striking. For Bellamya and Melania a conductivity of $500-600 \times$ µmhos ($0.05-0.06°/_{oo}$) seems to be the limit and Bellamya has an optimum range of from $0.15-0.2°/_{oo}$ (Lévêque, 1972a). It must be pointed out that these salinities are well within the freshwater range and that the salinity of Lake Edward, which harbours a normal molluscan fauna, is over $0.7°/_{oo}$ (Ch. 13).

There is perhaps some limiting chemical factor peculiar to the water of Lake Chad which is associated with increasing salinity. Alternatively, the absence of molluscs from the north end may be a reflection not of present unfavourable conditions, but of some event in the recent past, such as very high salinity and exceptionally low water level, after which the molluscs have not yet become redistributed. On the other hand, Lévêque (1972a) observed rapid changes in some mollusc populations especially among the islands near Bol.

Of the three groups of animals the molluscs contribute most to the total benthic biomass. In the open water of the south basin they form 75–90% of the total. The benthos as a whole is most abundant in the open water zone of both basins and Hopson's estimate of the biomass of *Bellamya* and *Melania* off Malamfatori (over 1 000 kg/ha) is comparable with the maximum estimate by Lévêque (1972a) in the south basin. These figures will no doubt be modified later, but the undoubted concentration of benthos in the open-water zones of the two basins is presumably partly due to the abundance of organic detritus.

The survey of Dejoux *et al.* (1969) in the south basin showed a distinct overall seasonal fluctuation in the biomass of most benthic species. In general the populations come to a maximum from December to March, towards the end of the cool dry season. This is particularly clear with the oligochaets and insects, less so with the molluscs. It is curious that the chaoborid fly larvae, which are common in the dune-island zone, but rare on the east side of the lake, show the reverse seasonal periodicity, being most abundant in the warm wet season, June–September. It was suggested by Dejoux *et al.* (1969) that the changes in the temperature of the water are probably the main determinant of these seasonal cycles. Lévêque (1972a) found that changes in growth rate of some species of benthic molluscs are apparently correlated with seasonal temperature changes. The most abundant swarms of insects emerging from the lake are to be seen during calm spells in February when the temperature is beginning to rise. The mean water temperature fluctuates through at least 10°C, from 18°–20°C in January to 30°–32°C in May–August. This reflects the wide air temperature range in a region of relatively high latitude (13°N) on the edge of the Sahara Desert. In contrast the seasonal changes in mean water temperature of equatorial lakes, such as Lake Victoria, is no more than 1° or 2°C. It can also be said that the curve representing water level through the year is roughly parallel to that for total benthic biomass. The rise in the latter from September/October onwards might be a reflection of the increasing amount of food available from the Chari inflow and washings from the surrounding land, so that the maximum biomass is ultimately reached during the following dry season when the conditions are becoming less favourable and the decline is about to set in.

Though little similar work has been done on other African lakes as a basis for comparison, the quantitative estimates of biomass and production of the benthic molluscs by Lévêque (1972a, b) and Lévêque and Gaborit (1972) indicate that most of the sediments of Lake Chad provide a particularly favourable environment for them. The total biomass of benthic molluscs, though subject to great changes with fluctuations of water level, is apparently greater than that of any other group of animals, and even more than the aggregate of the zooplankton. They are a very important link in the ecosystem and play some part in removing accumulating ions, especially Ca^{++}, from the water. This abundance is presumably attributable both to the richness of the sediments and to the fact that littoral conditions (shallow and well-stirred water characteristic of the inshore regions of deeper lakes) extend over the whole of the open areas of the lake. It would, however, be unwise to assume that there are no other as yet unknown factors involved in this situation and it would be particularly interesting to discover why the benthic fauna of the similarly shallow Lake George is generally much poorer and includes very few molluscs. A possible reason for this difference is that most of the superficial sediment of Lake George is very soft and indeed liquid and is anoxic close to the surface, whereas the sediments of Lake Chad are mostly firmer and thus provide a better foothold for molluscs. In fact Lévêque (1972a) reports that they avoid regions with a very soft substratum. Some of the endemic molluscs of Lake Tanganyika have managed to adapt to this kind of habitat (p. 278).

Before 1960, published investigations on the fish were concerned primarily with systematics and distribution of species in relation to the past history of the basin. Most of this was summarised by Blache (1964) and a comprehensive review of the mainly native fisheries of the Chad basin is found in Blache and Miton (1963). Both of these include discussions of the ecology of the fish in the southern basin of the lake and in the rivers and flood plains to the south. For a short account of fisheries from the Nigerian shore see Hopson (1967).

There is no doubt that both the obviously abundant production of fish in the lake aand its affluents and the dramatic seasonal bursts of reproduction and growth are caused by the annual flooding of the inflowing rivers during the period from June to November. Owing to the extreme flatness of the Chad basin, the rivers, especially the Chari and Logone, overflow their banks and inundate vast tracts of country more than 5 000 km^2 and locally known as the 'Yaéré'. This eventually drains away down both the Logone and the previously dry bed of the El Beid. The latter then becomes an important link between the Yáeré and the lake, and great numbers of fish are caught in it as they migrate up and down. For an account of the fish and fisheries of the El Beid see Durand (1970). In the words of Blache and Miton, the country south and southeast of N'Djamena is a veritable '*Mesopotamie Tchadienne*'. It floods very rapidly. Dried organic and mineral matter, ashes from bush fires and the droppings from herds of animals that have accumulated during the previous five months of dry weather, begin immediately to decompose and to dissolve. There follows within a few days an explosive growth of phytoplankton soon to be followed by one of zooplankton. Later come the larger vegetation, grasses and flowering plants such as Polygonaceae and Alismataceae in the shallow water, in deeper water fixed and floating water plants such as *Pistia, Nymphaea, Stratiotes, Utricularia*, etc.

The fish start to move up the rivers as the water begins to rise and when the flooding starts they spread out in prodigious swarms over the inundated land and feed voraciously on this new-found abundance. The conditions are very favourable for breeding, with plenty of food and vegetation cover for the young. Many of the species have a varied diet and change their food as they develop and grow. There is phytoplankton and newly sprouting leaves and stalks for herbivores such as *Distichodus* and *Tilapia*, zooplankton for fish such as *Alestes*, and organic matter of all kinds for omnivores such as *Clarias*. Reproduction and growth are very rapid indeed. As an example, the herbivore *Heterotis niloticus* has been observed to reach over 500 g in weight within six months of hatching (Blache and Miton, 1962; Durand and Loubens, 1969).

As the floods subside towards the end of the year the fish move back with the currents into the river beds. This is the great time for fishing. Great numbers are trapped and caught in residual ponds choked with vegetation; some of the ponds are permanent, but many eventually dry up. It is in these that the lungfish *Protopterus annectens* aestivates in the dry mud, and even the airbreathing catfish *Clarias lazera* has been known to survive some time in almost dry mud.

Such seasonal inundation of dry land, with consequent bursts of production, are well known in many other tropical African river basins and around low-lying shores of lakes, but the great extent of the flooded areas and the predominant and decisive influence of the seasonal floods on what must be one of the most extensive and potentially productive inland fisheries in Africa, make these events in the Chad basin of particular interest. The total annual catch in the lake is at present between 30 000 and 100 000 tons.

Many species of fish are caught, but commercially the most important are *Lates, Heterotis, Citharinus* spp., *Labeo* spp. and *Distichodus. .Alestes baremose*, also an important fish, is mainly caught in the inflowing rivers. Its biology and growth have been investigated by Durand and Loubens (1969) and Lauzanne (1969). Its maximum length is about 35 cm, it feeds mainly on zooplankton and the abundant fishery is a reflection of the richness of the zooplankton. Durand and Loubens consider that there are probably two more or less separate populations in the south basin of the lake. One is apparently confined to the dune-island region, the other occupies the open water and migrates seasonally up the rivers (Chari-Logone and El Beid) feeding and breeding in the flooded areas. There appear to be two similar populations living in the open water of the north basin one of which migrate up and down the Yobe River (J. Hopson, 1972). Fishing is mainly by barrage traps and seine nets in the rivers and with gill nets in the lake. Growth studies in both basins have shown that the prominent rings on the scales are due to the annual arrest of growth during the cool high-water season which reaches a peak in January. Here is more evidence of the influence of the seasonal flooding and recession. The growth characteristics of forty-six species of Lake Chad fish were measured by Durand and Loubens (1969).

The Nile perch, *Lates niloticus* (Centropomidae), as in other parts of its range throughout the soudanian region, is the largest of all fish inhabiting the open water. It is the subject of recent study and review by A. J. Hopson (1972). The eggs and early stages were described for the first time by Hopson (1969c). Unlike most pelagic freshwater fish, they do not go inshore to breed but lay their eggs in the open water. The eggs and early larvae each contain a single large oil globule which serves to keep them afloat. Hopson discusses the occurrence and significance of oil globules in the eggs of other percoid fish. As with *Alestes*, the scales of *Lates* show well-marked seasonal rings denoting arrested grown during the cool season (Monod, 1945a; Hopson, 1968).

In conclusion, we may summarise the main features of particular interest that distinguish Lake Chad from the other large lakes of tropical Africa. It is centrally placed in an ancient shallow basin on the edge of the Sahara in a region that, unlike eastern Africa, has been tectonically more stable but has been subject to important changes in climate during the Pleistocene and up to recent times. Consequently there have been great fluctuations in the level of the lake and, more frequently and recently than with the other lakes, connections have been made with the neighbouring basins, whose faunas have found a meeting-place in Lake Chad. The apparent absence of endemic species of aquatic animals (except for one species of non-cichlid fish) from the rich fauna of a large tropical African lake in a closed basin is a very remarkable phenomenon. There are four possible reasons for this:

1. the very recent isolation of the basin;
2. the relatively few types of habitat available in the lake;
3. the already 'matured' and well-differentiated nature of the ancient soudanian fauna from which the fauna of Lake Chad has mainly been derived;
4. the instability of the environment due to the many and drastic fluctuations in level during the last few thousand years (environmental stability in relation to speciation is discussed on p. 146).

In spite of the present very low rainfall over most of the basin, the extremely high rate of evaporation and the absence of a surface outlet, the water in the 'normal' lake is fresh. This is due to the flushing out of the lake by a very large seasonal inflow from the heavy rainfall highland region to the south and a much smaller but more saline subterranean outflow under the sands to the north and northeast. The fauna has thus escaped the fate of that of some of the closed basins in the Eastern Rift Valley where a similar change of climate has reduced some previously large lakes to very saline remnants, e.g. Lakes Elmenteita and Magadi. Situated far enough to the south to receive water from the humid tropical highlands, it has avoided the permanent fate of the other Pleistocene lakes in the west central Sahara such as Araouane whose previous existence is known only from dry alluvial deposits and their aquatic fossils (Ch. 11). This matter is discussed further in the next section.

Faunal surveys to-date suggest a high productivity. This is presumably due to the very shallow water and the regular wind regime whereby the nutrients are rapidly recycled from the mud and are copiously renewed every year from the inflowing rivers. Well marked seasonal fluctuations in rate of reproduction and growth appear to be regulated by the annual flooding of the rivers and by seasonal changes of temperature to a greater extent than in the other large lakes of tropical Africa.

'Little' Chad (1973–)

From 1963 onwards the annual peaks of water level caused by the Chari floods became progressively lower, but there was no great change in the form of the 'normal' lake until the severe drought in 1972–73 which seriously affected the northern quarter of tropical Africa (Sahel) and reduced the inflow from the Chari River (Fig. 12.6). In 1973 the lake had shrunk to three tenuously connected areas of water (Fig. 12.1). At the end of 1975 the North Basin was dry. The lake was reduced to a remnant in the South Basin directly fed by the reduced annual flood

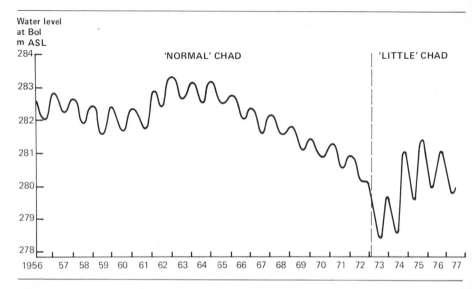

Fig. 12.6 Water levels of Lake Chad recorded at Bol. From 1973 onwards the north and south basins were separated and the area of open water was drastically reduced ('Little' Chad) (from Chouret, 1978)

from the Chari, which slowly percolated northwards through dense vegetation into the remaining patches of water between the islands of the Archipelago (Fig. 12.7b).

A striking feature was the very rapid growth of vegetation on the exposed sediments in the South Basin, and on the Grande Barrière. The dominant plants on the drier ground were the ambatch bush *Aeschynomene elaphroxylon* and the grass *Vossia cuspidata*, and in the more swampy regions *Cyperus papyrus*. The vegetation grew so luxuriantly that the Barriere became a dam between South and North Basins, which leaked only slowly. The development of the vegetation in the basin in 1974–76 was described by Fotius and Lemoalle (1976).

In 1974–75 the inflow from the Chari increased again to a volume previously recorded in the late 1960s (Fig. 12.6). Nevertheless, though the water level in the South Basin was then higher, the flow was so obstructed by the Barriere vegetation and reduced by excessive evaporation that only a small amount of water reached the North Basin during 1975. A few shallow patches of water remained and these dissappeared by the end of the year. The next season's flood was even higher but only a small amount of water crossed the Barrière.

It will therefore require exceptionally heavy floods to breach the Barrière and to bring the lake back to its 'normal' condition. Up to the end of 1977, at least, there waere no signs of a reversal of the recession. The Chari flood of 1976–77 was in fact somewhat lower than in the previous year. In the meantime it seems that Little Chad will persist as a small lake fed as usual by the annual floods from the Chari River with seasonal incursions into the Archipelago.

In spite of the absence of a surface outlet the salinity of the water in the North Basin of the 'normal' lake was always within the freshwater range, mainly due to a subterranean seepage to the northeast (p. 215). The effects of the recession after

1972 are thus of special interest. At a selected central point in the North Basin between mid 1973 and the end of 1974 the depth of water fell from 3·5 to 1·0 and the conductivity (K_{20}) rose from 900 to 3,500. During the same period the conductivity of water sampled near Bol among the dune islands did not exceed 570. These figures should be compared with those from the 'normal' lake in Fig. 12.3 and with those from the most saline of the truly 'freshwater' lakes in Table 5.1. It seems therefore that, in constrast with Lake Chilwa (p. 360), excessive salinification was not a major feature of the recession and would appear only in the final stages of dessication in the North Basin.

Iltis (1977) made a comparative study of the phytoplankton populations of the 'normal' lake in 1971–72 and of the 'little' lake in 1974–75. In the former there were in all about 120 species of algae. The four main regions of the lake – North and South Basins and North and South Archipelagos between the dune islands supported populations of algae distinct from one another in respect of species composition, though water was moving between them especially during the annual rise in level. In all regions the Cyanophyceae were dominant.

During 1974–75 the populations in the then virtually separated regions of the lake diverged still further. On the whole their density increased and there was a greater proportion of diatoms than formerly. The only apparent reflection of increased salinity was the appearance in the North Basin of certain species normally associated with moderately high salinity, such as *Oscillatoria platensis* (Cyanophyceae). The shoreward waters of the Archipelago were reduced to separate ponds containing a characteristic phytoplankton dominated by *Euglena spp.* (Chlorophyceae). As would be expected, the phytoplankton in the South Basin, in direct connection with the Chari and its annual floods, was least affected by the recession.

Estimates of chlorophyll *a* and of the rates of primary production by the phytoplankton were made during 1968–76 in the South Basin, the Archipelago near Bol and in the North Basin by Lemoalle (1975 and unpublished pers. comm.). During 1968–72 ('normal' lake, Table 7.1) the rate of primary production at Bol oscillated every year between similar, but slowly rising, limits corresponding to the steady fall in mean lake level (Fig. 12.6). The higher rates in mid year were contemporary with the seasonal low water levels (Fig. 12.4).

In 1973 and 1974, with the drastic fall in level and near isolation from the South Basin, the mid year rate of production at Bol was more than doubled and the minimum rate of the end of the year was much lower than in previous years. At the end of 1974 water flowed again into the Archipelago and with the rise in water level the rate and pattern of primary production became similar to what they were before 1973.

Measurements at a central station in the North Basin were made from the end of 1973. From then on the rapid fall in level was reflected in a rise in chlorphyll *a* from 4·3 mg/m³ in March 1973 to 2 700 mg/m³ in December 1974 and finally, just before drying up, to 3 600 mg/m³. The maximum rate of primary production measured in the North Basin in 1974 was about 30 000 mg O_2/m³. hour. This figure is about ten times the maximum recorded in Lake George (Fig. 7.3) and thus of the same order as that measured in Lake Aranguadi as judged from the relative rates of carbon synthesis per m² in Table 7.1. As with other characteristics, production in the remnant of the South Basin, still in contact with the Chari inflow, was less affected.

The clear inverse correlation between rate of primary production and water level could result from the increased density of the algae and higher concentration of HCO_3^- due to evaporation, and to the more rapid wind induced circulation of algae and nutrients in shallower water.

The depressing effects on the fisheries in the Chari delta of the exceptional fall in the volume of flood-water from the river in 1971–73 were discussed in detail by Quensiere (1976). The effects of the recession on the very rich fish fauna of the lake itself have been catastrophic, though very interesting. Benech *et al.* (1976) described the changing conditions and their effects on the fish during the critical period 1972–74 in the North Basin and in the Archipelago region which suffered most from the recession. The open water in the much contracted South Basin, directly affected by the annual (though much reduced) floods from the Chari River, has been ecologically less changed. In normal times the fishery is

Fig. 12.7 Lake Chad (*a*) in January 1908, (*b*) in November 1975 (from Chouret, 1970)

very productive and supplies a large area to the south and southwest and far into Nigeria. The effects of the recession on the fish trade along the many transport routes is described by Stauch (1977).

The drastic reduction in the volume of water ,and the complete drying out of the North Basin in 1974 destroyed great numbers of fish. Of the chemical changes in the water resulting from contraction by evaporation (e.g. precipitation of calcium and magnesium and increase of alkalinity) the most harmful seems to have been the stirring and consequent reduction in dissolved oxygen. The great increase in numbers and rate of production of the phytoplankton increased the amount of material consuming oxygen. There was a rapid reduction of dissolved oxygen at night and on very cloudy days when respiration exceeded photosynthesis. In Lake George (p. 88) the daily winds normally keep the water in circulation through the 3–4 m column frequently enough to prevent a dangerous fall in oxygen. Only rarely do very heavy wind-storms stir into the deeper water and bring up highly reducing organic matter which consumes the oxygen and kills

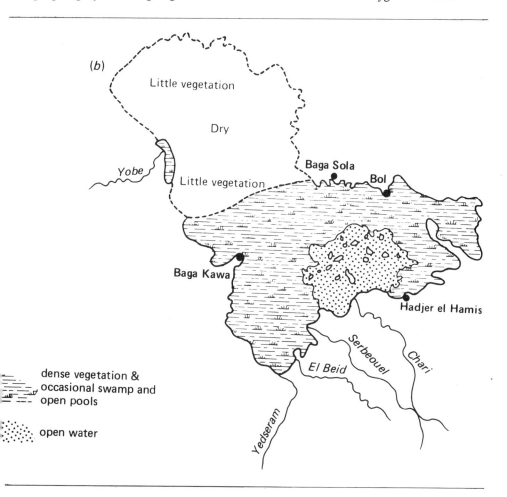

fish. In such shallow productive waters deoxygenation can thus result *either* from stagnation in the absence of wind *or* from deep stirring by very heavy winds. The shallower and more exposed the water the less powerful is the wind required to cause deoxygenation.

In 'little' Lake Chad such conditions have favoured frequent deoxygenation. In particular in 1972–74 the very shallow water in the North Basin was often deoxygenated by heavy wind storms. Immense numbers of fish died on these occasions and clogging of the gill filaments with suspended particles was an additional hindrance to adequate respiration.

Heavy storms seem to have accounted for several fish-kills in the Archipelago region near Bol in 1973. During 1974 much of the dense vegetation that had covered the exposed sediments had died and later decomposed when re-submerged by the annual flood water from the Chari. This was another cause of deoxygenation.

Records of fish catches during this period have shown a progressive diminution in numbers and the ultimate disappearence of several species. As expected, those species that are known to be especially sensitive to low levels of oxygen such as the Nile perch *Lates niloticus* and the tiger fish *Hydrocynus spp.* were the first to be missed by the fishermen.

Since Lake Chad has certainly suffered many such recessions in the past, some even more severe, we cannot be too pessimistic about the future, though the present state of the fishery is depressing enough. A succession of very high floods will however be needed to break the vegetation barriers and restore the 'normal' lake. It has been suggested (*pers. comm.* J. Lemoalle) that a very great loss of water by evapotranspiration from the vast expanse of emergent vegetation may be a significant hindrance to the recovery of the lake.

Anyone interested in the history of the lake and its fauna will wonder how the very rich soudanian fish fauna has managed to survive the recessions that are known to have occurred in the past. It is however generally agreed (Servant and Servant, 1970) that the severe desert conditions bringing sand-dunes (now partially submerged) across the present lake basin occured 12 000 years B.P. This was well before the final period of the Great Saharan Lakes when Lake Chad was probably in faunal connection with other soudanian waters (p. 198). Subsequent recessions, like the present one, have therefore never been so serious as to destroy the many species of soudanian fish. Sufficient permanent water must always have remained in the South Basin and lower Chari River to harbour representatives of all the species. Not only an adequate amount of water but also a variety of habitats were required for their survival.

13

Lakes Edward and George

The Rwenzoris, the most extensive of the high mountains of equatorial Africa, form an elongated block almost 100 km long at the base, rising from the Semliki plains south of Lake Albert (Mobutu Sese Seko) and extending southwestward to near the northern shore of Lake Edward (Fig. 3.4). Their mode of origin is a matter of controversy, but they seem to have been forced up as a single block in association with the formation of the Rift Valley. They form the southern wall of the main valley, down which the Semliki River flows from Lake Edward to Lake Albert, and separate it from a branch Rift containing Lake George. The highest peak, Mount Stanley, is over 5 000 m high and there are four separate groups of glaciers and snowfields, the lower limit of permanent snow being at about 4 500 m.

The great Ituri Forest of Zaïre crosses the Rift Valley and reaches the northwest slopes of the Rwenzori range which attracts an abundant rainfall (about 1 700 mm per year) brought partly by westerly winds across the humid Zaïre basin. The northwestern (Zaïre) slopes are more precipitous and the catchment area smaller than on the southeastern (Uganda) side of the watershed. More permanent rivers, whose volume varies greatly with the season, flow down the southeastern slopes into Lakes Edward and George than in the opposite direction into the Semliki River. This partly accounts for the abundance of standing water in the Edward-George basin where the annual rainfall is generally rather low (650–900 mm).

The Rwenzori (previously the Queen Elizabeth) National Park of Uganda surrounds Lake George and the eastern half of Lake Edward. The western half of the latter, including the outflowing Semliki River, is encompassed by the Parc National de Zaïre. This whole vast region of National Parks from Lake Albert to Kivu – the Great Rift Valley with its lakes, game plains and precipitous escarpments, the glaciated Rwenzoris and partially extinct Virunga volcanoes, the tropical rainforests of the Semliki valley, the mountain forests above 3 000 m and the alpine highlands above 3 800 m – present some of the most dramatic and beautiful scenery in Africa and, with its great variety of organisms and of conditions of existence, is of extreme interest to land and water ecologists.

The disposition of these two lakes is very peculiar. Lake George lies on a plain in a branch of the Rift Valley between the Kichwamba Escarpment and the Rwenzoris (Fig. 13.1). Its main inflows are the Rivers Nonge, Mbuku and Bumlikwesi and a branch of the Mpanga descending from the Rwenzoris. The other branch of the Mpanga is the reversed section of the Katonga River, which flows into L. Victoria, and drains the eastern escarpment (Fig. 14.1). The rivers from the mountains, after reaching the plain, filter through a large area of swamp before reaching Lake George whose surface area is about 290 km^2 and whose depth is nowhere more than about 3 m. It is connected to Lake Edward by the Kazinga Channel which is 40 km long and has a maximum width of less than 1 km. A casual observer from the shore of the channel would be unable to detect a flow in either direction. Only under the bridge at Katunguru, about half-way along its course, where the width of the channel has been artificially contracted to about 30 m, is it obvious that the water is generally flowing away from Lake George. The fact that the water of Lake George is fresh, with a conductivity (K_{20}) of approximately 200 which has been maintained for at least forty years (Beadle, 1932b), is evidence that there is enough flow-through to balance evaporation. There is no indication of a subterranean outlet. That there is a net outflow down the channel is further confirmed by the fact that the water of Lake George, which

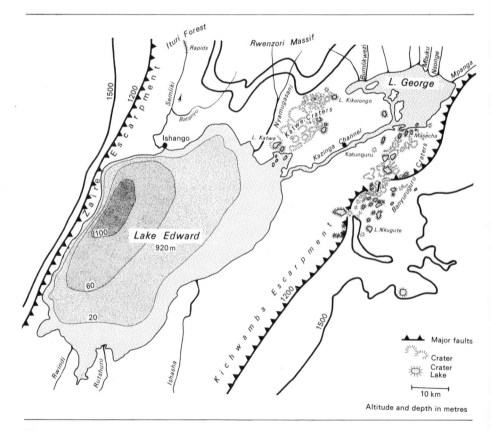

Fig. 13.1 Lake Edward and Lake George

is much less saline and more productive of algae and so visibly greener than that of Lake Edward, extends almost to Lake Edward. There is a rather abrupt boundary between the two waters marked by a change of colour, chemical composition and plankton. This boundary moves back and forth along the last 3 km of the channel according to the relative rainfall and evaporation in the basins of the two lakes. The force and direction of winds no doubt also play a part. The oscillation of the boundary between the two waters near the end of the channel was first noted in 1931 (Beadle, 1932b). The fall along the western half of the channel is less than 0·3 m in 24 km (Bishop, 1969). To the hydrologist Lake George is thus a kind of bay, though with a curiously constricted neck, of a much larger lake (Edward) whose outflow is the Semliki River. It is, however, obvious that the channel was formed under conditions different from the present – a matter to which we shall return later.

Very different is the character of Lake Edward, which is one of the Great Lakes of the Western Rift. Its length is about 65 km and the maximum width is 38 km. The deepest region (about 112 m) is a trench only 5 km from the western (Zaïre) shore from which the escarpment rises precipitously to highlands exceeding 2 500 m in altitude (Fig. 13.1). The eastern side of this trench is much less steep and rises with an almost uniform gradient for more than 30 km under water to the Uganda shore.

The main inflows to Lake Edward are the River Nyamugasani, which drains the southwestern end of the Rwenzoris, and the Ishasha, Rutshuru and Rwindi Rivers from the Kigezi and Rwanda highlands and the Virunga volcanoes in the south. The annual contribution from the Kazinga Channel is probably small compared with that from the rivers. The amount of water flowing through the lake, exclusive of evaporation, can be seen at the outflow via the Semliki River at Ishango in the northwest which is 30–40 m wide. The water leaves the lake as a rapid and turbulent stream about 3 m deep over rocks and boulders. It is so clear that the hippopotamus can be observed under water and large numbers of *Barbus* are seen facing the current. During its 250 km course to Lake Albert the Semliki drops about 300 m (920 to 620 m). Most of this drop is down several kilometres of rapids within the dense and inaccessible forest through which the river flows for about half of its course (Fig. 13.3). It is joined on the way by the streams which drain the northern slopes of the Rwenzoris and those from the opposite escarpment, and finally emerges on the Semliki plains as a more sluggish and winding river, and enters Lake Albert across a delta of its own making. The Semliki has all the appearances of a very young river which has as yet made little impression on the valley down which it flows. Geological evidence suggests that Lake Edward may only recently have overflowed by this route. Previously it appears that the upper section of the river may have flowed *into* Lake Edward and was subsequently captured by cutting back of the lower section which flowed into Lake Albert (de Heinzelin, 1955).

We must now consider the Pleistocene and recent past history of the Edward basin in the light of the geological and palaeontological evidence and of the composition of the present fauna. As in the Albert basin, three main periods in the Pleistocene are distinguishable from characteristic sediments which indicate a succession of three lakes.

1. Early Pleistocene sediments, whose fossil fish and molluscs are almost identical and which are probably contemporary with the Kaiso series of the Albert basin (p. 175). These beds are tilted upwards towards the west and are exposed along the northeast shore of Lake Edward and in the banks of the Kazinga channel within two kilometres of the mouth. They are exposed again on the Zaïre shore of the lake. As in the Albert Kaiso beds, most of the molluscs are now extinct, but it is remarkable that there are also the remains of *Crocodilus niloticus* and of the soudanian fish *Lates* and *Hydrocynus* which, though now living in Lake Albert and the Nile, are absent from the present Lakes Edward and George (Fuchs, 1937, 1939; Adam, 1957; de Heizelin, 1955, 1957).

The sediments are in general fine-grained, which suggests that the inflows into the early Pleistocene lake were relatively slow-flowing down rather slight gradients before the major rift faulting had started. The fossils are not distributed throughout the Kaiso beds but are concentrated in several lens-shaped bands of ironstone at different levels, some of which are almost entirely composed of molluscs with a few bones of fish and crocodiles (Fig. 13.2). These bands have been interpreted as shallow lagoons or bays subject to desiccation and to subsequent rather rapid flooding and deposition of further sediment which has preserved the fossils. It seems that the Kaiso lake was for a period very shallow and subject to occasional reduction to pools in which the animals were concentrated and died. It might, however, be mentioned that in Lake Tanganyika there are certain regions, such as near the south end in 20–30 m of water, where the bottom is largely com-

Fig. 13.2 Kaiso (early Pleistocene) fossil beds (dark bands) on the banks of the Kasinga Channel (see Fig. 13.1)

posed of the shells of molluscs. Prominent among these is *Neothauma*, a genus characteristic of the Kaiso beds but now surviving only in Lake Tanganyika. They have presumably been concentrated in this region by the prevailing water currents. Such concentrations of fossil molluscs are therefore not necessarily indicative of desiccation in shallow semi-isolated lagoons. Bishop (1969, p. 32) suggests that, rather than desiccation, the same kind of disturbance may have caused both the apparent periodic destruction of large numbers of molluscs and fish seen in the successive ironstone bands and the occasional mass mortalities fish now well known in tropical inland waters. These are caused by sudden usually violent weather stirring deoxygenated water to the surface. Another suggested cause is the volcanic discharge of toxic substances into the water, as may have occurred in Lake Edward in mesolithic times (p. 236). The final and complete extinction of the early Pleistocene Kaiso lakes with their characteristic molluscan fauna in both Albert and Edward basins was presumably due to the same climatic and tectonic causes, and it is very unlikely that fish, such as *Lates* and *Hydrocynus*, survived conditions which exterminated the molluscs.

Whatever the conditions during the Kaiso period, they were certainly widespread, since beds of the same character with a very similar fauna have been found at several points in both Lake Albert and Edward basins.

2. The mid-Pleistocene 'Semliki series' sediments which cover most of the Kazinga plain between Lakes Edward and George and are widespread in the Semliki valley. These sediments are much coarser-grained than the Kaiso which, according to Bishop (1969), indicates a major outburst of faulting during this period with consequent renewed erosion by rapid flow down steep slopes. No fossils have yet been found in the Semliki sediments in the Edward basin, and we have no direct evidence of the fauna of this (second) Pleistocene lake.

3. The late Pleistocene to recent period during which the most important event was the eruption of volcanic ash from the Katwe and Banyuruguru craters on either side of the Kazinga channel (Fig. 13.1). The course of events during this critical period is suggested by the superimposed series of beds at Ishango near the mouth of the Semliki River. These date from the Kaiso (early Pleistocene) to mesolithic (8 000–10 000 years ago). The mesolithic beds have yielded human remains, stone tools and fishing harpoons* and, more important in the present connection, a kitchen midden containing the bones of *Lates niloticus* and *Barbus bynni*, both now absent from Lake Edward but living in Lake Albert and the Nile (de Heinzelin, 1957; Greenwood, 1959; Cole, 1954, p. 248). These striking facts point to two main conclusions concerning the history of the aquatic fauna of the Edward basin, though more confirmatory evidence would be welcome.

First, if the soudanian fish of the Kaiso Lake Edward were, as argued above, exterminated before the mid Pleistocene, they must have reinvaded the basin between that time and the mesolithic. There may have been a passable connec-

---

* de Heinzelin (1964) discusses the importance of the Ishango mesolithic culture and develops an interesting thesis concerning the diffusion of techniques for making fishing harpoons from Ishango, down the Nile and northwestwards through Chad and into the Sahara as far as Araouane, where they were used in the neolithic lake.

tion between the Edward and Albert basins during the time of the mid Pleistocene (Semliki period) lake after the restocking of Lake Albert from the Nile following the extinction of the previous Kaiso Lake Albert (p. 174). It is worth noting that from recent studies of past glaciation on the Rwenzoris there is clear evidence of repeated advances and recessions of the glaciers during the Pleistocene. According to Livingstone (1962, 1967), carbon dating of sediments from a small lake (Mahoma) at 3 000 m indicates that the last advance reached its maximum about 15 000 years ago. This was before the final reduction of the mid Pleistocene lake and the subsequent volcanics, and we must presume that the rainfall in this region was greater, the temperature lower and the lake levels consequently higher than at present. The climatic as well as geomorphological conditions for a connection between the Albert and Edward Basins, that was passable to fish, may have existed at that time.

Second, the undoubted extinction of some of the soudanian fish, as well as the Nile crocodile, in the Edward basin was contemporary with the onset of the volcanics about 8 000–10 000 years ago. The mesolithic beds at Ishango are immediately overlayed by volcanic ash. We are tempted to speculate on a causal connection. Enormous quantities of ash containing, perhaps, soluble and toxic constituents were emitted and spread over most of the basin. This might well have been fatal for many fish, especially species such as *Lates* which seems to require clear and well-aerated water. If these eruptions were contemporary with the contraction of the lake through tectonic or climatic causes* the destruction could have been very great. This is supposed to have happened in Lake Kivu during the Virunga volcanics (p. 271). Whatever the cause, the certain fact remains that several species disappeared during the volcanic period.

The Kazinga Channel, with its present fluctuating direction of movement but with a net flow into Lake Edward, is one of the curious phenomena of this region. It would seem that a rather fast-flowing stream during a relatively short period is required to cut a well-defined and steep-sided bed through the very soft unconsolidated alluvial deposits. According to Bishop (1969) the early stages of cutting occurred during the final exposure of the mid Pleistocene (Semliki) lake sediments and before the onset of the volcanics. Its winding course has been caused by transverse faults. It could be imagined that, during the contraction of the Semliki lake, the floor of the shallow Kazinga–Lake George gulf began to emerge, leaving perhaps swamps and small lakes draining into Lake Edward. To confine this drainage into a single channel would seem to require a rather sudden drop in the level of Lake Edward, perhaps by faulting at the mouth of the Semliki. The rapid run-off could then cut a channel through the old Semliki lake sediments until the new level of Lake Edward was reached. From the present inadequate evidence other equally speculative theories could be proposed.

Both the early and the mid Pleistocene lakes spread into the plains in which Lake George and the Kazinga Channel now lie. But Lake George in its present form seems to have originated after a period of desiccation during and immediately following the volcanics (*c*.4 000 years ago). Studies of the ^{14}C age, the chemic-

---

* The existence of traces of ancient soils at the top of the mid Pleistocene sediments and under the volcanic ash on the Kazinga plain (Bishop, 1969) points to a contracted lake at the beginning of the volcanics.

al composition (Viner, 1977) and the diatom floras (Hawarth, 1977) of sediment cores provide evidence of this. Below a depth of about 2 m into the sediments is the surface of a clay and volcanic ash layer which is taken to mark the end of the volcanic period about 4 000 years ago which was followed by a short dry period suggested by a shallow layer of sand. Since then this lake has had a continuous existence and has deposited a series of layers of increasing organic content which show a progressive change in the diatom flora from *Melosira* and *Fragilaria* to *Nitzchia* and *Synhedra* as dominant forms. This transformation is apparently correlated with a change from oligotrophic to eutrophic, conditions. It may also be deduced from the diatom zonation that the present flora and conditions in the lake have persisted for about a thousand years.

The absence of crocodiles (*Crocodilus niloticus*) from the Edward basin is another curiosity which has encouraged speculation.* Since no remains have been found dating from later than Kaiso (early Pleistocene) we must assume that they disappeared with the rest of that fauna in the mid Pleistocene. A more immediate problem is raised by their failure to repopulate Lake Edward after the Semliki connection was established. The rapids in the Ituri forest are understandably a barrier to the upward migration of the soudanian fish such as *Lates* and *Hydrocy-nus* which are found below them. The crocodiles were also particularly abundant in the lower Semliki until recently decimated by hunters. They might, however, be expected to avoid the rapids and walk past them on land.

Unfavourable composition of the water of Lake Edward and lowered tempera-ture of the Semliki by the inflows from the Rwenzoris have both been suggested to explain the absence of crocodiles, but without supporting evidence. But fish, of which there are plenty in Lake Edward, are more susceptible than crocodiles to water composition, and there is nothing to suggest that Lake Edward is not a very favourable environment for all aquatic life. Measurements made in 1931 (Beadle, unpublished) on five of the streams descending from the Rwenzoris, including one of the largest (River Butangu), showed that their temperature rose to well over 20°C before they joined the Semliki, which, just above the rapids, after it had already received the water from these streams, had a temperature of 25°–26°C.

Anyone, however, who has visited the rapids (Fig. 13.3) which in one section flow as a roaring torrent, audible from a great distance, in a vertical-sided rock cleft about 10 m wide and more than 1 km long, and has had to make his way through the dense rainforest which overhangs the banks and cliffs in a tangle of branches, bushes and lianas, will at once think of another possible explanation. Since the rapids are apparently insurmountable by both fish and crocodiles, the latter have no alternative but to migrate through many kilometres of thick forest. Apart from the problem of food there would be little incentive to make such a journey with no opportunity for basking in the sun which, for crocodiles, appears to be a physiological necessity.

* A single live specimen of the small crocodile *Osteolaemus tetraspis* was found by Temple-Perkins (1951) in a stream in the forested hills south of Lake George. It is a forest-breeding species known from the hill-streams of the Zaïre forests. This specimen was perhaps the last survivor in the Ed-ward basin where it has presumably been exterminated and its habitat disturbed by hunters and settlers.

Fig. 13.3 Rapids on Semliki River in the Ituri Forest, Zaïre (see Fig. 13.1).

A study of the past and present fish fauna of the basin thus suggests that since the early Pleistocene the aquatic fauna has twice been partly, but apparently not wholly, destroyed. The nature of these catastrophic events is doubtful, though tectonic and climatic factors were both probably involved in the mid Pleistocene, and there is suggestive evidence that volcanic eruptions played a part at the end of the mesolithic period.

The present fauna of the two lakes is such as might be expected to result from such a chequered history. The total number of species of non-cichlid fish is about thirty. Some of the genera characteristic of Lake Albert and the Nile are absent (e.g. *Polypterus, Hydrocynus, Distichodus* and *Citharinus*), and certain entire families are missing (Characidae, Centropomidae, Schilbeidae, Mochokidae, Mastacembalidae, Malapteruridae) (Worthington, 1932; Trewavas, 1933). There are in addition nine other species which are not in the lakes, but are confined to the inflowing streams. A few typical soudanian fish now living in Lake Albert and the Nile have somehow survived or have managed to re-enter the lake. Such are *Bagrus docmac* and *Sarotherodon niloticus*. A remarkable case which is difficult to explain is *Sarotherodon leucostictus* found in both Lake Edward and Albert, but nowhere else. Its biology has been studied by Lowe (1957).

Until recently about twenty-three species of cichlid fish had been recorded from Lakes Edward and George. Many of these are closely related to species in Lake Victoria and five were reported to be identical in both lakes (Trewavas, 1933; Greenwood, 1965b, 1966, and 1973, p. 3). From this it had been concluded that there was a very recent connection by which these species had moved from Lake Victoria to Lake Edward, and that some, but not all, have since diverged from the parent stock. The existence of a former connection is consistent with the geological evidence. It is now generally thought that Lake Victoria originated after the middle of the Pleistocene as the result of upwarping of the western side of the basin which caused 'back-ponding' of the previously west-flowing rivers (see Ch. 14). The newly formed lake would probably have continued for some time to drain by the Katonga valley westwards into the Edward basin. Further uprising finally stopped the outflow and reversed the Katonga which now flows into Lake Victoria. According to this theory the connection between Lakes Victoria and Edward existed as recently as the late Pleistocene – a matter of a few thousand years ago. The mesolithic beds at Ishango (p. 236) apparently show that much of the fish fauna of Lake Edward was destroyed during the volcanic eruptions 8 000 – 10 000 years ago. Victorian fish may therefore have arrived in the Edward basin after that time, and the evidence from both ends, though still rather uncertain, appears to support a very late connection which could have provided the route.

On the other hand many more species of cichlids have now been found in Lake Edward, and the total number there and in Lake George is likely to approach 100, most of them endemic (pers. comm. P. H. Greenwood). Moreover, further examination of fresh specimens of the five species supposedly identical with species in Lake Victoria has shown that three of them (*Haplochromis macrops, ishmaeli* and *guiarti*) are different. The other two (*H. nubilis* and *Astatoreochromis alluaudi*) are certainly found in both lakes, but they are also widespread elsewhere in this region of Africa. The cichlid fish of the Edward and Victoria basins have therefore diverged further from each other than was previously thought. It would perhaps be more consistent with the evidence to suggest that the divergence began at an early stage during the early Pleistocene uplifting of the western edge of the Victoria basin when the Victoria and Edward basins were sufficiently interconnected to have a more or less uniform fauna, but were separated enough to prevent interbreeding between the populations in the two incipient lakes.

The relation between the faunas of Lakes Edward and Victoria has been discussed partly to demonstrate the difficulties in fitting together all the known biological and geological facts to give a coherent and consistent historical picture of a region which has undergone such drastic and complicated changes. Many uncertainties remain and even the geological background cannot yet be regarded as fully established.

The composition of the fish fauna of Lakes Edward and George suggests that these lakes are in the process of recovery and recolonisation after some partial extinction. A study of the past and present Mollusca gives the same impression. The early Pleistocene (Kaiso) molluscs mostly disappeared, and since the mid Pleistocene the composition of the molluscan fauna came progressively to resemble that now existing. The number of species which have survived is, however, considerably smaller than had existed in previous periods during the Pleistocene, and is fewer than in other large African lakes at the present time (Mandahl-Barth, 1954; Adam, 1957).

General surveys of the invertebrate fauna of Lake Edward were made by

Damas (1937) and by Verbeke (1957a). In Verbeke's Table 37 (p. 162) the number of species of turbellarian worms, insects and Crustacea, though greater than in Lake Kivu, is not very different from that in Lake Albert. Very many species are common to both lakes. Compared with fish the invertebrate groups, other than molluscs, are in general more mobile out of water or are better adapted with resistant stages for passive transport. It is therefore not surprising that Lake Edward, in which conditions are now favourable for them, is well populated by invertebrates. But we must presume that, like the fish and molluscs, they were seriously reduced in the recent past.

Some concerted production studies on Lake George were made during 1966–71 under the auspices of the International Biological Programme. The results were reviewed and discussed by members of the team at a Symposium (Greenwood and Lund, 1973) and by Burgis (1978). The work on water circulation and primary production has been discussed in Chs 6 and 7. The zoological work will be reviewed briefly followed by some general conclusions concerning the ecosystem as a whole. Photosynthetic production by the phytoplankton is clearly the major source of organic matter in this lake. Except in the mouths of the inflows and around the edge of some of the islands, the inshore waters are devoid of submerged macrophytes (Lock, 1973). This is probably due to the absorption of most of the light in the upper metre by the very dense phytoplankton. Also, in the absence of rocky shores, there is no large source of food from attached algae and associated organisms (*Aufwuchs*) which is an important food in many other lakes. Of the high gross production on the part of the planktonic algae (Table 7.1) a very large proportion, probably well over 95% of the energy so fixed, is dissipated in respiration by the plankton community (phyto- and zooplankton and micro-organisms).

In addition to providing inorganic nutrients for photosynthesis by aquatic plants, the inflows from rivers and flooded land must bring in some organic matter synthesised by land plants and by the emergent vegetation in the great areas of swamp which choke the main inflows from the north. The size of the contribution from these sources is unknown. Another source of organic matter as well as of nutrients that may be significant is the large population of hippopotamus in this region. These big animals feed on land at night and lie around defaecating and urinating in the water during the day, thus transferring to the water considerable quantities of material photosynthesised on land.

Of the zooplankton the most abundant species is the herbivorous cyclopoid copepod crustacean *Thermocyclops hyalinus* (Burgis, 1969, 1970, 1971b, 1974). In all, two species of Copepoda, four of Cladocera and sixteen of Rotifera were commonly found in the open water. As would be expected, there were differences in the abundance and composition of the zooplankton between the open and inshore waters and the swamps draining into the north end of the lake. During 1967 to 1970 there were seasonal increases in the total number of individual animals mainly of the Copepods. These changes were most marked in the open water where the numbers rose to a peak at the beginning of the rains in September/October and in February/March. The increases thus occurred during the dry seasons and their connection with climatic or hydrologic events is unexplained. Nevertheless, these seasonal fluctuations in numbers are very small compared with those reported from temperate lakes where there is a great outburst in spring and a catastrophic reduction in winter.

A more detailed study was made of *Thermocyclops hyalinus*. From such data as the age structure of the population, egg numbers, rate of development before and after hatching and the biomass from population density and from micro-estimations of the carbon content of individuals, a tentative estimate was made of the rate of recruitment and loss in the lake and of the gross production of this species at several sites and by calculation in the lake as a whole. Such estimates, and the methods and assumptions upon which they are based, however, are subject to error. Important in this connection, and a subject about which we know very little, is the nature of the food of the zooplankton, some members of which, like *Mesocyclops leuckarti*, are carnivorous.

Of the benthic fauna, other than micro-organisms, the oligochaet worms are an important component, but it is difficult to assess their position and importance in the production cycle leading to fish because their bodies have no skeletons and rapidly disintegrate in the gut of predators, and so escape detection. The most obvious benthic animals are the larvae of chaoborid and chironomid lake flies (McGowan, 1974). These, as in other lakes, periodically emerge from the water surface in large swarms (Fig. 4.2) in which *Chaoborus* spp. are the most numerous. The early larval instars are confined to the mud, but the final stages migrate upwards mainly during the night and catch and devour the Crustacea and Rotifera in the zooplankton (Fig. A3). They are themselves eaten by certain fish (Table 13.1) particularly when they swarm to the surface as emerging pupae.

The general stability of the lake's fauna is well demonstrated by the benthos (Darlington, 1977). Regular collecting from nine selected stations during six months (a complete seasonal cycle) showed the near constancy in space and time in the abundance and composition of the main benthic species. The only significant regional difference was between (a) the open mid lake where the mud is most exposed to the daily stirring by winds, and actively swimming larvae of *Chaoborus* spp. and the ostracod crustacean *Cyprinotus* sp. were dominant, and (b) the inshore region which is more protected from winds and the mud is more compacted by debris from the emergent macrophytes (papyrus, etc.). Here is an environment better suited to burrowing animals such as chironomid larvae and oligochaet worms. These two regions of course grade into one another.

A striking feature of the benthos is the scarcity of molluscs which are apparently represented only by one species of gastropod *Melanoides tuberculata*, and this has a spasmodic distribution and is nowhere abundant. Other species are found among the littoral vegetation and contribute to the diet of the mollusc-eating fish (Table 13.1). This is in great contrast to the similarly shallow Lake Chad in which molluscs account for most of the benthic biomass (p. 222). We can only speculate on the causes of this difference. There is a much greater variety of sediments in Lake Chad (p. 220). In Lake George a soft flocculent organic mud covers most of the bottom, sandy surfaces being relatively small in area. There is a comparatively small number of species left in the Edward-George basin after recent extinctions, but the molluscan fauna of Lake George is much poorer even then that of Lake Edward.

Discussion on the ecology and production of the fish and fisheries of Lake

Table 13.1 The food preferences (but not the exclusive foods) of the common species of fish in Lake George, from Dunn (1972) modified with later information from P. H. Greenwood. The taxonomic status of several of the species of *Haplochromis* is still in doubt.

Main food preference	Group A: Adults larger than cr. 15 cm (total length)	Group B: Adults smaller than cr. 15 cm (total length)
Phytoplankton	*Sarotherodon leucostictus* †**S. niloticus*	*Haplochromis nigripinnis*
Macrophytes		H. schubotzi
		H. eduardianus
Aufwuchs		H. limax
Insects (larvae and/or adults)	†**Barbus altianalis*	*Aplocheilichthys pumilus*
		Barbus kersteni
		B. neglectus
		Ctenopoma muriei
		Haplochromis angustifrons
		H. elegans
		H. schubotziellus
		H. wingatii
		H. aeneocolor
		Marcusenius nigricans
Molluscs	†**Protopterus aethiopicus*	Haplochromis mylodon
		H. taurinus
Zooplankton		*Aplocheilichthys pumilus*
		A. edwardensis
		Haplochromis pappenheimi
		Young stages of many species
Fish	†**Bagrus docmac*	
	†**Clarias lazera*	
	Haplochromis squamipinnis	

* numerically important
† commercially important
Note. H. pappenheimi, previously known from Lake Edward, but recently found in Lake George, is the only zooplankton-feeding cichlid so far found in these two lakes and may well be an important link in the food-chains.

George has been published by Dunn (1972) and Gwahaba (1974, 1975, 1978). In addition to the cichlids approximately twenty species of fish have so far been reported from the lake. Only five grow large enough to be commercially important. These are all soudanian species *Sarotherodon niloticus, Barbus altianalis, Protopterus aethiopicus, Bagrus docmac* and *Clarias lazera*. They are dependent on the primary production in different ways. The most important commercial fish, *Sarotherodon niloticus*, feeds directly on the plankton, filtering out particulate matter by means of special structures on the gill arches described by Greenwood (1953) in *T. esculenta* of Lake Victoria. As already mentioned (p. 51), it has recently been established that, contrary to previous opinion, *Sarotherodon niloticus* and *Haplochromis nigripinnis*, and presumably other phytoplankton-feeding cichlids in African lakes, consume and assimilate the blue-green algae and in particular *Microcystis* which is the major constituent of the phytoplankton in Lake George. Much of the most abundant product of photosynthesis thus passes directly into the fish.

The food preferences of the fish of Lake Edward were studied by Verbeke (1959b) and Hulot (1956). Those of Lake George are summarised in Table 13.1. It can be seen that there are a large number of small fish, as well as one of the larger commercial species, *Barbus altianalis*, that feed mainly on insects and particularly on the larvae and pupae of chironomid and chaoborid flies. None of the small species fish listed in Table 13.1 is of direct commercial value, but most of them are preyed on by the three large fish-eating species. Together they appear to exploit all the main sources of food in the lake, much of which is thereby transferred to the large predators, especially *Bagrus docmac* and *Clarias lazera*, and these are both important commercial fish.

In conclusion it may be said that the cycle of energy and materials in Lake George involves an exceedingly high rate of gross primary production on the part of the phytoplankton supported by a very rapid recycling of nutrients. This in turn is maintained by a very frequent circulation of water between the mud and the euphotic zone. It is probable that this high rate of production has persisted with little seasonal variation since the origin of the lake in its present form and climatic regime. Though the net primary production is very low by comparison with the gross, enough is passed on to support a very productive fishery. A distinctive feature of this type of lake, in which it also differs from the very productive Lake Chad, is the direct consumption of the phytoplankton by large populations of the cichlid fishes *Sarotherodon niloticus* and *Haplochromis nigripinnis*, the former being the most important commercial fish in the lake. The phytoplankton also enters the production cycle via the most abundant member of the zooplankton, the herbivore *Thermocyclops hyalinus*, which, in turn, is consumed by the carnivorous *Chaoborus* larvae and by several species of fish and by the young stages of many more. There are, of course, many other pathways from algae to fish than are implied in Table 13.1, such as through the benthic animals and the insects, molluscs and other organisms inhabiting the littoral vegetation. But the productivity of the fishery depends primarily on the direct consumption of the phytoplankton.

Apart from the losses through the respiration of all the organisms in the ecosystem, any attempt to construct an energy and materials budget for the lake as a whole must take account of losses in the outflow, in the swarms of emerging lake flies which fall onto the land and in the fish removed by fishing birds as well as by man. Whatever the quantities involved in the various gains and losses, the residual net resources of energy and materials have been enough to support a rich fishery that reached its peak in the early 1960s (Hickling, 1961, p. 185).

Of all the tropical lakes so far investigated Lake George is distinguished by the overall stability of the biomasses of its organisms and of the rate of production. The relatively small fluctuations that have been recorded over six years oscillate round an annual mean which is very constant. In deeper tropical lakes marked seasonal changes are caused by alternating windy and calm seasons which affect the circulation of nutrients, and in some shallow lakes (e.g. Chad) by the seasonal rains within the catchment which produce a massive inflow of nutrient-rich water as well as flooding surrounding land. With sufficiently frequent winds at all seasons the circulation of nutrients is more or less continuous in the shallow water of Lake George. However, what is more remarkable is the very constant

Fig. 13.4 Lake George and the foothills of the Rwenzoris with approaching storm

water level which has varied by only about 0·1 m during 1966–72 (pers. comm. A. B. Viner). There have been periods, as in the early 1960s, when the levels of all the great lakes were much higher than now and Lake George flowed into the crater lake Kikorongo for a few weeks (Fig. 13.1). But this involved less than a one metre rise in level. It is obvious that the level of Lake George is partly controlled by that of the very much larger Lake Edward which must thus act as a kind of 'buffer'. The inflows to both lakes are mainly derived from high montane forest regions, especially on the Rwenzoris, where the flow is partly maintained by the high rainfall, which is not subject to very great seasonal fluctuations. It is probable that the steep fall of the River Semliki behind the rock still at the outlet from Lake Edward maintains an unimpeded outflow which can rapidly compensate for an increased inflow and thus acts as a natural spillway. The ecological stability of Lake George is thus controlled by a combination of fortuitous circumstances. The delicacy of the balance has been demonstrated by the disastrous consequence (fish-kills) of occasional abnormal calm periods followed by outbursts of violent wind.

14

Lake Victoria

Though first found by Europeans later than Lakes Tanganyika and Malawi, Lake Victoria was, at least until the First World War, better known, and was, indeed, the only African Great Lake that many people outside Africa had heard of. This was partly due to the dramatic circumstances of its discovery by Speke in 1859 and 1861, bringing the final solution to the problem of the sources of the Nile that had fired imagination through most of recorded history. The subsequent much publicised visit in 1875 by Stanley, the great journalist-explorer, both confirmed Speke's statements and brought the lake and the peoples living around its north and west coasts to the notice of a very wide public. As the reservoir that maintains the basic flow of water down the Nile, the lake immediately became of great interest to hydrologists.

Before the arrival of the Europeans, and for some time after, the lake provided the best means of transport for men and goods between what are now Kenya, Uganda and northwestern Tanzania. A voyage by canoe along the inshore waters of the north and west coasts of the lake was easier and safer than the exhausting journey on foot which, among other hazards, involved the crossing of innumerable swamp courses that drain into the lake.* Many of the early missionaries to Uganda arrived, like Stanley, by canoe. Though for a long time past the Arabs had been trading in this area for ivory and slaves, it seems that they did not put their sailing dhows onto the lake until about 1880. With the great expansion of trade between the three countries bordering the lake, the dhow became increasingly important and remained a serious competitor of the steamship up to 1940. But the short-circuiting of the railway to Kampala in 1952 and the improvement in the rail-steamship services finally reduced the dhow traffic to insignificance. The new motor roads and airports during the past thirty years have further reduced the relative importance of lake transport, though the absolute quantity of men and goods carried in steamships is greater than ever, and two freight-train-carrying ships have recently been launched.

This brief summary of the history of Lake Victoria as a means of communica-

---

* Bishop Willis, a missionary in Uganda during the early years of this century, is said to have had boots made in London punched with holes to let the water out.

tion has been presented partly to emphasise its great importance to East Africa during most of the past hundred years. An interesting account of this subject up to the mid-1950s has been written by Ford (1958). Garrod (1961) has reviewed the history of the fishing industry on the lake and its fluctuations in response to technical innovations, legal regulations and developments in transport and marketing.

The hydrology of the lake thus become of interest at an early stage, and the first bathymetric survey was completed in 1900. Regular measurements of water level have been recorded since 1899 (Fig. 14.3).

Fish had always been an important food for the tribes living along the coast, especially for the Luo around the Kavirondo Gulf who were noted as expert fishermen and boatmen. Several individual biologists had collected material from the lake and the general character of the fauna and flora was known before 1920. But the first systematic biological investigation was the fisheries survey in 1927–28 by Graham (1929). This was done from a small steamboat and, besides investigating the fish and fisheries over the whole lake, some hydrological data relevant to ecology (temperature and some chemical characteristics) were collected. Little more was done until after the Second World War, but from about 1948 onwards our knowledge of the ecology and fisheries of the lake increased rapidly with the founding of the Lake Victoria Fisheries Service and of the Fisheries Research Laboratory at Jinja, both partly financed by the three countries bordering the lake. Some investigations were also undertaken by the Government Fisheries Departments and more fundamental research relating to the origin and biology of the lake was done by the Uganda Geological Survey and by the scientific departments of Makerere University at Kampala.

In spite of all the interesting and important information that has been collected, it is more difficult to present an integrated picture of the origin and essential hydrological and ecological features of Lake Victoria than of most of the other Great Lakes of Africa, about which a smaller amount of detailed information is available. This difficulty no doubt arises on the one hand from the complicated and still doubtful nature of the geological and climatic events that have determined its history and, on the other hand, from its immense size and its much-indented shoreline with many large and shallow swampy bays which occupy a high proportion of the total surface area and are ecologically very important but different from each other and from the open water. Generalisations on the ecology and productivity of the lake as a whole are not likely to be helpful.

The origin and past history of Lake Victoria has been the subject of some controversy. Since the pioneer geological work of Wayland (summarised in 1934) there has been much written on the subject, but the principal established facts and their various interpretations can be followed from a few recent publications which provide references to all the relevant literature – Bishop (1969) for geology; Bishop and Posnansky (1960) for archaeology; Kendall (1969) for climatic and ecological history from the study of sediment cores; Worthington (1954) and Greenwood (1951a, 1965a)) for distribution and speciation of fish, and Temple (1969) for a recent interpretation of combined geological and zoological evidence. Most readers, after studying this literature, will, I believe, conclude that a wholly convincing interpretation has yet to be presented, and then only in the light of more geological evidence and perhaps with a greater understanding of the process of speciation in cichlid fish. This is, however, far from a purely academic subject, since the history with which we are concerned is very recent. An understanding of the climatic, tectonic and biological events since the mid-Pleistocene would not only explain the present situation, but would probably give some indication of future trends and might enable us, if not to control them, at least to prepare for them. The great economic importance of Lake Victoria to East Africa and to other countries in the Nile basin gives some point to this argument.

There is good evidence of an earlier lake, or system of lakes, whose fauna is represented by fossils in sediments of Miocene age from Rusinga Island and from the neighbouring mainland near the coast of the Kavirondo Gulf at the northeast corner of Lake Victoria. This was named 'Lake Karunga' by Wayland (1931). According to Kent (1944) a lake was formed during the early Miocene by volcanic blockage, and included the area of the present Kavirondo Gulf and extended over the Kisumu plains to the east and some distance beyond Rusinga Island to the west (Fig. 14.1). Kent suggested that this lake was finally drained westward by faulting and erosion during the mid Pleistocene. The fish fauna was undoubtedly 'soudanian' in character since the Miocene fossil beds contain the remains of both *Lates* (the Nile perch) and *Polypterus*, neither of which are indigenous to the present Lake Victoria, but are typical of the soudanian water systems from the Nile to West Africa (Greenwood, 1951b). Though it is unlikely that *Lates* would survive a reduction of the old lake to residual swamps, this could hardly account for the extinction of *Polypterus*, which has a lung and is a normal inhabitant of swampy regions. On both geological and biological grounds therefore it is difficult to avoid the conclusion that the Miocene lake system was finally and completely drained and that the present lake arose *de novo*. What happened to these fish when the Miocene lakes disappeared, and why did they not survive in the west-flowing rivers at that time crossing the future Lake Victoria basin? Why did they not continue to live in Lake Victoria?

The major points at issue, some of which are still subject to controversy, are: (a) the faunal connections of the Miocene lake system and the causes of its disappearance together with its fish, (b) the time and mode of origin of the present Lake Victoria and of its previous connection with the Lake Edward basin; (c) the origin of its fish fauna; (d) whether climatic changes have played a significant role in determining the history of the lake and the evolution of its fauna, in addition to earth movements which have without doubt been a major influence.

One of the difficulties is raised by the very large gap in the geological evidence. Recent dating by the potassium-argon method has shown that Lake Karunga existed at least 22 million years ago, that the main mammalian fossil beds on Rusinga Island are more than 18·5 million years old, and that the latest Karunga fossil fish from the Miocene Rusinga beds must be at least 16·5 million years old (the age of the overlying volcanic lavas). Since that time until the Pleistocene (within the past 2 million years) we have as yet no stratigraphical evidence from which we can get a picture of the events between the decline of Lake Karunga and the birth of Lake Victoria (Bishop *et al.*, 1969, van Couvering and Miller, 1969).

Lake Victoria is situated across the Equator at an altitude of 1 240 m. It is one

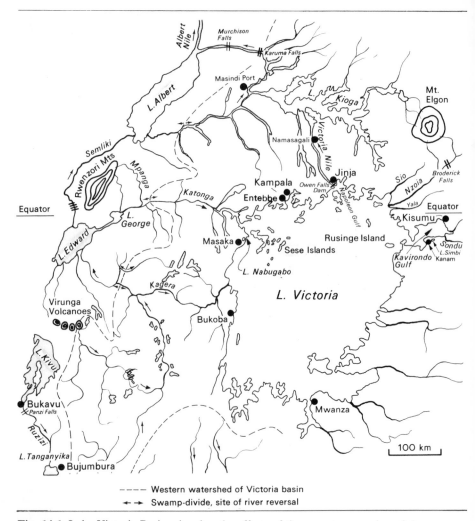

Fig. 14.1 Lake Victoria Basin, showing the effects of the recent upwarping of the western side of the basin which reversed the upper sections of the previously west-flowing rivers and flooded the basin to form the present lake. The overflow to the north then flooded the upper branches of a river to form Lake Kioga which overflowed the escarpment (Murchison Falls) to join Lake Albert and the Albert Nile.

of the largest lakes in the world with a surface area of about 75 000 km², and lies in a shallow basin between the two uplifted ridges along which the major rifting has taken place. Its origin is thus quite different from that of the large Rift Valley lakes, compared with which it is relatively shallow with a maximum depth of about 80 m.

Whatever the extent and connections of the former Lake Karunga, the configuration of the land and water systems in this region during the Miocene, before the major uplift and faulting, was totally different from now. The best interpretation of all the data would seem to be that, during the Miocene, Pliocene and part at least of the Pleistocene, the main drainage was by a series of rivers flowing

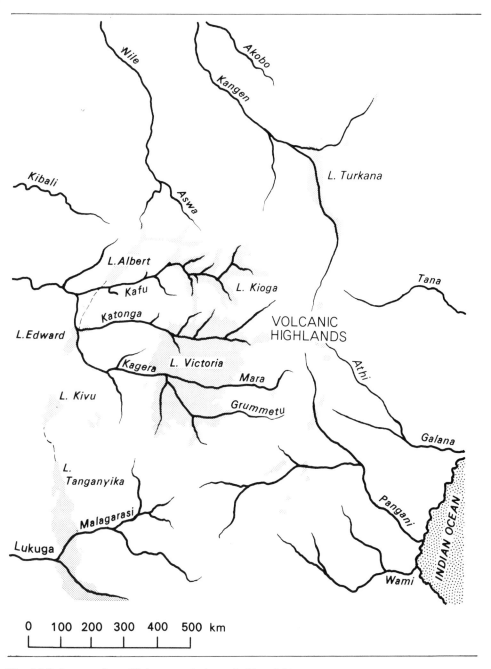

Fig. 14.2 Supposed pre-Pleistocene drainage in East Africa

westwards from high land in what is now western Kenya, across the regions of the future Victoria basin and Western Rift (Fig. 14.2). The ultimate destination may, as now, have been the Nile.

The hydrology of the Lake Victoria basin shows some very peculiar features (see Fig. 14.1). For example, the angles at which their tributaries join them suggest that the rivers Kafu and Katonga are flowing westwards to Lakes Albert and George in the Western Rift. In fact, they are both flowing to the east. There is, however, no doubt that at one time they flowed to the west and that their direction has been reversed by upwarping of the land on the western edge of the basin with which the faulting and formation of the Western Rift were associated. This explains, among other features, the remarkable fact that on the course of each of these two valleys there is a swampy watershed from which the water flows in both directions (Fig. 14.1). To the west the courses of the Rivers Nkusi and Mpanga are short and steep down over the escarpments to Lakes Albert and George. The Mpanga now receives most of its water from the Rwenzori Mountains. From the swampy watersheds the eastward courses of the Rivers Kafu and Katonga are very long and become progressively less steep as they cross the flatter country on the floor of the basin and become for the most part slow-flowing swamp courses draining into Lakes Kioga and Victoria. It is clear that these wide valleys choked with swamp vegetation, which are so characteristic of the northwestern shores of Lake Victoria, could not have been made by the present sluggish water-courses. They must originally have been cut by the rivers that flowed at a greater pace from the east. The Kagera River, which is now the major permanent inflow to Lake Victoria and could therefore be regarded as the 'source' of the Nile, is also, though less obviously, a reversed water course. There are 'swamp-divides' at two points along its western watershed (Fig. 14.1). The present Kagera River, through the uprising of the Virunga volcanoes in the late Pleistocene, now receives large quantities of water from them and from the Rwanda highlands, water which previously flowed to the north (Fig. 15.1).

Bathymetric surveys of the lake have revealed a pattern of contours giving some indication of previous east–west valleys across the region (Fig. 14.2). (See also Bishop and Trendall, 1967.) Deep cores into the sediments and geophysical measurements should confirm that they are the remains of river valleys which flowed to the west before the lake covered them.

The latest views on the origin of the present lake are summarised and discussed by Bishop and Trendall (1967), Bishop (1969), Temple (1969) and Kendall (1969). It is now generally agreed, and supported by much geological and archaeological evidence, that the Victoria basin was formed as part of a complex of earth movements which began in the Miocene and eventually resulted in the faulting of the Western Rift Valley. This involved the gradual uplifting of a ridge of land on the western side of the present basin. These movements began seriously to affect the hydrology of the region in the mid Pleistocene, and much of the previous east–west drainage was reversed by the rising land on the west so that the major rivers, Kagera and Katonga, were 'ponded back' to form branching and probably at first swampy lakes similar to the present Lake Kioga (Fig. 14.1). Further flooding with continuing uplift caused these two lakes to join, and Lake Victoria was born. It is possible that the uplifting was not rapid enough to prevent the rising lake from overflowing for a long time to the west by the Katonga valley. Lake Victoria might in fact have been draining into the Edward basin in the Western Rift well into the late Pleistocene – a matter of great significance in relation to the composition of the present fauna of Lake Edward (p. 238). Bishop

(1969, p. 107) estimates that this outlet was finally blocked by the continued uplift 25 000 to 35 000 years ago. The lake was then at a very high level, over thirty metres above the present lake, and flooded much of the present valleys of the Kagera and Katonga. The past 20 000 years has been a period of relative tectonic stability in this area, but the level of the lake has dropped in three steps as shown by raised beaches at about eighteen metres, twelve metres and three metres respectively above present lake level. Only the lowest of these has been carbon-dated at about 3 700 years B.P. from charcoal left from human occupation of a small water-worn cave at this level near Entebbe (Brachi, 1960). The formation of beaches implies a period of stability at each level. The fall from the initial high level has been interpreted as due to the establishment of the northern outlet at Jinja by the back-cutting of one of the tributaries of the original upper Kafu River (now Lake Kioga), and the stepwise fall in lake level as the result of successive down-cuttings of the outflowing river.

Speculation on the history of the eastern side of the Lake Victoria basin must take into account the discovery (Trewavas, 1937) of nine specimens of a tilapia (*Sarotherodon*), all with four anal spines, among the fossils collected by Louis Leakey in the early Pleistocene beds at Kanam near the southern shore of the Kavirondo Gulf (Fig. 14.1). Four spines are characteristic of tilapias of the eastward-flowing rivers and are unknown in species west of the Rift Valleys. It has been reported in only a single individual of each of the two indigenous species in Lake Victoria and of the Lake Edward species *S. leucostictus*. Many hundreds of specimens have been examined. The fosssils were assigned to *Tilapia* (*Sarotherodon*) *nigra*, a native of the east-flowing Athi, Tana, Juba, Webi Shebeli and other smaller rivers in the area as well as to the river that feeds Lake Natron in the Eastern Rift (Fig. 3.4). It is now considered to be a subspecies of *Sarotherodon spilurus* (Trewavas, 1966b).

These fossils suggested to Leakey and to Kent (1942) that the watershed between the east- and west-flowing rivers in the early Pleistocene was further west than now, at least in the region of the present Kavirondo Gulf. The later stages in the subsidence of the Victoria basin may have shifted the watershed to the east and reversed the upper reaches of the east-flowing rivers. This is a speculative suggestion to account for the presence of *Sarotherodon spilurus niger* at Kanam during the early Pleistocene. More fossil and other geological evidence would be helpful.

It has unfortunately been difficult to assess from geological evidence the part played by changes of climate in causing these great fluctuations in the amount and distribution of standing water in the Victoria basin during the Pleistocene. This is because earth movements have been so extensive and continuous during most of the period that the effects of climatic change cannot be separately distinguished. Even during the relatively stable period of the past 20 000 years the progressive fall in the level of the lake has usually been attributed solely to cutting down of the northern outlet. But it is difficult to accept as a certainty without more direct evidence that on three separate occasions there was a rather sudden cutting down at the outlet, presumably through softer rock, to cause the drop from one beach level to the next. No evidence on this point has apparently been got from the outlet itself. In view of the more convincing evidence of great climatic fluctuations and of their effects on lake levels in the nearby Eastern Rift in Kenya, and of the predominantly climatic causes of the appearance and disappearance of large, standing waters in other parts of tropical Africa and especially in the Sahara during the same period, the importance of climatic changes in the recent history of the Victoria basin cannot be dismissed without further evidence.

The recent rise in the level of Lake Victoria (as of the other Great Lakes) during the early 1960s brought the water to within a metre of the lowest beach level. This gives a hint of what a relatively small change of climate can do, though it is, of course, certain that the level has never risen above the three-metre beach during the past 3 700 years, except perhaps for a very short time, because the latter would then have been washed out. As already mentioned, water levels can, in principle, be altered by changes both in rainfall and in temperature, the latter by controlling the rate of evaporation.

The best prospects for a reconstruction of late Pleistocene climates in the Victoria basin would seem to come from the study of lake sediment cores. The physical features of the sediments give an indication of the conditions under which they were deposited, the chemical composition and the types of diatoms in the various layers can reveal something of the composition of the water and of productivity, and the pollen discloses the nature of the land flora and thus the kind of climate of the region at the time of deposition. The layers can be dated by the ^{14}C method.

Kendall (1969) examined two cores from Pilkington Bay near Jinja and one from the open lake near Entebbe. The first two comprised sediments deposited during the past 15 000 years and thus covered most of the period of the final contraction of the lake from the eighteen-metre beach level. A large amount of information was extracted and deductions were made concerning conditions which are not disclosed by the geological evidence from exposed sediments. In the present connection it will be enough to note the following conclusions. The level of the lake was too low to reach the outlet from about 14 500 to 12 000 B.P. and again for a short period around 10 000 B.P. This is deduced mainly from the copious precipitation of carbonates in the sediments, indicating an increase in salinity of the water as the result of evaporation in a closed basin. The changes in climate that are inferred from this are supported by the pollen analyses which show an absence of forest vegetation before and its appearance after about 12 000 B.P. and a decline around 10 000 B.P. This indicates, as would be expected, periods of low rainfall contemporary with low lake levels. Between 10 000 and 7000 B.P. evergreen forest came back and developed more abundantly than at any other time during the period under consideration. Thereafter semi-deciduous forest began to take its place and it seems that the last 5 000 years has seen a transition to present conditions in the basin. In view of the probable effects of forest clearance by man it is not possible to interpret with confidence the pollen data for the past 3 000 years. Kendall considers that the three raised beaches at eighteen metres, twelve metres and three metres were formed during the past 12 000 years. This implies that the highest of the three levels was contemporary with and at least partly due to the high rainfall, and was reached after a dry period had reduced the lake to a very shallow, and perhaps saline, remnant in a closed basin. But, as he points out, this level could not have been reached with the outlet at the present height, which subsequently must therefore have been cut down. The wet and dry periods inferred from the Lake Victoria sediment cores correspond rather well with those deduced for the same period from investigations of sediment cores, exposed sediments and from the movements of glaciers in other parts of East Africa (references in Kendall, 1969; Butzer *et al.*, 1972).

The picture that is beginning to emerge, though still obscure in several particulars, is one in which the uprising of the western half of the basin has played the predominant part in the formation of the lake, but changes in rainfall and/or temperature have been a contributory causes of the considerable fluctuations in lake level, at least during the past 15 000 years.

It is likely that most, if not all, of the ancestors of the fish at present in Lake Victoria (other than recent introductions) lived in the pre-Pleistocene west-flowing rivers that were later flooded to form the lake. As in the other two Great Lakes, Tanganyika and Malawi, there has been a burst of speciation on the part of the cichlid fish in response to the change from river to lake conditions. But the special interest of Lake Victoria is that this has apparently happened more recently and, perhaps, more rapidly (if we accept the geological history outlined above), and with, at first sight, less opportunity for ecological isolation in different types of habitat.

It seems that the cichlid fishes are at present particularly plastic and more prone to speciation than other groups of African fish, and their breeding habits are such as could conceivably lead to reproductive isolation even in a relatively uniform environment. Nevertheless, it is generally supposed that some kind of spatial isolation of sections of the populations must have occurred in the past to account for the prodigious outburst of speciation resulting in nearly 200 endemic species and 4 endemic genera of cichlids in Lake Victoria. More than 150 of these are species of the genus *Haplochromis* (Greenwood, 1974a). There are three main types in which the skeletal and muscular apparatus of the mouth and phraynx is much modified from that of the supposedly primitive and generalised insectivorous condition, which is also represented. There are plant eaters, mollusc eaters, the fish eaters, and those adapted for the curious habit of extracting and eating the eggs and young larvae that are carried in the mouths of the brooding females of most of the cichlid species. Moreover, the presence of 'species flocks' of *Haplochromis*, each containing several closely related species with apparently identical feeding habits, is generally regarded as explicable only as the result of some degree of separation in the past and subsequent reunion after divergent speciation (Greenwood, 1965a). On this assumption and in the light of the available geological knowledge, there have been several suggestions regarding the conditions during the Pleistocene that could have caused fragmentation of the water in the Victoria basin. Reduction of the lake to a number of separate sections during a dry (interpluvial) period in the mid-Pleistocene (Worthington, 1954) was proposed before more recent geological research had suggested a later origin for the lake itself. Since then, as already indicated, there has been a tendency to attribute all the major changes in lake level to earth movements and erosion of the outlet. Temple (1969) suggests that in the early stages of 'ponding back' the flooded west-flowing rivers would have become, at least for a time, separate branching lakes before they finally joined together. This is very likely and could conceivably have provided conditions favouring some divergent speciation of the riverine cichlid fishes. But, as we have seen, recent investigations on sediment cores from the lake (Kendall, 1969) suggest that during the last 15 000 years of relative tectonic stability there have been two major changes in lake level that were associated with climatic changes. Prior to 12 000 B.P. the lake may have been reduced to a rather saline remnant without an outlet. Between then and 10 000 B.P. it rose, possibly to the eighteen-metre beach, and fell again to below the outlet, finally rising to the three-metre beach level and has remained below this for the past 3 700 years. Though progressive cutting down of the outlet via the Victoria Nile must have regulated the final fall, the major rises in level and the falls below the outlet during this period must have been determined primarily by changes in climate. It is, of course, the very low levels that can cause fragmentation of a lake and provide conditions that may favour speciation. Kendall suggests that high salinity during the low-level periods may have driven fish into the inflowing streams and isolated several populations.

Though much of this story remains very obscure, we can now at least conclude that the past 15 000 years has seen some very great fluctuations of water level in both directions with consequent effects on the area and distribution of standing water in the basin. In addition to this we may assume that Lake Victoria has always been subject to the lesser changes of level due to fluctuations in rainfall, such as have been recorded during this century (Fig. 14.3). These have caused considerable alterations in the area of the land under water and the formation of new, though temporary, patches of water isolated for varying lengths of time. In view of the apparent rapidity of genetic change in the cichlids under favourable conditions, we cannot dismiss the possibility that such relatively minor environmental changes may have had some evolutionary consequences in the past and that some at least of the divergence among the cichlids may have occurred more recently than has hitherto been supposed. We are, however, unlikely to get nearer to solving this problem until we know more about the genetics of the cichlids. How far is reproductive isolation with consequent divergent speciation in these fishes dependent upon the kind of spatial separation in the past for which we have been seeking evidence?

There is one interesting situation in the Victoria basin that appears at first to be particularly relevant to this problem. Just off the western shore of the lake, and about 90 km southwest of Entebbe, is the small and shallow swampy Lake Nabugabo about 20 km^2 in area of open water, and with a maximum depth of about 5 m (Fig. 14.1). It was cut off from the main lake by a sand and shingle bar which was blown across the mouth of a bay by the prevailing southerly winds. It is fed by four streams that drain from the west through swamp-choked valleys. The water from Lake Nabugabo seeps underground through sandy soil to the north into a swamp which leads into the Katonga Bay of Lake Victoria (Fig. 14.1). The particular interest of the cichlid fishes of Lake Nabugabo was first pointed out by Trewavas (1933) and they have since been studied by Greenwood (1965b). Of the nine species of cichlids, four are also found in Lake Victoria. The remaining five species, all of the genus *Haplochromis*, are endemic to Nabugabo, but are closely related to species in the main lake, and it must be

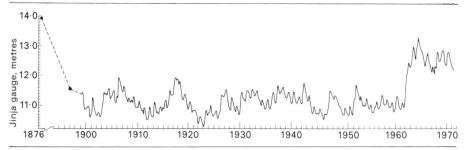

Fig. 14.3 Lake Victoria water levels between 1876 and 1970 (from data supplied by the Commissioner of Water Development, Entebbe)

assumed that they have been derived from a common stock. The sand bar is at the same level as, and therefore contemporary with, the three-metre raised beaches that have been dated at 3700 B.P. It follows that the divergence from the parent stock of *Haplochromis* spp. has occurred within the relatively short period of 3 700 years. In Lake Nabugabo we have, therefore, a demonstration both of the kind of situation that favours such changes and of the rate at which they can occur. It must, however, be emphasised that, though the visible differences between the endemic species of *Haplochromis* in Nabugabo and their relatives in the main lake are in some cases as great as between the latter and related species in the same flock, yet we do not know how far the Nabugabo species have gone toward reproductive isolation. This could, in principle, be discovered from a comparative genetical investigation. The amount of divergence is, of course, small compared with that which produced the main feeding types in Lake Victoria, for which a much longer time was presumably required. Nevertheless, we cannot doubt that these small changes are steps on the course of speciation. But there is no indication that there are any special conditions in Lake Nabugabo to which these fish are becoming adapted by selection, though the peculiar composition of the water is probably partly responsible for the virtual absence of Molluscs. The fish which normally eat molluscs, such as the lungfish *Protopterus aethiopicus* and the cichlid *Astatoreochromis alluaudi*, must get most of their food from other sources. Greenwood (1965b) has shown that the latter is adaptable in this respect, and in situations where molluscs are scarce its pharyngeal mill, specially adapted for snail crushing, may be modified for a more general diet. The possible chemical and biological causes of the absence of molluscs and crabs from Lake Nabugabo are discussed in Ch. 5 (p. 66).

When Lake Victoria was first formed it is likely that the rate of change of the relatively unspecialised cichlid fishes from the rivers was faster than it was more recently in Nabugabo, where conditions were more restricted and not greatly different from those in the lake. Moreover, the Nabugabo cichlids were derived from species that had already been adapted to lake conditions. This point is further strengthened by the example of Lake Lanao on the island of Mindanao in the Philippines which was formed by volcanic blockage about 10 000 years ago (Myers, 1960). In this lake the original cyprinid fish, confronted with the new lacustrine conditions, diverged at a prodigious rate and produced eighteen endemic species and four endemic genera. According to Myers there is only one riverine species (*Barbus binotatus*) from which all the Lanao cyprinids could have evolved. It is widespread in the region and very variable. Apart from recent introductions, the cichlids are absent from that part of the world. The example of Lake Lanao perhaps is more relevant to the early history of the cichlids in Lake Victoria than is the situation in Lake Nabugabo (see also discussion in Lowe-McConnell, 1969).

The non-cichlid fishes have also changed and have produced twenty-nine endemics aut of a total of fifty species, and one endemic genus. But, as we should expect, there is much less divergence from the riverine stock than has occurred among the non-cichlids of Lake Tanganyika during a considerably longer period. In Lake Victoria only one non-cichlid, namely the endemic catfish *Xenoclarias eupogon* (Clariidae), shows a well-marked modification clearly related to a change in mode of life (Greenwood, 1958, Gee, 1975). Superficially these fish closely resemble the genus *Clarias*, especially *C. alluaudi*, which is also found in the lake and from which it was presumably evolved. But the dorsal air-breathing section of the branchial cavities and their superbranchial organs, which are characteristic of *Clarias*, have entirely disappeared, and the gills have correspondingly increased in bulk and surface area. *Xenoclarias* have been caught at the bottom of the deeper water where obligatory air-breathing, as in *Clarias* spp., might be a hindrance. This same modification has occurred in the clariid genus *Bathyclarias* in the much deeper waters of Lakes Malawi and Tanganyika (p. 300). It is not however easy to understand what conditions could have induced *Xenoclarias* in Lake Victoria to revert to purely aquatic respiration since there is no evidence that the lake has ever been much more than 30 m deeper than at present (p. 251). Moreover *Clarias mossambicus*, an air-breather, has also been caught at the greatest depths (Gee, 1968).

We must now consider the principal ecological features of the present Lake Victoria and its associated waters, and discuss some biological phenomena that are better demonstrated or have received more attention here than elsewhere. The lake can be described briefly as an enormous and relatively shallow expanse of open water bounded, especially along the southern, northwestern and northern shores, by innumerable shallow bays, many of which lie in the branch valleys of the pre-lake river system and which provide favourable conditions for emergent and submerged large vegetation (macrophytes). The most significant and permanent inflow is the Kagera River draining the high-rainfall area of the Rwanda and Kigezi Highlands bordering the Western Rift (Figs. 14.1 and 15.1). The next most important is the Nzoia River in the northeast, which comes from the large volcano Mount Elgon and the Cherangani hills in the Kenya highlands. This is, however, subject to much greater fluctuations in flow with the more extreme wet and dry seasons in the catchment area. From the east and south come smaller seasonal rivers draining rather dry and flat country including the Serengeti plains. The inflows from the north and northwest are mostly slow flowing swamp courses whose origin has already been discussed. The most important of these is the Katonga River. It is interesting to note that, if the uplifting that ultimately reversed this river during the late Pleistocene had extended a little further west so that the swamp-divide was about 20 km west of its present position, the Mpanga

River would now join the Katonga and much of the water from the northern Rwenzoris would reach Lake Victoria (Fig. 14.1).

Lake Kioga, which lies in the flooded branches of the old west-flowing Kafu River, now receives the outflow from Lake Victoria (the Victoria Nile) and drains northwards and falls over the low northern end of the rift escarpment (the Murchison Falls) and thence to Lake Albert and the Albert Nile. It was formed, like Lake Victoria, by 'ponding back' of the Kafu River, but the uprising towards the escarpment of the Western Rift above Lake Albert probably occurred later than the corresponding movements that formed Lake Victoria further south. Kioga was, for a time, at a higher level than at present and was then, perhaps, still draining westward to Lake Albert via what is now the Nkusi River. Reversal of the flow of the Kafu may have started when the stream, now called the Lower Victoria Nile, had 'captured' the waters of the lake at the point shown in Fig. 14.1 as a gap in the watershed at the Karuma Falls (Bishop, 1969). The breakthrough of the northern outlet from Lake Victoria probably succeeded these events. Lake Kioga has essentially the same fauna as the inshore waters of Lake Victoria, since the Owen Falls at the mouth of the outlet, before they were submerged by the hydroelectric dam just below them, were little more than steep rapids and no barrier to the movements of fish. It is the Murchison Falls that are quite impassable and have preserved the isolation and the endemic character of the Victoria-Kioga fauna.

We cannot doubt that the major changes in water level that occurred in the late Pleistocene played an important part in determining the biological history of the lake and the nature of its present fauna. There have, however, been smaller fluctuations, though greater than the normal seasonal rise and fall of 0·2–0·4 m, that have been recorded during the past seventy years and have given opportunities for studying their ecological effects. It can be seen from Fig. 14.3 that from 1900 to 1960 there was an overall stability with several peaks at approximately the same level. There has been much discussion and some theories concerning the climatic changes that are reflected in the periodic high levels (see Lamb, 1966, pp. 183–6), with the object of finding a basis for predicting future levels in relation to inundation of land and port installations, the construction and operation of the Owen Falls hydroelectric station and the regulation of the outflow down the Nile. The picture has been radically altered by the unprecedented rise in level from 1961 to 1964 (Fig. 14.3). Though little attention had previously been given to it, there is no doubt, from scattered records made by missionaries and other early visitors to Uganda, that between 1875 and 1880 the lake rose to a height which, according to Lamb (1966, p. 184), was 0·5–0·7 m above the high level in 1964. By 1898 it had fallen to the general level maintained during the first half of this century, which is now seen to be relatively stable and drier interval between two periods of heavier flooding. The task of predicting the future has therefore become still more difficult. The belt of partly submerged dead trees along much of the shoreline, which in 1970 had not yet fully disintegrated, is a testimony to the length of time since the water had previously reached the 1964 level.

The magnitude of the ecological effects caused by fluctuations of water level depend upon the area of the land that is periodically flooded and desiccated.

Low-lying shores are clearly the most affected and even the normal seasonal mid year rise in level may temporarily inundate considerable areas of land. The organic matter on the newly waterlogged land decomposes rapidly and the released nutrients stimulate a high rate of production in the shallow water which then becomes a rich feeding ground for many invertebrates and small fish. Though the breeding seasons are much less marked than in temperate regions, most of the animals in the lake breed more vigorously during the major rainy period and subsequent time of high water level which occurs from May until July. The very young of several species of fish, especially of some cichlids, find shelter from predators, as well as food, in the peripheral shallows. The mature females of the air-breathing catfish *Clarias mossambicus* have been observed to lay their eggs in small flooded streams near the lake: the eggs have adhesive discs by which they are attached to the bottom or to some solid object. When the water level falls the young fish may be isolated in pools, but they are capable of making their way back to the lake over damp ground (Greenwood, 1955). The effects of seasonal high water levels on the ecology of the dense lake-fringing swamps, which are a major feature of much of the shoreline of Lake Victoria, are discussed in Ch. 18. It is then that lungfish *Protopterus aethiopicus* makes its breeding 'nest' in the papyrus and the young fish develop in comparative safety among the matted roots and vegetable debris.

In Chs 6 and 7 it was explained how the seasonal rise in primary production in the open water of Lake Victoria in June and July is caused mainly by the heavy southeast trade winds, which follow the rainy period and stir up the nutrients from below after a period of relative stagnation. The annual rise in water level and the flooding of low-lying shores make another seasonal contribution to productivity at about the same time, though the causes of the two phenomena are different.

That the extraordinary rise in level in the early 1960s would have some very marked ecological consequences was therefore to be expected. The peripheral flooded land was more extensive and remained under water for a longer time than during former peak water-level periods since 1900. Moreover, a characteristic feature of the low-lying swamp-fringed shores was the appearance of inland lagoons which were more or less isolated from the lake by a belt of dense swamp of varying width. These lagoons provided a new habitat, or at least a great expansion of a previously more rare one, which in aggregate around the entire lake shore must have been very extensive. The distribution of the temporary lagoons around Napoleon Gulf from which the Nile flows out of the lake (Fig. 14.1) were studied by Welcomme (1970). This was in 1964 when the lake level was highest. They first appeared in 1962 and most had dried up by 1965. After 1964 those most distant from the lake became progressively more isolated. Some lagoons appeared among the flooded trees behind the swamp fringe, but more extensive were the 'papyrus lagoons' – flooded land behind the papyrus belt. The latter were very shallow and conditions in them resembled those in a tropical fishpond. With no shade from direct sun and a very rich flora of algae, especially diatoms, there were much wider diurnal fluctuations in temperature and dissolved oxygen than in the open water of the lake. It was probably owing to these special conditions and because, to reach them, fish must pass through a belt of dense swamp, that only a few species managed to colonise them. During the later stages in the most isolated lagoons only the two cichlids *Sarotherodon leucosticus* (a recently introduced species) and *Haplochromis nubilis* were found. The amount of food available to the successful invaders was, however, very great – especially algae, vegetable debris and insects – and the fish multiplied to dense populations and began to 'runt', that is they reached maturity at a small size, a phenomenon well known in overcrowded fishponds. The peripheral lagoons thus favoured the multiplication of certain species of fish that were easily caught, and provided an interesting demonstration of the selection of species by a particular environment. Nevertheless, as the lagoons lasted only two or three years they could hardly have played a significant part in the speciation of the lake's fish: a very much longer period would be necessary for them to have a greater influence.

The biological consequences of the flooding of marginal land during the high-level period of 1962–65 were nevertheless very great, but the production of fish in this and the other large lakes must depend considerably on the normal annual rise in level. Particularly striking examples are the lower Chari valley in the basin of Lake Chad (Ch. 12), and the controlled annual fluctuations of water level in the manmade lakes that have an important influence on organic productivity (Ch. 20).

Though many of the fish, including most of the cichlids, of Lake Victoria live and breed in the lake itself, there are a number of others, particularly of the cyprinids, characids and siluroids, that ascend the rivers when in flood, breed in suitable places upstream and return with the young fish to the lake as the level drops. The breeding seasons of these anadromous (ascending) fish are thus determined by the rainy seasons and are more clearly marked than those of the fish that live permanently in the lake where conditions are more static and uniform. Ascension of rivers for breeding occurs in all the large African lakes (e.g. Lake Chad and the Chari River), but some investigations on the anadromous fish of Lake Victoria will be mentioned here (Whitehead, 1959). These were made during the period from 1955 to 1957 on the rivers, especially the Nzoia (Fig. 14.1), which enter the northeast corner of the lake. During the floods there are productive fisheries on this and other smaller rivers in the Nyanza Province of Kenya. The most ingenious of the fishing methods is the *Kek*. This is a staked reed barrier across the river with gaps which lead into channels ending in one-way traps. These can be adjusted to catch either ascending or descending fish. The fishermen sit day and night on raised platforms taking the fish from the traps.*
About 2 000 tons of fish per annum were caught from 1955 to 1957 in the Nyanza rivers during the up and down migrations. The total annual harvest from all the rivers affluent to Lake Victoria must be very great, and an understanding of the biology of the anadromous fishes is thus of considerable practical importance. Any scheme for exploiting a river as a water supply for irrigation, for hydro-electricity or for domestic purposes must ensure that the fish migrations are not obstructed in either direction.

According to Whitehead (1959) thirteen species of fish in Lake Victoria are without doubt anadromous. Among these are seven species of *Barbus*, *Alestes*

---

* For a general account of the fishing methods used in the Kenya rivers, see Whitehead (1958).

jacksoni, *Labeo victorianus*, *Schilbe mystus* and *Clarias mossambicus*. Another eight species were thought to be at least occasionally anadromous but their habits were not sufficiently well known. More recently Okedi (1969, 1970) has shown that some individuals at least of five small species of mormyrid (elephant snout) fish ascend the north coastal rivers to breed. The number of partially anadromous species is probably greater than at first thought. Greenwood (1970) has drawn a distinction between two subspecies of *Barbus radiatus* in Lake Victoria. *B. r. radiatus*, which has a wide distribution in eastern Africa, is found mainly in the inflowing rivers and streams, whereas *B. r. profundus*, known only from Lake Victoria, is apparently confined to deep water (18–65 m) and presumably breeds in the lake. The rest, including all the cichlids, remain in the lake for breeding. There are a few, such as *Barbus amphigramma* and a few *Haplochromis* spp. that are primarily river fish, most of them spending their whole lives in the rivers. *Alestes baremose* in Lake Chad has two populations, one anadromous, the other staying in the lake to breed (p. 224). This has not yet been demonstrated of any fish in Lake Victoria, though there are apparently populations of the otherwise anadromous *Labeo victorianus* that live permanently in the upper reaches of the large rivers (Fryer and Whitehead, 1959).

When the rivers begin to rise and to flow more rapidly into the lake, the anadromous fish congregate in large numbers around the mouths. Whitehead (1959) recorded some of the changes that occurred in the Nzoia between dry and wet seasons – water level, rate of flow, turbidity, pH and temperature. He concluded that level and rate of flow are the main stimuli to upstream migration. The degree of change in the other factors varied greatly from one year to another. As with fish migrations elsewhere, the causal relations are no doubt very complicated and probably involve internal hormonal reactions related to the development of the gonads. There are in general two flood periods in the Nyanza rivers (March–May and August–September), but they vary from year to year and are sometimes joined to form one long flood period. In the last case the migrations tend to be spread over the whole period, which suggests a predominant influence of river flow. The length and accessibility of the Nyanza rivers determines the time spent and the distance travelled upstream. In 1955 the Yala River was inaccessible owing to blockage by papyrus swamp at the mouth. The Nzoia can be ascended by large fish as far as the Broderick Falls about 110 km upstream. The Sondu on the other hand is blocked by a falls at 8 km. The fish that penetrate to the highest possible levels are those, such as *Barbus altianalis*, which breed in the upper rocky stream bed where the flow is rapid. Most of the fish, however, breed at lower levels in lateral flood waters which are very productive of aquatic plants and invertebrate animals and thus provide abundant nourishment for the young fish. Among these are *Labeo victorianus* and *Schilbe mystus*. There are a few, like *Clarias mossambicus* and *Alestes jacksoni*, which may ascend smaller streams that are in flood for a short time, and breed over flooded land near the lake shore. In addition, the riverine species, such as *Barbus neumayeri* (named *B. portali* by Whitehead, 1959) and some river-bound populations of *Labeo victorianus*, migrate both up and down during floods in search of suitable breeding grounds up side-streams or over flooded land. Less is known about the return of the old and young fish to the lake or of the stimuli that start the downward movement. It can be imagined that the drop in level and the chemical effects of stagnation, such as deoxygenation, on the dwindling flood waters may play some part.

It is tempting to speculate on the evolutionary significance of the anadromous habit among some of the fishes of Lake Victoria (discussion in Whitehead, 1959, and Corbet, 1961). There is general agreement that these fish were derived from those of the rivers that preceded the lake. Many species have remained unchanged and are found elsewhere in rivers. It might therefore be supposed that some of the previous river species continued to breed in the rivers affluent to the lake, but extended their feeding grounds to the lake itself when the river levels dropped in the dry seasons, returning to the rivers to breed in flood periods. This habit involves adaptation to greatly changing conditions and may account for the small degree of speciation among the anadromous fish compared with the lake-bound fish. None of the species so far known to be anadromous is endemic to the Victoria basin. The environment of the lake-bound fish is very much more stable and has enabled new species to become established through isolation from one another behind relatively small ecological barriers. The cichlids are the supreme example. This argument, though highly speculative, is consistent with the idea put forward in Ch. 8 (p. 146) that the generally greater number of species in the humid tropics than at higher latitudes is due, at least in part, to the stability of the equatorial environment. The small amount of speciation among the anadromous fishes of the lakes is an exception which apparently supports this thesis.

From our knowledge of the feeding habits of the fishes of Lake Victoria it is clear that the chaoborid and chironomid flies occupy an important position in the ecosystem. This is certainly also true of most of the other great lakes, but some interesting observations have been made on these insects in Lake Victoria and it would be appropriate to discuss them briefly at this point. The periodic appearance of lake-fly swarms is well known on many lakes. What appear to be gigantic clouds of smoke, sometimes of over fifty metres in height and occasionally more than a kilometre in length, rise from the surface of the lake (Fig. 4.2). The majority of the swarms on Lake Victoria appear a few days after the new moons. They are mainly composed of chaoborids, especially *Chaoborus anomalus* (Appendix A3), with a lesser number of chironomids of the genera *Chironomus*, *Tanypus* and *Procladius* (Macdonald, 1953, 1956). The water seems to boil with struggling fish devouring the pupae as they rise to the surface to emerge; the swarms are often followed by flocks of white-winged black tern, kites and other birds taking advantage of the abundant feast. Of the myriads of emerging flies only a small proportion succeed in mating and in laying eggs on the surface of the water. The swarms are often blown onto the land by the onshore winds during the afternoon and are attracted by lights at night (Fig. 4.2). Though non-biting, the flies are small enough to penetrate mosquito netting and nocturnal invasions of houses are very unpleasant; allergic reactions have been recorded in some people. The enormous numbers of flies that die on land are, of course, lost from the lake, though whether this is a significant drain on the lake's productivity remains to be determined. Direct exploitation of adult lake flies as human food, as on Lake Malawi (p. 304), has apparently not been practised on Lake Victoria.

As already mentioned, the most abundant species of lake fly on Lake Victoria is *Chaoborus anomalus*.* The transparent and almost invisible larvae capture and feed on the zooplankton, particularly Crustacea and Rotifera. For this they are provided with strong prehensile antennae and powerful mandibles (Fig. A3a). The first and second instars are themselves planktonic, but the third and fourth spend the day in the bottom mud and swim up towards the surface during the night. The chironomid larvae, on the other hand, are confined to the bottom mud where they feed on detritus, that is, decomposing particulate organic matter and associated micro-organisms. Macdonald (1956) found 2 000–2 500 chaoborid and about 1 000 chironomid larvae per square metre of mud surface in Ekunu Bay near Jinja. A few measurements elsewhere suggested that the population density over much of the lake is of the order of thousands per square metre. The total biomass of lake fly is therefore prodigious. The chaoborids are an important link between the zooplankton and the fish. The chironomids, by their burrowing activities in the mud, assist in the general process of decomposition and in stirring the soluble nutrients into the water above. As the prey of fish they return organic matter in the sediments more directly into the ecosystem.

Though many of the fish take lake fly larvae and pupae as part of their diet, the elephant snout fish, *Mormyrus kannume*, is the most persistent consumer of chironomids, sucking them from the mud by means of its tubular mouth and pharynx (Fig. B1c). Since this is an important commercial fish, there is here a short food link between mud and man. Species of another mormyrid, *Gnathonemus*, and several cichlids of the genus *Haplochromis* are also avid consumers of chironomids.

That some, though not all, aquatic insects show a periodic pattern of emergence related to the phases of the moon is now well established (Corbet, 1964), and the lake flies of Lake Victoria provided one of the first examples of this remarkable phenomenon (Macdonald, 1956; Corbet, 1958; Tjønneland, 1958a, b; Fryer, 1959d). It has for a long time been recognised in many animals in the sea, where the tidal effect of the moon's phases provides at least one of the possible causes. A similar lunar periodic emergence was discovered for the mayfly larva *Povilla adusta* in Lake Victoria by Hartland-Rowe (1958). Simultaneous emergence presumably facilitates mating, but the causal link with the early phase of the moon is still very mysterious and a satisfactory explanation is yet to be found. If it is the increasing intensity of light at night, this can at most be an external trigger, and there may be others which release pressures that have been building up in the maturing larvae. There is, however, at least one ecological consequence: the amount of food available to insectivorous fish fluctuates with a monthly rhythm and is reflected in changing stomach contents of the fish. During one of the peak periods a single specimen of *Mormyrus kannume* was found to contain about 5 000 larvae (Macdonald, 1953).

The ultimate consequences of the introducing of new species of fish into Lake Victoria are yet to be seen, but it has provided an interesting demonstration of the adjustment of an ecosystem to change in species composition. During the early 1950s four tilapias (Cichlidae) were introduced. In 1955 the Nile perch

* For descriptions and distribution of *Chaoborus* spp. in the east African lakes, see Verbeke (1957b).

Lates niloticus (Centropomidae) was introduced into Lake Kioga and a few years later was found in Lake Victoria.

The biology of the two tilapias endemic to Lake Victoria was studied by Lowe (1956) and Fryer (1961), and the history of the tilapia fishery of the lake prior to the introductions is summarised by M. J. Mann (1970). Though both species live mainly in the inshore waters, *Sarotherodon esculentus* favours more sheltered regions than *S. variabilis* which is common along more exposed shores and in the mouths of rivers where there is more movement of water. Both species feed on the phytoplankton and on algae in the bottom deposits.

Originally there was no fish feeding principally on the submerged macrophytes (water weeds) that are abundant in the shallow inshore waters especially in the bays, and to exploit this important source of organic matter was the main objective in introducing new tilapias. In fact, only two of the four introduced species are habitual consumers of macrophytes. These are *T. zillii*, a soudanian fish originally found in Uganda only in Lake Albert, and *T. rendalli* (formerly known as *T. melanopleura*), a close relative of *T. zillii* from southern tropical Africa. Both of these are particularly suited to culture in ponds where macrophytes grow more readily than algae, and hand-feeding with the leaves of land plants is easily done. The other two introduced species both feed on phytoplankton and bottom deposits. They are *Sarotherodon niloticus* which is widely distributed in the soudanian region and in most of the Rift Valley lakes, and *S. leucostictus*, a native only to Lakes Albert, Edward and George (Lowe, 1957).

These fish were put into the lake at various points between 1951 and 1961, and their subsequent fate and biology was followed so far as it was practicable to do so. Some conclusions from these investigations were published by Fryer (1961) and by Welcomme (1966, 1968). It was to be expected that at some points there would be competition between adult fish for food and breeding sites and among the young stages for space and food in the sheltered shallow water in which they grow ('nurseries'). It seems that the lake has so far provided such an abundance of these requirements that, with slightly different preferences and a certain flexibility, these species have come to occupy sufficiently different niches as to avoid serious competition with each other or with the two endemic species. The numbers of the new species have progressively increased, especially of *T. zillii* which by 1966 was almost as numerous as *S. variabilis*. The great rise in water level in the early 1960s certainly contributed to this result by providing more vegetable food and sheltered shallow water in the flooded areas. The fishery as a whole has so far benefited from these introductions, but ecological equilibrium has not yet been re-established and further reports are awaited with interest. The presence of hybrids from *nilotica x variabilis* and *zillii x rendalli* crosses has been reported. The situation is thus becoming complicated, but very interesting.

A further complication has arisen because of the introduction from Lake Albert of the soudanian predator fish *Lates niloticus*, the Nile perch. This is the largest of all African freshwater fish. The heaviest recorded was over 140 kg from the Upper Nile, but fish over 45 kg are common. Its size, strength and good flesh make it attractive both to commercial and to sporting fishermen. The proposal to introduce it into Lake Victoria was originally made with the object of exploiting more effectively the organic reserves of the lake. In particular there are vast numbers of small cichlid fishes, mainly *Haplochromis* spp., that are caught only on a very small scale and many of them inhabit the deeper open water where they have so far escaped the fishermen. It was assumed from knowledge of *Lates* in its native habitats, such as Lakes Albert and Turkana, and in some smaller lakes, such as Lake Kioga, into which it had previously been introduced, that these smaller cichlids would become the principal item in its diet and that the *tilapis* would escape large-scale predation.

As might be expected there was considerable opposition among biologists to a measure the success of which was very far from certain and which might result in great damage to the important tilapia fishery. What if, in the conditions of Lake Victoria, it were the tilapias that became the principal food of the introduced *Lates*? This was the main objection to the proposal, but among others it was also pointed out that the lake's resources would be less efficiently exploited after the addition of another link in the food chain (e.g. Fryer, 1960c).

In 1960, however, five years after its introduction into Lake Kioga, it was recorded off the north shore of Lake Victoria. Its mode of entry is unknown, or at least undisclosed. It is unlikely to have come from Lake Kioga through the turbines of the Owen Falls Power Station, but it might have escaped from fish ponds which drain into the lake. Soon after this, it was deliberately introduced near Jinja and Kisumu.

During the twelve years since its first appearance *Lates* has become well established. A comparison of its feeding habits in its native Lakes Albert and Turkana and in Lakes Kioga and Victoria shows some interesting changes (Hamblyn, 1966; Gee, 1969). Though the young of the tilapias are certainly devoured, its principal food in Lakes Albert and Turkana is *Hydrocynus* and *Alestes* (Characidae), *Engraulicypris* (Cyprinidae) and the crustacean prawn *Caridina*. On the other hand, in Lakes Kioga and Victoria they have so far fed mainly on *Haplochromis* spp. (Cichlidae) and several species of mormyrids. Not only have the numbers and average weight of *Lates* in Lake Victoria progressively increased, but they have been feeding more vigorously than their relatives in Lake Albert. This is shown both from examination of stomach contents and by the higher 'condition factor' (ratio weight/length).

The abundance of available prey in Lake Victoria is perhaps due both to the numbers of *Haplochromis* and to their lack of previous experience of such a large and agile predator. Selection, we might suppose, will eventually adapt them better to the new situation and a higher proportion of those that are left will more easily avoid predation. It is to be expected, therefore, that the numbers and condition of the *Lates* will reach a peak and then decline to a position of equilibrium. In the meantime the Nile perch has become an important item in the fishery and there are no signs as yet that the Tilapias fishery has thereby suffered, but we cannot be entirely confident of the ultimate situation.

Mention should be made of the almost total destruction during the past twenty years of the Nile crocodile (*Crocodilus niloticus*) in Lake Victoria, as in many other tropical African waters. They have been shot mainly for their skins though, apart from the danger to human life, they can trouble fishermen by smashing the gill-nets while taking the fish. The adult crocodiles feed largely on fish, but their previous importance as a predator in Lake Victoria is difficult to assess. They have been protected, though much poached, in the Victoria Nile below the Murchison Falls. Studies of their biology have been made by Cott (1963) – feeding and breeding behaviour, growth, predators (lizards, birds and mammals) on the eggs and young, and the effects of poaching and of tourism. This information and that collected by Modha (1967) on Central Island, Lake Turkana, will greatly assist in the conservation of this species.

Another development that may well have important ecological consequences is the proposed large-scale trawl-fishing of small cichlids in the deep offshore waters. It had been known for some time from experimental trawling that there are very large stocks of the endemic *Haplochromis* spp. and related genera outside the normal inshore fishing grounds. A proposal to start a fish-canning industry in Uganda based on large-scale trawling of *Haplochromis* stimulated further investigation of these fish, of means of detection (e.g. by echo-sounder) and of methods of trawling. Begun in 1955 by the East African Fisheries Research Organisation at Jinja, the investigations have been continued from 1967 as a project of the United Nations Development Programme based at Jinja (see *Ann. Reps. E. Afr. Fish. Res. Org.*, 1968 onwards; Regier, 1971). For an account of the exploratory bottom-trawling done under this project, see Bergstrand and Cordone (1971).

The distribution, numbers and species composition of the *Haplochromis* appears to be determined mainly by depth of water and nature of substratum (pebbles, sand, mud). Different species are adapted to feeding on vegetable matter, detritus, insect larvae, molluscs and smaller fish, or combinations of these. As well as migrating horizontally, some species move diurnally between surface and bottom. It seems, however, that deep trawling yields the largest catches. As

Fig. 14.4 Typical landing for canoe-caught fish, near Jinja, north shore of Lake Victoria

already stated, there are a very large number of species and, apart from the general nature of their foods, we have practically no knowledge of the biology of most of them.

In 1973 it was planned that the production of fishmeal should be the main objective and that the industry should be based on both Ugandan and Tanzanian ports. The catches would include a small proportion of larger fish suitable for direct human consumption. On the basis of investigations to date, it has been suggested that up to thirty-seven trawlers could eventually be involved to provide annually about 15 000 metric tons of *Haplochromis* giving 3 000 tons of fishmeal (Regier, 1970). From information so far available it is not possible to assess with confidence the effects of this large operation on the stocks and species composition of the *Haplochromis*, nor, indeed, on those of other species of fish in the lake. The industry will therefore have to be built up in stages with repeated assessment of the ecological situation. Its development will be followed with interest by ecologists as well as by fisheries experts and economists. It is particularly important that the present inshore canoe fisheries (Fig. 14.4), which are based on a very large number of small-scale units around the whole lake, and are economically and socially very valuable, should not be allowed to decline through the economic or ecological consequences of this operation. A more efficient inshore fishery from private canoes and small boats and a mechanised large-boat fishery in the deep water must be able to exist together without competition (discussion in Jackson, 1971).

15

Volcanic-barrier lakes: Lake Kivu

The effects of recent geological events on the landscape and on the water systems and their faunas are nowhere more obvious than in the Western Rift between Lakes Edward and Tanganyika (Fig. 15.1). The eruption in the late Pleistocene of the Virunga volcanoes threw a barrier of lava and ash across the valley some 100 km south of Lake Edward which had catastrophic effects on the hydrology of the region.

Prior to these eruptions much of the drainage into the Edward basin from the south was derived from the high watershed in the Rift Valley near the present town of Bukavu and from the western slopes of the Rwanda and Kigezi Highlands. Fig. 15.1 shows quite clearly what subsequently happened. The eruption of seven major volcanoes across the floor of the valley completely blocked the northward flow of what was then the upper Rutshuru River, and the water was held up and flooded the valley to form the present Lake Kivu.* The level continued to rise until the water flowed back over the watershed to the south cutting a gorge through an older volcanic barrier and falling over the Panzi Falls into the Ruzizi River and so down into Lake Tanganyika.

The streams from the western slopes of the Rwanda and Kigezi highlands were also blocked and were flooded to form Lakes Bulera and Luhondo which then overflowed back to the south and ultimately reached Lake Victoria via the Nyavarango and Kagera Rivers (Fig. 15.1). The eastern branch of the Rutshuru River narrowly escaped complete blockage, but its uppermost valleys were partially obstructed by a stream of lava from the small isolated group of volcanoes at Muko. This was the origin of Lake Bunyoni. The northern edge of the main lava field from the Virunga volcanoes reached the middle section of this stream and obstructed it at a point behind which the valley was flooded to form Lake Mutanda. The overflow, though deflected to the north by the advancing lava, continued on its course to the Rutshuru River and to Lake Edward. The lava from Nyamulagira has flowed westwards through a gap in the escarpment and blocked some

* There was probably a smaller lake in this valley before the Virunga volcanics (see below). Owing to the inadequate geological and palaeontological evidence there have been several different theories of the early history of Lake Kivu (see Marlier, 1958b). The one presented here will not be universally approved.

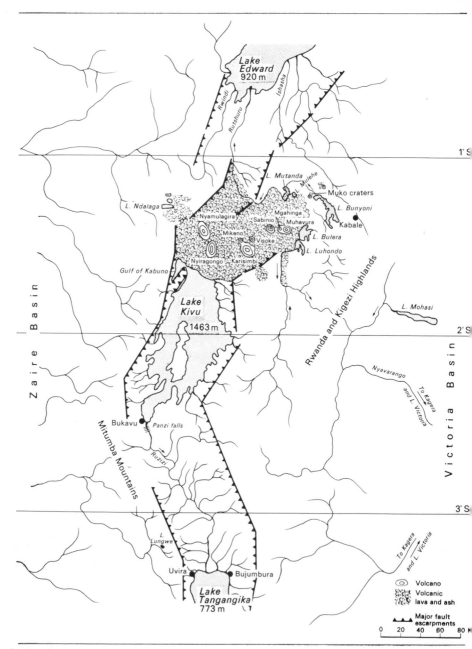

Fig. 15.1 The Western Rift Valley between Lake Edward and Lake Tanganyika with the Virunga Volcanoes north of Lake Kivu

small steep-sided valleys which previously drained to the east. This gave rise to the small group of Mokoto Lakes one of which, Ndalaga, was examined by Damas (1937). The outflows from these disappear beneath the lava field and it is presumed that they eventually drain into Lake Kivu.

The general appearance of these lakes is what would be expected from their origins. They lie in steep-sided branching valleys that show clear signs of having recently been 'drowned', the tops of the higher ridges projecting above water as islands and promontories. In all cases the obstructing lava flow is very obvious. In form and setting they are exceedingly beautiful, and the view across Lake Mutanda with the changing mist and light on the volcanoes in the background cannot adequately be described or photographed (Fig. 15.2).

Though not actually formed by a volcanic dam across a valley, many of the volcanic crater lakes are ecologically in a similar situation. For example, Lake Nkugute south of Lake George (Fig. 13.1), whose water circulation was discussed on p. 102, is a deep volcanic crater on the floor of a small valley near to its head. It is filled with water by seepages from the hills surrounding the head of the valley and discharges in a stream down the valley below. Like the barrier lakes, it has arisen by volcanic action on the course of a stream.

Of the eight Virunga volcanoes the two at the western end of the chain are still active (Fig. 15.1). Nyiragongo (3 470 m or about 2 000 m above Lake Kivu) erupted in 1976 after quiescence for more than seventy years during which it was capped by a cloud of vapour intermittently illuminated by explosive outbursts which were confined within the apical crater. Nyamulagira (3 056 m) with its several associated vents has been erupting at intervals since 1894 when this region was seen for the first time by a European, von Götzen. The most violent of these eruptions occurred between 1938 and 1940 when a river of lava more than 100 km² in area flowed southwards in waves and eventually poured into Lake

Fig. 15.2 Lake Mutanda and the Virunga Volcanoes Muhavura and Mgahinga. The lavaflow from these volcanoes blocked the valley and formed Lake Mutanda (see Fig. 15.1)

Kivu near the northwest corner. The Gulf of Kabuno-Kashanga was almost cut off from the main lake but a connecting channel no more than 150 m wide remains. These contemporary eruptions are the latest spasms of the volcanoes which first became active perhaps 100 000 years ago and, within the last 20 000 years, by damming the valleys, created the lakes with which we are now concerned. The largest of the lakes in the southern foothills of the European Alps, including Maggiore and Como, have this same general form. They too are recently flooded valleys but caused by damming by masses of rock debris (moraines) excavated, pushed down and left by the retreating glaciers at the close of the last ice age.

To the biologist the main interest of Lake Kivu is the relative poverty of the fauna and its ecological immaturity. These features can be related to the past history and present hydrology of the lake (discussed below). The fish fauna is mainly an attenuated version of that of Lake Edward and is unrelated to that of Lake Tanganyika into which it now drains. This is to be expected from the geological evidence which, as we have seen, suggests that Lake Kivu was formed by volcanic damming of a valley previously draining into the Edward basin. The same families that are missing from Lake Edward, but which are well represented in Lake Tanganyika (Characidae, Centropomidae, Schilbeidae, Mochokidae, Mastacembalidae) are absent, but there are also no species of the families Lepidosirenidae, Mormyridae, Bagridae and Cyrinodontidae which are represented in Lake Edward. In fact, there are only sixteen species in the lake itself; they belong to the families Cyprinidae, Clariidae and Cichlidae. In addition, there are another sixteen species of the same three families in streams and pools within the Kivu basin (Poll, 1939a, b). Three of the Lake Kivu species are also found in Lake Tanganyika. Of these *Clarias mossambicus* and *Barbus serrifer* have a wider distribution, but *Barilius moori* (Cyprimidae) is otherwise found only in Lake Tanganyika and is therefore the only species of the very rich fauna of Lake Tanganyika which has without doubt managed to surmount or circumvent the Panzi Falls.

The origin of the soudanian elements in the fauna of Lake Tanganyika will be discussed later (Ch. 16), but it should be noted that from the present distribution of species of fish there is no evidence for a faunal connection, prior to the Virunga volcanics, between the Tanganyika and Edward basins via the Ruzizi and Kivu valleys. Marlier (1953) has shown that the fish fauna of the affluent streams of the Ruzizi valley is composed of typical Tanganyikan species, quite distinct from those in the streams of the Kivu basin. Boutakoff (1937) had previously found geological evidence of a north-flowing stream in the Ruzizi valley prior to the late Pleistocene volcanic blockage. Marlier's work, and, indeed, the composition of the Kivu fauna, suggest that it could not have flowed from Lake Tanganyika. It might have arisen from a watershed just north of the latter and thus have been included in the pre-volcanic Edward basin. We cannot avoid the conclusion that, apart from *Barilius moori* which probably migrated quite recently, there has been no movement of fish between the Tanganyika and Edward basins since the Tanganyikan fauna began to evolve in isolation as long ago as the Pliocene.

During the relatively short time since Lake Kivu was isolated by the Virunga volcanic barrier there has been some speciation on the part of the fish. Of the cichlids all six *Haplochromis* spp. are endemic and the *Sarotherodon niloticus* has diverged enough to be accounted an endemic subspecies *regani*. As an ecosystem, however, the lake is poorly developed, and several niches are unfilled (Verbeke, 1957b). There is, for example, no major predator on the *Haplochromis* which are feeding mainly on algae and insect larvae, which, in turn, feed on algae and detritus. This part is played by such large fish as *Bagrus* in Lakes Edward and Victoria and *Hydrocynus* and *Lates* in Lakes Tanganyika, Albert and Turkana. In the absence of mormyrids (elephant snout fish) the invertebrates, especially the insects, in the mud under the peripheral oxygenated water are probably incompletely exploited. Another ecological deficiency is the absence of an effective predator on the zooplankton, much of which is irrevocably lost in the sediments below the deep water. This niche is filled by the 'sardine' (*Stolothrissa tanganîkae*) in Lake Tanganyika, *Aplocheilichthys* spp. in Lakes Edward, Albert and Victoria, and by *Alestes baremose* in most of the soudanian waters. Suggestions were made by Verbeke (1957b) for the introduction of a suitable species in the hope that more organic matter would thereby be converted to edible fish and less lost in the mud.

The deficiencies of the ecosystem of Lake Kivu are partly due to the small range of types of fish present. Nevertheless, the fauna is much richer than that of the other volcanic barrier lakes in this region (e.g. Luhondo, Bulera, Bunyoni and Mutanda) and of Lake Nkugute, all of which, before recent artificial introductions of fish such as *Sarotherodon niloticus*, contained only one species of *Clarias* and a very few of *Haplochromis*. This would suggest that, though most of these lakes were formed by blockage of hill streams whose normally sparse fish fauna they have inherited, Lake Kivu was preceded by a lake in effective faunal connection with Lake Edward and that it was enlarged and deepened by the Virunga volcanic barrier.

The recent seismic (echo) sounding by Degens *et al.* (1971b) show that the thickness of the sediments in the north basin, which is the deepest part of the lake, exceeds 500 m. They estimate that there was probably a lake there as far back as the Pliocene. Since the surface of these sediments is only about 60 m above the present water surface of Lake Edward, the level of the old Lake Kivu, before the appearance of the Virunga volcanic dam, may have been not much above that of Lake Edward. If so, the water connecting them could have flowed sufficiently slowly to allow the exchange of lacustrine species of fish. The fauna of the present Lake Kivu might have been derived from the partial destruction of a previously richer lake fauna. It has been suggested by Poll, (1939a, b) that chemical effects of the Virunga volcanics exterminated the species of fish which, it is assumed, were previously there. Unfortunately there is no palaeontological evidence of the mid-Pleistocene fish fauna of the Kivu basin, such as we have for Lake Edward, which has encouraged a similar interpretation for the very recent disappeareance of some of the fauna (p. 236). The solution to these problems may one day be found in the bottom sediments of Lake Kivu, and the foregoing very speculative suggestions may be put to the test.

There is no doubt that even now the peculiar hydrological conditions in the lake which are summarised in Table 6.1 are restricting the evolution and potential productivity of the ecosystem. Owing to the permanent stagnation of the high-density water below about 70 m most of the organic matter is locked in the bottom mud and is prevented from recirculating. But there are other related circumstances which are restricting the abundance of living organisms. The shape of the basin, with its very steep sides, is such that the great majority of the bottom is apparently covered by more than 70 m of water and this is perpetually deprived of oxygen. The 70 m contour is so close to the shore that there is only about 12% of the total area of bottom which is in contact with oxygenated water. The benthic fauna is restricted to the strip from above the 70 m contour up to the lower edge of the vegetation zone at about 8 m. It is relatively scarce and poor in species, consisting mainly of chironomid fly larvae and tubificid worms which are not exploited by any of the fish except *Barilius moori* which devours the chironomids as they finally pupate and rise to the surface. Nearer the shore the organic mud becomes progressively more sandy and is encrusted and consolidated by a calcareous deposit covered with a growth of algae, which is unsuitable for benthic burrowing animals. Most of the shore is rocky and the submerged rocks are similarly encrusted. The origin of this precipitate is clear. The salinity of the water is high even at the surface (about $1°/_{oo}$, see Table 5.1) and has a correspondingly nigh concentration of calcium bicarbonate which is continuously precipitated as carbonate through the removal of CO_2 by the very abundant photosynthesising littoral algae. The filamentous *Cladophora* (Chlorophyceae) is the commonest alga on the rocks and calcareous shallow bottom. Submerged higher plants, such as *Potamogeton* and *Ceratophyllum*, as well as emergent reeds (e.g. *Scirpus*, *Cyperus* and *Phragmites*) grow rather sparsely in the shallow water (Verbeke, 1957a).

The majority of the invertebrate fauna is found in this narrow littoral zone but, owing to the calcareous consolidation of the bottom and to the relative scarcity of rooted and floating vegetation, the fauna is poor in numbers and in variety of species compared with the littoral fauna of Lake Edward. Nevertheless, most of the invertebrate groups characteristic of a lake littoral are represented – Nematoda, Oligochaeta, Turbellaria, Cladocera, Ostracoda, Copepoda, Insecta, Hydrachnida, and gastropod molluscs. It is here that the majority of the fish are to be found, e.g. *Sarotherodon*, *Haplochromis*, *Clarias*, *Barbus* and *Barilius*. More detailed information on the distribution of the littoral fauna in relation to the substratum and type of vegetation can be found in Verbeke (1957a).

The sediments laid down off the mouths of streams are kept free of calcareous crust by the inflowing water. According to Verbeke, they form a reservoir of benthic animals which are much more abundant than elsewhere, and they are important for the breeding and larval growth of some of the fish, e.g. *Barbus altianalis*, *Sarotherodon niloticus* and *Haplochromi graueri*. Without these estuarine sediments the fauna of the lake as a whole might be even more restricted.

The pelagic zone, which is the deep water beyond the littoral region, covers about 90% of the lake surface. It is habitable by animals only down to the limit of oxygen at a depth of about seventy metres which is determined by the lower limit of circulation from wind action at the surface. Though the species of algae in the euphotic zone have been determined (van Meel, 1954, pp. 487 ff.) nothing is known of the rate and the seasonal variations in primary photosynthetic production. It is, however, probable that it is moderately high in view of the abundance of the zooplankton which must be feeding directly or indirectly on the products of photosynthesis. Measurements made in 1953 by Verbeke (1957a) of the volume of zooplankton below a unit area of surface (Fig. 15.3) showed that through most of the year the biomass was similar to that of Lake Edward. It increased during the windy period of August–September though much less than in Lake Edward. This increase presumably followed an intensification of photosynthetic production caused by the stirring up of some of the lower nutrient-rich water. These seasonal fluctuations of zooplankton were confirmed by Kiss (1959).

In the absence of zooplankton-feeding fish and, indeed, of any fish, in the pelagic zone, there is a more or less closed circulation of nutrients from the lower levels of the stirred water through the phyto- and zooplankton and back to the lower water by falling and decomposition, the rate of the cycling being increased during the windy periods. Some nutrients and organic matter are not doubt brought into the pelagic zone by the inflows, and some are continually sinking irrevocably into the uncirculated bottom water and mud. But the size of the standing crop of zooplankton suggests that this cycle could be profitably exploited by the introduction of a zooplankton-feeding fish. For this reason both species of 'sardine' from Lake Tanganyika *Stolothrissa tanganikae* and *Limnothrissa miodon* were introduced in 1963. There was no further news of their fate until

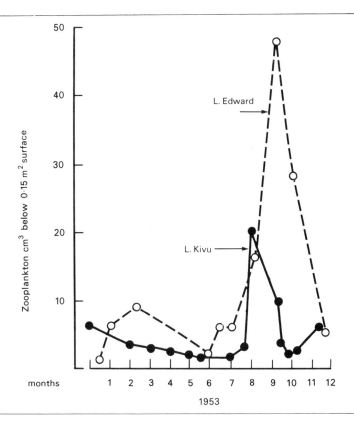

Fig. 15.3 Changes in volume of plankton in one year in Lakes Kivu and Edward (Verbeke, 1957a)

1976 when a visiting scientist noticed them in the market at Bukavu. An investigation in the following year discovered a small sardine fishery based on five landing sites. Only *Limnothrissa* however were among the catches nor could any *Stolothrissa* be found in the lake (pers. comm. R. L. Welcomme). The future of this fishery will be followed with interest.

Reference has already been made to the occasional mass destruction of fish which has been reported at irregular intervals from many tropical lakes. These are clearly attributable to unusually violent stormy weather bringing water to the surface from a normally stagnant oxygen-free and often H_2S-charged lower layer. It would seem that the hydrological conditions in Lake Kivu are favourable for such an event, but no reports of this have apparently been published. It is possible that owing to the very steep density gradient, the stability of the lake is sufficient to resist disturbance by the most violent weather. On the other hand, such catastrophes may have been partly responsible for destroying much of a previously richer fauna.

It is worth noting in this connection that the lava flow from Nyamulagira, when it reached the northwest corner of the lake in 1938, apparently obliterated the entire fauna of the Gulf of Kabuno, which has subsequently been repopulated through the connecting channel (Verbeke, 1957a). The Virunga volcanic barrier might therefore have had some catastrophic effects on the fauna of a previous lake which, as suggested above, may have existed on the course of a river draining into Lake Edward.

Though Lake Kivu is apparently an unproductive lake with a fishery of little economic importance, a study of the reasons for this condition are most interesting and instructive. The violent volcanic activity in the recent past is probably responsible both for the poverty of the fauna and for the present peculiar hydrological conditions which are restricting a potentially higher productivity.

Of the other volcanic barrier lakes of this region Lakes Bulera and Luhondo in Rwanda have been the most thoroughly studied (Damas, 1954–55). They show some peculiar features. Formed by blockage of two adjacent branches of the same river by lava flows from the volcano Muhavura, they are now connected together (Fig. 15.1). The drainage, previously to the northwest into the Edward basin, is now reversed. Lake Bulera flows directly over the Ntaruka Falls into Lake Luhondo, which discharges into the Kagera River system and thus into Lake Victoria.

The peculiarity of Lake Bulera is that, though very deep (max. 173 m), it was oxygenated to the bottom at the time of Damas's investigations. His suggestion that this condition is maintained by cold water from neighbouring hill country underflowing the warmer surface water has already been discussed in Ch. 6. All the other lakes under discussion, except Ndalaga, are moderately deep (40–70 m), but shallower than Bulera, and are all apparently permanently stratified and anoxic below 10–20 m in their deepest regions. This is characteristic of small relatively deep tropical lakes which are well sheltered from wind by surrounding hills (Ch. 6).

The inflow into Lake Luhondo, on the other hand, comes from the already heated surface water of Lake Bulera. It is apparently permanently stratified. The almost uniform composition from top to bottom of the lower anoxic water led

Damas (1954–55) to conclude that there was some kind of continuous but slow circulation of the whole column. This suggestion has been made here for other permanently stratified tropical lakes (Ch. 6).

Damas made quantitative estimates of the standing crop of phyto- and zoo-plankton in the volcanic barrier lakes of Rwanda as well as in Lakes Kivu, Edward and in some artificially dammed lakes in Katanga. A comparative review of this data is to be found in Damas (1964). The estimates were made by counting algal cells and individual animals.

In conclusion we may say that, because of their origin from blocked hill streams, most of the many volcanic barrier lakes in the region between Lakes Edward and Tanganyika have very few species of fish and tend to be ecologically immature with some niches unocciupied. Lake Kivu is exceptional in that it probably derives from a previous lake which was enlarged and deepened by volcanic blockage. It consequently has a more varied and lacustrine assemblage of fish species, though the fauna may have been somewhat depleted by the late Pleistocene volcanics. The continuing discharge of saline water into the bottom of the basin has produced a permanent dense and anoxic mass of water below the oxygenated surface layer to which all aerobic life is confined and from which great quantities of nutrients are presumably being lost to the lower layer. Owing to the very steep density gradient the upward recycling of nutrients from the lower water into the euphotic zone must occur on a relatively small scale. In Lakes Tanganyika and Malawi the density gradients are comparatively slight, and it is suggested in Ch. 6 that recycling of nutrients from the anoxic lower water, aided by a system of internal waves, is probably an important factor in productivity. Water has in fact been pumped from the anoxic depths to the surface of Lake Kivu and an expected increase in algal productivity was the result (p. 129).

The volcanic barrier lakes are of special interest in the study of the evolution of the African lakes and their faunas because their age is intermediate between that of the very young manmade lakes (Ch. 20) and that of most of the other large lakes which were formed before the late Pleistocene.

16

Lake Tanganyika

The more we get to know about the two greatest of the African Rift Valley lakes, Tanganyika and Malawi, the more interesting and exciting they become. Their enormous size and depth, the grandeur of the surrounding mountains (Fig. 16.2) falling precipitously to rocky shores and sandy beaches, their dramatic geological history and present hydrology, and, above all, their unbelievably rich and beautiful endemic faunas – all combine to give them an aesthetic and scientific attraction so far appreciated by relatively few people.

When in 1858 Burton and Speke reached Lake Tanganyika, of which they had heard from Arab traders, they were disappointed to hear that it did not, as they had hoped, flow out to the north into the Nile basin (Burton, 1860; Speke, 1863). Apart from this negative discovery relating to the then burning question of the source of the Nile, the most important scientific result of their visit came ultimately from a collection of mollusc shells from the beaches, which had attracted Speke by their very unusual appearance. The unique fauna of the lake was thus brought to the attention of European zoologists, and there followed a long debate on its origin which is not yet fully resolved.*

Lake Tanganyika lies on the floor of the Western Rift about 100 km south of Lake Kivu at the much lower altitude of 773 m. It is 650 km long with an average width of about 50 km. The maximum depth of the north basin is 1 310 m, that of the south basin 1 470 m. Most of the coastline is precipitous, and for long stretches the escarpment falls directly into the lake. On the west side of the south basin the steep escarpment continues under water and drops to 1 470 m below the water surface within about 4 km of the shore. Only at the extreme north and south ends is the underwater gradient at all moderate and even there the 100 m contour is less than 10 km from the shore.

The main inflows are the Ruzizi from Lake Kivu in the north, on which the Panzi Falls provide a faunal barrier between the two lakes, and the Malagarasi from the east, a swampy river draining the flat lands south of the Victoria basin. By comparison the other inflows have small catchments and many are seasonal.

---

* A brief historical account of the major biological investigations on the lake up to 1950 is given by Capart (1952a).

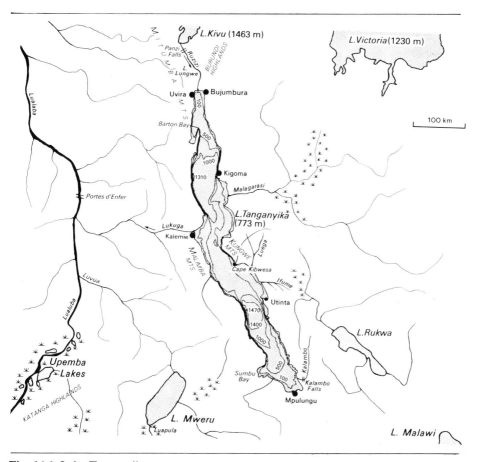

Fig. 16.1 Lake Tanganyika

On the west side is the outflow, the Lukuga River, which, when first seen by Cameron in 1874, was blocked. A subsequent rise in lake level had caused a break in the accumulated plug of swamp vegetation when Stanley visited it in 1876, and in 1879 it was seen by Joseph Thompson to be in heavy flood. In the early 1890s the flow had again almost ceased.†

Unlike Lake Victoria, whose main source of water is from direct rainfall on the surface, most of the rain in the immediate region of Lake Tanganyika is trapped by the high escarpments. A characteristic feature of the rainy season is the progressive building during the day of long banks of storm clouds above the escarpments until they finally discharge with much thunder, lightning and torrential rain which streams down the rocky ravines into the lake. Even when the Lukuga River is open the major loss of water is by evaporation from the surface, which is certainly also true of Lakes Victoria and Malawi and probably of most tropical lakes. On the other hand, though the outflow is of minor importance in the overall water budget at the present time, the salinity of the water, though high (over $0.5°/_{oo}$, see Table 5.1), is certainly not as high as would be expected in a

† For a recent discussion of the water level fluctuations see Camus (1965).

Fig. 16.2 Lake Tanganyika. The eastern escarpment and the Kungwe Mountains seen from a boat on the South Basin.

lake with no outlet. The water of Lake Turkana, which has been in a closed basin for a few thousand years, is about five times as saline (Table 5.1). In spite of the even higher salinity of the inflowing Ruzizi River (c. $0\cdot65^{o}/_{oo}$), the flow-through during the past few thousand years since the establishment of the Luku-ga outlet has been sufficient to prevent a great accumulation of salt.

Around the turn of the century, the first discussions concerning the origin of the lake and its fauna were dominated by the striking theory that it is the rem-nant of an ancient sea which was connected in the Jurassic with the Atlantic across the Zaïre basin. This theory, which was held with conviction by several competent biologists, was based mainly on the apparent marine or 'thalassoid' form of some members of the fauna. It originated from Böhm's discovery in 1883 of the medusoid coelenterate ('jellyfish') *Limnocnida tanganikae* and was sup-ported by Günther (1898) (Fig. 16.4), but was most strongly suggested by the appearance of some of the prosobranch gasteropod molluscs, which has originally arrested the attention of Speke. Some of these, notably *Spekia, Stanleya* and *Edgeria*, are globular, smooth and very thick shelled with a well-developed foot and thus adapted to hold fast to rocks which are strongly pounded by the waves. Such well-marked adaptation is common among littoral molluscs on rocky seashores, but had not previously been encountered in lakes. Others with long spines from the shell and protruding siphons (e.g. *Paramelania* and *Tiphobia*), characters otherwise only known in certain marine gasteropods, inhabit the large areas of very soft mud down to the limits of oxygen (150–200 m) where the supporting skeletal structures and a long siphon might well be advantageous (Fig. 16.3).

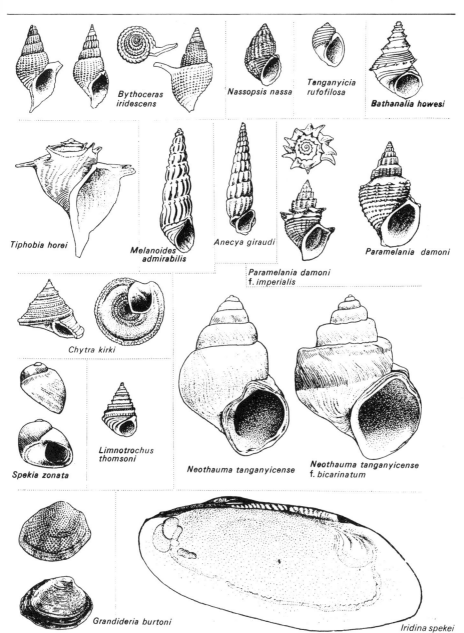

Fig. 16.3 Some endemic molluscs from Lake Tanganyika (Leloup, 1952)

J. E. S. Moore, who was the most notable exponent of the marine relict theory, added the potomonid crabs and some other groups of animals to the list of supposed marine relicts. If the two species of clupeid fish, which belong to a predominantly marine family, had been known they would presumably have been regarded as very strong evidence. Moore's *The Tanganyika Problem* (1903) is a fascinating discussion of the geological history of the lake and of the origin of its fauna in the wider setting of the East African lake region, based on his expeditions in 1895 and 1898. It is all the more interesting for being perfused with a theory which was later to be discredited. We now know that all the 'thalassoid' species are related to freshwater forms found elsewhere in Africa from which they have diverged during a long period of isolation in adaptation to special conditions in the lake. The argument was first set out by Cunnington (1920) and further work has confirmed this general conclusion. There is no geological evidence to support a former marine connection, and the extensive sediments in the Zaïre basin were mostly deposited in fresh water during the Pliocene.

It has more recently been suggested, e.g. by Beauchamp (1939, 1940, 1946), that the form of the thalassoid molluscs might have been induced by the present peculiar composition of the water, in which the ratios Cl/SO_4 and Mg/Ca are similar to those in seawater, or by a formerly higher salinity approaching that of the sea. The highly endemic nature of the present fauna is attributed to selection in the past for resistance to high salinity, alkalinity or to the peculiar ionic composition (e.g. high relative concentration of magnesium), and that even now the composition of the water is such as to demand special adaptations. This theory was extrapolated into the past by Fuchs (1937) in suggesting that the presence of some 'thalassoid' molluscs such as *Neothauma*, a genus now living in Lake Tanganyika, in the Kaiso (early Pleistocene) beds of the Albert and Edward basins is evidence of a higher salinity of the water at that time.

Though the water may well have been more saline in the past (see below), it cannot have exceeded a salinity tolerated by a large and varied freshwater fauna and, therefore, can never have reached a level much above $5°/_{oo}$, or 15% of seawater (see Ch. 5). Lake Turkana, as already noted, is well on the way to this condition with a salinity (ca. $2.5°/_{oo}$) about five times that of Lake Tanganyika, and with a higher alkalinity and pH. Yet it still provides a very favourable environment for freshwater organisms, none of which show even the beginnings of 'thalassoid' features. The waters of Lakes Albert, Edward and Kivu are also more saline and alkaline than Lake Tanganyika and the Mg/Ca ratio is high in all four lakes. Moreover, there is no evidence whatever that either high salinity or particular ionic ratios would have any such effect on freshwater molluscs nor, indeed, that the form of the marine molluscs, to which they have been compared, has anything to do with the composition or salinity of the seawater. It would seem more likely that we are concerned with adaptations to certain kinds of physical environment, e.g. waver-beaten rocky shores and large areas of very soft organic mud, which are common in the sea but rare in inland lakes. Lake Tanganyika is exceptional in having provided these conditions on a large scale without a break for a few million years (see discussion in Kufferath, 1952, and Leloup, 1950a, pp. 238–52). The suggestion that the apparent absence of planktonic Cladocera is due to the high level of magnesium (Cunnington, 1920), inspired by the absence of Cladocera from the open seas, has been discounted by Hutchinson (1933) who showed that some Cladocera are tolerant of far higher Mg concentrations and by Beauchamp (1939) who demonstrated that Tanganyika water is not toxic to Cladocera species taken from marshes near Kigoma. Cladocera have subsequently been found in the lake, but are restricted to shallow water (Harding, 1957) for reasons which have yet to be discovered.

The geological history of the lake and the origin of its fauna are still in some particulars the subject of controversy (see Willis, 1936; Cahen, 1954; Poll, 1950). During the Miocene, before the major rifting, the drainage across the area was to the west, and the Malagarasi together with the Lukuga probably formed one of the main rivers. When the lake first appeared as the result of the large rifting movements during the Pliocene, those rivers that were transected by it were thus draining into the Zaïre. The largest of these, the Malagarasi, was interrupted by the lake into which it then began to flow. This episode has been dramatically corroborated by the discovery in the Malagarasi River of species of fish which do not live in the lake but are found in the Zaïre basin on the other side of the lake, e.g. *Polypterus congicus*, *P. ornatipinnis*, *Sarotherodon niloticus upembae* (p. 167), *Tetraodon mbu* and *Distichodus maculatus* (Fig. B3). These are mostly riverine, not lacutrine, fish and there seems to be no reasonable explanation other than that they represent the old early Pliocene riverine fauna of the Zaïre basin which was isolated in the Malagarasi by the formation of the lake (Poll, 1946).

There may initially have been two lakes on the sites of the present north and south basins. There followed a very long period, probably more than a million years, during which there was no outlet. The fauna originally inherited from the previous river system was thus isolated in a complex of quite new habitats to which there was plenty of time for adaptation. The length of the period of isolation is reflected both in the total number of endermic species of fish (176 out of 214) and especially in the number of endemic species of cichlid fishes (130 out of 134). The figures for percentage endemism in both the cichlids (98%) and non-cichlid fishes (57%) of Lake Tanganyika are remarkably close to those from Lakes Victoria and Malawi which have also had a long history of isolation (Lowe-McConnell, 1969). What distinguishes Lake Tanganyika and reflects its greater age and period of isolation is the wider divergence in the speciation of the non-cichlids (Table 8.2). There are as many as eight endemic genera. Apart from one (*Xenoclarias*) in Lake Victoria, no endemic genera of non-cichlid fishes are to be found in any of the other African lakes, though the cichlids have produced 30 in Lake Tanganyika, 20 in Malawi and 4 in Victoria. Lake Tanganyika is remarkable also for the extensive speciation among animals other than fish. Most of the invertebrate groups have several endemic species, and in some cases genera. For example, the ratio endemic:total known species of bivalve molluscs is 5:13 (Leloup, 1950b), of gastropod molluscs 37:60 (Leloup, 1950a)*, potamonid crabs 3:4 (Bott, 1955), of planktonic copepod Crustacea 11:13 (Lindberg, 1951). It is a curious fact that the only calanoid copepod, *Diaptomus simplex*, which is the main constituent of the zooplankton, is not endemic. Another anomaly, already mentioned, is the absence from the deep water plankton of cladoceran Crustacea. They are found only in shallow water especially near the mouths of rivers, and of the six known species none are endemic (Harding, 1957). One of the most interesting adaptations is shown by the endemic species (and genus) of trichopteran (caddis fly), *Limnocetis tanganicae* (Marlier, 1955, 1962) (Fig. 16.5). It skims over the surface of the deep water like a gyrinid (whirligig) beetle and is attracted at night by the lights of the *ndagala* fishermen (p. 291). Seven or eight mil-

---

* For a comparison of the molluscan faunas of Lake Tanganyika and Malawi see p. 301.

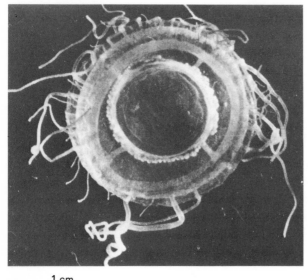

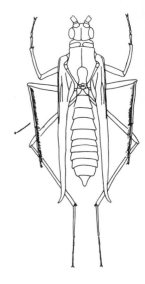

Fig. 16.4 *Limnocnida tanganikae*, view of under-surface, from Bouillon (1957). The 'freshwater jellyfish' of Lake Tanganyika and neighbouring smaller lakes. It differs from the species in Lake Victoria (*L. victoriae*) in its longer tentacles and in reproducing asexually by budding from the edge of the manubrium (mouth) as shown. *Limnocnida* spp. have been found in many African lakes, big and small. It appears spasmodically in large swarms in the plankton. For the life history, which includes a fixed 'hydroid' stage, see Bouillon, 1957 and Beadle *et al.*, 1960

Fig. 16.5 *Limnocetis tanganicae*, a flightless endemic trichopteran (caddfly) living on the surface of Lake Tanganyika (from Marlier, 1960

limetres long, it has lost its powers of flight, its wings being reduced and its limbs adapted for support and movement on the water surface. It also swims under water. The nymphs have been seen, but nothing is known of the breeding which presumably occurs inshore. As Marlier (1962) points out, flightless adaptation to surface pelagic life is rare among the Trichoptera, but there are species similarly adapted in Lake Baikal in Siberia and Lake Titicaca in South America, both of which are very large lakes with a long history of isolation. There are a few species of Diptera and Hemiptera that have managed to colonise the surface of the open sea in this way.

Among the vertebrates other than fish there are two endemic snakes which are almost permanently aquatic and feed on fish. The small colubrid *Glypholycus bicolor* follows the swarms of sardines (Fig. B2) upon which it feeds, and the larger water cobra *Boulengeria annulata stormsi* (Elapidae) lives during the day among the rocks on the shore and goes into the water at night to catch fish (Poll, 1952b).

Though a few more endemic species and perhaps genera of animals will be found in this and other African lakes, there is no doubt that the number in Lake Tanganyika greatly exceeds that in any other. This is surely evidence of great age and a very long period of isolation.

Though long isolation of an ancient Miocene fauna has certainly been responsible for the general features of the fauna of the present lake, there are certain details which must be added. These can best be discussed in relation to the fish. There are several soudanian species in the lake which are identical with those now living in the Nile basin. There are three possible explanations (see discussions in Poll, 1950, 1953, and Marlier, 1953).

1. It is possible that they are members of the original widespread Pliocene riverine fauna and have remained unchanged since then. *Bagrus docmac*, *Ctenopoma muriei*, *Malapterurus electricus* and *Sarotherodon niloticus* are considered by Poll (1950) to belong to this group.

2. Some, such as *Hydrocynus lineatus*, are soudanian species that now live in the Zaïre basin and may have come into the lake very recently (lake Pleistocene) when the Lukuga outflow was established, which probably also gave access to the lake for a few species which had evolved in the Zaïre River, e.g. *Alestes macrophthalmus* and *Citharinus gibbosus*.

3. Marlier (1953) points out that several of the soudanian species in Lake Tanganyika, such as *Bagrus docmac* and *Ctenopoma muriei*, are capable of traversing shallow swampy waters. These species also live in Lake Victoria from which, he suggests, they may have come during one of the very wet periods in the late Pleistocene. The watershed between Lake Victoria and the Malagarasi is very low and a not very great rise in water level could have provided a shallow and swampy route for the passage of such fish.

It was at one time thought that the presence of modern soudanian ('nilotic') species of fish in Lake Tanganyika could only be explained by the previous existence of an outlet to the north into the Nile basin. As already mentioned in relation to the origin of Lake Kivu (p. 270), the fluvial deposits described by Boutakoff (1937) in the upper Ruzizi valley apparently indicate that there was a river flowing northwards in the mid late Pleistocene just before the drainage was reversed by the uprising of the Virunga volcanoes. Whatever the correct interpretation of these deposits, it is surely impossible that a north-flowing river during the late Pleistocene could have given access to the lake for soudanian species without at the same time allowing some of the, by then, well-evolved Tanganyika fauna to escape to the north. The latter certainly did not happen. Marlier (1953) compared the fish in the affluent streams of the Ruzizi valley with those in the inflows to the Kivu basin. He found that the two faunas were quite different. The Ruzizi inflows have characteristic species which are absent from the Kivu streams and vice versa. These findings indicate that there was no recent faunal connection via the Ruzizi valley between Lake Tanganyika and the drainage system to the north which is part of the Nile basin. There are, however, fossil lake beds halfway up the course of the Ruzizi containing molluscs at present living in Lake Tanganyika, thus indicating that the lake very recently extended some distance up the valley. These are the only fossil beds so far known in the Tanganyika basin, but they are unfortunately too recent to reflect the earlier history of the lake (Cahen, 1954, pp. 348–54).

The conclusion, therefore, is that most of the fauna of Lake Tanganyika, which is mainly endemic, is clearly derived as a consequence of long isolation from the Miocene fauna of rivers that were previously draining into the Zaïre basin, but there have been later additions whose origin and time of arrival are still in doubt. Some of the unchanged soudanian species may derive from the original fauna, some may have arrived quite recently across the Victoria–Malagarasi watershed or up the Lukuga river. The difficulty is to conceive of a route by which soudanian species could have entered the lake without Tanganyikan species leaving it. The very recent Lukuga outlet has without doubt allowed both Zaïre species to enter the lake and a few lake species to pass into the Zaïre. This is the only route by which we are certain that recent exchanges have taken place in both directions between the lake and outside waters. Some species, such as *Hydrocynus goliath*, *Distichodus fasciatus*, and *Labeo lineatus*, which must have entered the lake by this route, seem to have arrived so recently that they are still in the process of extending their range. Since only large specimens have so far been found, it is possible that they have not yet established breeding grounds in the lake, their numbers being augmented from time to time from the Lukuga River (Coulter, 1966).

It is generally agreed that the long period of isolation was brought to an end by volcanic blockage in the late Pleistocene which reversed some of the drainage from the Rwanda and Burundi Highlands. There may previously have been a highland lake with an outflow to the north into the Edward basin which was blocked by the volcanic lava flow, the water level rising to form the modern Lake Kivu and eventually overflowing to the south over a falls and down the Ruzizi River to Lake Tanganyika (p. 271 and Fig. 15.1). This, it is generally thought, raised the level of the lake and started the Lukuga outflow.

It cannot be doubted, in view of the great climatic changes in the Pleistocene, that the level of the lake has fluctuated greatly during its long history as a closed basin. Nevertheless, as already pointed out, it could never have contracted so much that the salinity of the water rose much higher than that of Lake Turkana or the very rich endemic freshwater fauna would have succumbed. The echo-sounding survey made by the Belgian Expedition (1946–1947) revealed some interesting features in the bottom profile and sediments which may relate to the history of the lake prior to the final major rise in level (Capart, 1949). The valleys of the main inflowing rivers along the east coast continue under water and can be traced to a depth of about 550 m. The echo-tracing reproduced in Fig. 16.6 was made along a line between two promontories on the east coast of the south basin. It shows a cross-section of two such valleys. Capart concluded that these could have been formed only at a lower water level when the bottom down to the present 550 m contour was exposed to the air and would be effectively eroded by the rivers. The soundings also showed some details of the sediments. Down to 100–150 m there was evidence, confirmed by dredging, of a firm single layer of sandy silt. From 150–550 m there were echo-traces of up to eight alternate layers of hard and soft sediments. Below 550 m the number of layers decreased and below 850 m there was no evidence of any but a single layer of very soft material. Capart interpreted the hard layers as evidence of recurring periods of exposure to the air of the bottom down to the 550 m contour thus providing apparent support for his interpretation of the subaquatic valleys. It is perhaps significant that the altitude of the 550 m contour (220 m) is approximately that suggested by Willis (1936) as the general level of the land before the formation of the Rift.

Some doubt has been thrown on Capart's conclusions by Livingstone (1965) on

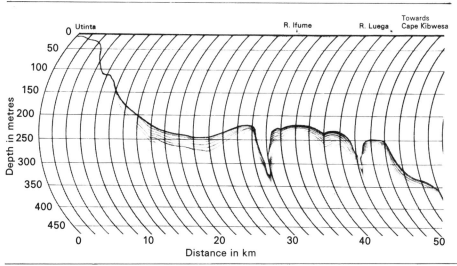

Fig. 16.6 Lake Tanganyika. Echo soundings along east coast between Utinta and River Leuga (Fig. 16.1) (from Capart, 1949)

the basis of a single short (10 m) sediment core taken in 440 m of water near the south end of the lake. This included one hard layer at about 5 m down the core which was found to be composed of volcanic ash laid down in water. Its age, from carbon dating, was 11 000–12 000 B.P. These findings show that this one hard layer, at least, has no necessary connection with a former low lake level. Livingstone makes the suggestion, previously rejected by Capart, that the sub-aquatic valleys have been eroded under water by cold density currents laden with sediment flowing down from the surrounding high country. Underwater erosion of this kind is known to occur along sea coasts and in some large European lakes. It is certainly necessary to explain why the valleys have persisted under water without being filled with sediment. Livingstone's findings suggest that the bottom down to 440 m has been submerged continuously for more than the past 12 000 years. A drop of water level to the 550 m contour would have divided the lake into two halves joined by a narrow channel, and with the water surface at 700 m there would be two quite separate lakes. Neglecting the possible complications caused by earth movements we could be reasonably certain that the lake has been divided, perhaps more than once, in the past. It has been suggested (e.g. by Livingston, 1965) that repeated separation of the two half-lakes could have favoured the high degree of speciation on the part of the fauna.

One of the species of cichlid fishes of the rocky littoral, *Ophthalmochromis ventralis*, has two distinguishable varieties, *O.v.ventralis* and *O.v.heterodontis*, living at the north and south end of the lake respectively. Between these populations are some intermediate forms. Poll and Matthes (1962) interpret this as evidence of divergence within two previously separated basins. The two varieties have now met and are hybridising. Whether or not this is the correct explanation, indications of this kind are scarce and there is no good evidence that the great variety of available habitats and the very long period of isolation of the lake as a whole have not in themselves provided sufficient stimulus to divergent speciation. Another apparently similar example is the separate northern and southern distribution of the two subspecies *melanostigma* and *macrops* of the cichlid *Callochromis macrops* (Poll, 1956; Fryer and Iles, 1972, p. 530). There seems to be no possibility of invoking such complete and temporary fragmentation of Lake Malawi to account for the extensive speciation which has occurred there during a probably shorter period of isolation.

These investigations have been discussed not because any definite conclusions can as yet be drawn, but because they point very clearly to the kind of work that should now be done over a wide area of lake on the disposition, composition and age of the sediments. Thereby we can reasonably expect to gain a much clearer picture of the history of the lake and, perhaps, valuable information on the geological and climatic events in the recent past.

The main characteristics and composition of the present fauna have been known for many years, as shown by the great amount of information summarised by Cunnington in 1920. The number of recorded species has been much increased since then, especially by the Belgian Expedition in 1946–47. Our knowledge of the ecology of the lake is, however, as yet rudimentary and for practical reasons relates mainly to the fish. There is one feature of Lake Tanganyika, as of Kivu and Malawi, which has clearly had a profound ecological effect. This is the enormous volume of permanently stagnant, anoxic and H_2S-charged water below about 200 m. The reasons for this have already been discussed (p. 95). The total volume of the lake is approximately 30 000 km^3, but only about one-quarter of this contains enough dissolved oxygen to support life other than the anaerobic micro-organisms which are permanent inhabitants of the anoxic lower water and the larvae of lake flies which spend part of their time there. Consequently, the benthic fauna is almost entirely restricted to the small fraction of the bottom that is below less than 200 m of water.

In this respect a comparison with Lake Baikal in Siberia is of great interest (see Brooks, 1950; Kozhov, 1963). Baikal is remarkably similar to Tanganyika in area, depth, long period of isolation and mode of origin on the floor of a rift valley. It differs mainly in its geographical location and consequent north-temperate continental climatic regime with severe winters and warm summers. Owing to the seasonal temperature reversal and complete overturn, the bottom water even at the greatest depth (1 741 m) is permanently well supplied with oxygen, with a temperature below about 50 m of less than 5°C at all seasons. In addition to the bottom water and mud at medium depths, corresponding to the deep benthic zone of Lake Tanganyika, Lake Baikal has therefore a very extensive and habitable abyssal zone which has probably persisted in its present condition for a few million years. During the maximum glaciation in the late Pleistocene an ice-sheet reached at least the north end of the lake, the surface of which must have been frozen for long winter periods if not continuously (Flint and Dorsey, 1945; Kozhov, 1963). The fauna could more easily have survived in the abyssal region where the temperature would have been close to that now prevailing (5°C). But the shallow littoral zone and the surface water generally, to which much of the fauna of Lake Tanganyika is confined, were presumably obliterated by the Pleistocene ice. Consequently, it is the deep water and bottom mud in which most of the famous endemic invertebrate fauna of Baikal is to be found, a region which is uninhabitable in Lake Tanganyika below about 200 m. The most remarkable members of this fauna are the amphipod Crustacea of the family Gammaridae (entirely absent from tropical Africa) of which there are 240 species, and all but one are endemic and comprise 34 endemic genera (see Kozhov, 1963, Table XIV, for more data about the fauna). Baikal thus differs from Tanganyika mainly because of the very different climatic regime now and during the Pleistocene. But the groups of animals which compose the faunas of the two lakes are also different because they are derived from two very different pre-Pleistocene faunas. Unlike the situation in Lake Tanganyika there are some undoubtedly marine elements in the fauna of Lake Baikal including, among others, the polychaete worm *Manayunkia* and some species of the gammarid Crustacea. But these are probably derived from immigrants from the sea rather than from the fauna of a previously marine Baikal (see Kozhov, 1963, Ch. 6). Geological recent marine immigrants are widespread in the inland waters of northern Europe where, unlike East Africa, there have been great changes in the distribution of sea and land since the Mesozoic and particularly during the Pliocene and Pleistocene.

General accounts of the main ecological regions of Lake Tanganyika have been given by van Meel (1952, with reference to the vegetation), Leloup (1952, invertebrates) and Poll (1950, 1952b, fish).

1. Inflowing rivers

Investigations on the ecology of the inflows to the Kivu and Tanganyika basins were made by Marlier (1951–54, 1953). Many of these are short, seasonal torrents from the escarpments; they are especially common on the west shore. The aquatic flora and fauna are poor and the fish are temporary visitors from the lake. The water is of very low salinity compared with the lake and usually much colder. The fauna and flora of the permanent inflowing rivers is by comparison very rich, but according to Leloup (1952) their invertebrate faunas are almost entirely different from that of the lake, so abrupt is the change from the riverine habitat to the almost oceanic conditions in the lake.

Of the permanent inflows the greatest single contribution is, according to Gillman (1933), made by the Ruzizi. Its water, being derived partly from Lake Kivu, has a salinity ($0\cdot6-0\cdot7^{\circ}/_{\circ\circ}$) actually higher than that of the lake, and tends to sink under the surface. Though the uppermost branches of the Kivu-Ruzizi drainage system are mostly seasonal mountain torrents with a poor fauna, the lower reaches, and particularly the Ruzizi valley, have a typical riverine invertebrate fauna (Marlier, 1951–54). There are a number of lake fish, such as *Barilius moori* which ascend the Ruzizi to breed especially in times of flood, and there are several others, such as some species of *Barbus*, which are more or less confined to the river.

The River Malagarasi, which enters the lake through an extensive swampy delta, provides a type of habitat well known around Lake Victoria, but which hardly exists elsewhere in the Tanganyika basin. Open channels traverse great areas of dense swamp composed of *Cyperus papyrus*, *Typha*, *Carex*, and other emergent plants. There are also extensive stretches of deeper and slow-flowing water choked with water weeds such as *Potamogeton, Ceratophyllum, Chara* and *Utricularia*, and in places with a floating mat of water lilies (*Nymphaea*), water chestnut (*Trapa*), the water fern (*Azolla*) or the Nile cabbage (*Pistia*) (van Meel, 1952). In contrast to the dense papyrus and reed swamp, these situations provide a relatively stagnant but nevertheless well-oxygenated and very productive environment for a rich fauna of invertebrates and fish. On a much smaller scale similar conditions are found in the lower reaches and estuaries of the Ifume and other small rivers. A greater contrast with the conditions in the lake could hardly be imagined. Though the mouth of the Malagarasi is not discernible from the lake its presence is betrayed at some distance by the brown colour and shallowness of the water and the occasional patch of swamp vegetation established on a raised portion of the submerged alluvial fan brought down by the river. Floating islands of papyrus are sometimes seen well out into the lake.

The fine organic mud of the submerged deltas of the permanent rivers, which spreads out some distance into the lake, provides a particularly favourable environment for several species of the light-weight bivalve molluscs, e.g. *Grandidieria*, and the spiny prosobranch snails such as *Tiphobia* and *Bythoceras* (Fig. 16.3).

2. Littoral region

Owing to the clarity of the water enough light for the photosynthesis of plants rooted on the bottom reaches a depth of about twenty metres, and this is taken to define the limit of the littoral region. Since it is also the region which is most disturbed by winds and waves, rooted vegetation is actually established only where there is sufficient shelter from mechanical disturbance as in the bays of Burton, Kigoma, Sumbu and in small inlets and creeks. In the shallow parts of such sheltered waters the forest of water weeds such as *Ceratophyllum*, which grow from the muddy bottom, supports a rich invertebrate fauna; among them the Mayfly (ephemerid) larvae and crabs are most prominent and serve as the main food of some of the littoral cichlid fishes and of the young stages of *Lates* spp. These, when mature, move out and become the major predators in the pelagic region. In the open but still relatively shallow (40–80 m) and often muddy waters of some bays, such as Sumbu Bay in the south, the bottom is densely populated by molluscs and crabs and is a very favourable habitat for certain fish such as *Chrysichthys* spp.

Along much of the shore, however, the mountains fall precipitously to the water's edge, where gigantic blocks of rock lie in a disordered mass and are spread under the water as far as the eye can see. Except on the most exposed surfaces where few organisms can maintain a hold against the violence of the breakers, the submerged rocks are covered with a furry coat of brown and green algae which harbours many small invertebrates such as Crustacea, rotifers and insect larvae. Here are to be found the strong-shelled 'thalassoid' prosobranch snails such as *Paramelania, Spekia* and others (Fig. 16.3) and about forty endemic species of cichlid fishes. Between and under the rocks are cracks and crevices in which crabs, prawns and many of the fish take refuge during the day. This kind of habitat, at least on a large scale, is unique to Lakes Tanganyika and

Malawi and, except for the absence of tides, resembles coral reefs and rocky seashores more than any other inland-water habitat. The behaviour and ecological relations of the fish can be studied by underwater observation. A few species, previously unknown, have been discovered in this way. A most remarkable phenomenon are the small, brilliantly coloured cichlid fishes (e.g. *Limnotilapia* and *Petrochromis*) nibbling like flocks of sheep at the carpet of algae on the rock faces. But there is a great range of feeding habits reflected in the types of teeth and shapes of mouth. Some are mainly carnivorous (e.g. *Lamprologus*) and feed on crustacea, insects or even young fish. The very complex relations between the species of browsing cichlids, with the implications for their evolution, has been more thoroughly studied in Lake Malawi (Ch. 17). Marlier (1959) has, however, described several subspecies of *Trophius moori* which are distinguishable from each other by their colour patterns. These he calls 'microgeographical subspecies'; they appear to be actively diverging at the present time. One form has become sufficiently differentiated to be assigned to a new species *T. duboisi* – a conclusion supported by observations on its behaviour in aquaria with members of the parent species.

Though long stretches of rocky shore are a characteristic feature of the Lake Tanganyika littoral, they are interspersed with long sandy or stony beaches. These provide less varied habitats than the submerged rocks, but are populated by characteristic invertebrates and fish (Leloup, 1952; Poll, 1952b). Except very close inshore, the submerged sand is mixed with mud and provides favourable conditions for bivalve molluscs such as *Grandideria* and *Iridina* (Fig. 16.3) and *Coelatura* which collect the particles of organic matter, algae, etc., that are stirred up by the turbulent shallow waters. The submerged, stony shores harbour a rich invertebrate fauna of oligochaets, leeches, planarians, larvae of Diptera and Trichoptera and aquatic Hemiptera, crabs and prawns. The fish of this region are mostly cichlids which feed on the animal and plant material on the bottom (one cichlid, *Tylochromis polylepis*, feeds on molluscs), and their generally rather sombre colouring makes them almost invisible even in the shallowest water. Two species, however, *Sarotherodon tanganicae* (Cichlidae) and *Barilius moori* (Cyprinidae), feed at the surface on plankton but are mainly confined to inshore waters.

In the littoral region of both Lakes Tanganyika and Malawi there is thus a very great variety of habitats. The main sections, rocky and sandy, are discontinuous and alternate with each other. This seems to provide a degree of isolation between populations of the same species and is probably one of the stimulants to the great burst of speciation in the region. But within one major habitat, e.g. rocky shore, there is a wide range of sub-habitats or niches which may have stimulated divergence leading to ultimate reproductive isolation and speciation. This gives exciting opportunities for resarch on ecological, behavioural and genetical aspects, in a situation which appears to be actively evolving at the present time.

3. The benthic or sublittoral region

This is defined as the area of lake bed below about 20 m depth of water, which is out of range of effective light and is not directly disturbed by winds and waves.

Its lower limit (at a maximum of about 200 m) is determined by the total absence of oxygen. It is deeper in the windy season (June–August) than at other times of year, and is subject to rhythmic fluctuations caused by internal waves (see Ch. 6). A glance at the map in Fig. 16.1 will show that this region covers a large area. It can be compared with the submarine continental shelf, though it differs from this in that, with increasing depth, conditions become progressively less favourable for aerobic organisms and it borders on an abyssal region which is practically devoid of animal life.

The surface of the bottom sediment is composed of soft mud which, below about the 100 m contour and in shallower water opposite the mouths of the larger rivers, is almost pure organic matter with a consistency of thick soup. This material is inhabited by many invertebrates, such as chironomid fly larvae, copepod and ostracod Crustacea, and molluscs. Leloup (1950b, p. 69) shows how the form of the bivalve mollusc *Grandideria burtoni* (Fig. 16.3) varies with the consistency of the mud in which it lives. In very soft mud the shell valves are much expanded and the general shape is globular, and this presumably helps to prevent sinking into the anoxic lower mud. The same form is adopted by some other bivalves in this habitat such as *Brazzaea anceyi*. The prosobranch snails are typically those with projecting spines, such as *Tiphobia horei* and *Bythoceras iridescens* (Fig. 16.3). *Paramelania damoni*, too, shows a variation of form in that those living in the very soft mud of the benthic region have more pronounced spines, which suggests that these are also concerned with prevention of sinking (Leloup, 1950a, p. 198). The crab *Platytelphusa armata* is a common scavenger in this region.

All these animals depend directly or indirectly upon the organic detritus which is produced in the water above or is brought in by the inflowing rivers. There are many species of fish that are mainly confined to the bottom. Endemic species of cichlids are the most abundant especially those of the genus *Trematocara* and *Hemibates*. Some species feed directly on the detritus, others select a diet of insect larvae, oligochaets and small crustacea from the mud (Poll, 1956).

Coulter (1966, 1967a) set gill-nets on the bottom (120 m) of the deep benthic region of the southeast arm of the lake, and when lifting the nets took water samples from the same level for measurement of oxygen. Some species of fish were caught in water with very little oxygen (below 2% saturation) and some even in apparently anoxic water. *Lates mariae* was one of the species recorded. This raises the interesting possibility that these fish have become in some manner, physiologically or metabolically, adapted to low oxygen levels and even, for a time, to anoxia.* The probable persistence for a geologically long period of a stagnant anoxic lower layer, whose upper boundary was frequently fluctuating, might have provided suitable conditions for such adaptations if the fish were genetically capable of producing the appropriate variations. There is no doubt that the Nile perch *Lates niloticus* of Lake Albert cannot live in anything but rather well-oxygenated water (Fish, 1955b).

---

* See p. 321 for an example of a cichlid fish apparently adapted by means of a very high haemoglobin level for periodic visits into the lower anoxic water of a small crater lake.

4. The pelagic region

This has features which are as remarkable and characteristic as those of the littoral. It comprises all the surface water above the anoxic abyssal region, that is to say to a depth of 100 m to 200 m according to the position and season (Ch. 6). Its horizontal boundary is indefinite and pelagic animals are commonly found in the water over what has been defined as the benthic region. It therefore occupies a much greater area and volume than the other regions except the abyssal (see Fig. 16.1).

With the exception of Lake Kivu the zooplankton of the other lakes is directly consumed by certain species of fish. But in most lakes this is not as important a link in the food chains leading to the fish as the direct consumption by fish of phytoplankton, macrophytes, insect larvae or organic detritus. In Lake Tanganyika, however, there are two small endemic genera and species of the herring family (Clupeidae), *Stolothrissa tanganikae* and *Limnothrissa miodon*, both usually about nine centimetres long, which have so successfully exploited this source of food that they have multiplied to prodigious numbers and are themselves the major item in the fishery (Fig. A4.2). Through a hierarchy of predators, the zooplankton has become the basic nutriment of an abundant and varied population of pelagic fish.

As in some other parts of the world, clupeids have invaded the large rivers of tropical West Africa, including the Zaïre, in which there are now several related species. Some found their way into Lake Tanganyika, but into none of the other large lakes. Since they are so well integrated into the economy of the lake, it seems more likely that they derive from the ancient pre-Pliocene riverine fauna that they have entered the lake comparatively recently with the opening of the Lukuga overflow. The presence of two endemic species of isopod Crustacea *Lironeca enigmatica* and *L. tanganikae*, parasitic on the two clupeid fish, poses the same problem (Fryer, 1968a, b). Another species of the genus, *L. expansa*, is know from the Zaïre basin, and there are others which are purely marine. No parasitic isopod Crustacea have as yet been reported from any other freshwater in tropical Africa. Since the clupeid fish in the West African rivers have presumably come from the sea, it is likely that these parasites have come by the same route. But we need to know more about their occurrence, distribution and life histories. For many of the unsolved problems relating to the origins of the Tanganyika fauna mid-Pleistocene and earlier fossil evidence could be decisive. This may well have been buried beneath the present lake since the final rise to the level of the outlet.

Limnothrissa lives mostly in shallow inshore waters and at certain seasons is caught off the beaches in seine-nets. *Stolothrissa*, the Tanganyika 'sardine',* is confined to the pelagic region at the edges of which it is often with *Limnothrissa*. If feeds exclusively on zooplankton which is mainly composed of copepod Crustacea, the most abundant species being the calanoid *Diaptomus simplex* (Lindberg, 1951). There are also a few species of rotifers and atyid prawns, and frequently

---

* *Ndagala* (in Swahili). The tribes around the lake each have their own name for this important fish. For a study of the breeding biology of both species see Ellis (1971).

swarms of the medusoid coelenterate *Limnocnida tanganicae*, the freshwater 'jellyfish', are seen at the surface (Fig. 16.4). As in other lakes the bulk of the zooplankton migrates diurnally, being concentrated at between 50 m and 120 m during the day, while at night they are within 10 m of the surface. The sardines follow these movements with the result that the surface water, which in mid lake is wonderfully clear and seems to be devoid of life during the day, may towards evening become almost turbid and seething with massive shoals of sardine. The larger predator fish are seen darting around among them. The native fishermen have taken advantage of the powerful attraction of light, and there was originally a brushwood fire on a platform in the bows of the canoe. Attracted by this light the sardines swarm around the canoe in such numbers that they appear to fill all the available space. They are caught with very large fine meshed scoop-nets mounted on long poles. Under most favourable conditions, at the height of the season and during the early part of the night, it is merely a matter of ladling out the water which is full of fish. Towards dawn, the catch progressively declines as the zooplankton, and with them the sardines, begin to leave the surface.

This technique has now been modernised, though light is still used, but now it has been made less hazardous by the introduction of kerosene vapour lamps, and large-scale commercial fishing is done from the Zambian south end of the lake and from Burundi in the northeast. A number of boats with powerful lights are towed into position by powered craft in the evening and brought back in the early hours to the processing plant. Large ring-nets are used. At the height of the season (June to October) when the fish are most abundant and all the private and commercial boats are out, more than a hundred points of light may be seen at one time scattered over the surface of the south end of the lake. The sardines are mostly dried in the sun and transported over a wide area, the main market being the Copper Belt in Zambia. They are at present by far the most important of the commercial fish in the lake. In 1966 it is estimated that, excluding catches for private consumption, about 5 500 metric tons fresh weight of fish were taken from the Zambian waters (less than a tenth of the total area of the lake) of which about 4 300 tonnes were sardines.†

All of the species of pelagic fish are endemic. Besides the *ndagala*, there are two much less common species which also feed on the zooplankton. These are *Lamprichthys tanganicus*, an unusually large species (up to 10 cm) of the family Cyprinodontidae and *Engraulicypris minutus* (Cyprinidae) of about the same length. It is to be noted that all four of the zooplankton-feeding species are of about the same size. In addition, the fry of the large predator fish at one stage feed on the zooplankton and may be found among the others.

The principal predators on the sardines (*ndagala*) in the pelagic region are *Luciolates stappersii* (Fig. B2) and *Lates microlepis*, both centropomids of 40–70 cm in length and closely related to the much larger Nile perch *Lates niloticus*, and the cichlid *Bathybates minor* (about 20 cm). Other large predators such as *Lates mariae* and *angustifrons* migrate between different zones in search of prey and are caught at times in the deep benthic water as well as near the surface. The

† Calculated from *Fisheries Statistics* (Natural waters), 1966. Republic of Zambia Central Statistics Office, Lusaka.

biology of *Luciolates* and the *Lates* species in Lake Tanganyika and the effects of recent large-scale commercial fishing on the stocks of these important fish have been studied by Coulter (1976). The populations are being reduced and altered in relative species composition, and at least one species (*Lates microlepis*) is in danger. But it is thought that with careful management the high rate of reproduction in the very favourable environment could maintain the fishery at its level at the time of the investigation.

The sardine fishery has been singled out for special mention because it is unique to Lake Tanganyika (though now introduced into Lake Kariba, p. 372, and Lake Kivu, p. 273), for its commercial importance, for its special biological interest and as an indication of the productivity of the lake. It has sometimes been assumed, on the basis of experience with lakes in temperate climates, that the pelagic region of the large permanently stratified tropical lakes, of which Lake Tanganyika is the most notable, are relatively unproductive or 'oligotrophic'. This matter was discussed on p. 000 in the light of two recent sets of measurements of primary productivity. There is no doubt of a high rate of production in the upper pelagic water of organisms that are the basis of the ecosystem, including the fish. There remains however some doubt concerning the manner in which plant nutrients are sufficiently rapidly circulated from below and to what extent undecomposed organic matter may be brought up from the sediments to contribute directly to secondary production via bacteria and protozoa. As already pointed out, the clarity of the surface water during the day is deceptive. Most of the organisms are then at greater depths. The volume of water occupied by them (down to 100–200 m) is very great and the euphotic zone exceeds 20 m. It perhaps differs from shallower productive lakes, not in level of production per unit area of surface, but in that the whole process is 'diluted' through a much greater volume of water. It is thought by some (e.g. Leloup, 1952, pp. 86–8) that such apparently high productivity of the open waters must be based to a significant degree on airborne organic dust transported perhaps from as far as the arid subtropical regions of northern and southern Africa. The nature and magnitude of the various basic sources of organic production will no doubt ultimately be defined by further research.

A brief comparision between Lakes Tanganyika and Malawi is made at the end of Ch. 17.

17

Lake Malawi

Over 600 km long and about 770 m at its deepest, Lake Malawi (formerly Lake Nyasa) is, after Lake Tanganyika, the largest and deepest of the Rift Valley lakes (Fig. 3.4 and 9.4). It has the same general form and appearance as Lake Tanganyika, being flanked by mountain ranges which in many places, especially in the northern half, fall as precipitous escarpments to the shore and continue under water with undiminished gradient to great depths.

Though the lake had been known to the Portuguese (and called 'Marawi') for more than a century previously, its northerly extension remained to be determined (see Fig. 2.1). It was David Livingstone who, between 1859 and 1863, first explored it thoroughly and brought it to the notice of the outside world. Apart from his own *Narrative of an Expedition to the Zambezi* (1865), his journeys and the subsequent turbulent history of the lake region have been many times recorded. A recent general account up to the independence of Malawi is Ransford (1966).

One of Livingstone's companions, John Kirk, later an important political figure as British Resident in Zanzibar sent to London a collection of fish from the lake and thus aroused interest in the very peculiar fauna. It was only a few years previously (1858) that Speke with his collection of mollusc shells had done the same for Lake Tanganyika.

There is no doubt that Lake Malawi poses some equally fascinating hydrological and biological problems, but, except for the fish, less is known about it than about Lake Tanganyika. Among the reasons for this is certainly the dramatic, though later exploded, theory of the marine origin of some of the fauna of Lake Tanganyika which stimulated much interest and investigation in the early part of this century (p. 278). Nor has Lake Malawi had the benefit of a large, well-equipped and widely ranging expedition comparable with the Belgian Expedition to Lake Tanganyika in 1946–47. Nevertheless, we now understand some of the major features of the hydrology of the lake which have been outlined in Ch. 6. Most work has, however, been done on the fish and their ecology; these will be discussed in this chapter.

Geological and palaeontological evidence is inadequate for definite conclusions, but rifting movements had certainly begun long before the Pleistocene during the early part of which it was thought by Dixey (1938) that the lake first appeared. It is therefore younger than Lake Tanganyika, but has been long enough isolated from other water systems to have produced an endemic fauna almost as remarkable as that of Lake Tanganyika. Unlike the latter there is only one deep basin (in the north), but in most important respects the two lakes have much in common. They are both subject to a similar climatic regime involving alternation between a wet and relatively calm season and a period of heavy winds and somewhat lower temperatures. This cycle acting on a basin of similar shape and north–south orientation has apparently caused the same kind of seasonal changes in water circulation in which wind-induced internal waves play an important part in circulating nutrients between the mass of anoxic water below about 200 m and upper layers in which most of the organisms live.

The outlet by the Shire River to the Zambezi, like the Lukuga from Lake Tanganyika, is intermittent, though unlike the Lukuga, only one cycle has occurred during the past century. It is clear from Livingstone's narrative that around 1860 there was much water in the then navigable upper Shire. From measurements by Livingstone and other missionaries between 1860 and 1890 a tentative curve was drawn by Pike and Rimmington (1965) to represent the changes in lake level before the systematic records were started (Fig. 17.1). From 1896, apart from a seasonal fluctuation of 0·4 to 1·7 m, there was a progressive fall to a minimum in 1915 at which time the outflow ceased. From then on the level rose steadily to a maximum in 1935 when, in the wet season, it was about 6 m above the minimum in 1915. But it was not until this high level was reached that the outflow was resumed. The reason for this delayed break-through was the building up of sand bars across the bed of the outlet after the flow had ceased. The blockage was augmented by silt deposited in the bed by the small tributary streams during the wet seasons and was consolidated by the growth of vegetation. During the rains, the upper section of the river bed sometimes actually drained

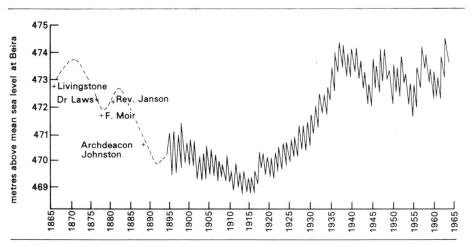

Fig. 17.1 Lake Malawi water levels 1865–1963 (redrawn from Pike and Rimmington, 1965)

back into the lake. Consequently a much higher lake level was required to dislodge these obstructions than was needed to maintain an outflow once established. The effective level was reached in 1935, the barriers were washed away and the outflow has continued until now, the lake level having remained relatively high (Dixey, 1924; Cochrane, 1957).

The catchment area of the Malawi basin (Fig. 3.4) is very much smaller both absolutely and relative to the lake area than that of Lake Tanganyika which includes the Kivu basin and the Malagarasi plains. The inflows are mostly rather short watercourses from the escarpments and nearby mountains and their volume depends directly on the rainfall in the immediate region. The hydrology is delicately balanced and the level of the lake and the volume of the outflow react rapidly to changes in local rainfall between wet and dry seasons as well as to longer-term fluctuations. These changes transform the inshore habitats. Except where the escarpments drop direct into the lake, there are coastal plains of varying width which are alternately flooded and exposed. It has been suggested that these changes have been one of the stimuli to the abundant speciation among the cichlid fishes (Trewavas, 1947; Lowe, 1953). Some recent changes in lake level are clearly marked on the rock shown in Fig. 17.2. The existence of raised beaches at least 100 m above the present lake level indicate that the outlet was once at a much higher level and has since been eroded down or, what is very likely in this unstable region, has subsided (Dixey, 1924).

The salinity of the lake water is comparatively low (K_{20} 200–230) being about one-third that of Lake Tanganyika. This difference is no doubt partly due to the absence of a major inflow of more saline water like the Ruzizi into Lake Tan-

Fig. 17.2 Monkey Bay, Lake Malawi. Recent higher level marks up to about 1·5 m above the water

ganyika. There is also no evidence that Lake Malawi was ever a hydrologically closed basin except during the short periods when the outlet was obstructed. On the other hand the very large proportion of endemic species of fish suggests that the lake has been faunally isolated for a very long time. The upper Shire River as far as Matope (100 km) has a slight gradient and spreads out on the way into the very shallow Lake Malombe. For the next 80 km there are a series of cataracts, the Murchison Rapids, with a total drop of about 280 m. This is the barrier which now prevents exchange of species between the lake and the Zambezi via the lower Shire. If the lake has never been a hydrologically closed basin for a long period, such a barrier must have existed for most of the lake's history. The lower Shire (about 230 km long) flows over a relatively level plain and spreads into swamps in its lower reaches. During stoppage of the lake outflow between 1915 and 1935 these swamps dried out and were brought under cultivation, with consequent serious social disruption when they were again flooded.

There are more than 245 species of fish in the lake which is a greater number than in any other African lake including Tanganyika (214 so far recorded). But it is interesting to note that the number of families represented (7) is only about half that in the other great lakes (Lowe-McConnell, 1969, Table 1).

The previous connections between Lake Malawi and the neighbouring water systems are very obscure and were probably complicated. It is evident, however, that none of the Malawi fish has come directly from the north. The fauna of Lake Tanganyika has totally different relations, and such prominent fish as *Lates* (Centropomidae), which are soudanian in origin, and the Clupeids which came from the Zaïre basin, are absent from Lake Malawi. On the whole the Malawian fish are most closely related to those of the Upper Zambezi (above the Victoria Falls) and less to those of the lower Zambezi into which the lake flows via the Shire River – so effective is the barrier presented by the Murchison Rapids.

There are some 15 species common to Lake Malawi, the Upper Zambezi as well as to the shallow swampy Lake Bangweulu which drains to the north into the Zaïre (Worthington, 1933). Lake Bangweulu contains only about 40 of the more than 400 species in the Zaïre basin from which it is at present faunally isolated by the Johnston Falls on the Upper Luapula (Fig. 9.3). Some of these have, probably recently, managed to make their way into the Zambezi basin perhaps over the Luapula-Kafue watershed (Bell-Cross, 1965, and Ch. 8, p. 134). In fact there is a close relation between the fishes of the Upper Zambezi and Kafue and those of the western Angola rivers on the one hand and those of Lake Bangweulu and the Upper Luapula on the other. The intervening watersheds are mostly flat and swampy (Ricardo, 1943; Trewavas, 1964; Poll, 1967). It is more difficult to understand how some of them reached Lake Malawi. There may have been a temporary connection across the watershed between the Chambesi and the streams flowing into the northwest corner of the lake (Figs. 9.3 and 9.4). Apart from this probable route of entry by the ancestors of many of the fish, close affinities between some non-cichlid lake fish and species now living in east coastal rivers (e.g. the Ruaha) suggest another route which is now blocked perhaps by reversal due to earth movements (pers. comm. E. Trewavas). The immigrant fish have evolved into a greater number of new species than is found in any other African lake though, as mentioned above, the number of families involved is comparatively small. As in the other long-isolated lakes, Victoria and Tanganyika, it is the cichlids that have diverged most. There are more than 190 species of which at least 97% are endemic. Of the 50 to 60 species of non-cichlids about 65% are endemic (Lowe-McConnell, 1969, and Table 8.2).

The broad ecological zones in Lake Malawi have been discussed by Bertram *et al.* (1942), van Meel (1952, pp. 181–94) and Jackson *et al.* (1963). They are, in general, similar to those in Lake Tanganyika.

1. The inflowing rivers are inhabited by fish which are mostly not endemic, some being restricted to the rivers, others seasonal migrants from the lake during the breeding period when the water is high.
2. In the lake itself there are the following main zones:
   (a) the estuaries of rivers and sheltered bays;
   (b) sandy shores;
   (c) rocky shores;
   (d) the relatively shallow (to about 20 m) but open inshore waters;
   (e) the pelagic region or upper 100 m or so of surface water offshore;
   (f) the benthic region comprising the mud surface and the water just above it over all the lake where the bottom is lower than about 20 m (euphotic zone) and above the limit of oxygen (150–200 m); and
   (g) the abyssal zone below 200 m, mainly in the north basin, where the water is permanently anoxic.

It is interesting to note that not only do the main ecological zones in Lakes Tanganyika and Malawi have the same characteristics, but that their resources are exploited by the fish in much the same manner. The species involved are, however, different, some niches being occupied by members of different families in the two lakes. This is particularly evident in the pelagic zone where, as in Lake Tanganyika there is an abundant zooplankton composed of about nine species of copepod and cladoceran Crustacea, of which the most abundant are *Diaptomus kraepelini* and *Mesocyclops leuckarti* together with larvae of chaoborid flies (Jackson *et al.*, 1963). These as a group exploit both the phytoplankton and the organic matter in the bottom mud. There is some indication that the plankton is generally more abundant in the southern shallower half of the lake (Jackson *et al.*, 1963, p. 45). This would certainly be expected since the nutrients would be more effectively circulated in shallow water. In Lake Tanganyika the zooplankton in the pelagic zone is mostly consumed by one species of fish, the sardine *Stolothrissa tanganicae* (Clupeidae). The other zooplankton feeders, *Engraulicypris minutus* and the young of some other pelagic fish, are far less numerous. In Lake Malawi, on the other hand, the equivalent zooplankton consumer, *Engraulicypris sardella* (Cyprinidae, local name 'usipa') is only one of several species which feed in this way. It is a small fish around 10 cm in length which is important for food and as bait for larger fish. It is probably the only completely pelagic fish in the lake in that it breeds in the open water of the pelagic region rather than inshore. But at certain times large shoals come close to the shore, where they are attracted at night by flares and hand-netted, and are caught in seine-nets on the beaches during the day.

The other zooplankton consumers in the pelagic zone are a group of about sixteen endemic species of genus *Haplochromis* (Cichlidae) which move about together in large shoals in the surface waters. They are distinguished by their protrusible mouths and are collectively known as '*utaka*' by the fishermen; they are mainly fished by the native open-water seine-net. They are most abundant in the shallow water above submerged rocky prominences where there is an upwelling of nutrients. Most fishing for *utaka* is done at not very great distances from the shore (Fig. 17.3). The latest systematic work on the *utaka* is by Iles (1959);

Fig. 17.3 Lake Malawi. A haul of *utaka* – mostly endemic cichlid species caught within a few kilometres of the west coast near Monkey Bay

they were also discussed by Trewavas (1935), by Lowe (1952) and by Fryer and Iles (1972). Like some of the cichlid species in Lake Victoria, they form part of one of the large *Haplochromis* 'species flocks' in Lake Malawi, an associated group of closely related species which have evolved in isolation in the lake. The zooplankton feeders are preyed on by several fish including the catfish *Bagrus meridionalis* and some species of *Ramphochromis* which are fast swimming thin-bodied cichlids with jaws and teeth adapted to predation. These predators are themselves of considerable economic importance.

Of particular interest are the fishes of the deep benthic region and the water around the lower limit of oxygen which varies according to position and season between 200 and 300 m. Though there have been no simultaneous oxygen measurements and fish catches, such as those done by Coulter in Lake Tanganyika (p. 290), several species of fish have been caught below 200 m and some near to 300 m. It is likely that these fish spend at least some time in water containing extremely little oxygen. A few species of the pelagic cichlids (*utaka*), the catfish *Bagrus meridionalis* (Bagridae), the elephant snout fish *Mormyrus longirostris* (Mormyridae), *Synodontis nyassae* (Mochokidae) and *Haplochromis heterotaenia* (Cichlidae) have all been taken from these depths (Jackson *et al.*, 1963, p. 58). Of special interest are the catfish of the family Clariidae which are well known from shallower waters in tropical Africa and are noted for the accessory air-breathing organs above the gill cavity (superbranchial organs) whereby they can live in stagnant waters and even under the anoxic conditions in some swamps (p. 320). In Lake Malawi there are four species of the genus *Clarias* which are not endemic to the lake and are found mostly in the associated rivers and swamps. The remaining ten species are of the genus *Dinotopterus* (*Bathyclarias*)* and are all endemic. They are found in various habitats, but at least four species have been recorded down to almost 300 m (Jackson *et al.*, 1963). The superbranchial organs, though valuable in shallow water from which the air above is easily accessible, are obviously of no use at great depths, whatever the level of oxygen. Experiments with several species of *Clarias* have in fact shown that, like the lungfish *Protopterus*, they cannot get enough oxygen through their gills which have a much reduced surface area. Aerial respiration has become a necessity, even in well-aerated water. It is therefore not surprising, but nevertheless very interesting, that in the genus *Dinotopterus* there is a tendency for reduction of the superbranchial air-breathing organs and a corresponding increase in the surface area of the gills. In three of the deep-water species in Lake Malawi, *D. foveolatus*, *filicibarbis* and *rotundifrons*, the superbranchials have entirely disappeared (Greenwood, 1961). The same has happened to species of *Xenoclarias* in Lake Victoria (p. 256 and Greenwood, 1958). This is, of course, not an adaptation to the level of dissolved oxygen, but to the inaccessibility of the air and is a reversion to purely aquatic respiration. Whether these fish are in any way adapted to a scarcity of oxygen by, for example, modification of their haemoglobin or of their respiratory metabolism, remains to be investigated.

As in Lake Tanganyika, the rocky shores provide rich and varied habitats

---

* They were originally assigned by Jackson (1959) to a new genus *Bathyclarias*, but were later included by Greenwood (1961) in the genus *Dinotopterus* of which there is one species *D. cunningtoni* endemic to Lake Tanganyika.

occupied by a large number of fish beautifully adapted to different niches. Owing, however, to the work of Fryer (1959a) in the neighbourhood of Nkata Bay we know much more about this habitat and its fauna in Lake Malawi. The upper surfaces of the submerged rocks are covered with a carpet or *aufwuchs* of blue-green algae (mostly two species of *Calothrix*) to which many other algae are attached and which harbours a large number of invertebrates – harpacticid (Copepoda) and ostracod Crustacea, chironomid fly larvae, water mites (hydrachnids) and larvae and nymphs of other insects. The crevices under and between the rocks provide hiding places for the crab *Potamonautes lirrangensis* and refuges from predators for fish.

The rocky shores of Lake Tanganyika, as already noted, have provided conditions for the evolution of a unique assemblage of endemic molluscs. It is therefore astounding that they are virtually absent from the same habitat in Lake Malawi. This difference between the two lakes is obvious even to a superficial observer who has the opportunity to visit both. Fryer's suggestion that in Lake Malawi they are kept in check by the fish, among which there are several species capable of eating them, is difficult to accept with confidence until we know more about the ecological relations of the fauna of the rocky shores of Lake Tanganyika. There is however no doubt that the molluscan fauna of Lake Malawi is more restricted than that of Lake Tanganyika as shown in Table 17.1. In Lake Tanganyika the family Melaniidae (Thiariidae) has thirteen of the endemic genera of 'thalassoid' snails, including those which inhabit the rocky shores. In Lake Malawi this family is represented only by the non-endemic genus *Melanoides* with six species (five endemic), none of which has established itself on the rocks (Crowley *et al.*, 1964). We can only guess that the basic molluscan stock, from which the present fauna of Lake Malawi was derived, included no species with a genetic constitution capable of producing forms adaptable to the rigorous conditions on the wave-swept rocks. Whatever the explanation, it is at least striking evidence of the isolation of the two lakes from each other and of the very different origin of much of their faunas.

The genus *Neothauma* (Viviparidae) known from the early Pleistocene beds of the Albert and Edward basins (p. 235) has survived in Lake Tanganyika (Fig. 16.3). In 1964 Crowley *et al.* reported another species of this genus (*N. ecclesi*) endemic to Lake Malawi. This clearly posed a considerable problem, but more recently Mandahl-Barth (1972) has re-examined the molluscan fauna of the lake and has given a number of cogent reasons for concluding that this is an endemic species of the nearly related but not endemic genus *Bellamya*, namely *B.*

Table 17.1 The number of families, genera, species and endemic species of gastropod and Givalve molluscs recorded from Lake Tanganyika (Leloup, 1950a, b) and from Lake Malawi (Mandahl-Barth, 1972)

	Gastropoda				Bivalvia			
	Families	Genera	Species	Endemic species	Families	Genera	Species	Endemic species
Lake Tanganyika	9	36	60	30	4	12	14	5
Lake Malawi	6	9	27	17	4	5	11	5

ecclesi. Three of the four species of this genus found in Lake Malawi are endemic. For an ecological study of two endemic species of *Bulinus* (Planorbidae) see Wright *et al.* (1967). Live specimens of one of these, *B. nyassanus*, was dredged from the mud at about 25 m – the greatest recorded depth for a pulmonate mollusc in tropical waters.

There is a prodigious number of fish among the rocks – on the average some six to seven per square metre of surface water (according to Fryer, 1959a). They comprise about twenty-seven endemic species of cichlids and five of other families and nearly all of these appear to be confined to the rocky regions. They are locally known as '*mbuna*'. The cichlid species are adapted to at least six types of feeding, as reflected in their jaw mechanisms and teeth. The commonest type (e.g. *Labeotropheus fülleborni*) feeds on the algal *aufwuchs* by means of a wide ventral mouth provided with scraping teeth. Others suck up the algae with a very mobile mouth fitted with tooth pads. Several others, and two of the non-cichlid species, are equipped with protruding teeth which enable them to grab the small crustacea and insect larvae from out of the algal carpet. One species, *Haplochromis kiwinge*, is adapted, like the *utaka*, to catch the zooplankton in the surrounding water. Two species of rock cichlids are predators on other fish as well as on insect larvae and crabs. A most remarkable mode of feeding is practised by the cichlid *Genyochromis mento* which has a protruding lower jaw with which it pulls off the scales and pieces of the fins of other fish, especially of *Labeo cylindricus* (Cyprinidae). This curious habit has been recorded of fish in other African lakes (Fryer *et al.*, 1955).

The animals of the rocky shores thus form a closely knit community which seems to be exploiting in a most efficient manner the products of algal photosynthesis on the rock faces. It would be very difficult to measure and express quantitatively the productivity of this region at any trophic level, but primary production and the rate of turnover are likely to be high. The turbulence of the shallow water along the shore would ensure a continuous supply of plant nutrients from the mud beyond the rocks and some of the substances washed in by rain from the land could be exploited before they reached the deeper water. The direct economic importance of the *mbuna* fishes is understandably small owing to the difficulties of catching on a large scale among rocks. But the production cycle which starts with what might be called a large-scale natural 'plate culture' of algae on the rock faces, supplied with ample solar energy and circulating nutrients at all seasons, is probably providing a considerable excess of organic matter to be washed into deeper water and ultimately incorporated into some of the important commercial fish. It should be possible to investigate this point.

The sandy shores, which along much of the coastline alternate with stretches of rocks, provide a quite different environment and are populated by other species. Of the twelve species of cichlid fish recorded by Fryer (1959a) off sandy shores only one is common to both regions. The main primary producers in the water above the sand are the water weed *Vallisneria* and a great number of microscopic algae which encrust its flat tape-like leaves. Among these plants is an invertebrate fauna – copepod and ostracod Crustacea, the prawn *Caridina nilotica*, chironomid fly larvae and larvae and nymphs of Dragonflies (Odonata) and Caddis flies (Trichoptera) are the commonest types. But the most notable difference from the rocky shores is the presence of molluscs – four species of gastropod snails and three of bivalves were recorded by Fryer. The range and habits and mechanisms of feeding among the cichlid fish is consequently different from that of the rock species. *Haplochromis mola* feeds on molluscs for which it has large crushing pharyngeal bones. *Haplochromis similis* and *Tilapia rendalli* browse on the water weeds and *Sarotherodon saka* takes phytoplankton from the water and detritus from the bottom mud. Two species of *Lethrinops* dig chironomid larvae from the sand. Most of these cichlids seem to be confined to the sandy shores, but they are preyed on by several species which frequently come inshore from deeper waters.

Fryer (1959a) recognised an 'intermediate zone' between the two above types of shore composed of smaller rocks among patches of sand. But the flora and most of the fauna belong to the sandy region, though a few species of cichlid fish, e.g. *Cyathochromis obliquidens*, seem to be almost confined to the intermediate zone.

There is therefore a very abrupt change in fauna between rocky and sandy regions with little exchange between them. Nor is there any apparent exchange between two neighbouring rocky regions even when the intervening sandy stretch is quite short. Fryer (1959b) suggested that the clear separation of these two ecologically different habitats has lasted long enough for the separate evolution of two different faunas each with its assemblage of endemic species. Here is an apparently good example of allopatric speciation within an environment which seems at first sight to be devoid of barriers. The barriers are, however, ecological but appear to be as effective as a range of mountains.

We have been considering the independent evolution of the two faunas on the rocky and sandy shores. If the barriers between the two habitats are as rigid as they appear to be, we might expect that the fauna of each single patch of rock or sand would have diverged in isolation and that each would have its own assemblage of species distinct from that of every other patch. Few observations have so far been made on the differences between the same species of fish in the same habitat in different parts of the lake, but there are certainly some small though clear differences that may be genetic (Fryer, 1959a, p. 267; Fryer and Iles, 1972, Ch. 16). There are however no major geographical differences which might be expected if the barriers had always been as effective as they now appear to be. We are led to the tentative conclusion that, though the ecological isolation from each other of the two habitats has been maintained for a very long time, there is at present, or has been on occasions in the recent past, some exchange between the faunas of individual patches of the same habitat. Climatic and tectonic events have certainly caused great changes in water level and might well have altered the distribution of the different types of shoreline. Such events could have given opportunities for movements of species with the result that the now isolated individual populations in the same type of habitat, though showing some incipient divergence, are essentially the same all over the lake.

Another coastal habitat studied by Fryer (1959a) was a shallow bay of which the bottom is mainly sand with an overlying loose layer of organic detritus. The water level is naturally subject to great variations and perhaps for this reason, and because the water is frequently disturbed by winds, conditions are not stable enough for the establishment of true stagnant swamp vegetation. It is colonised by reeds such as *Phragmites mauritianus* and aquatic grasses such as *Vossia cuspidatus* which are in places closely packed. The conditions here seem to be similar to those outside many of the lake-fringing swamps in Lake Victoria and other lakes where there are banks of reeds and grasses in water that is more disturbed than in the swamp but is still shallow enough for emergent vegetation. Examination of one such bay in Lake Malawi, Crocodile Creek, showed a much greater number and variety of aquatic insects than in the other two habitats – dragonfly nymphs (Odonata), aquatic beetles (Coleoptera) and bugs (Hemiptera). There were also many more species of small Crustacea – copepods, cladocera and ostracods. The molluscs on the other hand (three species of gastropod snails and no bivalves) were less well represented. On the whole the invertebrate fauna was of the kind to be expected in rather calm water. Of the nine recorded species of fish five were also found on sandy shores. The fauna of this bay may differ in some degree from that of other similar habitats round the lake , but it probably has the main features of this kind of relatively calm water, which is nevertheless sufficiently disturbed and unstable to prevent deoxygenation through the accumulation of decomposing plant material.

An interesting appendage to the distribution of fish is that of the external crustacean parasites on the fish (Copepoda and Branchiura). Seventeen species have been recorded in Lake Malawi by Fryer (1968b). The distribution of these does not necessarily follow that of the host fish since the Branchiura and the larvae of the copepods are free-swimming at least for a period. The eggs might also be washed or transported some distance. The interpretation of the present distribution of the species is therefore not easy. There are several which are widely distributed in Africa – *Dolops ranarum* (Branchiura), *Lamprolegna monodi*, *L. hemprichii* and *Lernaea barnimiana* (Copepoda) are found in all the major African river systems. On the other hand, the long isolation of Lake Malawi is reflected in five endemic species of *Lernaea* (there are seven in Lake Tanganyika).

The fauna in the bottom mud has not yet been investigated in detail but Lake Malawi is noted for its periodic very large swarms of lake flies that are composed mainly of *Chaoborus edulis*. The biology of lake flies is discussed in Chapter 14 (see Fig. A3). The early larval stages of the chaoborids and all the larval stages of the chironomid midges are undoubtedly to be found in the mud over large areas of the bottom of Lake Malawi. The older chaoborid larvae are planktonic. All stages are important items in the food of some of the fish, especially of species of *Dinotopterus* (*Bathyclarias*) (Jackson et al., 1963). David Livingstone (1865) was the first to record that these flies (*nkungu*) are eaten by the people round the lake. They are still eaten – striking evidence of the prodigious number of flies in the swarms. They are attracted by lights and fall in heaps on the ground to be scooped up, boiled or roasted and made into 'cakes' about 3 cm thick and 10– 20 cm in diameter. Livingstone noted that they taste rather like caviare or salted locusts.

In conclusion, it would be appropriate to summarise briefly and compare the main features of the two great lakes Tanganyika and Malawi as far as our present knowledge permits. The former is the older, but they were both formed in the same manner as deep trenches in the floor of the Rift Valley. The Tanganyika trench cut across some rivers which were flowing westward into the Zaïre basin. Lake Malawi on the other hand was formed by faulting along the line of a previous valley which had been draining southward into the Zambezi basin. The original riverine fish faunas which became isolated in the two lakes were therefore different in composition, and this is clearly reflected in the present fish and mollusc faunas. Though there are now a large number of species of fish in Lake Malawi, there is a smaller range of families. Several of the typical soudanian families of fish as well as the clupeids, which are found in Lake Tanganyika, are absent from Lake Malawi. Most of the fauna of the former is derived from the north and west, that of the latter from the south. The greater age of Lake Tanganyika is apparently reflected in the eight endemic genera of non-cichlid fish. There are none endemic to Lake Malawi.

Both lakes have provided the same types of habitat in a manner and on a scale different from the other large lakes. To these habitats the original faunas responded with the evolution of a large number of endemic species. The biological resources of the two lakes have been exploited in much the same manner, but different species are involved. The pelagic zooplankton, for instance, is consumed mainly by one species of clupeid fish in Lake Tanganyika, whereas in Lake Malawi several species of cichlid and one cyprinid play this part.

Speciation among the inshore fishes of both lakes may have been stimulated by the fragmentation of the coastline into discrete alternating zones of rocks and sand with occasional shallow bays of shallow calm water. This and the fluctuations in water level and consequent alteration in the distribution of the inshore ecological zones may have favoured the evolution of the characteristic endemic faunas of these zones. As usual it is the cichlid fishes that have responded most actively. Since Lake Malawi has been the scene of the most dramatic and extensive speciation among the cichlids, rivalling even Lakes Victoria and Tanganyika, much discussion and controversy on the causes and progress of speciation have centred on Lake Malawi. The main points at issue are the origin of the different groups of species, the part played by environmental changes in the past, the significance of breeding and feeding habits, genetic variability in different species and the possible influence of predators. These matters cannot adequately be dealt with here but the main lines of the discussion can be followed from the publications of Worthington (1933), Trewavas (1935, 1947), Lowe (1953), Jackson (1961b), Fryer (1960a, b, 1965), Greenwood (1965d), Fryer and Iles (1955), the general review by Lowe-McConnell (1969), and the book by Fryer and Iles (1972).

In more recent times both lakes have received species of fish from the Zaïre basin, but by very different routes – Lake Tanganyika by direct invasion via the Lukuga outlet; Lake Malawi probably over the high swampy watershed from the Bangweulu basin. For this reason each lake has been invaded by different types of Zaïre fish.

As far as we can see from the evidence, Lake Tanganyika was a closed basin for a very long period, probably much more than a million years, before the opening of the Lukuga outlet. Lake Malawi, on the other hand, probably overflowed into the Zambezi intermittently during its whole history, but a faunal barrier below the outlet, now represented by the Murchison Rapids, seems to have existed for most, if not all, of the period. More could certainly be learnt of the past history of the lakes from deep sediment cores.

In both lakes there is apparently the same kind of stratification, and the seasonal movements of the water layers are probably similar. The lower oxygen-free and nutrient-rich water is being circulated into the upper layers, perhaps on a more massive scale than was previously thought. Wind-induced internal waves would play an important part in this.

Though no measurements have been made, there is little doubt of the high rate of production in the well-stirred waters of the rocky shores and shallow extremities of these lakes. Recent measurements indicate that the rate of gross primary production in the open pelagic water of Lake Tanganyika is about as high as in Lake Victoria (Table 7.1), but the possibility of other sources of organic matter have been suggested (p. 130). It is unlikely that the two lakes differ very much in this respect.

18

Tropical swamps: Adaptation to scarcity of oxygen

The conditions for life in tropical swamps are peculiar and very different from those in open lakes and rivers. The extent of the swamplands of tropical Africa is, however, enormous. They form a major habitat of great ecological and human importance. Their total area is difficult to assess, but it certainly amounts to many hundreds of thousands of square kilometres and possibly exceeds that of the open waters of all the lakes.

We can define a swamp as a stretch of water in which conditions are dominated by closely packed aquatic vegetation. In general, the water is shallow and slow moving, so that the characteristic swamp vegetation can be established. The growth of emergent plants is often so dense that vast areas of water are entirely hidden from view. There are, of course, extensive swamps in some temperate countries but, apart from differences in types of plants, the high rate of continuous growth and of decomposition at tropical temperatures produces peculiar conditions which will be discussed in this chapter.

The 'basin and swell' structure of the African continent inherited from the remote past (Fig. 3.1) has determined the location of all but one of the main swamp regions. Many of them are the remnants of larger areas inundated during the late Pleistocene. In the Araouane-Niger basin lie the great swamps of the Middle Niger extending for about 250 km above Timbuktu. They include a complex of small shallow lakes and are the remains of the very large Pleistocene Lake Araouane (Fig. 11.3). Much of Lake Chad, which lies in the very much larger Chad basin, is choked with swamp (Ch. 12). In the southern end of the Sudan basin the Upper Nile spreads out into the great areas of papyrus and grass swamps known as the Sudd, an Arabic word meaning a blockage. It would not be too fanciful to suggest that during the past two or three thousand years the Sudd has had a profound, though negative, influence on the history of eastern tropical Africa. It has formed an almost impenetrable barrier across the Nile valley which could otherwise have provided a relatively easy route of exchange across the Sahara between the Mediterranean peoples and those of the lake regions and highlands of eastern Africa. Such contacts, on any significant scale, were established only during the last century (Ch. 2).

Very great areas of swamp lie in the central Zaïre basin, especially in the region north of the confluence of the Zaïre and Oubangui Rivers. Lake Bangweulu, with its great expanse of associated swamp, lies at the head of the Luapula River in the southeast branch of the Zaïre basin. In the remaining large ancient basin of the Cubango-Kalahari are Lake Ngami and the Okavango swamps which cover an area of about 15 000 km² (Figs. 9.3 and 9.4).

Of more recent origin is the shallow basin of Lake Victoria between the two Rifts, formed by earth movements since the Miocene. It has already been explained how these movements cut across the previous east–west drainage system and how the old rivers on the west side of the basin have been uptilted and actually reversed so that they now flow eastwards into Lakes Victoria and Kioga down a very slight gradient (Ch. 14). The gently falling land surface between the Western Rift and the lakes is now drained by a branching system of river beds that are choked with swamp vegetation, and most of the branches of the Kafu

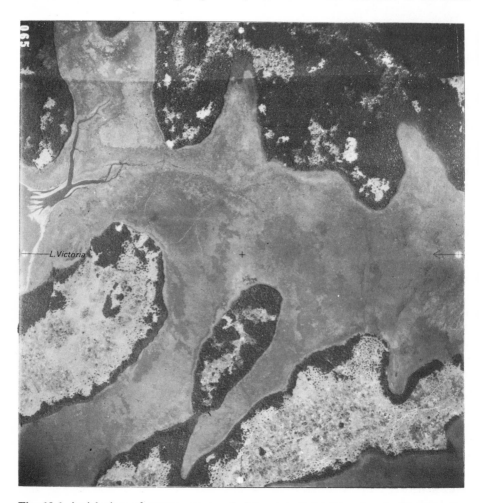

Fig. 18.1 Aerial view of swamp courses draining into Lake Victoria near Entebbe (by permission of Uganda Lands and Survey Department)

Fig. 18.2 Swamp course near Entebbe. Part of Fig. 18.1 seen from the ground

and Katonga Rivers flow westward before joining the main eastward streams (Fig. 14.1) – convincing evidence of reversal. From the air these branching grey-green swamp courses, some as much as two kilometres wide and extending into the invisible distance to the west, are a striking and characteristic feature of Uganda west of the lake shore (Figs. 18.1 and 18.2).

The uprising of the Virunga volcanoes in the late Pleistocene both captured a heavy rainfall and deflected into Lake Victoria, via the Kagera River, water which previously flowed into the Edward basin. This, in its middle course, traverses some 150 km of open and relatively level valley in southern and eastern Rwanda which is dotted with swamps and small shallow lakes. Thereafter the gradient increases and the river is confined in a narrower course. The lower section of the Kagera River, which provides the largest permanent inflow into Lake Victoria, is faster flowing and unimpeded by swamp.

Some remarkable examples of recently uptilted valleys are to be seen in the highlands of Rwanda and of the neighbouring Kigezi district of Uganda. There are several nearly level and steep-sided valleys choked with swamp (Fig. 18.3) which terminate in a sudden increase of gradient and in some cases with a waterfall.

Since swamps develop in shallow water they are subject to expansion and contraction in area with changes of water level due to fluctuations of rainfall. These changes, according to local conditions, may be rather small as in steep-sided valleys, or may involve enormous expansions of the area covered, as in flat open valleys and basins such as the Sudd region of the Upper Nile, Lake Kioga, Lake Chad, the Middle Niger and Gambia Rivers (Rzóska, 1974; Beadle and Lind, 1960; Blache, 1964; Blanc *et al.*, 1955; Johnels, 1954). As with lake levels, there is normally a regular seasonal fluctuation with periods of several years of greater expansion or reduction. Some swamps are seasonally or occasionally completely desiccated. During the period of recorded observations, and no doubt for the past several hundred years, certain areas of swamp have persisted and could thus be

Fig. 18.3 Papyrus swamp choking a highland valley in Kigezi, Uganda

called permanent. But the type of vegetation seems to depend mainly upon the short-term seasonal fluctuations, and a region which has been flooded continuously for two or three years is likely to be colonised by 'permanent' swamp plants, though in rare very dry seasons they may be totally destroyed. This brings us to the nature of the vegetation and its influence on the conditions in the water.

We are mainly concerned in this chapter with swamps with emergent vegetation of which papyrus is the best known, at least in eastern tropical Africa. In these, photosynthesising parts of the plants emerge into the air above the water which is thereby in varying degree shaded from sunlight and wind. They will be referred to as 'shaded swamps'. The word 'swamp' is also commonly applied to shallow waters supporting submerged water weeds and floating leafed plants. Since they are not shaded by emergent vegetation they will be called 'unshaded swamps'. Along the coasts and island shores of lakes such as Chad, Bangweulu, Victoria and Kioga and long rivers with shallow margins or in calm backwaters, dense papyrus and grass swamps often merge into the open water through an intermediate zone with floating leaves of water lilies (*Nymphaea*) and water chestnut (*Trapa*) and submerged weeds such as *Ceratophyllum*, *Potamogeton* and *Utricularia*. Here the water is well exposed to the sun and dissolved oxygen is abundant, and during the day it is often supersaturated by intense underwater photosynthesis. This rather narrow band of calm, well-aerated and nutrient rich water is of great ecological importance as it supports a large invertebrate fauna and is the feeding ground of many fish, especially the young and growing stages which also find protection from predators among the submerged vegetation. A large proportion of the Malagarasi swamps are of this character. They cover an area of more than 1 500 km^2 between Lakes Victoria and Tanganyika and drain into the latter. Conditions in the unshaded swamps are very different from those in the dense papyrus swamps to be described later. Carter (1955) studied the gradient in some chemical features along a transect from open water through water-lily swamp into the dense papyrus along the shore of Lake Victoria near Jinja.

Very acid swamps, where the dominant vegetation is the moss *Sphagnum*, are much rarer in tropical Africa than in cold temperate climates where large areas of 'blanket bog' are formed over high ground and are irrigated directly from an abundant rainfall. There are, however, two kinds of circumstances which have led to their formation in tropical Africa. In the craters of some extinct volcanoes, such as Mgahinga on the Uganda-Zaïre border (Fig. 15.1), and in some other very small basins where rainfall is high, the catchment area is extremely small and is leached of much of its soluble minerals. The water is thus very low in salts, particularly bicarbonate, so that the reaction is very acid (pH 3·5–4·5). Some species of *Sphagnum* are adapted to these conditions. They can also be found in more extensive drainage systems if the rocks of the area contain very low soluble salts. The water in the swamp courses draining from the west into Lake Victoria along about 120 km of coastline south of the Katonga River (Fig. 14.1) is of very low salinity. At one point the inflows are held up to form a small lake (Nabugabo, area about 25 km^2) behind a sand bar which has been formed by wind-induced currents in the main lake. Lake Nabugabo merges on the east side with an extensive *Sphagnum* swamp of several square kilometres. The water under the *Sphagnum* is very acid (pH 3·5–4·0) and its salt content is about one-fifth that of the main lake into which it drains under the sand. The important practical consequences of the low mineral content of the water is the virtual absence of molluscs and thus of the vectors of bilharzia from these inflowing swamps and from Lake Nabugabo and its associated sphagnum swamps. This type of bilharzia-free water is unfortunately rare in tropical Africa. Deficiency of calcium rather than of total minerals is likely to be the limiting factor since the fauna of Lake Nabugabo is not particularly poor in fish and invertebrates other than molluscs and crabs (Potamonidae) both of which are well represented in Lake Victoria. These have a higher demand for calcium than the other groups of animals.

We can now turn to the shaded swamps which provide an aquatic environment very different from any other. Papyrus (*Cyperus papyrus*) is the characteristic and most widely known swamp plant of central and eastern tropical Africa. One of the mysteries of African plant geography is the scarcity of papyrus west of Lake Chad, in which it is very abundant. It has been reported from small and isolated localities in West Africa – Lagos, Dahomey, Ivory Coast and Guinea, but it is apparently absent from Ghana and the Volta basin and even from the great swamp 'internal delta' of the Niger between Mopti and Timbuktu (Fig. 9.2, pers. comm. from F. N. Hepper). There are many swampy regions in West Africa that appear to be suitable for papyrus, but our knowledge of the conditions needed for its maintenance are inadequate as a basis even of a tentative explanation for its absence. Thompson (1976, p. 180) affirms that papyrus is rare in the waters of West Africa because of the large seasonal changes in volume of flow of the rivers. This cannot be accepted without comparative data on the hydrology of the Eastern and Western tropical African rivers. Even if this hydrological difference were in general established it would not necessarily be the *cause* of the interrupted distribution of papyrus. In fact, both the Gambia and the middle Niger are large perennial rivers which seasonally flood out into extensive and permanent papyrus-free swamps. The hydrological regimes, especially of the latter, seem to be very similar to that of the White Nile through the papyrus swamps of the Sudd. More detailed and widespread hydrological and biological information is needed to solve this problem. At one time papyrus grew in the Lower Nile valley where the ancient Egyptians made their papyrus parchment (hence the word 'paper') by hammering together strips of pith taken from the stems. Owing perhaps to intensive irrigation systems over the centuries it is not now found in the Nile Valley north of Khartoum. Until it was recently drained, Lake Huleh in northern Israel was the most northerly site for papyrus. It appears that for permanent establishment papyrus requires an almost continuously waterlogged soil or mud, though a free water surface is not essential. It propagates by the branching growth of the large rhizomes which are submerged in the waterlogged and therefore generally oxygen-free mud. Oxygen, however, reaches the tissues of the rhizome and associated rootlets by diffusion through a system of air spaces and tubes connecting them with the atmosphere through the stomata on the stem and terminal 'brushes'.

A stand of papyrus, with a number of other associated plants, will continue to spread by extension of the rhizome network so long as the soil is waterlogged and certain chemical and other conditions, which are not yet fully understood, are favourable. It spreads out from the shores of lakes and flooded valleys so that the rhizome network or mat is actually floating with an entrapped mass of decomposing organic matter and silt forming a kind of floating layer of soil. There is no limit to the depth of water above which it can flourish, and during the wet season the papyrus mat, which covers the entire surface of some of the western Uganda swamp valleys (Fig. 18.2 and 18.3), floats above several metres of water. Particles of partially decomposed plant material are continually falling from the mat through the anoxic water to form a layer of 'peat' at the bottom in which decomposition is virtually arrested. In the region of Lake Victoria this peat is no more than 2 to 3 m deep, but in the valley swamps of the Kigezi highlands of western Uganda it sometimes has a depth of 20 to 30 m This greater accumulation may be due to the lower temperature retarding decomposition, or perhaps to the shape of the uptilted, deep and steep-sided highland valleys which may form more effective sinks for the retention of peat deposits.

The lakeside papyrus swamps are limited in their outward spread by winds and currents which frequently tear off pieces of the mat. Flotillas of papyrus islands are a familiar feature of the inshore waters of Lake Chad, Lake Victoria and Lake Kioga after a heavy storm. In Lakes Kioga and Bangweulu and in the Nile Sudd projecting mats of papyrus and other vegetation and floating islands are sometimes large and secure enough to support a fisherman and his hut and even his cattle (Fig. 18.4). In regions where it is abundant, papyrus is used for a number of domestic purposes such as roof thatching, matting, fishing-floats, rafts, etc. Its general distribution is apparently partly limited by temperature, and in equatorial regions it is not found above about 2 000 m.

Besides papyrus, permanent swamps are populated by several other types of emergent plants including other species of *Cyperus*. In the East African swamps, for example, the grasses *Vossia cuspidata*, *Miscanthidium violaceum* and *Loudetia phragmitoides*, the bulrush *Typha*, the reed *Phragmites mauritianus* and the sedge *Cladium jamaicense* are common and often form dense and extensive stands in certain places. The distribution of each is presumably determined by special growth requirements not all of which are yet understood (Lind, 1956). The grass *Miscanthidium* is often mixed with papyrus, but in some situations, e.g. the valley swamps flowing into Lake Victoria, it is the dominant plant, papyrus being limited to a strip along the edge (Fig. 18.1). The dominant vegetation of temporary seasonal swamps is grass; the papyrus is confined to the edges of remnant streams and pools or to places where, owing to lack of drainage, the soil remains wet during the dry season.

There are many other plants associated with both papyrus and grass swamps, e.g. the fern *Cyclosorus* (*Dryopteris*) *striata*, climbing grasses and Convolvulaceae, *Polygonum*, spp., etc. About sixty species have been recorded in Uganda swamps (Lind, 1956).

There is also a characteristic swamp tree, *Mitragyna stipulosa*, which grows in waterlogged soils and around the edges of swamps. Like some other swamp plants, including the mangrove of tropical coastal salt swamps, it has pneumatophores or special respiratory organs that protrude above the mud and connect with the atmosphere the air spaces in the roots which are submerged in anoxic mud. In *Mitragyna* these are on arched portions of normal roots known as 'knee-roots' (McCarthy, 1962a, b).

The following account of the conditions in the water and mud of permanent papyrus swamps is based on work done in Uganda (Beadle, 1932a; Carter, 1955; Beadle and Lind, 1960; Visser, 1964a). Thompson (1976) gave a more botanically oriented account of conditions in the Uganda swamps, with discussion on the nutrition and growth of papyrus, which is extremely rapid, and on the exploitation of swamplands. The peculiar conditions in swamp waters are primarily due to the rapid decomposition of large quantities of plant material. Photosynthesis and growth of the upper parts of the plants that are exposed above the water are very rapid. These ultimately die and fall onto the surface of the mat, sink and become waterlogged, so that the mass of vegetable matter undergoing anaerobic decomposition is being continuously and copiously augmented.

In dense stands of papyrus the terminal brushes form a 'closed canopy' so that, as in a tropical rain forest, the interior is shaded and in the daytime cooler than the air above. Though measurements of light intensity have not yet been made, it is probable that the scarcity of aquatic algae and weeds in the open pools in the mat is due partly to insufficient light. For this reason there is little or no photosynthetic production of oxygen in the water, which is also well protected from wind stirring. The only source of oxygen is therefore by diffusion from rather still air. On the other hand, the consumption of oxygen by decomposing organic matter is so rapid that except during flood periods, exposed water under dense papyrus may be devoid of oxygen to within a centimetre of the surface. The straight unbranched leaves of grasses and reeds provide less shade from the vertical sum, but under very dense stands of the tall swamp grass *Miscanthidium* deoxygenation can be almost as extreme as under papyrus.

Carter (1955) in his study of the transition between lake and papyrus swamp along the shores of Lake Victoria near Jinja showed that, though the dissolved oxygen in the outer *Nymphaea* and weed zone would reach saturation during the day, continuously anoxic conditions prevailed even near the surface 15–20 m inside the papyrus swamp. The distance from the edge of the papyrus at which oxygen was still detectable was increased with rising lake level during the rains. But in the dense interior of papyrus swamps, even during periods of heavy rain when there is often a considerable flow-through, it may be impossible to detect oxygen in surface samples except in regions where the current is rapid.

Apart from these extensive areas of anoxic water, the general level of dissolved oxygen in all shaded swamps is undoubtedly low. Measurements taken at the surface in many densely shaded Uganda swamps at different seasons have rarely shown more than 5% saturation and often nil. There is every reason to suppose that this is a general characteristic of all dense papyrus and to a lesser degree of grass swamps of tropical Africa, as it certainly is of the Paraguayan Chaco swamps in South America (Carter and Beadle, 1930).

Investigations of the large swampy Lake Bangweulu at the head of the Luapula valley at the southeast corner of the Zaïre basin (Figs. 18.4 and 18.5) have demonstrated similar conditions.* The main inflow is the River Chambesi which drains through the massive swamp (about 4 000 km² in area) southeast of the lake. Around the edge is a band of flood plain inundated in the rainy season from November to March, but progressively dried out from about July. Most of the swamp is, however, permanently flooded and populated in varying density by emergent swamp plants. It is intersected by deeper channels of open water which connect numerous small, open lagoons. The flow-rate in the channels is, of course, high in the rainy season and very slight towards the end of the dry season. A series of oxygen measurements in the channels and lagoons made by

---

* I am indebted to D. Harding and to A. P. Bowmaker for information on the general ecology of Lake Bangweulu, of which this is a brief summary of an aspect relevant to the present discussion. The work was done under the auspices of the then Joint Fisheries Research Organisation based at Samfya on the west shore of the lake. For a general geographical account see Debenham (1947).

Fig. 18.4 Lake Bangweulu. Floating fishing camp. Vegetation in foreground: *Vossia cuspidata; Phragmites* sp. in background (photograph by A. P. Bowmaker)

Fig. 18.5 Lake Bangweulu. Typical peripheral swamp which dries up in dry seasons, with *Nymphaea caemlea* and *Eleocharis dulcis* (photograph by A. P. Bowmaker).

Harding (pers. comm.) demonstrated the extent to which oxygen is extracted from the water as it traverses the swamp. The oxygen level in a given stretch of open water in the swamp depends on the rate of flow, the extent to which the water has seeped through the neighbouring dense shaded swamp, wind stress, and the intensity of photosynthesis. The oxygen saturation level in some channels during the dry season was in fact as low as 5%. There is no doubt that at such times, and possibly for much of the year, the water in the dense, shaded regions of these swamps, as in Uganda, is anoxic almost to the surface.

The fisheries of Lake Bangweulu are of considerable economic importance (Ricardo, 1939a, 1943) and the ecological effect of these oxygen conditions, as in other swamps, must be considerable. They control the seasonal movements of fish, other than the air-breathers, and the distribution of the invertebrates upon which some of the fish feed. Lake Bangweulu is certainly a very suitable site for an intensive study of tropical swamp ecology.

The water under these conditions is slightly acid and the pH of papyrus swamp water is usually between 6·0 and 6·5. Though some organic acids are present the acidity seems to be due mainly to carbon dioxide from organic decomposition. Some measurements made by Milburn and Beadle (1960) of carbon dioxide (dissolved plus bound as HCO_3^-) in a Uganda papyrus swamp gave the very high figure of 148 mg/1 at the end of a long rainless period. After three days of flooding it had dropped to 35 mg/1.

The waterlogged mat and bottom peat are not only free of oxygen, but are highly reducing. Samples taken from just below the surface of the mat will usually reduce and decolorise methylene blue, and redox potentials (Eh) of −100 mV have been recorded within 30 cm of the surface. Even the water in small open pools among the mat may on occasions have a potential as low as +100 mV at 10 cm below the surface and contain no measurable oxygen within 2 cm of the surface (Beadle, 1957).*

Since the organic matter is mainly carbohydrate, with little protein, the gaseous end-product is mainly methane (CH_4) which bubbles to the surface. In pushing through a papyrus swamp forest one's feet frequently sink into the mat and much gas rises to the surface. There is not normally enough hydrogen sulphide to be detectable by smell, though it is sometimes obvious for a short time after a sudden rise in water level following heavy rains. Under these circumstances a much larger amount than normal of fresh vegetable matter is suddenly submerged, the oxygen is rapidly exhausted and the sulphur compounds are reduced to hydrogen sulphide which, under the slightly acid conditions, escapes into the air. Inorganic sulphate in the water must also be reduced to H_2S, but the amount available is not usually enough to produce a detectable quantity. From Lake Kioga have come reports of frightening sheets of flame from the ignited gas (methane) stirred up by the shallow-draft steamboat which at one time plied between Namsagali and Masindi Port along channels through the papyrus swamps.

---

* The actual potential measured (with a bright platinum electrode) depends upon the nature of the substances taking part. It cannot be taken as a measure of oxygen concentration except that in natural waters a potential below +150 mV indicates the certain absence of dissolved oxygen. Oxygen-free swamp waters were in fact found with potentials up to +200 mV, rising to +300 to 500 mV when aerated (Beadle, 1957, 1958).

Visser (1963) found that the composition of the gases evolved from decomposing anaerobic and waterlogged papyrus stems is approximately – methane 60%, and carbon dioxide 30%, the remaining 10% being hydrogen, carbon monoxide and ethylene. Swamps are therefore in this respect different from the other extensive anoxic tropical environment, the stagnant lower water of some of the deep lakes in which H_2S is produced and accumulates in quantities which are probably toxic to many organisms. Apart from carbon dioxide, there is as yet no evidence that under natural conditions any of these gases, nor any other decomposition products such as volatile organic acids and humic substances, normally accumulate in sufficient concentration seriously to affect animal life in swamps.

Between 1950 and 1958 the Hydrobiological Research Unit of the University of Khartoum investigated the Sudd swamps of the Upper Nile (Rzóska, 1974, 1976a, pp. 202–13). There are about 40 000 km^2 of permanent swamp in the Sudd basin and about twice that area of land is inundated between September and November when the inflowing rivers are in flood. The swamp vegetation is very similar to that already described in the Lake Victoria basin with which the Sudd is connected by the White Nile. Three major habitats were recognised:
1. Open rivers and channels.
2. 'Standing' waters.
3. Flooded land.
The observations were made in the first two at points accessible by boat. Though the reducing effects of decomposition were apparent, no swamp water was found to be in the extreme anoxic condition typical of the dense papyrus swamps of the Lake Victoria basin. It is probable that the flowing water in the rivers and channels spreads out into the surrounding swamps sufficiently to prevent complete deoxygenation of the water for a considerable distance on all sides. There must surely be great areas of swamp not accessible by boat in which extreme anoxic conditions are to be expected, especially during the low-water season. This can be confirmed only by direct observation. When, however, the composition of the water entering at Bor is compared with that of the water 500 km to the north where it leaves the main swamps from Lake No (Fig. 11.3), the effects of reduction due to decomposition in the swamps are very obvious – lower oxygen and pH, higher carbon dioxide and iron, and lower sulphate (Talling, 1957a). The last is presumably due to reduction to sulphide which ultimately escapes as H_2S, though, as in the Uganda swamps, this is rarely detectable by smell. This overall effect on the through-flowing water occurs in spite of vigorous photosynthesis by the abundant phytoplankton in the open rivers and 'standing' waters, which also support rich populations of zooplankton, whose ecology and distribution are discussed by Rzóska (1974). This is surely evidence of highly reducing conditions in a large proportion of the Sudd swamp waters.

The chemical exchanges between tropical swamps and the water that flows through them have recently received some attention and could be of some practical importance. In 1931 it was noted that a papyrus swamp blocking a small river in Western Uganda removed phosphate from the water passing through it (Beadle, 1932b). Talling's observations of the chemical effects on the outflowing Nile water of the anaerobic conditions in the Sudd swamps have been mentioned in the previous paragraph. Gaudet (1975, 1976, 1977) studied the exchanges of inorganic ions, phosphate and nitrate between the living papyrus plants, the floating mat, the peat and the free water in a small swamp blocking a stream flowing into Lake Victoria near Kampala. Estimation of the common elements in the various components of the swamp certainly disclosed exchanges between them. But the system is very complicated. Uptake and release of ions is obviously involved in the growth of papyrus and other aquatic plants and the living plants contain large amounts of the common elements, especially potassium and nitrogen. But much of the ions are also attached in some manner to the material of the mat and peat. They must also be actively taken up by the abundant micro-organisms as well as adsorbed by the organic molecules of the dead material. Gaudet observed that the flowing water washes out into the lake some of the organic debris that falls to the bottom under the floating mat (called 'peat' on p. 312). This probably provides much of the nutriment for the abundant plant and animal life in the well oxygenated water just outside the swamp (p. 311). In this manner the swamps may contribute something to the productivity of the lake by releasing into it solid organic matter synthesised from materials brought down by the river. It has been suggested that such an inflowing valley swamp might be made to process the waste organic matter and nutrients from the sewage system of a lakeside town and at the same time to enhance the productivity of the lake without pollution. We need however to know much more about the processes involved and the conditions required for optimal functioning.

Those peculiarities of the dense papyrus swamp environment that must obviously affect aquatic organisms are thus the result of the decomposition of vegetable matter – scarcity of oxygen, reducing conditions and a high level of carbon dioxide. These conditions, as already stated, are further encouraged by the dense forest of vegetation above the water which prevents violent wind stirring and reduces the incident light so that there is practically no photosynthetic production of oxygen in the water. The basis of organic production is therefore almost entirely photosynthesis in the aerial parts of the swamp vegetation. Dense papyrus and grass swamps, which provide the kind of conditions we have been discussing, probably cover a very great area of eastern and central tropical Africa. The adaptations by which many animals are able to spend some or all of their time in them are of great interest, and it is possible that we are here concerned with the kind of conditions in which the air breathing lung of vertebrates first evolved. The vast extent of the swamplands poses some important problems of land use, to solve which we should know as much as possible of the peculiar conditions in swamps and of the mechanisms by which animals and plants are adapted to them.

The quantity and variety of aquatic animal life in the interior of dense permanent swamps is undoubtedly much less than in the *Nymphaea* and weed zone of lake and river margins, but the fauna is richer than a cursory examination would suggest, even during the dry season when the water is most stagnant and the typical swamp conditions are most marked. During rainy seasons when the water level may rise, many animals not found there at other times become temporary inhabitants of these regions, mainly peripheral, where the effects of decomposition have been reduced by flooding. However, we are now concerned with the fauna of the stagnant interior where extreme swamp conditions prevail. The following groups of animals have been found to be common in the mat or in anoxic pools of water in the dense papyrus swamps of Uganda:

Ciliate Protozoa		e.g. *Blepharisma undulans*
Nematode worms		*Dorylaimus* spp.
Oligochaete worms		*Alma emini*
		Stuhlmannia spp.
Mollusca	Bivalvia	*Pisidium fistulosum*
	Prosobranchia	*Pila ovata*
	Pulmonata	*Biomphalaria sudanica*
Crustacea	Cladocera	*Pseudosida bidentata*
	Copepoda	*Mesocyclops leuckarti*
	Ostracoda	undet.
Insecta	Diptera	mosquito and chironomid larvae
	Coleoptera	undet.
	Ephemeroptera	nymphs undet.
Fish		*Protopterus aethiopicus*
		Polypterus bichir
		Clarias lazera
		Ctenopoma muriei

The species quoted for Protozoa, Nematoda, Oligochaeta and Crustacea (for Cladocera see Thomas, 1961) are the commonest that were recorded, but several others were found, some of them not identified.

Lack of dissolved oxygen can obviously be overcome by breathing atmospheric air from above the water surface. Many of the aquatic insects and their larvae (Hemiptera, Coleoptera, mosquito larvae) are air-breathers and are thus adapted in this respect for life in swamps. Because of the importance of malaria and other mosquito-borne diseases more is known about these insects than about the others found in swamps. About eighty species breed in the East African swamps of which about forty species are found in the permanent valley swamps in which deoxygenations is most intense (Goma, 1960a, 1961a). Certain species, fortunately including the most virulent of the malaria vectors *Anopheles gambiae*, do not breed in the dense interior of shaded swamps. This may be due in part to an aversion on the part of the female to laying eggs on densely shaded water. But experiments in planting larvae in papyrus swamps have shown that conditions in the interior are somehow inimical to the survival of the larvae of *A. gambiae*. Whether this is due to the absence of suitable food, to toxic chemical factors or to the absence of oxygen is not yet known (Goma, 1960b). It must be remembered that air-breathers must lose some oxygen by diffusion into the anoxic and very reducing water. For small animals with a relatively large surface area this might be a serious problem unless the permeability to oxygen of their outer membranes is very low. Whatever the nature of the limiting factor, an interesting and important practical consequence is that, unless special precautions are taken, swamp reclamation by drainage can provide more favourable habitats for the breeding of malaria vectors than were available in the original swamp (Goma, 1960b).

The accessory aerial respiratory organs of several African fish are very well known. The lungs of *Protopterus* and *Polypterus* and the special organs in the gill cavities of most species of the catfish genus *Clarias* and of the anabantid fishes (e.g. *Ctenopoma*) are illustrated in many textbooks. To these must be added the airbladder-lung of *Gymnarchus niloticus* (Gymnarchidae) and *Phractolaemus ansorgei* (Phractolaemidae), the suprabranchial cavity of *Ophiocephalus obscurus* (Ophiocephalidae). The coiled suprabranchial organ of *Heterotis niloticus* (Osteoglossidae), widespread over the soudanian region and West Africa, is apparently concerned with the detection and selection of food particles. It is possible, however, that the airbladder, which is open to the oesophagus, has an aerial respiratory function (d'Aubenton, 1955). The airbladder, which is a hydrostatic organ in most teleost fish, has been adapted by infolding and vascularisation of part of the inner surface in several tropical fish in other continents. But *Gymnarchus* and *Phractolaemus*, together with *Amia calva* (America), have airbladder-lungs that are entirely filled with a spongy mass of alveoli as in a mammalian lung (Thys van den Audenaerde, 1959).

All the genera of African air-breathing fish are common in the swampy regions of the Zaïre basin except *Gymnarchus* which is a soudanian fish. In the Zaïre, as well as in West Africa and the Chad basin, there lives the remarkable freshwater flying fish or 'butterfly fish', *Pantodon buchholzi* (Pantodontidae) (Fig. A4.3) whose air-bladder is adapted as an organ of aerial respiration. It skims over the surface of the water like the flying fish of tropical seas (Greenwood and Thomson, 1960; Poll and Nysten, 1962). More specialised accounts of air-breathing organs in fish are to be found in Marlier (1938), Carter (1957) and Bertin (1958). Some evolutionary implications are discussed by Hughes (1966) and investigations of the respiratory physiology of *Protopterus aethiopicus* have been made by Lenfant and Johansen (1968) and McMahon (1970) (review in Johansen, 1970). Whatever the conditions under which these organs were originally evolved, they now enable the fish to live in the interior of dense swamps where the water is mainly anoxic. A further advantage is gained by a haemoglobin adapted to function at low oxygen and high carbon dioxide levels.[*] They are not, however, confined to swamps and *Protopterus aethiopicus* and *Clarias lazera*, for example, are common in the well-aerated open water of many lakes. The gill surface of *P. aethiopicus* is so far reduced that even in air-saturated water it is asphyxiated if prevented from reaching the surface for air. *C. lazera*, on the other hand, can get enough oxygen through its gills alone in well-aerated water, though the rate of ventilation is increased (Abid-Magid, 1971). The secondary reduction of the suprabranchial aerial respiratory organs and the increase in the surface area of the gills of some species of *Dinotopterus* (Clariidae) in the deep water of Lake Malawi has already been noted (p. 300). They have been interpreted as a reversion to purely aquatic respiration in fish previously adapted to life in oxygen-deficient water (Greenwood, 1961). A similar secondary reduction of the suprabranchial organs has occurred in *Sandelia capensis*, an anabantid fish in the South African Cape region (Barnard, 1943). It is generally agreed from the geological evidence that the environment during the Devonian, in which some of the bony fishes first

---

[*] The respiratory properties of the bloods of some species of fish in Uganda, in relation to the conditions in their usual habitats, was studied by Fish (1955b).

developed lungs, was probably similar to a tropical swamp and characteristically low in oxygen. But it would seem unlikely that these swamps were then in open communication with large areas of well-aerated water such as are now provided by the African lakes and that aerial respiratory organs would have had a greater survival value than they now have in the lake regions of tropical Africa.

A well-known example of temporary anaerobic existence, though the mechanism is not yet known, is the larva of the lake fly *Chaoborus*, which has often been mentioned in previous chapters. It lives during the daytime in the bottom mud of lakes and migrates towards the surface at night to feed on the zooplankton in the upper layers. Some species are commonly found in the permanently stratified lakes described in Ch. 6, whose lower water is devoid of oxygen. They have no respiratory pigment and a study of their respiratory metabolism would be of great interest. They are an important food for certain fish, but in stratified lakes with an anoxic lower layer they are vulnerable to most predator fish only when they visit the oxygenated upper layer at night or when the pupae come to the surface to emerge. A very interesting adaptation has recently been discovered by Green *et al.* (1973) in the volcanic crater Lake Barombi Mbo in west Cameroon. This is permanently stratified and devoid of oxygen below about twenty metres (see also p. 139). *Konia dikume*, one of the eleven endemic species of cichlid fishes in this lake, feeds mainly on the larvae of *Chaoborus* which, in turn, prey on the rotifers in the zooplankton that are confined to the oxygenated upper layer (Green, 1972). The mean concentration of haemoglobin in the blood of *K. dikume* (over 16 g/100 ml) is very much greater than that of the other ten species of cichlids which range between 5·5 and 8·7 g/100 ml. Though the oxygen capacity of the blood was not measured, this must surely provide an extra store of oxygen and should enable the fish to spend some time below the anoxic boundary and thus to prolong its otherwise restricted period of feeding on *Chaoborus*. Unlike the other species of fish in the lake, *K. dikume* was caught only in deep water and the organic debris and sand-grains as well as larvae in the stomach indicate that they actually feed on or near to the bottom (Trewavas *et al.*, 1972, p. 82). Though there is indirect evidence that some of the pelagic fish in Lakes Tanganyika and Malawi visit the region of lowered oxygen near the anoxic boundary (p. 290), this is the first recorded example of a freshwater fish physiologically adapted to spend some time in water quite devoid of oxygen by means other than an aerial respiratory organ filled with air at the surface. In view of this remarkable adaptation evolved in a comparatively short time, why have cichlid fishes failed to accomplish this particular feat in other stratified lakes such as Tanganyika and Malawi? Perhaps there are some, as yet unnoticed, that have done so. There are in fact several species of marine fish as well as crustaceans and squids that spend most of their time at a depth of several hundred meters in the north-eastern tropical Pacific Ocean where the level of oxygen is permanently 0·1–0·3 mg/l. They visit the well oxygenated surface waters for one or two hours every night. But the bloods of most of the fish examined have a surprisingly low affinity for oxygen. It is therefore difficult to avoid the conclusion that they derive some of their energy from an anaerobic metabolism. So far, however, direct proof of this is lacking (Douglas *et al.*, 1976).

Several air-breathing aquatic snails live in the interior of swamps. In East Africa the commonest are the large amphibious snail *Pila ovata* (Ampullariidae) whose mantle cavity is divided into water and air-breathing compartments, and the pulmonate *Biomphalaria sudanica* (Planorbidae), one of the vectors of human intestinal bilharzia (*Schistosoma mansoni*). Both these species are common also in other well-aerated waters. Apart from its air-breathing, *P. ovata* does not seem to be especially adapted to swamp life and its blood contains no haemoglobin. The respiratory physiology of *B. sudanica* has been studied by J. D. Jones (1964). Compared with the European *Planorbis* (*Biomphalaria*) *corneus*, it seems to be only slightly better adapted to low oxygen conditions in that its haemoglobin has a somewhat higher affinity for oxygen, which enables it to make use of the pulmonary oxygen at a higher level of ambient carbon dioxide. It would appear that, like the insects but unlike the fish, air-breathing in snails, which fits them to live in swamps, has been inherited from a previous life on land.

An interesting adaptation to air-breathing is shown by the swamp worm *Alma emini* which is very abundant in East Africa in the floating mat of papyrus swamps and in the organic mud at the edge of ponds, etc. Its feeding habits are similar to those of the earthworms and vast quantities of mud are cast onto the surface. It undoubtedly plays an important part in the breakdown and decomposition of plant material. It is also common and ecologically important in tem-

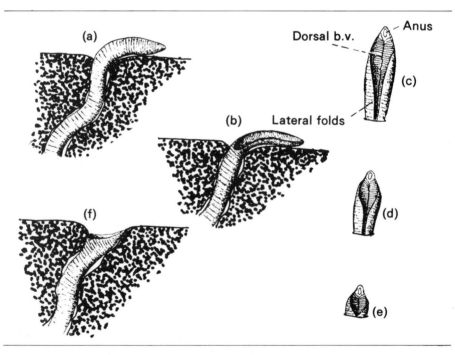

Fig. 18.6 *Alma emini*. Exposure to the air of the posterior dorsal surface and formation of 'lung'. (*a*) Extrusion of hind-end. (*b*) Hollowing of dorsal surface. (*c*)–(*e*) Dorsal view of extruded hind-end. Stages in the formation and closure of lateral folds and retreat into mud. The dorsal blood vessel and numerous lateral connexions are clearly visible in this region. (*f*) Final position after retreat into mud with tubular lung open to the air (Beadle, 1957)

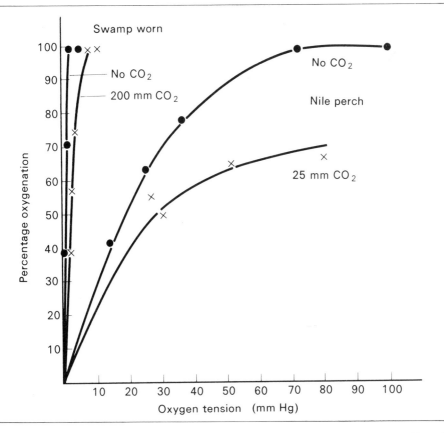

Fig. 18.7 Oxygen dissociation curves of the haemoglobin of *Alma emini* and the effect of CO_2, compared with the Nile perch (*Lates albertianus*) which inhabits the open well-oxygenated waters of Lake Albert (Beadle, 1957; curves for *Lates* taken from Fish, 1955b)

porary seasonal swamps which, in the dry season, provide grazing for wild mammals and cattle. It then retreats downwards and remains dormant in the still damp lower soil (Wasawo and Visser, 1959). In the wet swamp its habitat is both anoxic and very reducing, but it obtains atmospheric oxygen by exposing the hind end above the mud. The dorsal surface of this, which is highly vascular, is flattened out and then folded up to form a temporary 'lung' in which bubbles of air are drawn down into the mud (Fig. 18.6). Its haemoglobin is well adapted in that its affinity for oxygen is extremely high and it is unaffected by a very high level of carbon dioxide (Fig. 18.7). When submerged it can remove most of the oxygen from a bubble of air and can supply it to the tissues working at a very low oxygen level (Beadle, 1957). Further physiological work on this worm by Mangum *et al.* (1975) has confirmed the main conclusions set out above. The various aspects of the respiratory process were however subjected to more rigorous analysis. The high affinity for oxygen and the low Bohr effect of CO_2 were quantified, and the rate of oxygen uptake by the 'temporary lung' was found to be 50–60% of that through the entire body surface. It is remarkable that the very closely related *Drilocrius* sp. which inhabits the swamps of tropical South America be-haves in an apparently identical manner (Carter and Beadle, 1931).

On the other hand, most of the animals living in swamps do not breathe air and might therefore be suspected of having a partly anaerobic metabolism. It should be remembered that, though the great majority of species of animals are dependent upon free oxygen as the final hydrogen acceptor, the immediate source of energy is an anaerobic breakdown of carbohydrate yielding acids such as lactic, which are toxic in high concentrations but are subsequently oxidised. Oxygen is thus ultimately needed, though most species can live anaerobically for some time at least and, like the diving mammals, can accumulate some lactic acid which is later broken down when oxygen is available. In theory, permanent anaerobic life should be possible for an animal provided that the toxic acid end-products are removed from the body, or in some way rendered harmless other than by aerobic oxidation. There must also be plenty of available carbohydrate, because anaerobic breakdown is partial and yields less energy than complete oxidation.

There are, of course, several micro-organisms found in the swamp water and mud which are responsible for various stages of anaerobic decomposition – dinit-rifiers, ammonifiers, sulphate reducers, etc. (Visser, 1963), and these are per-manently independent of molecular oxygen. In spite, however, of a considerable amount of observation and experiment, there is no certain evidence that any mul-ticellular animal is capable of permanent anaerobiosis (von Brand, 1946; Beadle, 1961). Several species of ciliate protozoa are found in anoxic swamp water and mud, but experiments on one of these, (*Blepharisma undulans**) showed that survival in complete absence of oxygen is limited to about three days, whereas another species (*Bursaria* sp.*) found in open pools at the edge of swamps, dies as soon as the oxygen is exhausted (Beadle and Nilsson, 1959). Whatever the nature of the adaptation in *Blepharisma*, it is only temporarily effective and the animals must presumably rise from time to time to the oxygenated surface film. It may be that anaerobic respiration can continue until the animal has exhausted its reserve of carbohydrate which can be renewed only with energy provided by aerobic oxidation. More remarkable is the small nematode worm *Dorylaimus* sp. also found in the anoxic mud of papyrus swamps (Banage, 1966). This animal reacts to the experimental removal of oxygen by expending less energy on movement and apparently becomes active only at intervals. Some specimens survived more than two months of anoxia. In this case perhaps, by expending less energy, the carbohydrate reserves are much more slowly consumed than by the more energe-tic ciliate *Blepharisma*.

The swamp worm *Alma emini*, though breathing air in the manner already described, is normally immersed in mud that is highly reducing. Under these circumstances some at least of its tissues are probably functioning at a low oxygen level, and it can in fact survive for several days, but not permanently, when deprived of all oxygen. Biochemical investigations by Coles (1970b) suggest that the characteristic aerobic metabolism involving the citric acid cycle is used. But even in the presence of oxygen some of the carbohydrate is broken down anaero-

---

* These have recently been renamed *Blepharisma japonicum* and *Neobursaridium gigas* (pers. comm. J. R. Nilsson).

bically, yielding acetic, propionic and methyl butyric acids as end products (aerobic glycolysis). When oxygen is excluded these acids increase and traces of butyric and of a branched C_6 fatty acid (but no lactic) are produced. It is interesting to note that this two-way type of metabolism has been found in some nematode worms, such as *Ascaris*, parasitic in the intestines of vertebrates where the oxygen supply is very low and spasmodic (Smith, 1969). Though no animals have yet been shown to survive permanently in the complete absence of oxygen, the case of *Alma* demonstrates a metabolic adaptation to permanently low oxygen levels and to prolonged, though limited, periods of complete anoxia. Further research would no doubt show that other swamp animals and those inhabiting the anoxic lower water of some lakes have a metabolism adapted in this and perhaps in other ways to these conditions.

Adaptation to an environment, however, necessarily involves all phases of the life history and the eggs and early larval stages must somehow survive until the special features, such as accessory respiratory organs, have developed. Little is known about the life histories of many of the inhabitants of shaded swamps, though it seems that most reproduction occurs during the rains when the level of oxygen is raised at least in the surface water and particularly around the edge of the swamps.

The breeding habits of the three species of lungfish (*Protopterus*) have been studied – *P. dolloi* in the Zaïre basin (Brien *et al.*, 1959; Brien and Bouillon, 1959; Poll, 1959), *P. annectens* in the Gambia (Johnels and Svennson, 1954) and *P. aethiopicus* in the fringing swamps of Lake Victoria (Greenwood, 1958). These authors have also reviewed previous work on the subject. The 'nests' of *P. annectens* and *P. aethiopicus* are variable in form, but they are all holes excavated by the fish into the mud or wet soil at the edge of the swamp or into the matted roots and vegetable debris just under the water surface when the water level has risen during the rainy season. The cavity so made communicates in various ways with the surrounding water. A few thousand eggs are laid in the cavity and are guarded by the adults until some time after hatching. The larvae at first have filamentous external gills which must function as aquatic respiratory organs until the lungs are developed, after which they begin to leave the nest. According to Johnels and Svennson (1954) the male of *P. annectens* ventilates the nest by movements at the entrance, but no measurements of oxygen were made to test the effectiveness of this action. Otherwise a rather low level of oxygen would be expected within the nest. But only one set of measurements has been made – those by Greenwood (1958) in a nest of *P. aethiopicus* which has not been seen to ventilate the water. The concentration of oxygen near the surface of the water in the nest varied between 0·3 and 1·7 mg/l, or about 4 to 20% saturation,* over a period of two weeks. The temperature ranged from 20° to 25°C which would involve a relatively high rate of metabolism. It would be most interesting to discover whether the respiratory metabolism during the early stages operates, as in the swamp worm, at least partly, through anaerobic pathways which would become redundant after the lungs begin to function.

---

* 100% saturation is taken as in equilibrium with air. No precautions were taken to avoid the reducing effects of organic matter on the method used (Winkler). The actual oxygen levels could have been as much as 0·5 mg/l higher than these (Beadle, 1958).

The breeding habits of *P. dolloi* are different from those of the other two species, and would appear to be adapted to the peculiar conditions in the riverine swamps of the Middle Zaïre basin. When the river level falls in the drier season the surrounding flooded land is not wholly drained and extensive shallow swamps remain. The lungfish excavate a tunnel into the mass of vegetation and mud in which they make a 'nest' communicating with the air through a kind of chimney. The water in the tunnel in which the eggs are laid would probably, without this ventilator, be deprived of oxygen. The young fish are ready to take to the open water when the level rises once more.

One of the functions of the 'nests' of *P. annectens* and *P. aethiopicus* is presumably to prevent the eggs from falling into the lower anoxic water and mud where they would no doubt perish from anoxia. The small anabantid fish *Ctenopoma damasi* makes a floating nest in which the eggs are supported by a mass of foam made by the fish (Berns and Peters, 1969). This species and *C. muriei*, both of which are found in swamps in East Africa, lay eggs containing oil globules large enough to cause them to float. When the larvae start to move the oil is distributed in two sacs on either side of the body which serve as hydrostatic organs until the airbladder is later filled with gas and the oil is absorbed (see also Peters, 1947). This is however not a feature confined to swamp fish. Several marine and freshwater fish in all parts of the world have buoyant eggs which float among the pelagic plankton and are thus prevented from falling into bottom sediments. In many cases the buoyancy is due to oil globules, as in the egg of the Nile perch, *Lates niloticus* (Hopson, 1969c). A floating foam nest is made also by the African river pike *Hepsetus odoe*, a fierce strong-toothed predator that lives in the swamps of some West African rivers and in the Zaïre and Zambezi basins (Fig. A.4.3). The eggs are collected, cooked and eaten by the fishermen on the Gambia River – an unusual source of human food in Africa (Svensson, 1933).

The dense emergent swamp vegetation provides a habitat for a large variety of terrestrial animals – birds, tree-frogs, insects and even a few mammals. The papyrus forests are inhabited by a characteristic assemblage of species, but very few are confined to swamps. Even elephants make their way in and feed on some of the swamp grasses, and hippopotamus make tracks through the vegetation in the evening and early morning on their way to and from their terrestrial feeding grounds. These aquatic tracks are well worn and often used by fishermen as a means of access by canoe to the open water.

The sitatunga (*Limnotragus spekei*), a relative of the bushbuck, is especially adapted to swamp life. It is a good swimmer and spends most of its day in dense swamp from which it issues at night to feed on the neighbouring grasslands. Its cloven hooves are much elongated so that its weight is widely spread over the insecure substratum.

There are many birds that rest or feed, and sometimes nest, in shaded swamps, but for very few is this an especially preferred habitat. A boat moving quietly along the outer edge of the fringing swamps of Lake Victoria will surprise a large number of birds such as the purple gallinule (*Porphyrio alba*), squacco heron (*Ardeola ralloides*), goliath heron (*Ardea goliath*) and the long-toed lily trotter or jacana (*Arctophilornis africanus*), and occasionally a finfoot (*Podica senegalensis*) or even a small flock of pigmy geese (*Nettapus auratus*). All these will be feeding outside the papyrus swamp among the floating water lilies (*Nymphaea coerulea*), water chestnuts (*Trapa natans*) and submerged water weeds such as *Ceratophyllum demersum*.

The most characteristic of swamp birds is the shoe-billed or whale-headed stork (*Balaeniceps rex*). This grotesque animal, more than a metre in height, with a massive bill shaped like an ungainly boot, can sometimes be seen standing motionless on a papyrus island like a sentinel (though its bill is resting on its breast), waiting for a suitable fish.

During the day swamps are rather quiet apart from the occasional calls of herons and gallinules and the twittering of pied kingfishers (*Ceryle rudis*). But as night falls there begins a deafening chorus of cicadas and crickets punctuated by the tinkling of tree-frogs in the emergent vegetation and the croaking of toads in the pools along the edge.

Homo sapiens, the most versatile of animals, has become adapted to life in many apparently unfavourable environments. In some of the extensive swamp regions of tropical Africa the life led by the fishermen could truly be described as amphibious. Certain peoples, such as some subtribes of the Dinka in the Sudd region of the Upper Nile in the Sudan, are dependent upon swamp fish for survival at least during the long dry season. The fishermen live in the swamps for several months, often on floating papyrus islands, and spend much of their time in the water, fishing with spears and traps. They are better integrated into a natural ecosystem than most human beings at the present time.

The Jonglei Investigation Team (1954, Vol. I) described an interesting seasonal regime practised by some sections of the Dinka, Nuer and Shilluk tribes on the Upper Nile. In the wet season (May–September) they live in semi-permanent villages on the higher ground above the flood plains, keep their cattle in byres and cultivate millet, maize, vegetables and tobacco. At the beginning of the dry season (June–December) they move down with their cattle onto the intermediate open grasslands and, when these dry up, they proceed to the riverain flood plains and the swamps around the main river channels which then provide the only available pasturage. At this time fishing becomes the principal source of sustenance (see Rzóska, 1974).

Such annual movements are characteristic of peoples in all parts of the world that suffer from severe seasonal droughts. This is an example, to be found elsewhere in Africa (e.g. Lakes Chad, Kioga, Bangweulu, Rukwa, etc.), in which adaptation to swamps has become an integral part of the life of the people and necessary for their survival. The above investigations based on Jonglei on the White Nile were undertaken primarily because of a proposal to cut a canal from Jonglei northwards to rejoin the While Nile between its confluences with the Jeraf and Sobat Rivers (Fig. 11.3). This was to short-circuit the Bahr el Jebel to provide a shorter and unobstructed navigation channel and, by deflecting some of the water entering the Sudd, to reduce the loss by evaporation which is at present about 50%. The project has recently been revived though the canal is now to be cut between Bor and the mouth of the Sobat River so as to avoid all the permanent swamps. The hydrological consequences and their effects on the ecology and human activities in this region are still in doubt. Owing to the enormous area of the permanent swamps and the extreme difficulty in penetrating the interior very little is known of their ecology. Of all the world's inland water systems this is surely the most difficult to investigate.

The swamps of the Okavango Delta near the southern edge of the tropics in northern Botswana (Fig. 18.8) are of special scientific and practical interest. Since most of the country is very arid and the Delta drains into the northwest corner of the Kalahari Desert, the one apparently abundant source of water has naturally inspired some hopes. Geological, hydrological and biological investigations have recently been intensified with a view to exploiting the water and to improving the productivity of the wetlands of the Delta. Present knowledge of this situation was reviewed in the proceedings of a symposium held at Gaborone in 1976 (Nteta and Hermans, 1977). Campbell (1977a) gives a vivid impression of this remarkable inland delta, outlines its biological and human history and discusses possible future developments.

The Okavango River is fed from the Angolan highlands via the Cubango and Cuito Rivers (Fig. 9.4 and 18.8). At Muhembo the river descends 150 km of well defined swampy valley and then fans out over a great expanse of the desert but is finally obstructed by a northwest–southeast fault about 250 km from Muhembo. In the delta so formed large quantities of sediment brought down by the river are being deposited and swamp vegetation covers most of the extensive flooded area. The fault is one of many that demonstrate the tectonic instability of the region and quakes are frequently recorded.

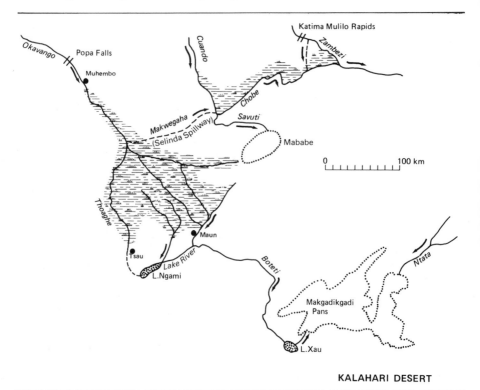

Fig. 18.8 The Okarango Swamp delta

The water is finally deflected by the fault into the southeast corner from which it flows both into Lake Ngami and by the Boteti River for about 200 km into the Makgadikgadi saltpans. The volume of these outflows varies with the seasonal rainfall in the Okavango catchment. Lake Ngami is sometimes dry and the Boteti occasionally ceases to flow.

The prehistory of this area is a matter of some controversy but it is generally agreed that, as in most of tropical Africa, there were wet and dry periods during the late Pleistocene. Evidence from cave deposits, ancient lacustrine sediments and land forms in the northern Kalahari point to a prolonged wet phase between about 17 000 and 14 000 B.P. (Grove, 1969; Cooke, 1977) At one time there was a large lake in the Makgadikgadi basin. Several opinions have been expressed concerning the previous drainage of the area and the obviously recent origin of the Delta. One of these suggested that the Okavango formerly flowed unhindered via the Boteti and Makgadikgadi to join the Limpopo River. Subsequently the Delta was formed by tectonic obstruction, and the resulting loss of water by evaporation from the swamps and the general lowering of the rainfall reduced the outflow to its present very low level. Even now, were it not for the enormous loss from the swamps (see below), a considerable lake would surely occupy the Makgadikgadi basin.

In historic times there have been great changes in the positions of the swamp channels and in the points at which they disappear as recognisable open water courses. Deposition of sediments raising the channel beds and occasional blockage by banks of vegetation, together with massive downflows of water during periods of heavy rain in the highlands, have frequently altered the hydrological pattern of the Delta and continue to do so. For example, early in the nineteenth century Lake Ngami was dry and a stream with a well-defined bed crossed the basin. In 1850 most of the flood waters of the Delta flowed out via the Thaoghe channel (now blocked) into Lake Ngami which was then much larger than at present (Campbell, 1977b). Both the area and the situation of swamp vegetation have correspondingly changed greatly during the past two centuries. The area of flooded swamp is now at a seasonal maximum in June–July and a minimum in December–January.

Another curious hydrological feature of the region is to be seen between the northern edge of the Delta and the Zambezi River (Fig. 18.8). The Cuando River descending from the northwest is obstructed by a low sand ridge and turns through a right-angle and flows, as the Chobe River, through swamps into the Zambezi. The Zambezi, below the Katima Mulilo Rapids crosses flat country, and in flood seasons overflows into the Chobe swamps, and thence returns by the Chobe River to its own course and down to the Victoria Falls. At the angle made by the Cuando and Chobe there is an occasional overflow (the Savuti) into the small swampy Mababe Basin. In addition to these complications, especially high floods down the Okavango overflow to the northeast via the Makwegaha (Selinda Spillway) into the Chobe River. This provides an occasional aquatic faunal connection between the Okavango and Zambezi basins.

The vegetation of the Okavango swamps includes most of the species typical of the swamps of tropical East and Central Africa (e.g. the Lake Victoria basin, p. 313). Their distribution in the different types of habitat is described by Smith

(1977). It is fortunate that the water hyacinth, *Eichhornia crassipes*, that has caused so much trouble elsewhere, has not yet reached the Okavango basin. The floating water-fern *Salvinia molesta*, which was an embarrassment during the early stages of Lake Kariba (p. 368) is now to be found in the Chobe River and the lower Cuando which, as already mentioned, are occasionally joined to both Zambezi and Okavango basins. Before long it will surely find its way into the Delta. More surprising is the absence of the floating Nile cabbage, *Pistia stratiotes*, indigenous to and abundant in most of tropical Africa.

The recent changes in the hydrology of the Delta discussed above have natural-ly been reflected in alterations in the distribution of the vegetation. *Cyperus papyrus* dominates in the permanently flooded areas, particularly around the wa-ter channels. That it is limited (according to Smith, 1977) to waters of low salin-ity (K_{20} 40−60) is interesting because this is not true of its distribution elsewhere (e.g. Lake Victoria, K_{20} 90−100, and Lake George, *c.* 200).* Though the rate of production of papyrus is, in terms of dry weight per annum (Thompson, 1977), extremely high, the economic value of this and other swamp plants is very limited (p. 000). There is evidence that agriculture has been practiced within the Delta only for the past 200−300 years mainly by exploiting the well-watered and rich soils from which the floods retreat in the dry seasons (Campbell, 1977b). Im-provements so far accomplished and possibilities for further expansion of irriga-tion are discussed by Thompson (1977) and the Ministry of Agriculture, Bots-wana (1977).

Stock-raising has had a chequered history owing to the tsetse fly, *Glossina mor-sitans*, transmitter of cattle trypanosomiasis (Nagana). Exploitation of the good grazing land within the Delta has thereby been prevented except for a period of years following an epidemic of rinderpest in 1896 which ravaged both the cattle in the surrounding lands and those wild mammals from which the fly otherwise gets its blood-meals. The consequent great reduction in the fly population per-mitted cattle-grazing within the Delta for some time, but the tsetse have now re-established themselves and stock-raising is again impossible in the interior.

Seventy to eighty species of fish are recorded from the Delta and Lake Ngami (Fox, 1977). This fauna is closely related to those of the upper Zambezi and Kafue Rivers (p. 000). In such an unstable and swampy environment, it is dif-ficult to make meaningful estimates of the sizes and composition of the various fish communities, but it is generally thought that the stocks are small compared with those in most other African swamp regions. The two Cichlid species *Sarotherodon andersoni* and *Tilapia rendalli* (*melanopleura*) appear to be the most abundant and may account for nearly 50% of the fish biomass. As expected, fish are scarcest in the dense and no doubt oxygen-deficient swamps and are more numerous in the open waters and especially those that are enriched with nutrients derived from nearby cattle grazing. There is so far no modern commercial fishery and local fishing by traditional methods is comparatively unproductive. Larger yields could no doubt be gained by controlled application of modern techniques and perhaps by pond culture combined with improvements in transport and marketing.

* The salinity of the inflowing Okavango River is low (K_{20} *c.* 30) but progressively increases down the Delta. Isolated pools in the dry seasons have been recorded as much more saline (K_{20} up to 3 000).

The water and vegetation of the Okavango Delta and its surroundings have naturally attracted many large mammals and birds whose continued existence depends upon the maintenance of the ecosystem (Patterson, 1977), and has so far been saved by the tsetse fly from serious human interference. The great diversity of terrestrial wildlife owes much to the interdigitation of two quite different types of habitat (Fig. 18.8). This is a unique feature of the Delta. Especially interesting are the amphibious antelopes *Limnotragus spekei* (the sitatunga p. 000) and *Kobus leche* (the red lechwe). The large terrestrial fauna could be better exploited for tourism and, by controlled culling, as a source of food.

It has been estimated that about 95% of the annual inflow (including the small amount of rain falling direct onto the Delta) is lost by evapotranspiration from the swamps (Wilson and Dincer, 1977). The remaining 5% leaves by the Boteti river towards Makgadikgadi, by the Lake River to Lake Ngami and by seepage into the ground water. This enormous loss to the atmosphere and the shallowness of the water obviously reduce very much of the value of the Delta as a source of water for use elsewhere. But some of the water has been used for small-scale copper and diamond mining, and for a domestic supply to the small town of Maun at the base of the Delta. None of these schemes involves very long-distance transference of water.

There are now plans for the future aimed at exploiting the water more effectively without destroying the ecosystem which, it is generally agreed, should be preserved. For example, clearing and deepening some of the channels would more rapidly draw water from the swamps during flood periods and so reduce the losses from evapotranspiration. This is the main objective of the Jonglei canals scheme for the Nile Sudd swamps (p. 317). Most of the permanent water lies in the northern half of the Delta (Fig. 18.8) from which it would have to be extracted and controlled. Since, however, the whole system is likely to remain unstable owing to earth movements, sedimentation and to wide fluctuations in rainfall any such measures would probably have to be frequently modified. For these reasons some have questioned the wisdom of investing a large sum of money in permanent works.

The special interest and attraction of the Delta derive both from its instability and from its position as an inflow into one of Africa's largest deserts after the Sahara, whose inhabitants, it is hoped, will benefit from scientific and ecologically orientated exploitation.

19

Temporary, saline and thermal waters: Lake Chilwa

In dealing with past history and the effects of climate in general and in relation to the different regions, the seasonal, medium- and long-term fluctuations of water level have been discussed (Ch. 3). The seasonal and medium-term changes in levels that have been recorded in historic times are clearly related to fluctuations in climate which determine the relation between inflow and evaporation.

We are here concerned with situations in which evaporation is periodically or continuously so much in excess of inflow that the salinity of the water may be raised by evaporation to a level which has profound biological effects, and in extreme cases the basin is dried up. In regions with well-marked wet and dry seasons, streams and small bodies of water that are well flushed out during the rains become dry without significant rise of salinity. But in closed basins desiccation is usually associated with accumulation of salt. For this reason and because adaptation of organisms to both of these conditions involves a physiological mechanism for dealing with a scarcity of available water, desiccation and high salinity have been brought together in this chapter.*

There are a few lakes with permanent outlets whose salinity is rather high. Lake Kivu is an example (Ch. 15). But here the salts are derived from surrounding and underlying volcanic rocks. This also occurs in many closed basins though, unless saline inflows are actually visible, it is sometimes difficult to decide whether a lake is gaining salts from sources other than by evaporation of freshwater inflows. Extraneous sources of salt are certainly important in most if not all of the salt lakes in the East African Rift Valleys. Most volcanic crater lakes have very small catchments and their water may derive directly from rain, and for that reason may be of very low salinity (Ch. 15). Some, however, are fed by underground seepages through soluble volcanic rocks and thereby become highly saline (Arad and Morton, 1969). The ancient salt industry of the Katwe crater lake in western Uganda (Fig. 19.2) has been maintained by a continuous seepage of salt water from below (Wayland, 1934). On the shores of Lake Magadi in the Eastern Rift there is a modern plant for the extraction of soda exported for mak-

---

* A geographically wider ranging discussion of this subject can be found in Cole (1969).

ing glass. Though this lake has very greatly contracted from what was a fresh-water lake in the late Pleistocene, the present chief sources of the alkaline salts are the saline hot springs at many points around the shore which have been estimated to contribute more than 1 000 tons of soda per day. The climate is generally hot and dry and the brine is saturated and crystallised by evaporation except during an unusually wet season (Coe, 1966).

There are a few lakes in Africa that have no surface outlet but are nevertheless fresh; moreover they have certainly contracted from a very much larger volume during the past few thousand years. Examples are Lakes Chad, Baringo and Naivasha. In the case of Naivasha (in the Eastern Rift in Kenya) lack of direct evidence has encouraged much speculation and differences of opinion as to the cause of this curious phenomenon (discussion in Richardson and Richardson, 1972). For me, however, it is difficult to doubt that an underground outlet is an important, if not the only significant, cause. Other suggestions have been: chemical precipitation and removal by winds of salt-laden dust from the shores especially during low level periods (deflation). Nevertheless, we have to face the fact that the great majority of lakes in closed basins, and there are very many, have become saline by evaporation and these influences have failed to prevent it. For example, Lakes Nakuru and Elmenteita which, during the previous high level period, were joined with Naivasha in a single large and fresh lake (Fig. 3.6) are now reduced to saline remnants. Gregory (1921, p. 107) who discovered Lake Baringo (160 km north of Naivasha) in 1893 reported that the water was flowing 'freely' into the lavas at the north end. No one has apparently investigated this since that time. It may be noted that the freshness of the water is maintained in spite of some highly saline springs discharging into the lake (Fig. 19.4). There is a suggestion that Lake Turkana is less saline than would be expected from the amount by which it has contracted in volume since the level fell below the previous surface outlet to the Nile (p. 179 ff). The case of Lake Chad has received much attention from hydrologists, and here there is certain evidence of a large-scale subterranean seepage into the sandy soils along the north and east coasts. Though some of the other factors mentioned above have played a part, it is generally agreed that the freshness of the water of Lake Chad is maintained mainly by the subterranean outflow (Ch. 12).

In the Sahara Desert, as explained in Ch. 11, the scarcity of exposed water is due both to the great excess of evaporation over rainfall and to the disappearance under the porous sands of large quantities of water coming from the peripheral and central mountains during the rains. Fresh water is in general obtainable in the desert only by digging into one of the subterranean reservoirs. Most, though not all, of the naturally exposed water is saline in some degree, often approaching saturation. The amount of permanently exposed water inhabitable by organisms, that have no means of combating high salinity and desiccation, is very small.

In tropical Africa as a whole the amount of temporary and saline water is thus far from insignificant. It is mainly to be found in the arid northern and southern belts and in the dry savanna between these and the humid equatorial region. Saline waters are also common in the two Rift Valleys, especially the Eastern. Apart from its intrinsic scientific interest, the study of the ecology and the physiological adaptations of aquatic organisms to high salinity and desiccation is important in relation to fisheries and to some vector-borne diseases in those regions where the water tends to become saline or to dry up at intervals.

To the simple question – at what point does a water cease to be fresh and begin to become saline? – there is no simple answer. It depends upon the context. 'Saline' water has been variously meant to denote water:

(a) in which salt can just be detected by human taste;
(b) that is just too saline as a permanent source of drinking water for man and other mammals;
(c) that is too saline to support a typical freshwater fauna and flora.

It is impossible to attach any quantitative meaning to these definitions. To mention one confusing example – the water of Lake Turkana (Table 5.1) has an unpleasant taste and unfortunate alimentary efects on people not accustomed to it. But acclimatisation is possible, and it is the normal drinking water of the Turkana tribe who live and fish along the western shore. The lake has an abundant and typical freshwater fauna and flora, though there is some evidence that the species composition of the plankton is to some extent affected by the relatively high salinity (p. 183). Of all the African waters that support an abundant freshwater fauna Lake Turkana is exceptional in being detectably saline to the taste. Others, such as Lake Edward, have a relatively high salinity but still provide good drinking water (Table 5.1).

In arid regions the life and the very existence of man and other terrestrial animals is determined by the quantity, quality and distribution of available water. For successful adaptation to very dry conditions great economy in the use of water is essential. This is effected by mammals and other animals in various ways, behavioural and physiological, to reduce evaporation from the body surface and to control loss of water from the kidneys. The most efficient use possible is made of all available water from such sources as saline pools, dew, plants and from the oxidation of foodstuffs, especially fat (metabolic water). Several small desert rodents, including the African jerboa (*Jaculus jaculus*) and gerbil (*Gerbillus gerbillus*) manage to live and retain a normal body water level without drinking water at all, feeding solely on air-dried plant material (Schmidt-Nielsen, 1964; Potts and Parry, 1964, p. 302 ff.). The fennec (*Fennecus zerda*), a small carnivore of the Sahara, lives under extreme desert conditions and is known to be entirely independent of drinking water. It feeds both on plant matter and on rodents, lizards and insects which themselves must be independent of free water (Monod, 1958). Birds, other than marine species, being very mobile, can reach scarce water more easily than other terrestrial animals. Only a few desert species are known to drink saline water or, like some insectivorous birds, to dispense entirely with drinking water. These have powerful salt excreting kidneys. It is remarkable that the salt excreting nasal organ by which many marine birds can survive permanently away from anything but seawater, has not been developed in terrestrial birds, even in those living in deserts where the organ would be of obvious use. There is some doubt about the ostrich which has a large nasal gland, though its function as a salt excretor has not yet been demonstrated (Schmidt-Nielsen, 1964, Ch. 15).

Very arid countries and extreme deserts therefore support a much larger number of terrestrial animals than might at first be supposed, because many have become independent of exposed freshwater, some being able to extract water from very concentrated salt solutions, and others getting their water by means other than drinking. All have highly efficient devices for preventing loss of water.

Inland saline waters differ in several important respects from the sea. They are relatively very small and are often extremely isolated one from the other. Their salinity is subject to great changes and indeed may start fresh and end as a dry salt-pan within a few weeks. During the course of evaporation the composition changes owing to the different solubilities of the ions, calcium followed by magnesium being the first to be precipitated as carbonates. Not only is each water subject to these drastic changes but there are great differences in composition between individual waters due to the varied origin of the soluble salts.

Inland waters have been classified into certain general types on the basis of inorganic composition (Livingstone, 1963). To illustrate two contrasting types of saline waters that are widespread in Africa, some data are given in Table 19.1. These samples have approximately the same salinity, and both are higher than that of seawater ($35°/_{oo}$).

1. A small natural pond in the northern Sahara in a depression in the sand reaching to the water table, well exposed to the sun and wind for evaporation. It is a typical 'sulphato-chloride' water, with a not excessively high pH, though rising to over 8·5 through photosynthesis in the daytime. The predominant anions are chloride and sulphate, and calcium and magnesium are relatively high.

2. A highly alkaline lake in the Kenya Eastern Rift. The pH is very high as the result of the presence of carbonate as well as bicarbonate derived from the carbonatite volcanic rocks. Chloride and sulphate are correspondingly low. The insignificant quantities of calcium and magnesium result mainly from precipitation of high pH and salinity. This kind of water is typical of the volcanic regions of tropical Africa including the volcanic mountain masses in the Sahara (e.g. Tibesti).

The functioning of all living cells depends, among other things, upon the presence within them of water and inorganic ions maintained at concentrations, if not constant, at least within a rather narrow range. In multicellular animals cellu-

le 19.1

	Approx. Salinity %₀	Milliequivalents per litre								
		pH	Na⁺	K⁺	Ca⁺⁺	Mg⁺⁺	Alkalinity (HCO₃⁻ + CO₃⁻⁻)	Cl⁻	SO₄⁻	Cations	Anions
Sahara adle, 1943a) /33)*	50⁺	8·7 §	410	11·8	23	86	19	443	93·2	531	555
rn Rift e Elmenteita lling and ling, 1965)	43⁺	>10·5	410	9·8	<0·25	<1·2	289	148	3·0	420	440

l in a depression 140 km south of Touggouert (Fig. 11.3)
:ulated from density measured by hydrometer in the field (subject to some error) on the basis of the density of
wn concentrations of NaCl
:ulated from conductivity
asured at 14·00 hrs when photosynthesis by the abundant algae had probably raised the pH to a maximum.

lar activity depends also upon the maintenance of ionic concentrations in the surrounding body fluids that are different from but related to those within the cells. This situation obviously cannot be maintained by aquatic animals merely by sealing themselves of in an outer impermeable membrane preventing exchange of water and ions with the outer medium. There must be a continuous exchange between inside and out, at least through the gut, respiratory and excretory organs. Active life of the whole body depends in fact upon a through-circulation of water and salt. The processes by which a more or less constant internal chemical composition with respect to water and ions is maintained in an environment, which is always different and usually fluctuating, are known as osmotic and ionic regulation respectively (for a discussion of this subject see Potts and Parry. 1964).

The body fluids of most marine animals, except fish, are isotonic with seawater, but, nevertheless, some of the ions are regulated at a different level. Life in fresh water necessarily entails keeping the body fluids more concentrated, or hypertonic to, the surrounding water. Conversely, to colonise very saline inland waters a mechanism for hypotonic regulation is required.

Less is known of the physiological mechanisms for the regulation of water and ions in the bodies of animals in highly saline waters than of those living in freshwater or the sea (for general discussion and references see Beadle, 1969). In the present connection, we are interested mainly in the evolutionary and ecological implications. It is a curious fact that almost all the species of animals found in highly saline waters are derived from freshwater types. Freshwater animals maintain the concentration of their body fluids hypertonic to that of the water partly by active uptake of ions from the water through some of the body surface as well as in the food. This compensates for loss of salts and gain of water by diffusion and is reinforced by the low permeability of most of the body surface and by the elimination of urine that is more dilute than the blood. In highly saline water these processes must be reversed in order to maintain the body fluids hypotonic to the water.

The permeability of most of the body surface of the brine shrimp, *Artemia salina*, is extremely low. The excess of salt taken up by the gut is excreted by special cells in the gills and the animal is thus able to keep the concentration of its blood at a comparatively low level in water of salinity about five times that of seawater. The only other information we have about these mechanisms comes from a few saline water dipterous insect larvae, and these are clearly derived from freshwater forms. The mosquito larva *Aedes detritus* is common in coastal and inland salt waters of Europe and the Mediterranean region. Its body surface is very impermeable and the saline water is taken in by mouth and thus into the blood. A fluid is eliminated into the rectum by the malpighian tubes that is isotonic with the blood. Cells lining the rectum then reabsorb water from this fluid so that an excess of salt is allowed to pass out, and the concentration of the blood is held at a low level.* This and many other insect larvae have presumably come originally from the land, but via fresh waters. On land the rectal reabsorption of water is of special importance to insects in conserving water, especially in dry

---

* There is a discussion of osmotic regulation in mosquito larvae in Clements, 1963, Ch. 3.

Fig. 19.1 A highly saline and alkaline crater lake. One of a group of late Pleistocene volcanic craters in the Western Rift Valley between the northern shore of Lake Edward and the Rwenzori Mountains (Fig. 13.1). Lake Katwe (Fig. 19.2) is also in this group

climates. This mechanism might be said to have 'preadapted' them for life in saline water.

Since the regulatory mechanisms in fresh and saline water have to work in opposite directions we might perhaps have expected that adaptation to the former would preclude an animal from living in the latter. In fact, during the course of evolution, the hypertonic regulating mechanism of species from many groups of freshwater animals has obviously been reversed, i.e. become hypotonic, to function in very saline water. The Insecta (Diptera, Coleoptera), Crustacea (Phyllopoda, Cladocera, Copepoda, Ostracoda), Nematoda, Rotifera and Protozoa are all represented by species, having close relatives in fresh water, in waters of such high salinity that hypotonic regulation can hardly be doubted in those which have not yet been confirmed by experiment.

When the matter is considered more closely the freshwater origin of most saline water animals becomes less surprising. There are several species that are found in both fresh and very saline water. The invertebrates *Brachionus plicatilis* (Rotifera) and *Arctodiaptomus salinus* (Crustacea, Copepoda) are examples of many of such (Beadle, 1943a, b). The mosquito larvae *Aedes detritus* and *A. natronius* and the cichlid fish *Sarotherodon alcalicus grahami* can be kept successfully in fresh water, though they have never been found naturally in anything but water of high salinity. The well-known seasonal migrations between the sea and inland freshwaters on the part of several species of fish are good demonstrations of the transition. In the eel and the salmonid fishes the hyper–hypotonic switch has been demonstrated and investigated.

We cannot therefore doubt that most saline water animals have evolved from freshwater ancestors and that they have effected a reversal of the osmotic and ionic regulatory mechanisms. From our present knowledge of distribution in relation to salinity it can be said that almost all so-called freshwater animals are found in water of salinity below $10^o/_{oo}$. Experiments on acclimatisation suggest that freshwater animals with no means of switching from hyper- to hypotonic regulation, that is the majority, succumb when the salinity is raised to between 10 and $20^o/_{oo}$. In very few cases $20–30^o/_{oo}$ is the limit. It was with this in mind that about $5^o/_{oo}$ is suggested as the highest salinity that is likely to have been reached in Lake Tanganyika during the Pleistocene (Ch. 16). In view of the very rich, varied and ancient freshwater fauna of the lake the limit is more likely to have been less than $5^o/_{oo}$. The salinity of Lake Tanganyika is now about $0·5^o/_{oo}$; that of Lake Rudolf is approximately $2·5^o/_{oo}$.

With some hesitation, but for most practical purposes, we might therefore define fresh water as water of salinity below $5^o/_{oo}$ and a freshwater animal is one incapable of hypotonic regulation. But we must bear in mind that the spectrum is a continuous one and that no precise division can be made.

In view of the great variety in chemical composition of different saline waters and of the same water during the course of evaporation, the question arises as to whether special adaptations are needed for life in saline waters of peculiar composition. Information upon which to base a definite answer to this question is difficult to find partly because there are factors other than water composition that may be responsible for the presence or absence of species in particular cases. For example, though the salinity of both lakes is around $250–300^o/_{oo}$ the absence of fauna from the Dead Sea and the presence of several species, including *Artemia salina* and insects in the Utah Great Salt Lake, has been attributed to the relatively high concentration of bromide in the Dead Sea (Nemenz, 1970). Something relevant may, however, be got from comparing the data from four surveys in Africa, two of some highly alkaline saline lakes in the East African Rift Valleys all with pH over 10 and with salinities ranging from about 10 to 45°/$_{oo}$* (Beadle, 1932b; Jenkin, 1936), two from less alkaline sulphato-chloride waters of the northern Algerian Sahara, pH 7–9, salinity range 10–80°/$_{oo}$ (Gauthier-Lièvre, 1931; Gauthier, 1938; Beadle, 1943a). The difference is very striking (Table 19.2). The fauna of the alkaline saline lakes, though within a rather lower salinity range, is, by comparison, very poor indeed. Five species of Rotifera, one of Ostracoda, two of hemipteran insects and some unidentified chironomid larvae comprise the whole so far recorded fauna of Lakes Elmenteita (salinity 4°/$_{oo}$), Nakuru (45°/$_{oo}$) and Bogoria (35°/$_{oo}$) in the Kenya Rift. Crustacea other than Ostracoda were absent from all except a small very alkaline crater lake (salinity 35°/$_{oo}$) west of Lake Naivasha, where in 1931 the copepod *Paradiaptomus africanus* was found in very large numbers. This is a species widespread in freshwaters of East Africa, but seems to be the only copepod crustacean recorded and identified from any of the very alkaline saline waters (Lowndes, 1936).† In three of the lakes examined, Bogoria in the Kenya Rift (salinity 36°/$_{oo}$), Kikorongo (35°/$_{oo}$) and Maseche (45°/$_{oo}$) in the Western (Uganda) Rift, no zooplankton of any kind could be found in 1931.

Table 19.2

	Rift Valleys (*Jenkin, 1936; Beadle, 1932b*) *Salinity 10–45%*	*N. Algeria Sahara* (*Beadle, 1943a*)	
		10–45%	*over 50%*
Rotifera	5	6	2
Platyhelminthes			
(Rhabdocoela)	0	1 + undet.	+ undet.
Crustacea			
Phyllopoda	1	1	1
Cladocera	0	6	0
Ostracoda	1	3	0
Copepoda	1	8	5
Insecta			
Diptera larvae	2 + undet.	8 + undet.	2
Hydrachnida	0	3	0
Nematoda	0	+ undet.	+ undet.
Mollusca	0	6	0
Cyanophyceae	6	22 + undet.	5 + undet.
	16 + undet.	64 + undet.	15 + undet.

+ undet. denotes additional but undetermined species.

* The salinities of these lakes are, of course, subject to great variations. The data given here are based on water samples and collections made on single occasions. The salinities are calculated from either conductivity or density measurements made in the field and are therefore rough approximations. Lake Kikorongo, an alkaline and saline volcanic crater lake in western Uganda (Fig. 13.1), is occasionally flooded by an overflow from Lake George and is thus temporarily colonised by freshwater animals including fish.
† Copepods have been recorded, but not identified from the saline and alkaline springs of Lakes Magadi and Manyara.

The foregoing remarks refer to the water in the main bodies of the lakes. Into some of them there flow small streams and seepages of much fresher water providing a gradient of salinity and conditions in which some rooted vegetation can be established, e.g. alkaline salt tolerant sedges (Cyperaceae) such as *Cyperus laevigatus* (see Fig. 19.2). In the fresher outer regions of these fringing marshes there is, as expected, a typical freshwater fauna. But in the more saline water further out among the vegetation is a suitable habitat for a few animals, especially insect larvae, that are alkali salt tolerant and seem to find in these situations a better food supply and protection than in the open lake. Two species of mosquito larva, *Aedes natronius* and *Culex mirificus* are very common and often found in vast numbers in such habitats around some of the alkali lakes in the Rift Valleys. *A. natronius* is apparently endemic to the alkaline salt lakes of the Lake Edward-George basin (Hopkins, 1952).

The saline waters of the northern Algerian Sahara present an entirely different picture (Table 19.2). These waters are typical of the non-volcanic regions of the Sahara in being relatively neutral, though very high rates of photosynthesis may raise the pH to almost 9 during the day. But the major anion is Cl^- rather than CO_3^{2-} (Table 19.1).

The two surveys of saline waters in the Algerian Sahara showed that all the major groups of freshwater animals are represented. A summary of the numbers

Fig. 19.2 Katwe Salt Lake, Uganda (Fig. 13.1). The alkaline salt-tolerant sedge *Cyperus laevigatus* is visible (photograph Uganda Government Ministry of Information, Broadcasting, and Tourism)

of recorded species from a survey in 1938 (Beadle, 1943a) is given in Table 19.2, and compared with the figures from the East African alkaline lakes.

These figures must not be taken to represent accurately the number of species to be found in African neutral and alkaline saline waters. Further surveys over much wider areas will certainly add to the numbers. However, there is little doubt that the general conclusion will remain that the fauna of the near-neutral is much richer in species than that of the alkaline saline waters. Moreover the comparison shows that certain groups not found in the very alkaline waters (Rhabdocoela, Cladocera, Hydrachnida and Mollusca) are represented in the more neutral waters and that in the latter, when the salinity exceeds 50°/$_{oo}$ the fauna is richer in some respects (particularly in copepods) than in the alkaline waters of much lower salinity. The work of Labarbera and Kilham (1974) on the copepods in forty-eight East and Central African lakes records their presence only in waters of conductivity (K_{20}) less than 10 000, except for one species, *Paradiaptomus africanus*, at K_{20} 15 000. These waters, especially the more saline ones, are highly alkaline. In the neutral waters of the Algerian Sahara three copepod species (*Arctodiaptomus salinus*, *Cletocamptus retrogressus* and *Horsiella brevicornis*) were found in waters of K_{20} 80 000 to 100 000 (Beadle, 1943a). Very alkaline and saline waters have been studied in Iran (Löffler, 1953), in Turkey (Gessner, 1957) and in the U.S.A. (Cole and Brown, 1967). All show a large biomass involving very few species.

Artemia salina, the brine-shrimp, is cosmopolitan and the best known of several phyllopod Crustacea adapted to life in saline water, and able to survive complete desiccation in the egg. It is common in the sulphato-chloride waters of the northern Sahara and recorded from salinities exceeding 50°/$_{oo}$. Another Saharan phyllopod, *Branchinectella salina*, is apparently confined to less saline water. *Artemia* is often found in prodigious numbers, as in the Great Salt Lake, Utah, U.S.A. It is not widely known that in one remote spot in the Fezzan, Libyan Sahara, it is an important item of human food. In a small permanent salt lake, Gabra Aoûn, of salinity around 160°/$_{oo}$, there is a dense swarm of *Artemia* feeding on the rich algal crop. The local people, the Dawada, net them, dry them in the sun and make them into 'cakes'. These have been shown to contain all the amino acids essential for human nutrition (Monod, 1969). The industry was flourishing when the lake was visited by the first Europeans Oudney and Clapperton in 1822. The only other recorded example, at least in tropical Africa, of the exploitation as a source of human food of organisms other than fish in highly saline water is the harvesting of the blue-green alga *Spirulina platensis* in the saline and alkaline ponds in the desert of the Kanem region north of Lake Chad and around Ounianga Kebir southeast of the Tibesti Mountains (Pourriot et al., 1967; Iltis, 1968, 1969–1971, Iltis and Riou-Duwat, 1971). This small spiral filamentous alga (Fig. A1) is characteristic of extremely alkaline and saline water and is sometimes so prolific that the water is coloured bright green. Iltis recorded it from salinities up to 270°/$_{oo}$, though the optimum for growth was 20–60°/$_{oo}$. They are collected as a sludge, dried in the sun and sold in the markets and even exported to Nigeria. He also studied over four years the fluctuations of salinity in the Kanem ponds (Fig. 12.1) and the associated successions of zoo- and phytoplankton species (summarised and discussed by Iltis, 1975).

After man, the only known terrestrial consumer of *Spirulina platensis* in African saline lakes is the lesser flamingo (*Phoeniconaias minor*). This and the greater flamingo (*P. antiquorum*) are commonly seen wading in shallow very saline and even alkaline waters. In some, such as Lakes Elmenteita, Nakuru and Bogoria in the Kenya Rift, they are numerous enough to give the appearance from a distance of large pink sheets spread across the lake (Fig. 19.3). *P. minor* walks through the water with beak partly submerged drawing in water and filtering off the algae, which are often mostly *Spirulina platensis*, by means of filter plates in the mouth. *P. antiquorum* feeds on the mud from which it filters out larger organisms such as chironomid larvae and plant seeds (Jenkin, 1957). Nothing is known about their mechanisms for regulating salt and water in the body, whether the nasal salt glands which are well developed play a part as well as the kidneys, but Vareschi (1978) reported that the lesser flamingoes on Lake Nakuru drink fresh water from the inflowing streams, though at intervals of apparently not less than five days. This was a study of the population fluctuations of *P. minor* (which varied between about 32 000 and nearly 1·5 million) and of its main food organism, *Spirulina platensis*, with observations on its feeding behaviour and quantitative estimates of the algae consumed. The work of others on the biology of these remarkable birds on the saline lakes of Africa is also discussed.

Between 1953 and 1962 the cichlid fish *Sarotherodon alcalicus grahami* (*Tilapia grahami*) was introduced into Lake Nakuru from Lake Magadi. It is the only fish in the lake and has multiplied greatly and grows to a larger size than in Lake Magadi. It has changed its mode of feeding from grazing on the filamentous

Fig. 19.3 More than 20 000 flamingos crowded along part of the north shore of Lake Nakuru, Kenya. A few are to be seen in the open lake. The vast majority are lesser flamingos (*Phoeniconaias minor*) with about a hundred greater flamingos (*P. antiquorum*). A few pelicans can be seen in the bottom left-hand corner (from Vareschi, 1978)

blue-green algae in Lake Magadi to filter feeding on *Spirulina platensis* and is a major consumer of this alga. The most obvious effect on the ecology of the lake has been the appearance of more than fifty species of fish-eating birds of which the Great White Pelican is the main predator. This has added to the attractions of the Nakuru National Park (Vareschi, 1979).

Only one phyllopod crustacean has so far been recorded from very alkaline saline water in East Africa. This is *Branchinella ornata* in pools in a shallow volcanic crater near Katwe, Uganda, which is dry for much of the year (Beadle, 1969; it is otherwise known only from South Africa in waters of unrecorded salinity – Linder, 1941). At the time of collecting the water had a salinity of about $20^o/_{oo}$, alkalinity 100 meq/l and pH 10·5. The eggs of *Artemia* placed in this water hatched successfully but the larvae died almost immediately. From the large amount of ecological data from North America, Europe, North Africa and Asia it has generally been concluded that *Artemia salina* is not adapted to live in alkaline carbonate waters. A few records, however, from carbonate saline water in the U.S.A. of pH up to 10 have been overlooked (Cole and Brown, 1967). But the taxonomy of the genus *Artemia* has not yet been properly studied. With a distribution covering three continents it is likely to have diverged into genetically distinct races, if not higher categories, that are physiologically adapted to different types of saline water. It has not yet been recorded from the alkaline carbonate waters of Africa, and is undoubtedly absent from most of them. The only group of animals that are apparently equally well adapted to both habitats are the Rotifera. But one species only, *Brachionus plicatilis*, was actually recorded from both types of water.

It is not possible to analyse in the same manner the available data concerning the algal flora except for the blue-green algae (Cyanophyceae) which are shown in Table 19.2. But diatoms are exceedingly numerous in both habitats, and together with blue-greens sometimes form a gelatinous carpet on the bottom of very shallow water at the edge of both neutral and alkaline lakes. Lake Magadi in Kenya is a notable example. The shallow lagoons into which the alkaline hotsprings flow are carpeted with blue-green algae which are the main food of the fish *Sarotherodon alcalicus grahami* (see below).

Without experiments to supplement the ecological information, we cannot with certainty point to a factor responsible for the relative poverty of the faunas of very alkaline saline waters. It might be the direct effect of high pH or bicarbonate, the relatively low calcium (though the absolute concentration is as high as in some freshwaters, Table 5.1), or an unfavourable balance between ions.

There are several species of fish of the mainly freshwater family Cyprinodontidae that are found in inland saline waters. In the northern Sahara *Aphanius fasciatus* has been taken from a wide range of salinities up to over $40^o/_{oo}$. It must certainly be capable of hypotonic regulation (Beadle, 1943a, b). Several of the cichlid fishes can live in water of moderate salinity, but the most remarkable is *Sarotherodon alcalicus grahami** endemic to Lake Magadi where it lives in pools fed by the alkaline hotsprings (Coe, 1966). It is a small fish up to seven or eight centimetres long, found in water of salinity up to about $40^o/_{oo}$ (calculated from density measurements), of pH 10·5 and of temperature up to 40°C. Over 80% of the

---

* Formerly known as *Tilapia grahami*, now recognised as a subspecies of *Sarotherodon alcalicus.*, pers. comm. E. Trewaras.

salts are $NaHCO_3$ and Na_2CO_3. A study of the acid-base balance in the blood of this fish has revealed some very remarkable features (Reite et al., 1974) Johansen et al., 1975; Maloiy et al., 1978; Maetz and de Renzis, 1978). The water with a very high alkalinity of about 200 meq/l and a pH of over 10·0 contains virtually no free CO_2 (Fig. 7.1) – a condition opposite to that faced by the swampworm (p. 322) and quite different from all freshwaters. Measurements made on fish subjected to the minimum possible disturbance indicated that the blood bicarbonate is unusually high (about 6 meq/l) and the blood pH about 8.4. To maintain the phenomenally steep gradient of H^+ and HCO_3^- (over ten times) and of other ions across the exchange surfaces, especially the gills, requires extremely powerful regulating mechanisms. In addition, there must presumably be some means of neutralising the highly alkaline water taken up through the gut. Relevant in this connection are some experiments on the alkaline saline water mosquito larva Aedes natronius. When immersed in alkaline water containing a suitable coloured indicator the water is taken up into the foregut of the larva where its pH is rapidly reduced from 10·5 to about 8·0 (Beadle, unpublished). The nature of the acid involved remains to be discovered, but the pH of the blood is kept constant when the larva is transferred from alkaline to neutral saline water in which it can also live in the laboratory. Similar changes might be observed in the gut of S. a. grahami.

About 30 and 160 km south of Lake Magadi in the Eastern Rift are two other alkaline salt lakes, Natron and Manyara, each of which has the endemic species Sarotherodon alcalicus. In Lake Natron S. alcalicus are found in shallow creeks around the lake and are not especially associated with inflowing springs. They were collected by Coe (1969) from water of 30–40°/oo salinity. Their breeding behaviour, which is often very complicated and specific in cichlid fishes, was found to be similar to that of S. a. grahami.

In Lake Manyara there is another closely related species, S. amphimelas. This has been recorded from water of salinity as high as 58°/oo (Makerere Expedition Report, 1961). It has now been found in other small lakes in the same region – namely Eyasi, Kitangiri and Singida (pers. comm. E. Trewavas).

Natron and Magadi are at approximately the same altitude (600 m) and lake sediments have been found at about 12 m above the level of both these lakes (Fig. 3.4). The Magadi deposits have been dated at about 15·000 year B.P. Absence of faulting or tilting of these beds suggests that the area has been relatively stable since that time. It is therefore likely that Magadi and Natron were then joined in one large lake whose water was probably fresh. The Magadi beds contain fossils of a Sarotherodon sp. that is larger than, but apparently showing affinities with, S. a. grahami. This species presumably lived in the previous large lake and may have been the ancestor of both S. a. grahami and S. alcalicus. Whatever the cause of the reduction in size, the very shallow water of the present habitats of these two species could never be colonised by anything but very small fish (see Coe, 1966, 1969, for references).

The higher altitude of Lake Manyara (about 950 m) and the present scarcity of geological evidence prevent us from speculating on a former connection with the other two lakes. But we would suspect that all three species have a recent common ancestor which lived in fresher water and was able to develop a hypotonic regulating mechanism and a means of dealing with the extremely high alkalinity during the subsequent few thousand years of increasing desiccation (Hotdship, 1976).

There are other tilapias that are living in water of between 1 and $10^\circ/_{oo}$ salinity (and occasionally higher). Since they are some of the most important of African fish as human food and have been introduced into India and the Far East this resistance is a matter of some practical significance especially in arid regions. *S. shiranus chilwae*, a subspecies endemic to Lake Chilwa in Malawi, is a fish whose economic importance depends partly on its ability to withstand relatively high salinity and alkalinity. *Sarotherodon niloticus*, whose several varieties are distributed over the whole soudanian region and up into the Jordan valley, is often found, and can be experimentally cultivated, in water too saline for most purely freshwater fish.* The most adaptable of the large commercially valuable species seems to be *S. mossambicus*. It is widespread in the warmer parts of south-eastern Africa south of the Zambezi (see p. 149). It inhabits coastal brackish waters and inland waters both fresh and saline due to evaporation during the dry season. It has been experimentally acclimatised to seawater of $35^\circ/_{oo}$ salinity (Jubb, 1967). As expected it is a hypotonic regulator in seawater (Beadle, 1969, for references). Maetz and de Renzis (1978) have recently investigated its regulatory mechanisms.

Hotsprings are included in this chapter because they are often very saline. They are also of great interest as a demonstration of the adaptability of living organisms. In volcanic areas such as the East African Rift Valleys water may come from heated regions at great depth and emerge at a high temperature, even to near boiling point (Fig. 19.4). Where the volcanic rocks contain much soluble salt, the water is also saline. There are a number of micro-organisms – bacteria, fungi and algae (especially blue-greens) – that live at the constant and very high temperatures of hotsprings. The hottest springs known are in Yellowstone Park, U.S.A. There it has been shown that some bacteria grow at 85–88°C and that the upper limit for photosynthesis of the blue-green algae is about 73°C. It is remarkable that the optimum temperature for photosynthesis by a given sample of algae was the temperature of the water at the point in the outflowing stream from which it was taken. This applies even to those taken from 70°C and demonstrates in a striking manner both the stability of the temperature gradient and the adaptive effects of natural selection (Brock, 1967; Kahan, 1969). The 'Maji Moto' hotsprings flowing into Lake Manyara, Tanzania, issue at 74°C (Makerere Expedition Report, 1961; Njogu and Kinoti, 1971). Sulphur bacteria and blue-green algae were found living in pools at this temperature among clumps of the sedge *Cyperus laevigatus*.

There are no aquatic animals that are known to tolerate temperatures as high as those to which some micro-organisms are adapted, except as resistant eggs or desiccated later stages. But several fish and invertebrates have been found living at temperatures high enough to require special adaptations. Thus, in the Manyara

---

* A subspecies of *Sarotherodon niloticus* has recently been found in water from alkaline hot springs flowing into the Suguta River south of Lake Turkana in Northern Kenya (Fig. 3.4). The salinity of the water in which the fish were found was recorded as $25^\circ/_{oo}$ and the temperature up to 41°C (pers. comm. M. J. Coe). Further discoveries of this kind are likely to be made in the more remote and arid parts of the Soudanian region.

Fig. 19.4 Hot Spring near Lake Baringo, Kenya, surrounded by a gelatinous mass of blue-green algae

hotsprings, mosquito larvae (*Culex tenagius*) were collected from water at 42°C and some copepod crustacea (unindentified) at 44°C.* It has already been mentioned that the fish *Sarotherodon alcalicus grahami* can live premanently at a maximum temperature of 39°C in the Magadi hotsprings. For short periods they can tolerate up to 41°C. Striking evidence of this was given by a sharp line across the stream bed showing the upper limit to which the fish browse on the algae. At this point the temperature of the outflow had dropped to 41°C. A few dead but apparently undamaged fish floating downstream had presumably spend too long in the danger zone between 36° and 41°C (Coe, 1966).

Another tropical African example, that has been investigated, is the Ikogosi hotspring in western Nigeria (Rogers *et al.*, 1969). The temperature at the source is 38°C and the water is fresh with a composition similar to that of the ground waters of the region (conductivity K_{20} 98). The pH rises from 6·0 to 7·5 within fifty metres of the source, presumably due to release of pressure and discharge of dissolved carbon dioxide. The warm water supports abundant populations of bacteria and fungi as well as diatoms. The zooplankton is restricted to copepod crustacea (harpacticids and cyclopids) and rotifers.

The most interesting aspect of hotspring biology is the existence of organisms

---

* The temperature of small pools of water exposed to the sun, e.g. in the hoofmarks of mammals, may rise to great heights in the afternoon. Larvae of *Anopheles gambiae* were found at 39°C in such pools near Lake Manyara (Njogu and Kinoti, 1971).

at temperatures that cause the denaturation and coagulation of the proteins of those that live at more normal temperatures. The means by which these fatal changes are avoided and the enzyme systems are adapted to function at such high temperatures is a matter of great interest as an example of the biochemical adaptation of living matter to extreme conditions. The many hotsprings in tropical Africa, and especially in the Rift Valleys, provide good opportunities for the study, by observation and experiment, of the adaptations of organisms to extremes of high temperature and, in many cases, also of salinity and alkalinity.

Complete desiccation is the fate of many inland waters in all parts of the world. Whether or not a particular body of water dries up depends both upon the rate of evaporation, determined by temperature, humidity and winds, and upon the length of time during which evaporation exceeds inflow. It is obvious that the shallower the water the soon will it be exhausted. The frequency of drying up varies greatly. Small puddles, as in the hoofmarks of cattle, may appear and disappear several times in the year even in the humid tropics. Such small puddles, so long as they last seven to ten days, can provide breeding places for *Anopheles gambiae* and thus a focus for malaria. Large ponds and extensive shallow temporary swamps, such as border the Gambia River and on the Kano plains off the northeast corner of Lake Victoria, fluctuate seasonally between flooding and desiccation. In the more arid regions the dry season is longer and in the central Sahara there are basins in which surface water appears only rarely, in some cases only once in ten or twelve years. At the other extreme, climatic fluctuations may be such that the normal condition is with water, and desiccation occurs only at rare intervals from five to ten years in the case of Lake Chilwa (p. 354) to hundreds or even thousands of years in some basins, as suggested by the geological evidence.

It is therefore not surprising that most inland water animals and plants are adapted in one way or another to survive complete desiccation. There are several methods by which this is done.

1. Animals capable of living and moving out of water, e.g. some insects whose adults are aquatic such as water beetles and bugs, can abandon the site and find, if they are lucky, some more permanent water.
2. Some, such as certain molluscs, fish and amphibia, are able to take refuge in the dry mud by reducing their metabolism to a minimum and by sealing themselves in a covering which retards loss of water by evaporation. This is known as 'aestivation'.
3. Survival in a state of dehydration. In nature they are air-dried and not completely dehydrated, but many such organisms have survived long periods of extreme artificial dehydration (p. 352). The spores of bacteria and algae, plant seeds and the drought-resistant eggs of freshwater invertebrates, *Hydra*, Platyhelminth worms, and Crustacea Branchipoda, Ostracoda, and Copepoda are well-known examples. Less familiar are examples of survival as dehydrated embryos, larvae and even adults (e.g. some rotifers, tardigrades, nematodes and insect larvae). In these metabolism is virtually stopped and they are temporarily 'dead'. This condition is known as 'cryptobiosis'. A number of animals (e.g. some of the ostracod and copepod Crustacea, and of the oligochaet and turbellarian worms) survive dry periods in encysted post-egg stages in the

mud, but it does not seem to be known if any of these are in a desiccated 'cryptobiotic' condition. It is likely that most aestivate in cysts that prevent evaporation. Some examples of aestivation and cryptobiosis will now be discussed.

The seasonal dessication of the eggs of *Triops granarius*, a crustacean from temporary pools in the Sudan desert, was studied by Carlisle (1968). The dormant eggs are in diapause (partially developed) and are often subject to a temperature of up to 80°C in the dried sediment exposed to the sun. They can be heated to 99°C (but not higher) for at least sixteen hours without harm. It seems that exposure to a high temperature is in fact necessary for continued development and that a second generation cannot develop from eggs that have not been desiccated.

Both aestivation and cryptobiosis are practised by some mosquitoes in response to complete desiccation of the larval habitat. Some, such as the genus *Aedes*, and particularly the yellow fever mosquito *Aedes aegypti*, lay dehydrated drought-resistant eggs on the damp mud as the water disappears. Most of the Anophelines, however, are dependent upon the survival of the impregnated female during the dry period. The female *Anopheles gambiae*, the most important of the African malaria vectors, has been observed in the Sudan to aestivate in a dormant condition in sheltered crevices during the long, dry season. Both activity and the development of the ovaries are retarded by low humidity (Omer and Cloudesley-Thompson, 1968).

A well-known African aestivating animal is the lungfish *Protopterus*. As the last of the water is disappearing it burrows into the soft mud and secretes round itself a cocoon of mucus, waterproofed with a layer of lipoprotein. In this it lies motionless and folded on itself, head and tail uppermost, and breathes air through a small aperture at the top of the cocoon. The position of the burrow is betrayed by a flat cap of dry mud which is thin and porous enough for exchange of gas with the atmosphere. In these particulars its behaviour is identical with that of its near relative *Lepidosiren paradoxa* in the temporary swamps of Central South America. Fishing by digging is practised in both continents.

Aestivation is not, however, a physiological necessity for the lungfish as is the drying of certain seeds and of the resistant eggs of some aquatic invertebrate animals that cannot develop without a preliminary desiccation. It is practised only when circumstances make it necessary for survival. In the swamps connected with the Gambia River, which are regularly dry for three to five months every year, most of the *Protopterus annectens* aestivate during the dry season and are rarely found in the river at any time (Johnels and Svensson, 1954). But in the lake regions of East and Central Africa *P. aethiopicus* inhabits the open waters of the lakes as well as the permanent lake-fringing and valley swamps which continue to provide for them a normal habitat during the dry season. Only in certain places are there extensive flat stretches of country subject to annual flooding and drying in which *P. aethiopicus* is trapped and obliged to aestivate. Such are the Kano plains off the northeast corner of Lake Victoria, and the region of temporary swamps north of Lake Kioga. In contrast the basin of Lakes Edward and George, in which *P. aethiopicus* is particularly abundant, provides extensive permanent swamps but relatively little seasonally flooded country and, as far as is known, the lungfish do not aestivate in that region. In certain parts of the Kano plains, where the water table remains close to the surface, Wasawo (1959) discovered that *P. aethiopicus* survives the dry season by burrowing into the ground and excavating a small chamber in which the water collects. By resting in this subterranean 'pond' it can survive without secreting a cocoon to prevent evaporation.

The physiology of aestivation in *P. aethiopicus* was studied by Homer Smith and entertainingly discussed together with other related matters in his two books (1956, 1961). The rate of oxygen consumption is progressively reduced to about 10% of that of the active fish, the blood circulation is much retarded until the heartbeat falls to about three per minute and some water is lost by evaporation. Urine production ceases and nitrogenous waste, previously partly ammonia which is toxic, becomes entirely urea which accumulates in large quantities in the blood and tissues. Lomholt *et al.* (1975) observed that the cocoon extends as a respiratory tube into the pharynx which prevents the normal swallowing of air practiced by the active fish and by the Amphibia. Ventilation of the lung during aestivation is effected by expansion and contraction of the lung itself as in the higher vertebrates. This state of dormancy can be maintained in the laboratory for several years, but it is not directly due to the lowered metabolism, loss of water or to the accumulation of waste products, but rather, it seems, to an active inhibition which is removed within minutes by the appropriate stimulus. The latter is apparently normally provided by asphyxiation when the fish is immersed in water in which it is unable to breathe except by rising to the surface. A rapid reawakening is certainly necessary when the ground is flooded at the beginning of the rains, because it cannot survive in water by gill-breathing alone and anyhow the water and waterlogged soil may soon be deoxygenated. The rapidity of the revival suggests that the control mechanism involve hormones released into the blood by the appropriate stimulus. This phenomenon, which has wider physiological implications, remains to be investigated in detail. There are several small species of cyprinodont fish in the African and American tropics unique in their ability to survive in the egg for long periods of complete drought. It has been shown that during embryonic life there are three stages at which development may be arrested (diapause) and at these times the eggs can survive drying out of the mud in which they were laid (Myers, 1952; Peters, 1963; Wourms, 1963). The genus *Nothobranchius* is widespread in tropical Africa, and *N. taeniopygus* (Fig. B3g) has been recommended as an antimalarial fish since it feeds on mosquito larvae and is naturally adapted to temporary waters (Vanderplank 1941). The early embryology of 'annual fish' has been studied by Wourms (1964, 1972). The egg is presumably surrounded by a membrane (chorion) that retards evaporation of water, but water and gas exchange and other aspects of the metabolism of the embryo during aestivation have not yet been studied. It would be of great interest to discover the mechanisms by which embryonic diapause is induced and broken. This phenomenon is otherwise unknown among vertebrate animals unless the arrested development of the embryos of certain mammals, whose implantation in the uterine wall is delayed, can be regarded as a related phenomenon.

The tailless (anuran) Amphibia, the frogs and toads, are mainly aquatic and, at first sight, ill-adapted to anything but a very damp environment. The skins of most species are, in fact, quite ineffective in preventing loss of water by evaporation in a dry air, and standing water is essential for the eggs and larvae of nearly all species. Nevertheless, the Anura, but not the purely aquatic Urodela (newts, etc.), are one of the most successful groups of vertebrates in the wide range of conditions to which they have become adapted (Poynton, 1964; Bentley, 1966; Mayhew, 1969; Warburg 1972). They are common in all parts of the world except the polar regions where the temperature is continuously too low. Their overall distribution shows little signs of having been hindered by regions with very dry climates, and there are even a few species to be found in deserts. Such species owe their capacity for adaptation to periodic and prolonged drought conditions to several features:

1. Much water can be lost, often exceeding 50% of body weight, without harm. The tissues can remain healthy with the blood concentration greatly increased, even doubled. These are much wider limits than those tolerated by other vertebrates, especially mammals.
2. Burrowing into the ground in dry periods for which the limbs are often adapted. The Californian desert frog *Scaphiopus couchii*, like the lungfish, secretes a mucus 'cocoon' around itself in the aestivation burrow, which presumably retards evaporation.
3. The end product of nitrogenous metabolism in the more aquatic species is mainly ammonia, which is toxic but diffuses away rapidly into the water. In those adapted to drier conditions urea accumulates in high concentration in the body fluids during aestivation without harming the tissues.
4. Water is stored in the bladder and lymph spaces and used when required. Only the anuran Amphibia (frogs and toads) and the Chelonia (tortoises and turtles) can actually reabsorb water from the bladder (Hicks, 1975).
5. Dehydrated frogs can take up water very rapidly through the skin even from damp soil. This and the uptake of water from the bladder is controlled by a hormone from the posterior pituitary (Ewer, 1952).
6. In view of the often very transitory water for breeding, development and larval life may be much curtailed or accelerated. Several species of the genus *Breviceps* found in southern tropical Africa not only burrow to escape desiccation, but even lay their eggs underground providing them with enough yolk for complete development. They can thus dispense with larval life and are quite independent of standing water, though a period of high humidity is needed for activity and feeding.

It has recently been found by Loveridge (1970) that the African frog *Chiromantis xerampelina* aestivates in rather dry places and stores its waste nitrogen in the bladder as a suspension of solid uric acid, previously thought to be an end product confined among the vertebrates to the reptiles and birds. Moreover, it loses water to a dry atmosphere no more rapidly than a chameleon. In these respects it has become physiologically a reptile.

Frogs and toads, though generally so dependent upon water, are thus in some instances surprisingly capable of surviving periods of extreme drought. The biology and physiology of the many species found in the drier regions of tropical Africa provide an interesting field for further research.

Among invertebrates an interesting example of aestivation, which has been studied by Visser (1965) and by Coles (1969a), is the large ampullariid snail *Pila ovata* which is common in most inland freshwaters of tropical Africa. As already mentioned, its mantle cavity is adapted for both water and air breathing. Like the lungfish it is not confined to stagnant or temporary waters and is found in large lakes, even for example off the rocky shores of the islands in Lake Victoria, where neither air-breathing nor aestivation is required for survival. But in temporary swamps and ponds, as the last water disappears in the dry season, it burrows into the mud, closes its operculum and remains dormant until the water returns. The rate of oxygen consumption falls during aestivation and is ultimately maintained at about one-sixth that of the active snail. Specimens kept on the shelf for more than a year may revive within half an hour when put into water. The same effect is produced by covering the operculum with water, which ultimately causes the animal to emerge. As with the lungfish, the main cause of the termination of aestivation seems to be asphyxiation. Exposure to nitrogen stimulates opening of the operculum and increases the oxygen uptake on return to air. Oxygen consumption, however, is also immediately increased by as much as sevenfold by merely handling the dormant snail. It can be imagined that this mechanical trigger might be operated by the first heavy storm of rain and thus facilitate a more rapid response to the subsequent flooding. The mud, when waterlogged, rapidly becomes anoxic and it is urgent for the snails to emerge into the open water without delay. These sudden changes in metabolic rate again suggest a hormonal control mechanism. A substance has been extracted from the liver which stimulates the respiration of muscle slices (Coles, 1969a). This may prove to be one of the links in the control mechanism which adjusts the rate of metabolism between the active and dormant conditions.

It has recently been found that some of the snail vectors of bilharzia can survive several months of desiccation. In some situations this may be epidemiologically important. *Schistosoma haematobium* (urinary bilharzia) in tropical Africa is mainly carried by snails of the genus *Bulinus* and is commonly endemic in regions of temporary standing waters which disappear in the dry season. It has been shown not only that the snails can aestivate, but also that the schistosomes within them can survive and are infective when the rains return. There is also evidence that under certain conditions *Schistosoma mansoni* (intestinal bilharzia) can survive some months of drought within the snail host *Biomphalaria* spp. (discussion and references in Jordan and Webbe, 1969, pp. 30–33).

A lesser known, but in some respects even more remarkable, aestivating tropical African mollusc is the bivalve *Aspatharia*. Throughout the dry season it lies buried in the dry mud with the valves closed. In the flat country bordering the southeast shore of Lake Malawi there is a species of *Aspatharia* which aestivates under about 5 cm of dry mud at the level of the last flood water in the beds of streams that flow into the lake during the wet season. It can be kept indoors in a dry and dormant state for more than two years and will revive within half an hour when immersed in water (pers. comm. M. Kalk). Nothing is yet known of its respiratory physiology, but it is surprising that it can survive at all under these conditions. As a bivalve (lamellibranch) mollusc, it has no special means for air breathing, and the mantle cavity is sealed from the air during aestivation by a plug of hardened mucus in the exhalent siphon and remains full of water which, it must be supposed, would rapidly become anoxic. Though the shell-valves fit very closely together, which is surely necessary to prevent leakage of water from the mantle cavity, the edges of the mantle are not actually fused (pers. comm. G. Fryer). It is however unlikely that a significant amount of oxygen could penetrate by this route – a matter that could easily be settled by experiment. In spite of the very high temperature of the dry mud exposed to the sun the rate of oxygen consumption is presumably very low indeed, perhaps zero, and a mainly anaerobic metabolism may be in operation. It is known, however, that some striking changes take place in the kidney during aestivation (Cockson, 1971). A number of concretions appear that are composed largely of calcium carbonate and phosphate. The former may be a sink for respiratory carbon dioxide to be released when the water returns. But, since no uric acid was detected, the fate of the nitrogen is unknown and the nature of the metabolic adaptations needed for survival for so long in a closed box remains to be discovered. It is likely that, as with *Pila*, a metabolic inhibitor is involved.

The ability of some organisms to survive in a state of desiccated cryptobiosis is one of the most remarkable feats of living matter. Though known for a very long time, especially in seeds and spores, neither the nature of the process nor the reason why only some organisms possess this faculty is fully understood (Grossowicz *et al.*, 1961; Keilin, 1959; Hinton, 1968). Attempts to measure respiration and other life processes have shown that the rate of metabolism can, at least in some cases, be reduced to zero. When dehydrated in an artificial water-free atmosphere they can survive much greater extremes of heat and cold than in the normal state. The eggs of the brine shrimp *Artemia salina* can survive up to one and a half hours at 103°C (Hinton, 1954), and cryptobiotic stages of bacteria, algae and other organisms have survived after exposure to a temperature approaching absolute zero (Becquerel, 1950). The largest and most complicated animals known to survive dehydration are certain insect larvae especially of the chironomid flies. The larva of the chironomid *Polypedilum vanderplanki* has been found in shallow rock pools in Nigeria, Uganda and Malawi that are alternately flooded and dried by the powerful sun, leaving a few centimetres of dry mud (Hinton, 1960a, b). They must certainly have a much wider distribution. After several years of dehydration in the laboratory, they come to life and continue a normal life history less than an hour after return to water. In the dry condition they can survive temperatures down to −270 and up to +102°C. They are probably naturally exposed to temperatures at least up to 70°C in the black, dry mud. They are resistant to repeated wetting and drying to which they are exposed in their habitat.

Cryptobiosis has in recent years come to claim some attention both in relation to the artificial preservation of living materials and to the possibilities of extraterrestrial life, but its most obvious significance relates to the present distribution of aquatic organisms on this earth. There has been much speculation, unsupported by direct evidence, concerning the manner in which small aquatic organisms move from one isolated body of water or dried basin to another – for example by winds and on or in the bodies of birds and mammals. Observations and experiments which demonstrate that transport by these means actually occurs have been performed and discussed by Evans (1958–59) and Maguire (1963). It seems that certain freshwater algae that are widely distributed, such as *Asterionel-*

la formosa, may be transported in a viable condition only in damp mud (Jaworski and Lund, 1970). But we are far from understanding the means of dispersal of many freshwater organisms whose distribution can only be explained by the assumption of extra-aquatic transport.

The ecology of very short-lived bodies of water in deserts has been discussed by Rzóska (1961) on the basis of observation on pools near Khartoum. A most remarkable feature is the rate at which some species can develop and reach maturity to produce another generation of drought-resistant forms. Adult *Moina dubia* (Crustacea, Cladocera) and *Metacyclops minutus* (Crustacea, Copepoda) were found within two days of the appearance of a pool and resistant eggs were produced within four to five days (water temperature 31–35°C). Many of these pools are dry before the end of two weeks, but the water becomes progressively more saline and may be unfavourable for some species long before the water finally disappears. A very short life history is essential for life in such ephemeral habitats. Some observations on previously dry pools in the Paraguayan Chaco, South America, during the dry season showed that fully developed copepod Crustacea appeared within six hours of the first fall of rain (Beadle, 1927, unpublished). These could only have come from animals aestivating in the mud in an advanced stage of development, and perhaps in a cryptobiotic condition.

Certain animals are confined to temporary waters. Their eggs will not develop unless they are desiccated. Notable examples are found among the euphyllopod Crustacea of the orders Anostraca (e.g. *Artemia*), Conchostraca (e.g. *Estheria*) and Notostraca (e.g. *Triops*). Morris (1971) observed that the dry aestivating eggs of *Artemia salina*, which contain embryos in the gastrula stage, are fully hydrated and are respiring within an hour of immersion in water. There is, however, no further morphological development for another twelve to eighteen hours. During this period, but not later, the eggs can be repeatedly dehydrated and rehydrated without damaging the embryos. This faculty might well have survival value in avoiding the otherwise disastrous effects of small showers of rain during dry seasons, and it would be interesting to examine the aestivating eggs of other freshwater invertebrates with this point in mind. A survey in which the crustacean faunas of Saharan waters of varying permanence are compared, is to be found in Gauthier (1938). The work of Meijering (1970) on the freshwater Cladocera in pools on an island off the Dutch coast is relevant to this discussion. The salinity of these fluctuates greatly due to periodic flooding by seawater which destroys the adult Cladocera. The resistant eggs, however, survive indefinitely in saline water and hatch as soon as the salinity falls to a suitable level. This must also be a feature of the Cladocera in those inland waters in which salinity is greatly increased by evaporation. The resistant eggs can thus serve as a protection against drought, high salinity or high and low temperature.

It should be mentioned at this point that high salinity and desiccation are not the only hazards that some freshwater animals can survive in diapause enclosed in resistant cysts. Such resting stages of planktonic animals produced in response to low temperature and reduced light are commonly found in sediments below permanent freshwaters in temperate climates. They contain embryos in arrested development which emerge with rising temperature and light in spring (Stross, 1966). Little is known of this phenomenon in the tropics, but an interesting case has been examined by Moghraby (1977) in the lower Blue Nile. The rich planktonic fauna (copepods, cladocerans and rotifers) in the clear water of the low level river practically disappears when the big seasonal flood comes down. When the flood subsides and the water clears the zooplankton is rapidly regenerated. If silt from the bottom of the fast flowing river is settled in clear water in the laboratory and exposed to light at a temperature of 20–30°C several species of the above groups emerge – e.g. *Daphnia lumholzi* (Cladocera), *Mesocyclops leuckarti* (Copepoda) and *Asplanekna brightwelli* (Rotifera). These can also be raised from dry mud left in pools during the dry season. Conversely the adults were caused to produced resistant eggs by dispersing silt in samples of previously clear water. It seems then that a variety of unfavourable conditions can induce the production of resistant eggs containing diapausing embryos.

Since inland waters are by nature transitory and fluctuating in composition, the evolution and distribution of the world's freshwater faunas and floras have depended much upon the ability to adapt to these difficult features of the environment. The matters that have been discussed in this chapter are therefore of fundamental importance to limnology and in some respects are of direct practical value. This can be illustrated by the case of Lake Chilwa in Malawi which will be introduced at this point as a finale to the chapter.

Lake Chilwa

Lake Chilwa in Malawi is an outstanding example of a lake in a closed basin that is subject to large fluctuations in level and salinity. In historic times longer-term changes of much greater amplitude have been superimposed upon rather regular seasonal fluctuations. It supports an important fishery (Mzumara, 1967) endangered at intervals by drastic contractions of the water, and has been the object of recent investigations by the University and Fisheries Department of Malawi (Kalk, 1969, 1970).* It lies in a tectonic depression southeast of Lake Malawi (Fig. 19.5) and covers a relatively small area even at its recent maximum in 1962 (about 700 km² of open water).

Contrary to previous opinion (Dixey, 1938) present geological and biological evidence is against a former connection with Lake Malawi from which it is now hydrologically quite separate (Kirk, 1967). It is generally agreed that before the Pleistocene when the rifting began which ultimately formed the Shire Valley there was a river flowing northwards across the region of the present Chilwa basin draining into the Indian Ocean via the Rovuma River. In support of this deep boring has disclosed some fluvial sediments below the lacustrine deposits on the floor of the basin. Unwarping of the Shire Highlands associated with the rifting caused a relative sinking of the Chilwa-Chiuta region of the valley and in the basin so formed the water collected and overflowed as before to the north (Fig. 19.5). The origin of the lake is therefore somewhat similar to that of Lake Victoria (p. 250).

Though the origin of the thirty species of fish recorded from the whole basin is not yet decided with certainty, it is not surprising, in view of the above geological

---

* A comprehensive account of Lake Chilwa and its basin and of its geological, biological and human problems appears in a recent book (M. Kalk (ed.), 1979).

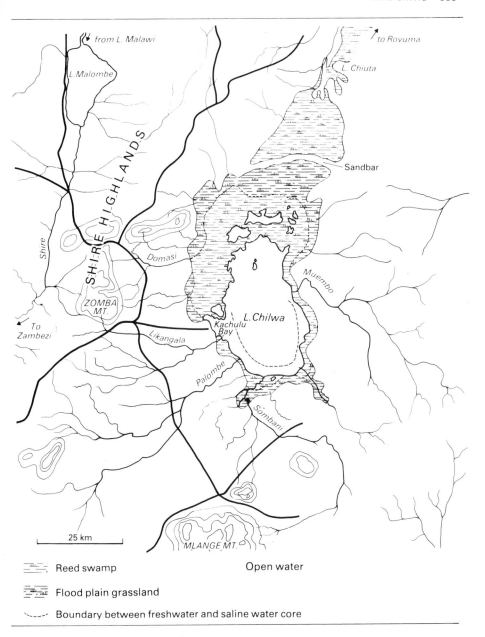

Reed swamp

Open water

Flood plain grassland

Boundary between freshwater and saline water core

Fig. 19.5 Lake Chilwa (redrawn from Kalk, 1970)

history, that they have closer affinities with those of the Rovuma and other east-flowing rivers than with those of Lake Malawi (Trewavas, 1966a; Kirk, 1967). One fish appears at first sight to have been derived as a subspecies, *chilwae*, from *Sarotherodon shiranus shiranus* which is endemic to the Lake Malawi basin; but this may not in fact be so since, according to Trewavas (1966b), the *S. s chilwae* in Lake Chiuta shows closer affinities with *S. rovumae* in the Rovuma River than with the subspecies in Lake Malawi. The taxonomy of this group of fish will presumably have to be revised.

Ancient beaches thirty, fifteen, nine and five metres above the highest recorded water level testify to some very large fluctuations in prehistoric times. It was at the thirty and fifteen metres levels that Chilwa and Chiuta must have been joined as one lake draining into the Rovuma and the Indian Ocean. The two lakes are now separated by a wind-blown sand bar. During the dry season water seeps back from Chiuta through the sandbar onto the flood plains north of Lake Chilwa (Fig. 19.5). The main inflows are from the Zomba Plateau in the west and Mlanje Mountain in the south. The seasonal changes in rainfall over the catchment are such that in most years there is a difference in water level of up to one metre between wet and dry seasons. Evaporation exceeds rainfall for about nine months in the year. Superimposed upon these annual changes are less frequent (every six to eight years) but wider fluctuations of rainfall with the result that the area and depth of the lake are on occasions drastically reduced (Figs. 19.7 and 19.8). The salinity and alkalinity of the water are then raised to a level that cannot be tolerated by most of the fish, and the fishery has to be abandoned (Morgan and Kalk, 1970).

At high level the water spreads outwards to form flooded marshes with a vegetation similar to that of seasonally flooded flat land around the East African Great Lakes. The swamps, which are most extensive at the north end of the lake (Fig. 19.5) are dominated by the bulrush *Typha domingensis* with thickets of ambatch *Aeschonomyne pfundii* (Fig. 19.6). Papyrus is remarkably scarce and is

Fig. 19.6 Lake Chilwa, flooded margin, showing the ambatch shrub, *Aeschonomyne pfundii*, the bulrush, *Typha domingensis* and water lily *Nymphaea* sp. (photograph by M. Kalk)

limited to the mouths of the Rivers Likangala and Domasi, where the water is not subject to great increases in salinity nor to severe desiccation. It is likely that either of these would prevent papyrus establishing itself (p. 312). Howard-Williams (Chs 7 and 13 in Kalk, 1979) showed by experiment that *Typha* is more resistant to raised salinity than the other macrophytes in the lake. He also studied other aspects of the ecology of the macrophytes especially the disappearance and recolonisation during a period of drought and reflooding. Conditions required for germination of the seeds are of course as important as those for maintenance of the growing plants and the dried saline mud during a recession is a very unfavourable soil for the germination of most plants. Another point of ecological importance is that *Typha* does not form a dense floating 'mat' of roots and decaying vegetable matter characteristic of papyrus. Movements of water between lake and fringing swamps are therefore more rapid and nutrients are transported more freely. The level of oxygen in *Typha* swamps is consequently generally higher than in dense papyrus and they thus provide a more favourable refuge for fish when the level of the lake falls.

During the past twenty years there have been three six-year cycles involving low levels of 1·5 to 1·7 m below datum. At still longer intervals between 1897 and 1968 there have apparently been several drops in level that exceeded 2 m and persisted for up to four years. The fishery consequently suffered very severely. In 'good' years, when the level is high, the Lake Chilwa fishery has provided about 50% of the total fish from all the waters in the country. In 1965 the lake provided nearly 10 000 metric tons of fish. In 1968, the catches had dropped to about 100 tons. The cumulative effect of this series of dry years on the level of the lake and the subsequent dramatic recovery with an increase of rainfall are shown in Fig. 19.7. In 1969 the total catch had already exceeded 3 000 tons (Eccles in Kalk, 1970). The social disruption caused by these violent fluctuations in the major local industry can well be imagined.*

Of the thirty species of fish recorded from the whole basin only three are common in the open lake and are the basis of the fishery – *Sarotherodon shiranus chilwae* (endemic subspecies), *Barbus paludinosus* and *Clarias mossambicus*. Their feeding habits were investigated by Bourn (1974). They migrate into the swamps when conditions in the lake become intolerable as in 1966–67. The peripheral less saline water in the swamps and river beds provide a refuge in times of recession and thus a reservoir of fish for natural restocking of the lake when the water rises again.

Though concerted investigations on the chemistry and biology of the lake were only started in 1966, it happened that late in 1968 the level reached the lowest recorded since 1920 and the fishery was brought to a standstill. It has therefore been possible to get some idea of the range of conditions to which the organisms in this lake may be subjected and the changes in the fauna and flora resulting from them. In 1967 *Barbus paludinosus* left the lake when the conductivity of the water had reached about 4 000 (see Table 19.3), *S. s. chilwae* at about 5 000, but *Clarias*, though reduced in numbers when the conductivity reached 4 000, did not entirely disappear until it had exceeded 10 000. During the first year of in-

---

* The history of the tilapia fishery is reviewed by Morgan (1970).

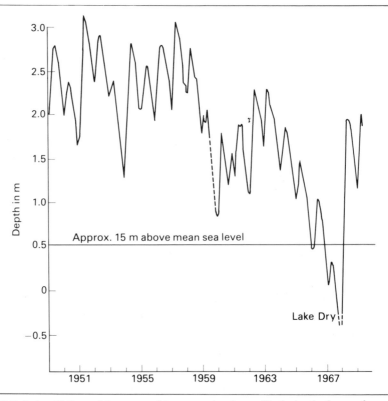

Fig. 19.7 Lake Chilwa. Mean monthly water levels at Kachulu (redrawn from Kalk, 1970)

creasing water level (1969) the fish populations were recovering much more rapidly than expected. About 3 000 tons of *Clarias* were caught in the lake and flooded marshes during that year. The *Sarotherodon* and *Barbus* were also increasing very rapidly, though net fishing was legally restricted until they had reached breeding size (M. Kalk, pers. comm.). This certainly suggests a large reserve of fish taking refuge during the drought in swamps and scattered pools around the basin.

Lake Chilwa also suffers from occasional mass mortalities of fish following abnormally violent winds around the turn of the year. As with the same phenomenon in Lake George, it is the air-breathing *Clarias* that escapes annihilation, and we must assume, until direct evidence is found, that death is due to asphyxiation following deoxygenation of the water when the highly reducing and oxygen-consuming mud is suddenly stirred to the surface.

At the present time the accumulation of salt in the lake is due mainly to evaporation of the inflowing freshwater and direct rainfall. The two small saline springs make a negligible contribution. We do not know whether this has always been so since its isolation from Lake Chiuta. The ancient intrusive rocks of the region are rich in minerals and may have contributed much of the salts in the lake.

The composition of the water resembles that of the alkaline lakes of the East

Fig. 19.8 Lake Chilwa. Low level, October 1967

African Rifts, the predominant ions being Na^+, HCO_3^-, CO_3^{2-} and Cl^-. The period, 1966–1970, during which observations have been made, began with extreme drought conditions with very low water level and high salinity. In 1969–70 there was a return to the previous higher level seen before the series of disastrously dry years beginning in 1964 (Fig. 19.7). The changes in composition of the water during 1966–70 can therefore be taken to illustrate the amplitude of the fluctuations to which organisms may be subjected in this lake. The approximate limits with respect to some factors of likely biological importance are quoted in Table 19.3.

The lowest concentration of ions was reached in the peripheral surface water two months after flooding. At the other extreme after prolonged drought there was a sprinkling of salt crystals on a black and slimy salt-saturated mud, and a small area of very shallow and saline water in the centre of the basin. During these few years conditions in the lake had ranged from those typical of a freshwater to the last stages in the evaporation of an alkaline saline lake. As already mentioned the latter extreme condition is reached only at intervals of many years. Betweenwhiles the seasonal fluctuations, though well marked, do not reduce the water to quite such a biologically disastrous condition. The figures in Table 19.3 do not refer to uniform and simultaneous changes throughout the whole lake, but indicate the range of conditions to be found within the lake during the period under consideration.

There can be no doubt that changes in salinity and chemical composition of the water accompanying the fluctuations in level are the major cause of the changes in the composition of the fauna and flora. But there are indications that other factors may also have an influence. This is suggested by the study by McLachlan

Table 19.3 Approximate maxima and minima of some characteristics of the open water of Lake Chilwa during the period 1966–70 (from Morgan and Kalk, 1970)

Temperature °C	21–37
Conductivity (K_{20})	276–16 720
Salinity (from K_{20})	0·3–16·7‰
pH	8·4–10·8
Na^+ meq/l	1·6–142
Ca^{++} meq/l	0·24–0·95
Mg^{++} meq/l	0·04–0·74
Cl^- meq/l	0·74–88
$HCO_3^+ + CO_3^=$ (alkalinity meq/l)	1·9–88

and McLachlan (1969) on the distribution of the benthic fauna during the wet season of 1968–69. The previous wet season had already begun to raise the level and decrease the salinity of the water. In the early stages of the 1968–69 wet season the diluted peripheral water had a conductivity of less than 500 (salinity c. 0·5‰) and supported a benthos composed mainly of chironomid larvae, with some larvae of dragonflies (Anisoptera) and mayflies (Ephemeroptera) as well as pulmonate snails and leeches. On the other hand, the central more saline region of the lake had a conductivity of over 2 000 (salinity c. 2‰) and was devoid of benthic animals. After a few weeks the whole lake was more or less uniformly mixed by winds with a conductivity of about 1 000 (salinity c. 1‰). At this time the entire insect benthic fauna of the lake, although not of the peripheral swamps, was apparently obliterated. This cannot easily be explained solely, if at all, as the result of overstepping a salinity or alkalinity limit. Lake Turkana, whose water has a similar composition and alkalinity, has a conductivity of over 3 000 and yet supports a rich and typical freshwater fauna, and that of the surface water of Lake Edward approaches 1 000. Both ecological and experimental work suggest that a salinity of about 5‰ (conductivity c. 5 000) is the upper limit for most freshwater animals (p. 72). It may be that in this case the increase of salinity in the peripheral water induced by the wind was too rapid for acclimatisation which might have been accomplished with a more gradual change. But there are other possible factors in the situation. Wind-stirring of the mud in the very shallow water might well release toxic substances and could cause some reduction in oxygen if not complete anoxia at the bottom for long enough to asphyxiate the benthic fauna. At such times the water becomes extremely turbid and the suspended silt may make the water impossible for these animals. These situations should be investigated in greater detail because very shallow waters in the tropics can be both very productive and particularly prone to catastrophic mortality when violently stirred by wind storms.

On the other hand, after a series of dry years such as preceded the very low level in 1967–68 increasing salinity was certainly a major factor in changing the composition of the fauna and flora and finally destroying most of them. In 1968 the central shallow remnants of the lake exceeded 16 000 in conductivity (c. 16‰ salinity). It appeared that, at the end, no other organisms remained but blue-green algae (Cyanophyceae) especially *Arthrospira platensis*. But, even then, there must have been other unfavourable factors at work since some other very alkaline waters of salinity up to 45°/$_{oo}$ support a richer fauna and flora (p. 339). High temperature in very shallow water must be mentioned as a possibility. Maximum temperatures of 40°C were recorded. The progressive changes in the water and sediments during the period of refilling (1968–70) were studied by McLachlan *et al.* (1972) and McLachlan (Chs 4 and 9 in Kalk, 1979).

Biologically disastrous fluctuations in level and chemical composition of the water, with occasional desiccation, are thus not confined to peculiar and small patches of water that are of scientific but of little direct economic interest. Lake Chilwa is an example of such a water that has recently been investigated and in which a significant proportion of a country's natural resources is involved. Lake Chad is another.

20

Manmade lakes

The most ambitious projects in applied hydrology and limnology are the large artificial lakes made by damming voluminous rivers. Though this is an ancient practice, it has been growing steadily in scale of operation and some of the manmade lakes that have appeared during the past twenty-five years are comparable in size with the largest natural lakes. For a long time past in the more arid parts of Africa, as in other continents, small dams have been thrown across river beds to store water for irrigation and for watering cattle during the long, dry season. Agriculture and stock-raising in the Republic of South Africa, where the climate is generally arid for long periods and there are no large natural lakes, depend very much upon reservoirs collected in this manner. The limnology of several of these has been investigated, e.g. Hartbeespoort Dam (Allanson and Gieskes, 1961). The recent great increase in size has followed from the demands of an expanding population for domestic and irrigation water and from the need for rapid economic development requiring a major source of power in the absence of an adequate supply of fossil fuels. Modern developments in hydroelectric, excavation and constructional engineering have made possible operations on a scale not previously contemplated.

The main purpose of these large schemes has generally been hydroelectricity, though in the case of the Nasser Dam on the Nile the regulation of water for the irrigation of Egypt is of even greater importance. The smaller seasonal reservoirs at Jebel Aulyia (White Nile) and Sennar (Blue Nile) were constructed solely for storage of water for dry season irrigation. There are also a numbber of small artificial lakes in tropical Africa that were constructed primarily for hydroelectricity; for example, there are three in the Upper Katanga Province at the southeast corner of the Zaïre basin which supply power to the copper mining region. These are the Lakes Mwandingusha and Koni on the River Lufira, and N'zilo on the Lualaba. Limnological work on these lakes by Magis (1962) has been mentioned (p. 90). Around 1950, at the time of the original planning and early stages of construction of the first of the very large African dams at Kariba on the Zambezi, neither the possible other uses nor all of the consequences of such a scheme were fully appreciated.

Like other specialists hydroelectric engineers have been apt to misunderstand or underestimate the importance of those aspects of a scheme that do not directly affect the main objective (megawatts) and are laudably motivated by technical enthusiasm as well as profitability. This exclusive attitude has not always been modified by local authorities who promote the projects. The fact is that applied ecology is more difficult to understand and to practice successfully than engineering science because, having a biological component, it is more complicated and the relation between cause and effect is less easy to establish. The ecological consequences of such large-scale interferences with the environment can be immediate or more often long delayed and irreversible. It should be added that ecologists have until recently paid most attention to the new lake itself and its immediate surroundings. The ecological effects on the river and country below the dam may be very important for agriculture, fisheries, wildlife and public health over a wide area even as far as the sea coast (Scudder, 1972).

The sudden appearance of an enormous lake in a region such as the Middle Zambezi valley which had previously been devoid of permanent surface water apart from the original river, is likely to have consequences and raise problems not previously encountered. There is the difficult social problem of the displacement and resettlement of the many thousands of people living in the region to be inundated. A positive benefit is a potentially prosperous fishery in the new large volume of calmer water in which nutrient-rich sediments settle and a great variety of habitats are provided for the fauna and flora. But the rapid development and maintenance of a highly productive ecosystem requires careful management in order to channel the production into species of fish of the greatest commercial value. During the first few years the production of fish usually rises to a maximum and then falls to a low level depending upon the intensity of fishing. The reasons for this are discussed in what follows (see also Petr, 1975). But exploitation of the fishery requires fishermen who are accustomed to the methods used in lakes. Such men are likely to be scarce in the area, and people who were previously agriculturalists may have to be recruited and trained.* Another benefit, that will doubtless be enjoyed by an increasing number of people in the future, is the recreational opportunities for fishing, sailing, swimming, etc.

New problems of public health will certainly arise (Waddy, 1975). In particular, the newly flooded land will inevitably provide more extensive habitats for the aquatic vectors of certain important diseases of man and domestic animals, such as bilharzia and liver-fluke (snails) and malaria and filariasis (mosquitoes). In the ultimately likely but not as yet imminent event of large-scale urban development at some points on the shore there will arise the unpleasant problem of water pollution due to domestic and industrial effluents. This is already beginning to claim some attention in the old-established and rapidly growing towns on some of the large natural African lakes and in the artificial reservoir Lake McIlwaine, near Salisbury, Zimbabwe (p. 390).

Even during the construction period, which may last for four or five years, some difficulties arise. Since the site is often remote from a centre with good

* For a discussion of the resettlement problems associated with the African manmade lakes see Scudder (1966), Chambers (1970) and Dufalla (1975).

communications a new heavy-duty road must be built and temporary housing and other facilities provided for the technical staff and labour force which, not counting their dependants, may number more than a thousand. The administrative, socal and medical problems involved in such an operation can well be imagined, and the effects on the region, perhaps one of special ecological interest and economic value, may be very harmful.

Another initial operation requiring special care, though mismanaged in at least one case, is the regulation of the rate of filling of the lake. The country below the dam is usually dependent in varying degree on the water from the river for irrigation of crops, stock-raising, fisheries and domestic use, and an adequate continuous outflow must be maintained. For this reason filling of most of the large artificial lakes has been spread over two to four years.

In most cases the land to be flooded has contained at least some ancient remains of historical or archaeological interest. This is now generally recognised, and surveys are normally made to record sites and to study and map those of particular interest before they are for ever consigned to the bottom of the new lake. The biggest, most expensive, and most publicised operation in archaeological salvage was the excavation, removal and re-erection of the Temple of Abu Simbel with its four colossal figures of King Rameses II (c. 1200 B.C.), which would have been submerged when the waters of the Nile rose as Lake Nasser behind the Aswan High Dam.

To the biologist the most interesting general aspect of manmade lakes is the transformation from riverine to lacustrine conditions. Some of the changes involved in the evolution of Lake Kariba could hardly, under the circumstances, have been foreseen. Such were the initial enormous growth of floating vegetation and the complete deoxygenation of much of the bottom water owing to the flooding of the terrestrial vegetation and organic matter in the soil. One of the objects of the biological investigations has been to provide a basis for predicting the probable course of events in future schemes of this kind. When it is realised that we are apparently witnessing, on a much accelerated timescale, the transformation of a river into a lake that marked the evolution of the Rift Valley lakes, these large-scale 'ecological experiments' are seen to have a much wider scientific interest. However, in order that the investigations should produce information of the maximum value, it is clearly very important that a thorough preliminary survey on the original river system should be made before the engineering works are started. There has been a tendency to allow too little time for a satisfactory initial survey.

A peculiarity of the large artificial lakes in tropical Africa is the annual drop in level (draw-down) purposely caused by the opening of the sluice-gates in order to make room for the incoming flood water in the rainy season. The drop amounts to several metres and as a result there is a seasonal alternation between inundation and desiccation of considerable areas of land. The ecological effects of these fluctuations are very important.

The most unexpected consequence, at least to a biologist, of a very large and deep new lake is the disturbance in the local tectonic equilibrium due to the accumulation of a great weight of water especially in a region of tectonic instability. Between 1961 and 1968 the epicentres of several earthquakes were located near the deep end of Lake Kariba (Fairhead and Girdler, 1969). Some of these quakes were estimated to be violent enough to have caused great destruction in a town, had one existed on that spot.

There have been several symposia devoted to manmade lakes – in London (Lowe-McConnell, 1966), in Ghana (Obeng, 1969), S.C.O.P.E. Report (1972) and in Knoxville, U.S.A. (Ackermann *et al.* 1973).

A considerable amount of limnological work has been done and published on Lakes Kariba and Volta. Preliminary surveys of the Kainji section of the River Niger prior to the closure of the dam have been published, but fewer details of the development of the lake are as yet available in print. General reports on Lake Nasser have been published, but so far there are no detailed publications from which the situation can be properly assessed. Consequently, of the four large artificial lakes, Kariba and Volta can be the most satisfactorily discussed. The much smaller Lake McIlwaine in Zimbabwe has been introduced as an example of artificial pollution, and it has been studied from that point of view.

Though there can be no clear-cut distinction between a manmade lake and a seasonal reservoir, an extreme case of the latter would be an impoundment, with a dam so widely opened each year that the original riverine conditions are temporarily restored. The Jebel Aulyia Reservoir (Table 20.1) on the White Nile above Khartoum has been chosen as an example of such an annual transformation, that has been investigated by biologists.

It has not been possible to keep in touch with all the latest hydrological and ecological developments in the large artificial lakes discussed in this chapter, nor to obtain enough worthwhile information concerning the more recent projects of this kind, e.g. Kafue in Zambia, Kunene in Angola, and others. It is however unlikely that any major problems will arise with subsequent projects that have not already been encountered and I hope that the following account of the first six of these lakes and of a few smaller reservoirs in Tropical Africa will provide a useful introduction to the subject. On the other hand experience has now shown that every new lake has its own unexpected peculiarities. Its ecological development and the potentialities for the future fishery cannot be predicted beforehand except in broad outline, though frequent monitoring during the early stages can indicate the final outcome.

Lake Kariba (Zimbabwe and Zambia)

Owing to the steep gradients and surrounding topography, the hydroelectric potential of the Zambezi below the Victoria Falls, like that of the lower Zaïre River, is colossal even by world standards. Several schemes have been considered and two have been completed – Kariba in 1958 and Cabora Bassa in 1974. A point of interest not encountered with other examples discussed in this chapter is the effect of a large artificial lake on the hydrology and ecology both of the river and of another similar lake subsequently made below it.

On closure of the Kariba dam in December 1958, the valley was rapidly flooded, and within a year the depth of water just behind the dam was about 60 m. By mid 1963 (four and a half years after closure) the depth at the same point reached its maximum of about 120 m. The lake fills the entire middle section

Table 20.1 Approximate data relating to some of the manmade lakes and seasonal reservoirs in tropical Africa

Lakes	Mean latitude	Altitude (m)	Maximum area (km²)	Maximum depth (m)	Ratio ann. inflow: volume	Maximum draw-down (m)	Date closure
Kariba							
(River Zambezi)	17°N	530	4 300	125	1:4	14	1958
Volta							
(River Volta)	7°N	92	8 800	80	1:4	3	1964
Nasser-Nubia)							
(River Nile)	23°N	183	6 000	80–90	1:2	20	1964
Kainji							
(River Niger)	10°N	155	1 250	60	4:1	10	1968
Kafue Gorge							
(River Kafue)	16°S	1 000	3 100	58	2:1	7	1971
Cabora Bassa							
(River Zambezi)	20°S	200	2 700	151	1:1	36	1974
Nyumba ya							
Mungu	4°S	670	180	41	1:1	3·5*	1965
Seasonal Nile Reservoirs							
Jebel Aulyia							
(White Nile)	15°N	377	600	12	8:1	6	1937
Sennar							
(Blue Nile)	13°N	422	140	16	50:1	16	1925
Roseires							
(Blue Nile)	12°N	480	290	50	16:1	13	1966

* The seasonal fluctuation in level. There is no artificial draw-down.

of the Zambezi valley, is about 300 km long, of mean width about 30 km and mean depth of about 30 m (Fig. 9.4).

Ecological research has been based on stations on both sides of the lake – in Zimbabwe the Nuffield Lake Kariba Research Station at Sinamwenda and the Fisheries Research Institute at Kariba, and in Zambia the Central Fisheries Research Institute at Chilanga.

Though the rainfall on the lake itself is low (400–800 mm annually) it is high in the upper Zambezi catchment area (800–1 600 mm). The annual inflow is therefore considerable and amounts to about one quarter of the volume of the lake. This ratio can be contrasted on the one hand with that of the large East African lakes such as Lake Victoria, approximately 1/100, and Lake Tanganyika, at most 1/1 500, and on the other to the small Nile reservoirs in which the annual inflow greatly exceeds the volume. The fact that the annual inflow/volume of Lake Chad (0·5–1) is actually higher than Kariba emphasises once more the peculiar hydrology of Lake Chad (Ch. 12).

The wide seasonal fluctuations of temperature associated with the relatively high latitude (mean 17°S), as with Lake Chad (13°N), are important in relation to productivity. The 24-hour mean air temperature at Kariba in winter (June–July 1965) was 17°C and nearly 10° higher in summer (28°C November–December 1965). The water temperature is also influenced, at least at the inflow end, by the relatively cool Zambezi water, but the entire range of lake water temperature is from 18° to 31°C. The temperature of the surface water fluctuates seasonally between about 20° and 30°C (Coche, 1968, 1974). This may be contrasted with the 2°–3°C seasonal change in the surface of Lakes Victoria and Tanganyika. It happens that the heaviest and most persistent winds are the southeast trade winds that blow during the cold season between April and August and so combine with the drop in temperature to cause the annual overturn.

Hydrological measurements were begun late in 1959 by Harding (1964, 1966). After a gap of several months, in 1964, they were resumed by Coche (1968). The first measurements near the dam in November 1959 showed a marked stratification with a thermocline at about ten metres, below which the water was devoid of oxygen and contained free H₂S (Fig. 20.1). During the following six months the thermocline descended gradually to about thirty metres in May. After this, with the lowering of air temperature and the onset of the southeast winds, the surface cooled rather rapidly and the whole column became homothermic, and a complete overturn in July brought oxygen to the bottom and dispersed the H₂S. With the calming of the winds and the rise in temperature in August stratification was re-established. This cycle, involving a period of maximum stability between December and April and an overturn in July or August, has been repeated in subsequent years.

The surprisingly rapid deoxygenation and sulphuration of such a large proportion of the deep water during the first year caused some anxiety for the fisheries and for the quality of the water to be released into the lower river through the turbines which draw water from below the thermocline. It was due to the sudden flooding of land which, by aqueous decomposition, brought into solution great

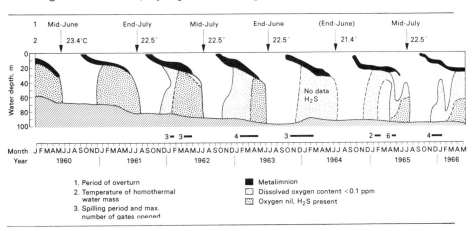

Fig. 20.1 Lake Kariba, deep water station. Progressive increase in depth. The initial deoxygenation and sulphuration of the water below 15 m during the warm and calm first half of the year was dispersed by the lower temperature and heavy winds during July–September. The extent and duration of deoxygenation decreased during the six years of observation (redrawn from Coche, 1968)

quantities of organic matter as well as soluble inorganic salts from the soil and rock and from wood ash left from the preliminary bush clearing in the valley. It was therefore of great interest and practical importance to find that, though the cycle has been repeated in subsequent years, both the extent and the duration of the deoxygenation have progressively declined (Fig. 20.1). These conditions are now confined for two to three months to the bottom of the deepest region and to a few small depressions in other parts of the lake bottom (Coche, 1968). The excess nutrients have presumably been 'fixed' in living organisms and in unde-composed organic matter preserved in the mud and some must have been lost in the outflow. It was therefore to be expected that an initial increase in concentra-tion of dissolved salts in the surface water was followed by a progressive decrease from 1958 to 1964 (Harding, 1966). This seasonal hydrological regime summa-rised in Fig. 20.1 was initially similar to that of a typical highly 'eutrophic' lake in a temperate climate, but thereafter it gradually became less productive. These hydrological changes were reflected in a general increase in productivity during the first few years, followed, if we regard the fisheries statistics as an index of general productivity, by a decline in 1963 towards an equilibrium at a lower level (see Coulter, 1967b, for a general summary of the events to that time). There must surely be considerable water movements associated with internal waves caused by the southeast trade winds, which have an important influence on pro-ductivity (Ch. 6), but they have not yet been investigated. Surface seiches have now been recorded. For more hydrological details and a discussion on fish pro-ductivity see Coche (1974).

During the first year there was an explosive multiplication of aquatic organ-isms: planktonic, floating and submerged plants, fish and invertebrate animals. This was to be expected over newly flooded land at a tropical temperature. But species differed in the rate at which their reproduction was accelerated (Balinsky and James, 1960; Boughey, 1963; Mitchell, 1969). The rapid increase of phyto-plankton soon became evident from the green patches in the surface water due to the blue-green alga *Microcystis*. The most dramatic and publicised event was the appearance and spread of vast floating sheets of the water-fern *Salvinia molesta* and, to a lesser extent, of the water cabbage *Pistia stratiotes* and the duckweed *Lemna*. By 1962 more than 20% of the lake surface was covered. Apart from the hindrance to navigation and fishing, there was a considerable reduction in dis-solved oxygen below the floating mats due both to shading and consequent reduc-tion in photosynthesis and to protection from wind stirring. Continued acceler-ated growth of these plants would certainly have created a serious situation, but from 1963 onwards they steadily declined and the floating mats of *Salvinia* are now much reduced and confined mainly to sheltered bays and the inlets of streams, especially in places where there was no preliminary bush clearing and the floating vegetation is caught up and anchored by partly submerged trees. There are also floating sheets of 'sudd' vegetation composed of emergent plants such as the sedge *Scirpus stoloniferus* and several grasses and other flowering plants, all rooted in mats of *Salvinia*.

It was not until 1964 that the submerged rooted vegetation (water weeds) be-gan to spread in the shallow waters, but now there are large areas round the shores where the bottom is covered with dense forests of weeds such as *Ceratophyllum demersum* and *Potamogeton pusillus*. This provides a favourable habitat for aquatic insects, snails and oligochaet worms and is an important link in the production cycle (McLachlan, 1969a). The expansion of a habitat suitable for snails is likely to increase the incidence of bilharzia (schistosomiasis) in nearby human population. A survey by Hira (1969) in the Siavonga area on the Zambian shore showed that a considerable number of the snails *Bulinus africanus* and *Biomphalaria pfeifferi* were infected with *Schistosoma haematobium* (urinary) and *S. mansoni* (intestinal bilharzia) respectively.

It is very difficult to decide what are the most important external sources of nutrients upon which the productivity of the lake must depend (Mitchell, 1973). Coche (1968) has published analyses of the common inorganic ions, nitrate and phosphate in the water from different parts of the lake and from the inflowing rivers and streams at different seasons during 1965. Since, however, the volume of water coming from the various inflows has not been measured, and must vary from year to year, it is not possible to assess quantitatively the relative contribution made by each to the productivity of the lake.

In spite of the enormous catchment area (Fig. 9.4) the water of the upper Zambezi has a rather low salinity (K_{20} 40–85). This may be due to absorption in the very extensive swamps in which the main tributaries collect before passing on to the Victoria Falls and to Kariba. The volume of the Zambezi is, however, so much greater than that of the other inflows that its overall contribution must be relatively high. Next in importance is the Sanyati River which discharges into the final basin before the dam. Its water is, on average, rich in salts and nutrients, but the volume of the annual discharge is very small compared with that of the Zambezi. The other rivers have small catchments, flow only during the rains and probably contribute relatively little.

There is another source of fertility from the seasonal changes of water level that are characteristic of manmade reservoirs and are often artificially regulated ('draw-down'). Near the end of the dry season in October and November the outflow sluices at Kariba are opened to make space for the flood waters from the inflowing rivers during the coming rainy season. The resulting fall in level has varied from year to year from three to eight metres. Along stretches of gently shelving shoreline this causes a considerable annual fluctuation in the area of inundated land which is alternately flooded and dried out. These temporary shallows, as in other lake basins, are a source of nutrients and a favourable environment for the reproduction and growth of organisms. Some chemical effects of flooding and drying have been studied by S. M. McLachlan (1970). During the dry season, after desiccation of the aquatic vegetation, the dry land is soon covered with rapidly growing plants, especially grasses, on which wild mammals graze and deposit their dung. When the lake rises again and the land is reflooded, this organic matter is quickly decomposed releasing salts and other plant nutrients. There is an initial rise in salinity of the water which later falls as these substances are diluted with offshore water. Water weeds such as *Potamogeton pusillus* and *Ceratophyllum demersum* are soon re-established. Floating mats of *Salvinia* develop on a large scale only over uncleared areas where projecting dead trees prevent them from drifting. The metabolism of the fast multiplying animals and micro-organisms is soon evident in a rise in dissolved CO_2, a lowering of pH and a reduction in dissolved oxygen.

Mitchell (1973) suggested that the prolific growth of floating vegetation would have fixed much of the nutrients suddenly brought in by the initial flooding of the land, thus reducing the loss by the outlet. This would subsequently have been more slowly released by decomposition and recycled in the ecosystem. If this were so, the excessive growth of *Salvinia* could be regarded as a beneficial buffer to the original inrush of nutrients.

The artificial seasonal rise and fall in water level, though not originally intended as such, provides a kind of 'fallowing' to grass and herbivores, as is sometimes practised in pond fish culture. The ponds are regularly drained and sown with crops that provide a 'green manure' for the next flooding (Hickling, 1962, p. 116). In the case of Lake Kariba the mammals also are contributing to the fertility of the lake in return for good grazing by accelerating the decomposition of the land vegetation. How significant is this contribution relative to that made by the inflows cannot yet be estimated, but it might be very important. On the debit side nutrients are being continuously lost in the outflow, by deposition in the bottom mud and by fishing. The turbines draw water from below 20 m where nutrients are accumulated during summer stratification.

These temporary shallow waters also provide nurseries for young tilapias which find in them both a good food supply and a refuge from predators, particularly the tiger fish *Hydrocynus*. When the water recedes they have a more permanent refuge among the reeds in deeper water (Donelly, 1969).

Thirty-one species of fish were recorded in the middle Zambezi before the dam was built (Harding, 1966). The most important commercial species are *Sarotherodon mortimeri Tilapia rendalli* (Cichlidae), *Hydrocynus vittatus* (Characidae), *Labeo altivelis*, *L. congoro* (Cyprinidae), *Clarias mossambicus* (Clariidae) and *Synodontis zambezensis* (Mochokidae). *Sarotherodon macrochir* was later introduced because it was thought that *S. mortimeri* might be slow to adapt itself to the new conditions in the lake. In fact, the latter has flourished and far outnumbers the other cichlid species in the commercial catches.*

Nine species of fish, well known in the Upper Zambezi above the Victoria Falls but not recorded for the Middle Zambezi, have appeared in Lake Kariba. There have been suggestions to account for this (Bowmaker *et al.*, 1978). Some species may have arrived accidentally with the introduced *Sarotherodon macrochir*. It has even been suggested that some may occasionally have survived the severe bashing when swept over the falls but were not suited to the conditions in the Middle Zambezi until the lake appeared (Bell-Cross, pers. comm.). This is rather difficult to believe in view of the obvious barriers to fish presented by some other much lower and less precipitous falls in Africa. A route which some consider more likely is through the Victoria Falls turbine inlet channels (Marshall, 1979). It has in fact been reported that some of the introduced Tanganyika sardine, *Limnothrissa miodon*, have survived the passage through the turbines of the Kariba

---

* For a well illustrated book on the common fish in the lake see Balon (1974), and Bowmaker *et al.* (1978) for a review of the history of the fish populations of the original middle Zambezi since the lake was formed. For a general account of the Kariba project see Kenmuir (1978).

Dam (Kenmuir, 1975). In this way hydroelectric projects might have effects on the distribution of fish in some African river systems.

Conditions in the original river fluctuated annually between two extremes. In the dry season the water was restricted to a relatively small stream confined in a narrow bed in which the fish became overcrowded and the predators, especially *Hydrocynus*, flourished at the expense of the populations of other species. During the rains the river overflowed and flooded large areas of valley bottom producing temporary stretches of very fertile shallow water favourable for breeding and for feeding of the young fish, and safe from predators. Without this periodic over-flowing the fauna of the river would undoubtedly have been much poorer.

The rise in the fertility and the great increase in volume of the water during the first year of the lake's existence was followed by an initial increase in the produc-tion of fish. There was no longer a period of overcrowding in the dry season so that a higher proportion of the younger fish avoided consumption by predators, and, as already explained, the artificial rise and fall of the lake have continued to provide rich and secure seasonal feeding-grounds for them.

Nevertheless, as might have been predicted, the original high production has not been maintained, mainly because it was caused by the initial sudden decom-position of organic matter from the newly flooded land which was thereafter per-manently submerged.

The large-scale felling and burning of trees and bushes over much of the area to be flooded was done to remove obstructions to navigation and to the setting of gill-nets. The fishermen, however, have found that, in spite of the difficulties, fishing is more profitable among the submerged tree trunks that were not re-moved. Their surfaces become encrusted with algae and associated organisms – chironomid larvae, nymphs of Trichoptera and Ephemeroptera and oligochaet worms (McLachlan, 1970a, 1974). The rotting wood is penetrated by the wood-boring nymphs of the mayfly *Povilla adusta* which feed on the algae. The com-bined surface area of the submerged trees may be more than twice that of the bottom from which they project. This mixed animal and vegetable growth is an important food for several fish especially *Distichodus mossambicus* and *D. schenga* (Citharinidae) whose jaws and teeth are adapted to scraping food from a flat surface. McLachlan (1970b) showed that, when fresh branches are cut and submerged, chironomid larvae begin to settle within twenty-four hours and in eight days the encrusting fauna and flora are well established and increasing rapidly.

The development of sediments over newly flooded soil and of the associated fauna of chironomid and trichopteran larvae and oligochaet worms was studied over a two-year period by McLachlan and McLachlan (1971) in the 'estuary' of the seasonal River Mwenda on the southern shore of the lake. The type of sedi-ment was determined by deposition of organic matter and inorganic particles from the river and from coastal erosion. Deoxygenation of the water below 20 m during summer stratification precluded all benthic fauna; nevertheless, they re-colonised these deep sediments after the overturn. As might be expected, the biomass of animals in the sediments was correlated with the amount of organic matter. But in the peripheral shallow water the presence of coarse particles of sand derived from shore erosion made the sediments a less favourable environment. It is interesting to note how very rapidly, after flooding, these encrusting and benthic faunas establish themselves.

Those trees that are in shallow water and are exposed during the seasonal low level are attacked by termites and boring beetles, with the result that they rapidly disintegrate and will disappear long before those that are permanently submerged. In the warm season, when the upper limit of the deoxygenated water rises, the fauna and flora on the tree stumps below this limit are destroyed or forced to move and they are then confined to the well-oxygenated shallower water. When the trees have finally disintegrated this important habitat will no longer exist. There will be further discussion of this matter in the section on the Volta Lake.

The change from riverine to lacustrine conditions has, of course, produced some new kinds of habitat to which the Zambezi fishes are not properly adapted. The most notable of these is the deep pelagic region in which the large quantity of zooplankton was not at first fully exploited for food by any of the fish, and to that extent production of fish was not as high as it could be. In Lake Tanganyika the two species of 'sardine' (*Limnothrissa miodon* and *Stolothrissa tanganicae*) have evolved in adaptation to this niche and are themselves very important commercial fish. In 1965 it was therefore decided to introduce *Limnothrissa* into Lake Kariba, and they were transported by air in aerated tanks. During subsequent years a few of these fish were reported from time to time, but late in 1969 large shoals were seen, and netting by night with vapour lamps, as on Lake Tanganyika, was begun. A detailed account of the fate of the introduced 'sardine' was given by Woodward (1974). They had occupied the open water of most of the lake and had been followed into the deeper water by their main predator, the tiger fish *Hydrocyon vittatus*. For no obvious reason they do not however seem to have produced such dense populations as in Lake Tanganyika and by 1971 the fishery was not as productive as had been anticipated. According to Woodward the value of this introduction is the more doubtful in view of the fact that the zooplankton feeder *Alestes lateralis*, previously restricted in numbers, had a population explosion shortly before the introduction of *Limnothrissa* which successfully competed with them and perhaps prevented them from occupying this important niche more successfully. It is however unlikely that these relationships had yet reached an equilibrium and in fact according to Marshall (1979) the sardine catches have considerably increased and the prospects for the fishery in general are much brighter than appeared in the early 1970s.

Lake Cabora Bassa

Though not the largest manmade lake in Africa, Cabora Bassa was designed to give the greatest output of energy (nearly 4 000 megawatts) for use both in Moçambique and in South Africa. (Fig. 9.4).

Some ecological work was done under contract from the Portuguese Government on the river before closure of the dam in 1974 and during the subsequent year by scientists from the Universities of Lourenço Marques in Moçambique and of Rhodes in South Africa. But the research has been severely restricted by the political events leading to and following the independence of Moçambique.

Compared with Kariba we therefore know much less of the detail of its ecological development and only that during the first year.

As already mentioned, the Middle Zambezi before the dams were built was a fast flowing river that in the dry season was reduced to a stream flowing between rocks and sandbanks. In the flood season it overflowed and inundated large expanses of the surrounding land except, of course, in the Kariba and Cabora Bassa gorges. These big seasonal changes prevented the establishment of large aquatic plants except in sheltered backwaters. The ecological changes in Lake Kariba itself have been discussed, but the regulating action of the Kariba dam on the river below greatly reduced the seasonal swings between massive floods and a restricted watercourse. The resulting more constant level of the river allowed the root aquatic vegetation to establish itself along the shores so that, when Lake Cabora Bassa began to fill, plants such as the sedge *Phragmites mauritianus* and the grass *Panicum repens* were already well advanced and spread rapidly. Their future will depend in part on the rate and amplitude of the regulated seasonal draw-down through the dam (Jackson and Davies, 1976; Bond *et al.*, 1978).

In view of the trouble caused by floating aquatic plants (especially *Salvinia molesta*) during the early stages of 'Lake Kariba their progress in Cabora Bassa was followed with considerable interest. The initial situation was however different in that some of the tributary rivers in this section, notably the Hunyani, already harboured the now notorious water-hyacinth *Eichhornia crassipes* and the water fern *Azolla nilotica* in addition to *Salvinia molesta* and the Nile cabbage *Pistia stratiotes* which inhabit Lake Kariba (Davies *et al.*, 1975; Jackson and Davies, 1976). The already well established emergent plants and the projecting remains of submerged trees and bushes held up large masses of floating plants that otherwise would have drifted down the lakes. The danger of a blockage in the narrow 37 m gorge leading to the dam, though serious, is apparently lessened by the east–west orientation of the lake, since the prevailing southeast winds have a large enough easterly component to sweep up the gorge.

Aquatic vegetation in such a lake, if confined mainly to inshore regions, is an advantage to the fishery. There is a source of food in the plants themselves, in the algae and invertebrates that encrust their surfaces and in the small free-swimming invertebrates that live amongst them. They also provide an essential refuge from predators for the young fish. If however the aquatic vegetation runs riot and covers a large proportion of the open water, as at first in Lake Kariba, both navigation and the fisheries are seriously hindered. The initial increase in floating vegetation was not infact as great as had been expected. There remains the difficult and important problem of human schistosomiasis (bilharzia), the vector snails of which are encouraged by dense aquatic vegetation. The disease is common in the region and energetic measures will be needed to control it.

The salinity, shown by conductivity of the middle Zambezi water, which varies with the season (K_{20} 40–80), was increased in Lake Kariba, and the outflow conductivity was found to be 90–100 (Coche, 1974). Hall *et al.* (1976) recorded 100–120 between Kariba and the future Cabora Bassa dam site. They concluded that the more saline water from the Kafue River (K_{20} 135–350) must partly account for this, and that this section of the river is therefore richer in nutrients than the Kariba section. But the total range of conductivities quoted above is rather small and is well within that of 'normal' freshwaters (Table 5.1) in which the major ions (measured as conductivity) are, as far as we know, always in more than adequate supply for the needs of plants and animals. The above salinity increase may or may not be accompanied by an increase in the consumable nutrients (as defined on p. 41) and possibly in trace elements, both of which have no detectable effect on conductivity but will stimulate production if their concentration is raised above a previously limiting level.

Bond *et al.* (1978) published in more detail the vertical and horizontal distribution of temperature, pH, oxygen and conductivity, mostly in the deep eastern end of the lake, during the first year after closure of the dam. In the windy season (May–July), apart from diurnal fluctuations near the surface, the temperature and oxygen levels were almost uniform to the bottom. During the calm period at the end of the year thermoclines developed with complete deoxygenation and appearance of H_2S below 30–50 m depth. This however was not so extensive as during the early stages of Lake Kariba and it was thought that seasonal stagnation of lower water would not be repeated for so many years in Cabora Bassa in view of the greater stirring action of the southeast trade-winds and the much higher ratio (1:1) of annual inflow to volume of the lake (Kariba 1:4). Unlike in the Kariba basin, there was no cutting and burning of trees and bushes to provide a rapid supply of nutrients from the ashes. The general productivity is not therefore likely to rise to an early very high peak, but in the long run it is thought that it will maintain a higher rate of production than Kariba. When Cabora Bassa has settled to a more or less steady condition it may be possible to compare its productivity with that of Lake Kariba and to decide whether it is basically richer in its nutrient resources.

Jackson and Rogers (1976) have described the main features of the fish fauna of the river before the impoundment and the changes that took place during the first eight months after closure of the dam. The most obvious difference from the early stages in Lake Kariba was the very large and rapid increase in the number of young fish that had avoided the predators such as *Hydrocynus vittatus* in the shelter of the already established luxuriant aquatic vegetation which also provided plenty of food. Thus, during the first eight months, the fish population had increased very rapidly.

Thirty-eight species of fish were originally recorded in the river, exclusive of its tributaries. Many of these, such as *Barilius zambesensis* and the smaller species of *Barbus* are characteristic of fast flowing rivers and soon disappeared from the lake. The cichlids (e.g. *Tilapia rendalli* and *Sarotherodon mortimeri*) were not increasing during the first year. Some of the siluroids, characids and cyprinids were mainly responsible for the explosive population growth. It is likely that several riverine species previously absent will by now have descended from the tributaries, especially the Kafue and Hunyani, and have become established in the lakes. The relative abundance of the species and the general ecological situation will certainly have changed greatly since 1975, and we await with interest the publication of more recent work. It has been recommended that the Lake Tanganyika 'sardine', which is established in Lake Kariba, should be introduced.

The ecological future of Lake Cabora Bassa will depend partly on the timing and size of the periodic drawn-downs and reclosures. These should be arranged to provide newly flooded land during the period of growth of the young fish. For engineering purposes there was a draw-down of 10 m soon after the lake was filled. This stranded and destroyed about 50% of the *Eichhornia* (Bond and Roberts, 1978).

There is clearly a great need, as with other similar projects, for close co-operation between engineers and ecologists in the management of this lake. This was unfortunately lacking in the early stages of construction and operation, which were planned with little serious consideration of the ecological consequence for agriculture, stock-raising, domestic water supplies, fisheries and wild-life, which affect the middle Zambezi valley, the lower river valley and the lands around the estuary For the views of ecologists on the conduct of this project, see articles by Tinsley, Davies and others in *African Wildlife*, Vol. 29, No. 2, 1975.

The Volta Lake (Ghana)

The closing of the dam at Akosombo about 100 km from the mouth of the Volta in May 1964 created the largest of Africa's artificial lakes (Figs. 11.3 and 20.2). Over 300 km long, its final surface area of 8 500 km² exceeds that of all the natural lakes except Victoria, Tanganyika and Malawi. Its maximum depth is about 75 m near the south (dammed) end.

In shape it differs from Kariba, which lies in the open valley of a single major

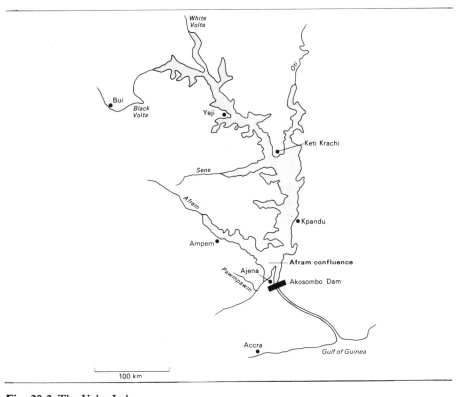

Fig. 20.2 The Volta Lake

river, the Zambezi, and has no very significant subsidiary inflows. A glance at the map of Volta Lake (Fig. 20.2) reveals at once that it has flooded a much branched river system. There is a striking resemblance in form, though on a much larger scale, to the volcanic barrier lakes of southwestern Uganda and Rwanda (Fig. 15.1). It can be said to provide a transition, or rather several transitions, from river to lake conditions reflected in a hydrological and ecological gradient from the inlet of each major river down the flooded side-arm into the main lake. As with Lake Kariba, the annual draw-down to compensate for the incoming flood water towards the end of the second rainy period in October and November exposes large areas of land especially in the savanna region around the northern half of the lake.

The limnological research has been done mainly under the auspices of the Volta Basin United Nations Research Project by the University of Ghana and by the Institute of Aquatic Biology of the Ghana Academy of Science. Summaries of the work during the first five years were published by Lawson et al. (1969), Entz (1969) and Biswas (1973). During the first two years biological investigations were confined to the gorge above the dam and to the more open waters of Ajena Bay just north of the gorge. Subsequently, it has been possible to extend the observations northwards and up the main arms.

With the example of Kariba in mind, it was with some anxiety and with much interest that biologists watched the early development of the lake and looked for signs of excessive accumulation of nutrients with a catastrophic deoxygenation and a growth of floating vegetation. For economic reasons there was no preliminary felling and burning of trees, except in two very small experimental areas, and in this important respect the initial conditions were different from those in Kariba.

In the event the growth of floating vegetation has been relatively restrained (Ewer, 1966; Lawson, 1967). During the first few months there was a considerable floating growth of the grass *Vossia cuspidata* and of the Nile cabbage *Pistia stratiotes* with unrooted masses of the weed *Ceratophyllum demersum* just under the surface and in shallow inshore waters. These growths could never have been called 'explosive' and have since been much reduced. The present extent of *Pistia* and *Ceratophyllum* is not greater than would be expected in a natural lake in this situation. They are mainly confined to the mouths of the inflowing rivers and to other sheltered situations along the shoreline, where they are of considerable ecological importance in harbouring a number of invertebrates that serve as food for some of the fish (Petr, 1968a). On the other hand, they have also provided a substrate and food for an increasing population of snails, especially *Bulinus truncatus rohlfsi*, one of the vectors of urinary bilharzia (*Schistosoma haematobium*). This disease has consequently spread and increased among the people living along the shores (Paperna, 1969, 1970; Inst. Aquat. Biol., 1969–70, p. 55, ff.). The situation has been aggravated by the immigration of fishermen from the sea coast who had abandoned their old fishing grounds when mechanised large-scale fishing was introduced, and found better opportunities on the lake. They brought their bilharzia with them from a region heavily infested with the disease. On the other hand the incidence of human sleeping sickness (trypanosomiasis) has at least temporarily been reduced by the flooding of sites (close to the edge of the original river) suitable for the breeding of the tsetse fly, *Glossina palpalis*.

It would be rash to suggest an explanation for this striking difference between the vegetational history of the two lakes. The abundance of soluble nutrient salts in the ashes from the initial large-scale burning of trees (not done in the Volta) may have given a stimulus to the growth of *Salvinia* in Kariba. It has also been suggested that *Salvinia*, unlike *Pistia stratiotes*, is an exotic plant introduced accidentally from South America and is thus freed from the competition, whatever that may be, that it would have faced in its own native habitat.

A general survey of the planktonic algae and main chemical features of the Volta River before the dam was closed was made by Biswas (1968). There was a seasonal increase of algae in the river, especially of diatoms, during the dry period from February to April.

Though the equatorial climatic conditions, especially temperature, are different from those of Kariba at the southern edge of the tropics, the hydrological development of the two lakes has followed a generally similar course. In the early stages after the closure observations on the deep water near the dam showed that a sudden increase in dissolved nutrients from the newly flooded organic matter provoked a great outburst of phytoplankton (mainly blue-green algae) and a consequent increase of oxygen in the surface water. The latter on occasions rose to as much as 300% (super-)saturation. After about four weeks the phytoplankton bloom was much reduced and the oxygen at the surface dropped to less than 20% saturation. Under the photosynthetic zone the oxygen concentration was still further reduced by organic decomposition and below about ten metres it was exhausted (Biswas, 1966a; Ewer, 1966). These conditions were obviously most unsuitable for fish and a very large number died during the first few weeks. The first stages of decomposition and deoxygenation were thus more severe than in Lake Kariba.

The depth to which oxygen penetrated at this time varied, as might be expected, from place to place and seemed to depend mainly upon the degree of exposure to winds. This was demonstrated on a transect made in February 1965 by Viner (1969) from the open water of the Afram confluence up into the more sheltered Afram arm of the lake (Fig. 20.2). At the upper end of the arm beyond Ampem the water was shallow enough to be stirred and oxygenated to the bottom.

Even during the first two years this stratification of the south end of the lake was partially dispersed with lowering of the oxygen boundary during the two periods of instability (August–September and January–February) discussed below. As the initial excess of nutrients was reduced by loss through the outflow and by fixation in organic matter deposited in the mud, so the depth of the oxygen boundary was progressively lowered. Eventually oxygen reached the bottom at all seasons in regions less than about twenty metres deep, and deoxygenation in the deeper water towards the south end occurred only during the calm seasons. Thus, by degrees, a regime became established involving a twice-yearly overturn in the southern half of the lake with a temporary penetration of oxygen to the bottom of the deep water. In the northern half there are indications of a single annual overturn (Fig. 20.3; Biswas, 1969). Conditions therefore became more favourable for fish.

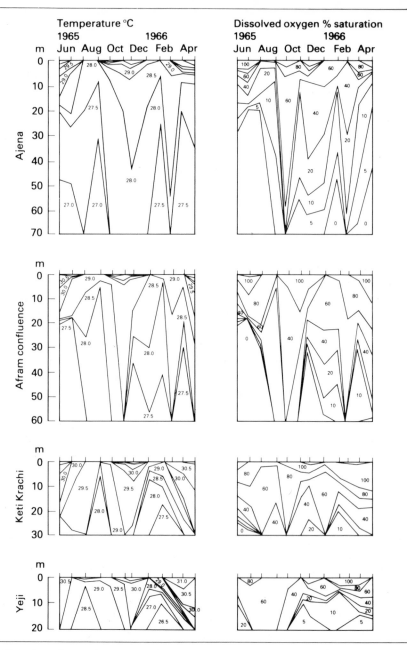

Fig. 20.3 Volta Lake. Seasonal changes in temperature and dissolved oxygen at four stations between Ajena in the south to Yeji in the north (see Fig. 20.2) (redrawn from Biswas, 1969)

This curious double seasonal regime, to which the lake has now settled, is due to the fact that it traverses three degrees of latitude (6·5–9·5°N) and crosses two vegetational zones – the forest in the south and the savanna-woodland in the north, which are subject to different climatic regimes (p. 37). The boundary between the two zones lies about halfway up the lake. The northern savanna region has a single annual seasonal cycle with a cool, rainy and windy period in August–October, which induces deep stirring of the water. For the rest of the year conditions are warm, dry and relatively calm. In the southern forest zone the regime is more complicated and involves two rainy periods, March–July and September–October. The lowest temperatures and the heaviest winds (from the sea) occur for a few weeks around August between the two rainy seasons when the skies remain cloudy. This is the main period of water mixing due both to lowering of air temperature and, more especially, to wind stirring. A second stirring tends to occur during the period of the much less violent wind (Harmattan from the north) in December and January when the temperature is high and the humidity is low. The work of Whyte (pers. comm.) on the nearby meteoric crater lake Bosumtwi, also in the forest zone, suggests that the observed lowering of the surface temperature and thus the instability of the water column at this time is due to evaporative cooling, and consequently an overturn can then be effected by a minimum of wind stress (p. 93). If, for any reason, the supply of nutrients to Lake Volta is reduced or increased, the degree and duration of deoxygenation in the deeper water is likely to increase or decrease accordingly. The present annual overturns and some deoxygenation near the mud surface during the calm seasons would seem to favour rapid decomposition of organic matter and the recycling of nutrients from the mud. The maintenance of the present apparently high productivity will in future depend upon the continued and adequate influx of nutrients from outside.

The fish fauna of the River Volta is rich and basically soudanian in character with some additional West African (Guinean) species (Daget, 1960b). Apart from the initial great destruction of fish in the lake caused by deoxygenation and possibly by some associated products of decomposition, most of the original river species appeared to flourish during the first year. Thereafter the character of the fauna began to change (Petr, 1968b, c; Evans and Vanderpuye, 1973; Lelek, 1973). The mormyrids (elephant snout fish), previously an important constituent of the river fishery, have quite disappeared from the lake except near the mouths of the inflowing rivers. Similarly, the characids are now represented in the lake by two (A. baremose and dentex) of the original many species of the genus Alestes. The reasons for these changes are not very clear. Apart from the destruction of habitual food and feeding grounds, the need for flowing water in which to breed is another possible influence.

On the other hand a number of species, formerly scarce in the river as judged from the fishermen's catches, have much increased in the lake. The three tilapia species of which are now the most important of the commercial fish are striking examples. In this case, the reasons are clear. The vegetable foods favoured by these three species – Sarotherodon galilaeus phytoplankton and attached algae, Tilapia zillii submerged grasses and detritus, Sarotherodon niloticus water weeds and algae – are all more abundant in the lake than in the former river.

Though the zooplankton has not been studied in detail, it has increased in the calmer and more favourable conditions in the lake, and is composed mainly of rotifers with relatively few crustacea. The one predominently zooplankton-feeding fish, *Heterotis niloticus* (Osteoglossidae) congregates in shallow water where the zooplankton is most abundant, and has noticeably multiplied.

During the period 1964–68, Petr (1970a, 1971) studied the chironomids (Diptera) whose larvae are the dominant members of the fauna in the bottom mud. The distribution and abundance of the various species seems to be determined by the physical and nutritional properties of the sediments, the amount of wave acttion and the oxygen level or associated chemical conditions, in the lower water. The initial great reduction in oxygen during the first year after the closure of the dam restricted them to the peripheral water where the depth is not more than about three metres. The subsequent general oxygenation, described above, has allowed them to colonise the sediments down to about twenty-five metres. The progressive changes that have occurred since 1964 in water level, oxygenation and in the establishment of a seasonal regime, have resulted in associated changes in the numbers, species composition and distribution of these larvae. They are now most abundant near the mouths of the inflowing rivers presumably because of the deposition in the sediments of particulate organic matter brought down by the rivers. If this is the explanation, we have here a demonstration of an apparently important contribution to the lake's resources of organic matter synthesised elsewhere.

An especially interesting feature of the lake is the appearance of a new, or at least very much extended, habitat provided by the submerged trees and other woody vegetation in shallow water. This has given a surface for attachment of algae among which there live a variety of small animals – crustacea, oligochaet worms and chironomid larvae – and the whole encrusting growth is eaten by several species of fish. In addition, the soft, rotting wood provides a suitable material into which the nymphs of the mayfly *Povilla adusta* bore. The tunnel so made is open at both ends with a membranous lining secreted by the nymph. Water is drawn through the tunnel by the movement of the gills and edible particles (algae and detritus) are filtered off by hairs on some of the limbs (Hartland-Rowe, 1958). The nymphs also emerge from their tunnels at night and feed on the organisms encrusting the surface of the wood. When mature they swim to the water surface and the adults emerge into the air. *P. adusta* is well known in all the tropical African lakes in situations suitable for the nymphs; e.g. rotting wood, decaying stems of papyrus, etc. But in the Volta Lake a particularly large amount of suitable material has suddenly appeared in the form of submerged trees, and the number of *Povilla* have greatly increased (Petr, 1970b).

The nymphs of *Povilla* are devoured by many species of fish. In the north of the lake the main predators are *Schilbe mystus* and *Eutropius niloticus* (Schilbeidae). In the south *Povilla* as well as other insects and zooplankton are consumed particularly by *Pellonula afzelini*, a clupeid closely related to the 'sardines' of Lake Tanganyika. This fish, about seven centimetres in maximum length, was previously rather rare in the river, but is now very abundant and, like its relatives in Lake Tanganyika, might become an important item of human food. But, together with *Povilla*, it is a link between the abundant algae and other organic matter encrusting the tree surfaces and free-floating in the water and the larger predatory fish *Lates niloticus* (Nile perch), *Alestes baremose* and the cichlid *Hemichromis fasciatus*. According to Petr (1970b) the tree-encrusting organisms occupy a very important position in the ecosystem, and the main intermediary between them and the fish is *Povilla adusta*. There is the interesting possibility that the present high production of fish from the lake owes much to the presence of submerged trees. The test for this thesis will emerge when the trees have ultimately rotted away. If, as a result, there is a marked decline in productivity, an imaginable remedy would provide quite a novel use for timber. There are signs, however, that the emergent and floating aquatic grass *Vossia* and sedge *Scirpus*, which are spreading in shallow sheltered water, may take the place of dead trees as a habitat for *Povilla* and the numerous encrusting organisms (pers. comm. from S. A. Whyte).

The workings of a hydroelectric dam involve the regulation of the outflow from the turbines and over the sluices so that the previous alternating annual regime of heavy flooding and relatively slight flow of the river below the dam becomes less pronounced and free from extreme fluctuations. It is to be expected that this change will have profound effects on the ecology of the lower river. The closure of the dam has in fact caused a dramatic increase in the growth of the submerged aquatic weeds *Potamogeton octandrus* and *Vallisneria aethiopica* (Hall and Pople, 1968). This is thought to result from preventing the annual heavy floods which had previously scoured out much of the weeds that had established themselves during the calm low water seasons. The expansion of weed has considerably affected the fauna of the river. Herbivorous fish such as the tilapias, which depend upon weeds for shelter and food, have multiplied. The snail *Neritina* sp. which feeds on *Vallisneria* has also increased. The populations of several other invertebrates must surely have been favoured.

Of particular interest is the bivalve mollusc or 'clam' *Egeria radiata* found in the lower reaches of most of the West African rivers where it is an important source of food (Purchon, 1963). The adults live in the submerged sandbanks, feeding on particulate organic matter taken in with the water inhaled through their siphons. The clam fishery on the lower Volta is conducted by women who collect them and, for the most part, sun-dry them. They also 'farm' them by collecting and planting young clams in sandbanks above their normal range, where they grow to maturity. In 1963 this industry was providing an annual cash income of over £100 000 (Ghanaan) and was employing 1 000–2 000 women (R. M. Lawson, 1963).

The life history of *Egeria* is peculiar.* Though the feeding and growing adults need only clean sandbanks free of vegetation, spawning and the survival of the motile veliger larvae require an increase of salinity to about $1^{\circ}/_{oo}$. The clams in the lower part of their range are exposed to this slight rise in salinity during the dry season when the junction between sea and fresh water moves further up the river. The veliger larvae, but not the later stages, apparently require a slightly higher salinity to maintain their salt-balance and thus to survive, though $1^{\circ}/_{oo}$ is still well within the 'freshwater' range and less than the salinity of Lake Turkana (Table 5.1). The life history is thus adapted to the conditions characteristic of the West African rivers with their strong seasonal floods. Spawning is triggered by the small rise in salinity that occurs when the down-current is least. The veliger

---

* The following information, some of it unpublished, concerning the life history of *Egeria* and the effects of the dam, has been supplied by W. Pople.

larvae are then able to maintain themselves and to move upstream in sufficiently calm water and to establish new colonies of growing clams.

Up to 1963, the dry season region of $1^o/_{oo}$ salinity, and therefore the spawning grounds, were situated at about 30 km from the mouth of the river. During the building operations in 1964 and 1965 the flood water was impounded above the dam and the downflow was continuously restricted. As a result the $1^o/_{oo}$ salinity region and the spawning grounds moved upwards to about 50 km from the coast. Since 1966 there has been a continuous outflow from the turbines and the dry season $1^o/_{oo}$ salinity zone has been pushed down to about 10 km from the mouth, where new spawning grounds have been established. From there larvae move upstream into fresh water and the young adults burrow into sandbanks whence they are transported up to the original 'farming' areas.

The construction of the dam has not, as had been feared, allowed the submerged vegetation to cover the sandbanks in which the clams feed and grow. It has caused the spawning grounds to move nearer to the mouth of the river. This was an unforeseen but fortunate outcome. If the gradient of the lower river bed had been such that the original spawning grounds were within 20 km of the coast, the entire population of *Egeria* might have died out through want of a suitable breeding place.

The case of *Egeria* has been introduced at this point as a notable example of the effects of dam construction on the fauna of the river below it. These effects can be as important as the biological changes in the new lake itself. The estuaries and brackish water regions of the large African rivers present many interesting biological problems ripe for investigation.

In many respects the development of the Volta Lake has followed the course expected from the investigation made before the construction began. The growth of the fishery has however exceeded expectations and a surprisingly large number of those people who were transferred to the settlement areas have returned to the lake shore to engage in profitable fishing (Butcher, 1973; Taylor, 1973). According to Kalitsi (1973) much greater benefits could have been gained but for 'our inexperience, defective planning and implementation, and lack of enterprise'. We hope that such praiseworthy self-criticism will be applied to other such schemes of this magnitude and importance, whose planners will be assisted by the accumulated experience. In fact, in spite of mistakes and omissions and in addition to providing hydroelectric power, the Volta Lake has surely much benefited the people of Ghana, and will continue to do so.

Lake Kainji (Nigeria)

Compared with the previous large manmade lake projects in tropical Africa, more preliminary work was done and published on the hydrology and biology of the section of the River Niger destined to form the Kainji Lake and on the impending problems of human resettlement (White, 1965; Imevbore, 1967, 1970; Visser, 1970; Scudder, 1966). Information on the development of the lake since closure of the dam in 1968 has been briefly summarised by Imevbore (1971) and Adeniji *et al.* (in Regier, 1971). More information was presented at a symposium on Lake

Kainji at the University of Ife in 1970 (Imevbore and Adegoke, 1975). More recent information of the fisheries is mentioned later. The relevant climatic and hydrological information is summarised by Lelek (1973).

This is the smallest of the first four large artificial lakes (Table 20.1). The dam is at Kainji, about 1 200 km from the mouth of the Niger (Fig. 11.3) and the lake extends northwards about 130 km to a little beyond Yelwa. It apparently occupies the site of a previous late Pleistocene lake. Its central section has submerged a large flat island (Foge), which had formerly divided the river into two branches, and it is here that it has spread laterally over relatively flat country and has its maximum width of about 30 km. At either end the lake is confined to a narrow steep-sided valley with a width of 1–5 km. The very rapid flow-through (ann. inflow:vol. 4:1) filled the lake in 2·5 months after closure, and the correspondingly very large draw-down (10 m) now results in a seasonal flooding and desiccation of a considerable area of land in the middle shallow section.

More than three-quarters of the area was cleared of trees by felling and burning to facilitate fishing. This might have been expected to provide abundant dissolved nutrients for an initial massive growth of aquatic plants. That, in fact, this did not happen has been attributed to the large annual draw-down and the consequent periodic exposure of much of the bottom of the peripheral shallow water so that submerged and emergent vegetation cannot become firmly established. It might be suggested that the rapid flow-through may have diluted and prevented a great accumulation of soluble nutrients. For further details of the development of the macrophytes, see Imevbore and Adegoke, 1975. There are however abundant invertebrates, especially the larvae of the mayfly *Povilla adusta* and chironomid fly larvae on submerged tree stems and such emergent vegetation as the grass *Echinochloa* sp. These provide food for fish but their numbers fluctuate with the seasonal exposure and submergence of the plants upon which they live, especially in the middle shallow section (Bidwell, 1976; Bidwell and Clarke, 1977).

Lake Kainji is situated in the transitional zone between the tropical dry savanna with its single-peaked rainy season and the humid tropical forest region with two annual maxima of rainfall. The number of peaks in the rainfall between May and October varies from year to year. But marked temperature stratification with deoxygenation of the lower water (below about 12 m) occurs only during the warmest time of the year at the end of the dry season and the beginning of the rains (March–May). During the rest of the year the winds are apparently strong enough to maintain deep circulation and oxygen to the bottom. The horizontal currents associated with the voluminous flow-through and the outflow near the bottom through the turbines contribute to the breakdown of stratification (Henderson in Imevbore and Adegoke, 1975).

The rapid flow-through and large draw-down are, of course, the consequence of the heavy seasonal floods of the River Niger and are of great ecological importance. The situation is complicated by the occurrence of two annual floods. The first, which brings the lake level to a maximum, comes in July to September and is due to rains in the local catchment within Nigeria. This is called the 'white flood' because it carries enough sediment to cloud the water. About six months later, and before the level has dropped to the minimum, comes the 'black flood' which is free of sediment and maintains the water at a medium level during

May–August. It is derived from rains fallen some months previously in the Upper Niger Basin.

The abundant population of fish have been for several years in a state of flux in adapting to the new lacustrine conditions. Over 100 species occur in the area (Banks *et al.*, McConnell, in White, 1965) and 82 species were recorded behind the temporary coffer-dam during the first stage of construction (Motwani and Kanwai, in Visser, 1970).

Apart from the provision of deep relatively calm water, the annual flooding of the surrounding land has released into the water both nutrients and particulate organic matter and has given food and shelter for young fish. There was an initial decline in the numbers of species previously feeding on the river bottom, especially the mormyrids, but the suspension of organic detritus at first favoured *Citharinus* and *Synodontis*, the numbers of which later declined. The tilapias, which where expected to increase, were for some time rather scarce but the subsequent increase of aquatic macrophytes and phytoplankton has favoured them and *Sarotherodon galilaeus* has become an important commercial fish. The appearance of a rich zooplankton has greatly increased the numbers of *Alestes baremose* and of the clupeids *Sierrathrissa leonensis* and *Pellonula afzeliusi*, and of their predators, particularly *Hydrocynus forskali* (Lelek, 1973; Otobo, 1974; Lewis, 1974; Blake, 1977a, b.) The prospects for a profitable fishery are therefore bright.

Lake Nasser-Nubia (Egypt and the Sudan)

The Aswan High Dam was completed and closed in 1964. The Nile's hydrological regime is summarised on p. 155. The main purpose of the dam is to trap and regulate the flood water from the Blue Nile arriving during September to November in order to provide hydroelectric power and to irrigate a much greater area of desert and for a longer period during the year. It is expected that the irrigated land will be increased from about seven to nine million acres. General reports on the early developments were published by Entz *et al.* in Regier, 1971; Entz, 1972; and Raheja, 1973. The limnological history of the lake up to 1974 is described by Entz 1976.

The lake extends for about 270 km between Aswan and a little south of Wadi Halfa in the Sudan (Fig. 11.3). The greater part in Egypt and the smaller in the Sudan are known as Lake Nasser and Lake Nubia respectively. It lies across the northern tropic and the relatively wide seasonal temperature fluctuations are accompanied by the wide, diurnal temperature swing and very low humidity, both characteristic of extreme desert. The mean annual rainfall is about 4 mm and the annual loss by evaporation from the water surface has been estimated at about 3 000 mm. It might have been thought that much water would be lost by seepage as the lake rises and floods the surrounding porous desert sand. This however is not likely to present a serious problem because the region is mostly underlain by Nubian sandstone interbedded with clay and thus is of low permeability. Such faults and cracks as have been examined are apparently well sealed with precipitated minerals, especially iron oxides (Wafa and Labib, 1973). The deposition of organic sediments from organisms produced in the lake itself will no doubt further decrease the permeability of the substratum.

These features together with the scarcity of terrestrial vegetation might be expected to produce an ecosystem very different from Lakes Kariba and Volta. Neither the initial outburst of production due to flooding of rich terrestrial vegetation nor the subsequent influx of nutrients from the annual inundation and desiccation of surrounding productive land, consequent upon the draw-down and refilling with the flood, were to be expected in this lake. In fact thee was no dramatic initial outburst of production with catastrophic deoxygenation of the water. Both these features characteristic of the tropical manmade lakes developed more slowly during the first few years. The major source of nutrients, and the only source of water, is the Nile itself with its load of nutrient-rich sediments brought down by the annual flood. Before the High Dam was closed this sediment reached the irrigated lands of Egypt and even the Mediterranean delta, which in fact is composed of Nile sediments. The new lake now acts as a settling tank since the flow rate of the incoming flood-water greatly decreases as it enters at the south end. The heaviest silt immediately falls to the bottom and progressively lighter particles are deposited as the water mass advances. The northern half of the lake is so far free of silt. Below Wadi Halfa in the south a mud delta is being laid down which extends further with every flood, reducing the depth of water and leaving an ever-increasing expanse of sediments exposed during the low water season.* Eventually the river is expected to cut a new and narrower channel through these sediments and the Sudan would then be provided with some very productive land easily irrigated from the river.

Egypt has however lost one of its most precious natural resources. The nutrient-rich sediments, annually dispersed with the flood-water over the irrigated land, have for several millennia maintained the fertility of the country. Without the sediments artificial fertilisers will surely have to be applied on an increasing scale. The Delta is so far the region most seriously affected. Very little water now reaches the Mediterranean, and no more sediments from the annual floods are deposited on the Delta. In the absence of the annual flushing out by the river water more of the coastal land is becoming saline by infiltration of seawater.† These are certainly the most serious consequences of the High Dam, but another is the increased evidence of water-dependent diseases such as bilharzia (schistosomiasis) and malaria which will require vigorous control measures (Farid, 1975). Other distressing, though interesting, medical and social problems were raised by the necessity to uproot and resettle more than 40 000 people previously living on the land now submerged. The details of this dramatic episode have been described by Dufalla (1975).

On the other hand, in addition to benefits from hydroelectric power and an increase in agricultural land, an important gain is a productive lake fishery. An artificially dammed lake is always more biologically productive than the original river, provided that the nutrients that accumulate in it are adequately circulated. The annual flood-water provides nutrients in solution but the abundant sediments that fall to the bottom decompose and yield much soluble nutrient materials. These are circulated into the surface euphotic zone by the currents gener-

* Entz (1978) describes the sedimentation processes in the lake up to 1973.
† The reduced outflow of freshwater into the sea is even affecting the marine fisheries in this region (George, 1972).

ated by the inflowing river, by wind action and by the seasonal alternate stratification and overturn of the deeper water. Consequently the productivity of the lake has progressively increased since its inception. A dense phytoplankton and accompanying zooplankton have now developed. Both are food for several species of fish. In the deposited sediments under the peripheral shallow water, especially in the more sheltered bays rooted emergent aquatic plants and floating weeds are establishing themselves. The mud, and the recently submerged sand have much extended the benthic communities and organisms such as chironomid larvae and oligochaet worms are flourishing. Finally, the rising waters have submerged great expanses of rocks which have been encrusted with algae and associated organisms (*aufwuchs*) and are browsed by some of the cichlid fishes as in Lakes Tanganyika and Malawi.

These developments have caused a great increase in the populations of fish. Latif (1976) lists 56 species of which 16 are commercially important. The inshore fish community differs from that in the deeper water. The inshore region is dominated by the phytoplankton-feeding cichlid *Sarotherodon niloticus* and the predaceous Nile perch *Lates niloticus*, the deeper water by *Alestes* spp. that feed, at least partly, on the zooplankton, and the predators *Hydrocynus* spp. The currently expanding fishery is based principally on *Sarotherodon* and *Alestes*. The recorded landings of fish per annum increased from 762 tonnes in 1966 to 12 257 in 1974 and at least 20 000 tonnes per annum are expected in the future, but careful management will be needed to maintain the yield.

Lake Nasser-Nubia is thus providing conditions for a productive fishery situated in the heart of an extreme desert and the rich sediments denied to the agricultural lands of Egypt are now at least in part being converted into fish. The ultimate balance between the profits and losses from this project will no doubt become clearer after a few more years. Rzóska (1976b) provides a short discussion and an assessment of the situation up to 1975.

Jebel Aulyia Reservoir (Sudan)

As one of the means for regulating the flow of Nile water into Egypt, which increases greatly in August–October with the flood from the Blue Nile, a dam was completed in 1937 across the White Nile at Jebel Aulyia 44 km south of Khartoum (Fig. 11.3). Every year the dam is closed in July and August when the flood reaches the confluence of the two Niles. The impounded lake is full by September and is progressively released as the Blue Nile subsides, until the dam is fully reopened in January (Fig. 20.5).

By the standards set by subsequent manmade lakes in Africa this one is comparatively small (*c.* 600 km² in area). It however raises the water level for more than 500 km upstream from the dam. Its peculiar biological interest arises from the annual closure and opening of the dam which causes this section of the river to become a temporary lake for a few months of each year.

This is of course one of many such annual reservoirs in tropical Africa, but during the 1950s some biological aspects of the annual fluctuations of this reservoir were studied by the Hydrobiological Unit based on the University of Khartoum. The effects on the water and its plankton were investigated in two ways:

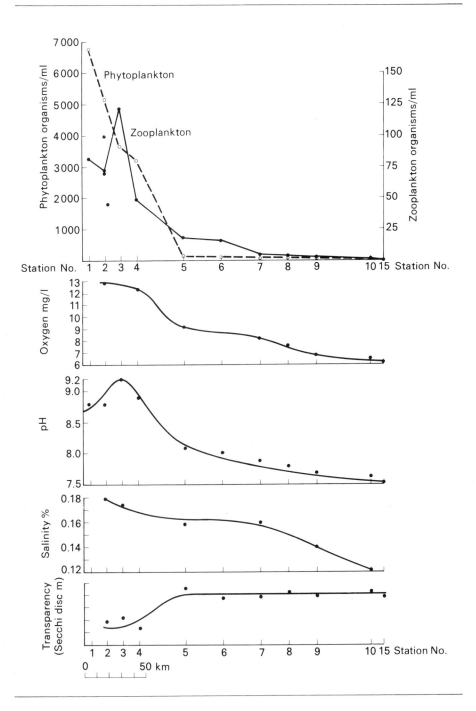

Fig. 20.4 A series of surface samples taken from below the Jebel Aulyia Dam to nearly 400 km above it in December 1951, i.e. in the middle of the full-lake period (selected and redrawn from Brook and Rzóska, 1954)

(1) by examining a series of samples taken along the length of the lake, when full, to demonstrate the transitional gradient from riverine to lacustrine conditions, and (2) by taking repeated samples from a point below the dam in order to record the changes in the down-flowing water due to the appearance and disappearance of the lake above the sampling point.

During a period of five days in December 1951, Brook and Rzóska (1954) took a longitudinal series of water and plankton samples from one point 20 km below the dam and from ten others along the length of the lake – a total distance of 396 km. The main results are summarised in Fig. 20.4. There was a dramatic increase, of more than a hundredfold, in the numbers of both phyto- and zoo-planktonic organisms between the furthest station upstream, with river conditions of flow and relatively small depth, and near the dam where the water was deeper and slow-flowing. There was also a change in the species composition of the plankton (not shown in Fig. 20.4). The blue-green algae (Cyanophyceae), very scarce at the upstream stations, were dominant at the lower 'lake' stations and represented mainly by *Anabaena flos aquae* and *Lynbya limnetica*. The common diatom *Melosira granulata*, present at all stations, was greatly increased in the lower 30–40 km. The crustacean zooplankton of the upper 'river' stations was exclusively composed of copepods. These were much more numerous at the lower stations, but most striking was the appearance in the 'lake' region of great numbers of Cladocera, chiefly *Ceriodaphnia rigaudii*, which were apparently absent from the upper river.

Rzóska (1968) took a set of samples from each of three stations within 5 km above the dam during October and December 1954 and October 1955. A number of samples taken in rapid succession on each occasion showed considerable variation in the numbers of each plankton species. These findings do not invalidate the general conclusions reached from the previous work, but they demonstrate the discontinuous horizontal distribution of plankton, which has been found also in permanent lakes, and complicates the problem of assessing the biomass of plankton in moving water.

Together with the longitudinal series of plankton samples discussed above, Brook and Rzóska (1954) took samples to show some of the associated changes in the composition of the water (Fig. 20.4). The gradient in photosynthetic rate corresponding to that in the density of the phytoplankton is reflected in the oxygen and pH curves, the former much exceeding saturation during the day near the dam. The decreased transparency of the water towards the lower end is also due to the increased density of plankton. The higher salinity is not so easily interpreted but was attributed to liberation of salts from the newly flooded land. It might also result from decomposition of the much denser plankton population. A high concentration of oxygen was found in the bottom water which indicates deep stirring and probably rapid recycling of the soluble materials from decomposition.

Measurement of plant nutrients was not included in this investigation but subsequent work in 1951–53 by Rzóska *et al.* (1955) showed that the density of the phytoplankton reached a maximum and began to fall before the lake was released by opening of the dam. This suggests a depletion of nutrients and indicates that, were the dam to remain permanently closed, the algal population would

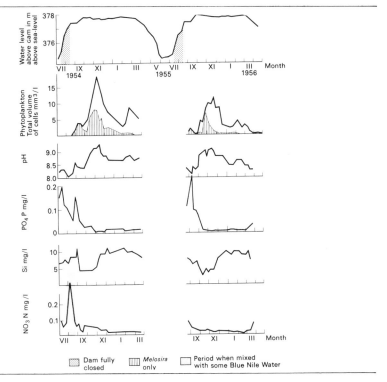

Fig. 20.5 A series of surface samples taken from one point below the Jebel Aulyia Dam during three seasons (1951–54) (rearranged and redrawn from Prowse and Talling, 1958)

reach an equilibrium at a lower average level – a sequence of events which subsequent was found in the large manmade lakes.

Two longitudinal series of samples in October 1954 and 1955 by Prowse and Talling (1958) confirmed these general conclusions and showed further that the plant nutrients silicate, phosphate and nitrate were generally reduced by the growing algal populations towards the lower end of the lake. They also took a series of samples during three seasons (1951–54) from a single point (Gordon's Tree) about 40 km below the dam in order to show the effects of the seasonal changes above the dam on the composition and phytoplankton of the outflowing water. The confluence of the rivers was near enough to affect the sampling point with some back-flowing water from the Blue Nile flood during about a month from mid July. Allowance must be made for this in studying the curves in Fig. 20.5 which summarise the major events during the two seasons 1945–56. The density of the phytoplankton begins to rise soon after the closure of the dam in July and rises to a peak in November, to fall by degrees to a previous low 'riverine' level in April. The growth of the phytoplankton is reflected in a rise in pH, a fall in phosphate and in the initial fall and subsequent rise in silicate. The latter corresponds with the growth and decline of the principal diatom *Melosira granulosa*. Successive measurements of photosynthetic rate through the water column during the period of the algal maximum (November–December 1953) gave rates ranging from 1·15 to 2·7 g C/m^2 day. This is of the same order as the rates found for the northern offshore water of Lake Victoria (Table 7.1), though the latter are maintained throughout the year. A single set of measurements in the lake above the dam in October 1954 gave a figure of 4·3 g C/m² day.

More details, particularly of the succession of the different planktonic species, can be got from the publications quoted above. It will be clear, however, that annual reservoirs, such as Jebel Aulyia, provide good opportunities for studying the initial stages in the transformation of a river to a lake. Unlike the permanent artificial lakes, these opportunities are never lost, but are repeated every year.

Sennar and Roseires Reservoirs (Sudan)

The Sennar Dam was completed in 1925 to hold a reservoir for the irrigation during the low-water season of the Gezira, a large area of country in the angle of the White and Blue Niles south of Khartoum (Fig. 11.3). The Gezira Cotton Growing Scheme has steadily expanded since its inception, and now covers an area of about 2 000 km². A second dam was later constructed (1966) further up-river at Roseires to augment the Gezira Scheme, to provide water for another irrigation area (Kenana) south of the Gezira and to generate electric power. Some characteristics of these two reservoirs are shown in Table 20.1. A limnological survey of the lower section of the Blue Nile before and after the closing of the Roseires Dam was made by the Hydrobiological Unit of the University of Khartoum, but the results are not yet published (see summary in Hammerton, 1972). As in the Jebel Aulyia reservoir there was a considerable rise in productivity as shown by the increased plankton populations and the improvement in the fishery as the result of the change from river to lake conditions. Certain other consequences have also been noted. The large African freshwater mussel *Etheria elliptica*, which is an important component of the fauna on the rocky bed of the river, are now not to be found in the reservoir. This is attributed to the massive deposition of silt. Another not unexpected result of the appearance of these, and other, Nile reservoirs has been the increase of insects that breed and develop in standing or slow-flowing water. In particular, the swarms of emerging chironomid flies (lake flies, see also p. 261) have greatly increased in the neighbourhood of the reservoirs. Though non-biting, they are extremely unpleasant and cause allergic symptoms, such as asthma, in some people. The flooded land around the reservoirs has provided prolific breeding places for mosquitoes. In addition the turbulent water flowing over the rocks from the sluice gates forms an excellent habitat for the small biting fly *Simulium damnosum*, the vector of the nematode worm causing onchocerciasis or 'river blindness' (p. 152). For information and discussions on the aquatic insect pests of the Sudanese Nile see Lewis (1966) and Rzóska (1964).

Lake McIlwaine (Zimbabwe)

The probability, indeed the certainty, that some tropical African inland waters will soon be subject to serious artificial organic pollution was raised in Chapter 19. This has been a problem for some years in South Africa, and the research done there is relevant to the tropics (e.g. on the Hartbeespoort Dam by Allanson and Gieskes, 1961). There is, in fact, at least one notable example in the tropics in Lake McIlwaine, a manmade reservoir providing the main water supply to the City of Salisbury, Zimbabwe (Fig. 9.4). The dam was completed in 1953, an initial limnological survey was made by Munro (1966) and recent investigations of the hydrological and biological effects of pollution were done by Falconer *et al.* (1970) and Marshall and Falconer (1973a, b). A brief discussion of this situation is therefore appropriate as an example and a warning of a problem that is undoubtedly appearing elsewhere in tropical Africa.

Lake McIlwaine is about 15 km long and formed by damming the Hunyani River (Fig. 9.4). It has an average depth of about 9·5 m and a maximum of about 27 m. It is unfortunately situated at a lower level than the city and 37 km to the southwest, and processed sewage effluents from the urban areas are discharged into the streams flowing into the lake. Owing to the very high content of plant nutrients, especially nitrate and phosphate, these inflows have caused the progressive eutrophication of the lake to a condition that could certainly be called 'polluted'.

The basic hydrological regime is determined by the relatively large seasonal changes of temperature due to the high latitude (17°55′S), by the seasonal heavy winds that are contemporary with the low winter temperatures from June to August, the seasonal fluctuations of inflow, and by the relative shallowness of the water. In 1968–69 the range of temperature at the surface was 21°–25°C in summer (November–March) and 16°–20°C in winter (June–August). Such conditions, as would be expected, are likely to cause a seasonal alternation of stratification and overturn, which is demonstrated by the temperature and oxygen data in Fig. 20.6, though the general impression is one of instability. A striking feature appeared during the apparently stable period in October–January when the lower half of the water column was free of oxygen. During this period the temperature at the bottom rose by 3°C (18°–21°), parallel with a rise at the surface. This indicates that, during this period, some deep stirring was taking place but not sufficiently rapidly to prevent the deoxygenation due to the intense decomposi-

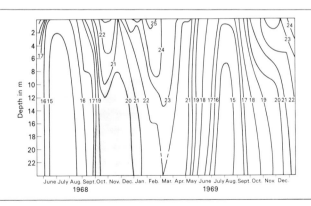

Fig. 20.6 Lake McIlwaine, near Salisbury, Zimbabwe. Isotherms (°C) showing seasonal alternation between stratification, with deoxygenation of lower water, and overturn which brings oxygen to the bottom. Stippled area denotes oxygen-free water. Redrawn from Falconer *et al.* (1970)

tion in the lower water. During the most turbulent and cooler period (April–July) mixing was rapid enough to raise the oxygen at the bottom to 20–50% saturation.

The isotherms in Fig. 20.6 suggest that the temperature and wind regimes would probably have lessened the volume of the deoxygenated water and its duration during the calm season if less of the dense and nutrient-rich water was entering with the inflows. The water column might have been less stable and resistant to disturbance from the occasional abnormal violent weather which is liable to occur at any time. The concentrations of nitrate and phosphate in the surface waters, though subject to seasonal fluctuations, are always rather high and there is no reason to suppose that the growth of algae is ever seriously limited by depletion of these nutrients. Analyses of bottom sediments have shown the very large amounts of phosphorus and nitrogen that are accumulated in them, especially in the central deeper channel of the original river.

The numbers of the phytoplankton of all species are continuously high but with a peak during the warmer summer period in December–January. Growth of phytoplankton seems to be controlled more by light and temperature than by nutrient supply. As is common under very eutrophic conditions the species are predominantly blue-greens (Cyanophyceae, Fig. A1), such as *Anacystis* (*Microcystis*) *aeruginosa* and *Anabaena flos aquae*, with the chain-like diatom *Melosira granulata* appearing during the main overturn in March–June.

The most obvious evidence of increasing eutrophication and the effects of pollution since 1960 has been the increased frequency of algal 'blooms' and the progressive general rise in the density of the algal populations. This has caused considerable trouble in the water filters, and it has been suggested that *Anabaena*, which is the dominant alga during the winter, may be responsible for outbreaks of gastroenteritis in children. This alga is known to produce substances that have toxic effects on man and domestic animals, and algal poisoning of cattle has been reported from the Hartbeespoort Dam (Allanson and Gieskes, 1961).

Studies of the bottom (benthic) fauna have shown, as expected, an uneven distribution of species of chironomid fly larvae and of oligochaet worms according to the type of sediment, level of oxygen, etc., associated with different depths of water above. But preliminary work has indicated that progressive increases have occurred in total numbers and in the species composition of chironomid larvae and oligochaet worms such as would be expected to result from increasing eutrophication (Falconer *et al.*, 1970). The most striking example is the greater increase in the numbers of the oligochaet *Limnodrilus hoffmeisteri* compared with *Branchiurus sowerbyi*.

Lake McIlwaine is as yet far from the appalling state of organic pollution that can be seen in some lakes and rivers associated with industrial cities elsewhere. It has been suggested above that the hydrological regime is retarding what must be the inevitable further deterioration if no active measures are taken to prevent it. It is to be expected, and is in fact known from elsewhere, that, after removal of the sources of pollution, conditions may not improve for a long time because the great store of organic matter in the sediments continues to be recycled and to support an excessive growth of phytoplankton. Whatever may be done to reduce the inflow of nutrients, it is therefore necessary to consider other remedies. A measure that would have additional benefits is to convert the phytoplankton into good edible food for man and domestic animals. This in principle could be done by suitable fish, such as those that feed direct on the algae and mormyrid species that feed on the benthic fauna, provided that the particular hydrological and chemical conditions are suitable for such fish. Research is needed to decide this. It may be possible, as has been done successfully elsewhere, to make use of the nutrients derived from sewage effluents for the culture of fish in ponds.

Lake Nyumba ya Mungu (Tanzania)

Nyumba ya Mungu ('The House of God') on the upper Pangani River was completed in 1965 and is a small reservoir by African standards (Table 20.1). It was designed for both irrigation and hydroelectricity and has provided a productive fishery. There are very many small reservoirs in Africa and their number is increasing, but Nyumba ya Mungu (Fig. 3.4) is briefly discussed here because it was investigated by a team (leader Patrick Denny) during three months in 1974 (July–September) and their findings were all published together (in Vol. 10, No. 1 of the *Biol. Journ. Linnaean Society*, 1978). Though seasonal changes could not of course be recorded, most of its characteristics appear to be such as would be expected from its dimensions and situation. There are however some features of special interest which will be emphasised.

The lake is about 30 km long and is fed by the Ruvu and Kikuletwa Rivers which drain the southern slopes of the Kilimanjaro and Meru volcanoes. Two swampy deltas are thereby accumulating at the north end. The water is mostly shallow (mean depth 6 m), but the bottom slopes gently to a depth of 41 m at the south end in front of the dam. There is no artificial 'draw-down' but the maximum depth is limited by a spillway from which it drops by a maximum of about 3·5 m in the dry season.

Its north–south orientation and the shallow water provide conditions for frequent deep stirring and circulation of nutrients by the northeast and southeast trade winds. The available evidence suggests that the stirring is maintained at frequent intervals throughout the year. Temporary diurnal temperature and chemical stratification (e.g. of oxygen) were noted in the more sheltered bays, but even the deeper water is unlikely ever to be subjected to serious deoxygenation. Some paired-bottle measurements indicated, as would be expected, a moderately high rate of gross production by the phytoplankton (mainly *Melosira* and blue-green algae). So far, however, the phytoplankton is not directly or indirectly the source of organic sustenance for most of the commercially important fish. Of the twenty species of fish recorded from the lake only one, *Sarotherodon esculentus* (introduced from Lake Victoria), feeds directly on the phytoplankton and appears in the catches. But the fishery is based mainly on the two cichlid species indigenous to the Pangani basin, *Sarotherodon pangani* and *S. jipe*, both of which graze on the algae encrusting solid surfaces, particularly the submerged stems of aquatic plants. *Tilapia rendalli* (the only other introduced species) consumes both the epiphytic algae and parts of the plants on which they grow. The vegetation of the lake supports a very large quantity of epiphytic algae which thus appear to be photosynthesising most of the organic matter ultimately harvested by the fishery.

Table 20.2 Some chemical features of the two inflows to Lake Nyumba ya Mungu

	Ruvu River	Kikuletwa River	Lake, south end
July–August 1974			
Conductivity (K_{20})	526	965	685
Alkalinity meq/l	3·6	9·2	5·6
pH	7·4–7·7	8·1–8·2	8·6–9·1

The lake is however still young and the pattern of the food web may change and other exploitable species may become dominant.

As usual with newly impounded lakes, productivity, as reflected in the fishery, increased greatly during the first few years and later settled down to a lower level. There has fortunately been no problem of blockages by floating vegetation. The water fern *Salvinia molesta* and the water-hyacinth *Eichhornia crassipes* are both absent, though the latter is well established and troublesome in the lower Pangani basin. A more ominous possibility in the future is the spreading into the lake of the deltas from the two main inflows from the north with dense growths of the large emergent plants, particularly the bulrush *Typha domingensis*. It is conceivable that a large proportion of the lake may ultimately be converted into a marsh overgrown with *Typha*.

A very interesting feature of the lake is the great difference between the salinities and alkalinities of the two inflows, at least one apparent ecological effect of which can be seen (Table 20.2). These two chemical properties of the waters are in the same range as those of Lakes Tanganyika, Albert and Edward (Table 5.1). Within this range however the Kikuletwa River is much more saline and alkaline than the Ruvu. The composition of the water at the south end of the lake suggests that the two inlets make roughly equal contributions to the water in the lake. The higher pH of the lake water is presumably due to photosynthesis by the more abundant phytoplankton. These differences are emphasised here because they are accompanied by a major difference in the vegetation. The Ruvu River as far as its entrance into the lake supports a rich growth of *Cyperus papyrus* and many other associated plants. On the other hand, the flora of the Kikuletwa is relatively poor, devoid of papyrus and is dominated by *Typha* which also replaces papyrus in the lake itself. In view of the known distribution of macrophytes in other African waters it is difficult to imagine that the difference in salinity and alkalinity can account for this. *Cyperus papyrus* flourishes in Lakes Albert and Edward. Nor do analyses of the major inorganic ions in the waters provide a possible clue. Nevertheless, there is as yet nothing to suggest that anything but a chemical difference is responsible. The absence of papyrus from some African waters has been correlated with a high fluoride level (Kilham and Hecky, 1973). This matter is emphasised here because it would seem to offer an admirable opportunity for the investigation (with experiments) of a seemingly clear-cut ecological problem, though it may prove to be more complicated. It might involve other organisms in addition to the macrophytes and have wider ecological implications for the African inland waters.

21

Some comments and conclusions

This book is not a comprehensive review of all the available information on this subject, though the references should enable anyone to track down all the details that have been published. It aims rather to present in outline what can reasonably be inferred from the evidence of the origin and history of the tropical African inland waters and their faunas and floras, to describe what is known of the cycles of events in the different water systems, particularly in relation to organic production, to discuss some interesting scientific problems raised, and to indicate the importance of these waters to the peoples of the area. The discussions have been conducted on a background of evolutionary concepts and limnological principles presented in the simplest possible manner. These principles were first established in relation to conditions in temperate climates, and their relevance to waters in the tropics has been discussed.

The reader will probably detect a certain tendency to avoid, if not some disrespect for, systems of classification that were formerly so prominent in limnological literature. This is particularly evident in relation to typing of lakes with respect to water composition, cycles of circulation and intensity of production. It is partly a matter of temperament, and I will admit to an unreasonable bias. At the same time I would submit that this is a fault on the right side which can help dispel the common illusion that a phenemenon with a label attached is thereby explained. Systems of classification based on well-defined criteria are useful, and, indeed, essential, but only as provisional arrangements of the data. They imply certain interrelations between the categories which further research may show to be untenable or at least of limited application. As an example of the contortions sometimes required to maintain the integrity of a classification – one of the suggested criteria for distinguishing a productive (eutrophic) from an unproductive (oligotrophic) lake was a relatively high or low surface/volume ratio respectively, which is certainly one of the factors affecting the rate of stirring of the bottom water and sediments. When, however, some eutrophic lakes were later found to have a low surface/volume ratio, they were said by some to exhibit 'bathymetric oligotrophy' (quoted by Russell-Hunter, 1970, p. 118). They obviously do not exhibit any kind of oligotrophy, and at that point one of the bases of the classification was clearly shown to be at fault. Such desperate attempts to preserve an untenable position are surely 'counterproductive' (to use a similarly disreputable and ugly expression), and do not encourage original thinking.

Another preliminary point to be made, which cannot be too strongly emphasised, is that all ecological situations, like the metabolism of an organism, however seemingly stable, are in *dynamic* equilibrium ('steady state'), and are invariably maintained by a balance between opposing forces. Even the inorganic features of a lake such as salinity, water level and the positions of the thermocline or anoxic boundary are controlled by a balance between the rates of processes that are acting to move them up and down. The concentration of dissolved nutrients, such as nitrate or phosphate, may at times be too low for detection by chemical analysis, though other evidence may show that it is being assimilated and recycled very rapidly. Biological features, such as the biomass and rate of production of a species and its relation to that of other species in the ecosystem, are controlled by a much more complicated constellation of opposing processes, organic and inorganic, that are correspondingly more difficult to unravel.

In attempting to draw general conclusions we are handicapped by the as yet inadequate amount of research done on certain important subjects especially in places from which more useful information could certainly be got. The data at our disposal are not uniformly reliable, and the conclusions that have been drawn from them must be accepted with varying degess of probability. Such disadvantages are however invariably associated in some degree with all ecological work which is faced with the great complexities of even the simplest ecosystem. For some of the phenomena under discussion we still do not fully understand the basic principles involved. For example, we have much more to learn about the mechanism of speciation and of the conditions, external and internal to the organism, that control the rate of speciation. Why are the cichlids apparently evolving more rapidly than other African fish at the present time, but not in all the water systems in which they are represented? There has been much speculation on this subject. There have been many suggestions, based on the chemical conditions in the different waters, to account for the curious distribution of certain organisms. We do not, however, yet know enough about the nature of the exchanges between freshwater organisms and their environment and of their special needs and tolerances to make more than very tentative guesses. More experimental work is needed, but backed by a combination of physiological and ecological knowledge and skills to ensure that the experiments are planned to provide answers that are relevant to the natural situation. Another important field, on which we need more background knowledge of the underlying processes, relates to the productivity of the African lakes. I refer especially to the study of water movements upon which the level and fluctuations of production largely depend.

Having made these precautionary and discouraging remarks, I will try to draw some very general conclusions that seem to be justified from the evidence, beginning with the historical aspects.

About twenty-five million years ago, in the early Miocene, a very long period was coming to an end during which the African continental surface had been comparatively stable. Erosion had reduced much of the landscape to what must have been a rather monotonous condition of immense plains, plateaux and smooth hills intersected by water courses draining to the coasts and into the ancient basins (Fig. 3.2). There were probably lakes from time to time in the basins and at least one large one (Lake Karunga in Kenya) in the higher country. From fossil evidence and from what can be deduced from the present distribution of species, the composition of the fish fauna of the inland waters was rather uniform over the whole of tropical Africa. Part of this fauna has survived without much change as the 'soudanian' fauna of the northern tropics. It may therefore be supposed that water connections between the basins had been fairly frequent, presumably through periods of heavy rainfall, erosion and river capture. There is no fossil evidence from before the end of the Miocene for the existence of any water system that was isolated for a long enough time to produce a large endemic fauna. The situation was drastically transformed by the subsequent uplifting of the highlands of eastern Africa with the associated fractures of the two Rift Valleys and the sagging between them to form the Victoria Basin. Into the sinks so formed the waters of all the Great Lakes, except Chad, have collected. The upper sections of a series of rivers previously flowing into the ancient Zaïre basin (Fig. 3.4) were interrupted by the Western Rift. The northernmost lakes that collected in this valley, together with Lake Victoria, eventually flowed northwards into the Nile and thus greatly extended the old Nile basin. The subsequent history of the water systems of eastern tropical Africa and of their interconnections was determined both by continued earth movements and by great fluctuations in the amount of surface water due to changes in rainfall and temperature.

The ecological result of these events was that a primarily river fauna and flora was faced with a great variety of new conditions in these very large lakes. The fish in particular reacted to the changes through the selection of genetic combinations adapted to the many new habitats, and thus new species were evolved. Moreover, the new lakes have been to varying degrees and for different lengths of time isolated from each other and from neighbouring water systems so that very remarkable endemic faunas have evolved within them. In each of the three lakes, Tanganyika, Malawi and Victoria, the endemic species are numbered in hundreds.

Owing to inadequate information and to our incomplete understanding of the mechanisms of speciation there are many inconsistencies and unsolved problems that are discussed in the appropriate chapters. But, on the whole, there is some discernible relation between the composition of the fish fauna and the extent of endemic speciation within a lake on the one hand and what is known of its past geological history on the other. The distribution of invertebrate animals, which for the most part are less waterbound than fish, does not as clearly reflect the past history of most of the water systems. But in the oldest of the lakes, Lake Tanganyika, there are many endemic species of molluscs, some crustaceans and even insects, and this must surely have resulted from a very long period of isolation.

In contrast to the revolutionary changes undergone by the old Miocene fish fauna in eastern tropical Africa, in the northern and western tropics it has remained closer to the original. There are many individual species of this fauna spread across the whole of the Soudan from the Nile, through Lake Chad and the Niger basin to the Senegal and Gambia Rivers, in waters in which endemic species are comparatively scarce, in spite of their present extreme isolation from each other by great expanses of arid country. In attempting to explain these partly contradictory situations we might point to the greater tectonic stability since the Miocene of the soudanian region compared with eastern tropical Africa and to the fact that the great fluctuations of climate during the past few thousand years have made and broken water connections between the main soudanian basins across what is now desert. The soudanian faunas have recently been less isolated than at the present time. In addition, the climatic fluctuations in the soudanian region may have prevented the water systems from providing stable enough environments for the establishment of many new species. We cannot, however, feel satisfied that all is now explained, particularly in view of the speed at which some species of fish, particularly of cichlids, are known to have evolved, and because a few highly modified species have in fact evolved in the Soudan, such as the cichlid *Gobiocichla wonderi* in the rapids above the internal delta of the Niger.

Apart from the Niger, the West African rivers that discharge southwards through the forests into the Gulf of Guinea have been more stable and longer isolated from the main soudanian water systems and have a correspondingly greater proportion of endemic fish. These include the clupeids which are derived from species that have invaded the rivers from the sea in geologically recent times.

The biological history of the Zaïre water system is different from that of the other ancient basins. Though for the most part lying outside the regions most affected by the post-Miocene earth movements, it has the greatest total number of fish species and of endemics. There are probably several reasons for this. In spite of great climatic changes in the central Zaïre basin the water system as a whole has remained intact at least from the Miocene, and has thus maintained an almost stable environment for a very long time. On the other hand, the faulting that has occurred within the basin, though on a relatively small scale, has been widespread. Through damming of the outflow by faulting a large lake was formed in the central basin which existed for a long time during the Pliocene before it finally drained away, and many falls and rapids appeared on the peripheral tributaries which became faunal barriers between the main river and the upper reaches, in which species have evolved in isolation. To these features we might add the immense distance traversed by the Zaïre River system which might itself contribute to the effective isolation of populations of fish. It is likely that the Zaïre River presents a greater number and variety of habitats for aquatic organisms than any of the other African basins. In addition, the large Pliocene lake would have provided other kinds of environment and thus probably contributed to the evolution of the present very large and varied fauna, which includes endemic species of fish belonging to many different families.

The volcanic barrier and crater lakes that have appeared in recent times, many of them within the past few thousand years, are mostly sufficiently isolated from neighbouring water systems to obstruct the movements of fish. That they have not, on the whole, been sites of much endemic speciation can be attributed to several causes. The faunas that originally populated both kinds of lake were generally poor in numbers and variety of species. This is to be expected of crater lakes, but the barrier lakes of East Africa were mostly formed by blockage of hill streams with a poor indigenous fish fauna. Lake Kivu is an exception to this, to account for which I have suggested that the blockage probably deepened an already existing lake, but its lacustrine fauna has been much attenuated perhaps by chemical contamination of the water from the volcanic lavas. The relatively small fish fauna of Lake Kivu has nevertheless produced a few new species and varieties of fish. Many of the crater lakes are impossible for fish because of their high salinity and alkalinity derived from leaching of soda-bearing lavas. It might also have been thought that small more or less circular crater lakes would present a very small range of habitats to which fish could adapt themselves, especially those that are sufficiently deep and sheltered to have a permanent mass of anoxic water at the bottom. The small crater lake Barombi Mbo in West Cameroon has, however, produced twelve endemic species of which eleven are cichlids thought to have been derived from two original species. But, apart from prolonged isolation, this lake does not seem to possess any of the features that might be thought especially favourable to divergent speciation.

It is clear that a very important factor in these situations is the nature of the fauna which originally happened to be in the lake or gained access to it at an early stage. Though we do not yet understand the genetic basis, there are some species of fish, especially among the *Haplochromis* group of the African cichlids, that have been in recent times, and probably still are, particularly adept at producing rather rapidly a wide range of genetic variations from which suitable combinations have been selected in adaptation to new situations. Considerable changes of structure, coloration and behaviour have been involved. The large cichlids *Tilapia* and *Sarotherodon* have been equally successful in colonising most of the tropical African lakes, but they have done this by, so to say, restricting themselves to fewer types of habitat. Fewer species have been evolved with much less structural difference between them compared with the *Haplochromis* group; in fact there are many geographical races of most of the species in different parts of their range which are not sufficiently different to be considered, on structural grounds, as separate species.

In the recently formed Lake Lanao in the Philippines it is the genus *Barbus* of the family Cyprinidae that has performed a similar feat of adaptive radiation and has colonised a great range of habitats. Many species of *Barbus* have evolved also in Africa especially in the rivers, but they have shown far less diversity than the African cichlids in the habitats to which they have been adapted and in their anatomical modifications. Whatever the explanation for these genetic differences in adaptability between groups of animals at certain times and in certain regions, they have played an important part in the recent evolution of the lake faunas. The well-known and numerous endemic prosobranch snails in Lake Tanganyika suggest that there are similar differences between groups of molluscs, though the relative scarcity of endemic prosobranchs in Lake Malawi poses a difficult question. However, we are not likely to get nearer to a solution of the remaining problems concerning the evolution of the lake faunas until we have more detailed facts about species and variety distribution and a greater understanding of geological history and of the physical and chemical events in the waters. Most of all we must know more of the biology and genetics of the organisms, especially the fish.

We are likewise faced with many gaps in our knowledge and understanding when we try to describe the patterns of inorganic and organic processes which contribute to the overall 'metabolism' of the tropical African lakes, and to assess 'productivity'. There are, however, certain very general conclusions that seem to be justified from what has so far been discovered. The relatively high and even temperature and illumination in the tropics, especially at very low latitudes (equatoria), are in themselves conducive to a high rate and annual total of *gross* primary production by aquatic plants. But the rate of breakdown through respiration by the plants themselves, and by the organisms that consume them, is correspondingly high. *Net* production of materials and energy available to fish is therefore not necessarily greater than in a temperate climate under otherwise similar conditions; at least there is no clear demonstration that this is certainly so. In theory the greater annual and long-term stability of most tropical lakes would be expected to lead to a more highly evolved community of organisms exploiting more efficiently the resources of the ecosystem. It must also be noted that, with the same percentage loss of energy and materials between the plants and the fish, the higher the original gross production the greater the absolute amount left over for the fish. This appears to be relevant to the situation in Lake George where the very small residual percentage of the initial extremely high primary production suffices to support a productive fishery.

On the subject of periodic fluctuations in the rate of production in tropical lakes we can express rather more confident conclusions. Unlike in temperate regions, temperature and illumination are continuously adequate over the year and are not responsible except near the boundaries of the tropics for seasonal changes in the rate of production. There are, however, two components of the tropical climate that in many situations show well-marked seasonal changes in intensity, namely rainfall and wind. The volume of the inflows to a lake are determined by the rainfall in the catchment. The inflows provide the main ultimate source of nutrients and sediments, but the extent to which increased downflows during the rainy season cause an immediate and significant rise in the rate of production depends upon the volume of the annual floodwater in relation to the volume of the lake, i.e. the extent to which the lake is 'washed through' by the floods. The well-marked seasonal regime in the shallow south basin of Lake Chad is determined by the annual heavy floods down the Chari and Logone Rivers by which the basin is almost flushed out. The same is obviously true of the seasonal artificial reservoirs, such as those on the Nile. Deep lakes of very large volume in relation to annual inflow, such as the large Rift Valley lakes and Lake Victoria, are not affected in this manner by the annual increased inflows except locally in the region of the river mouths. On the other hand, the annual floods cause a rise in water level in all lakes in which the depth depends both on the rainfall and on the nature off the outflow as well as on the rate of evaporation from the water surface. The consequent flooding of surrounding low-lying land can provide an important seasonal addition to the supply of nutrients. But its significance will depend upon the shape of the basin which determines the area of land subject to flooding in relation to the size of the lake. In artificial lakes the initial flooding of the land releases nutrients and, during the first year or two, stimulates production to a maximum from which it falls to a lower level, with fluctuations partly caused by seasonal flooding of surrounding land (Petr, 1975). The inshore fisheries of some of the large lakes, such as Lake Victoria, are probably supported to a significant degree by the annual flooding of peripheral land. The unusually high levels at longer intervals following a series of years with higher rainfall have certainly provided an increased supply of nutrients and additional suitable breeding grounds for fish in many lakes. The productivity of Lake George is however remarkable in view of the constancy of the water level.

In the majority of lakes the most important immediate source of nutrients is through recycling of the products of excretion and decomposition, the circulation of the former directly between the organisms in the upper layers and the transfer of the latter upwards from the lower regions by movements of water. The effectiveness of winds in causing these movements depends first upon their strength, duration and seasonality, second on the degree to which the lake is exposed to the winds by its shape, location, surrounding topography and surface area, third on the depth of water to be stirred, and fourth on the vertical density gradient due to differences in temperature or concentration of dissolved solids, which tend to resist stirring. For shallow and well-exposed lakes the winds may be sufficiently powerful and frequent, even during the calmer periods, to maintain a frequent circulation to the bottom with no marked fluctuations in the rate of production corresponding to seasonal changes in the wind regime.

At the other extreme are the small but relatively deep and well-sheltered lakes, such as some crater and volcanic barrier lakes described in this book, and those very large and extremely deep but wind-swept lakes, such as Tanganyika and Malawi. The depth of water relative to wind stress at the surface is such that only a fraction (in Lake Tanganyika a very small fraction) of the total depth is directly stirred from the surface. Though we have reason to suppose that there is some form of circulation within the lower water, there is not sufficient mixing with the upper layers to counteract the high rate of decomposition, and the lower water is permanently devoid of oxygen and accumulates nutrients and perhaps dissolved organic matter, but it is not equivalent to the 'hypolimnion' of a temperate lake in summer. The level of the anoxic boundary between these two main layers is held in position by a balance between, on the one hand, the rate of stirring from above and in the large lakes at least, through the action of internal waves and, on the other hand, the rate of decomposition below the boundary. This implies that the anoxic boundary is constantly being eroded and thus releasing nutrients into the water above. Moreover, the considerable fall in the level of the anoxic boundary during the heavy wind period (demonstrated in Lake Tanganyika) must be accompanied by a seasonal increased influx of nutrients into the upper water. In fact there are strong indications in Lakes Tanganyika and Malawi of increased algal production during the windy season due perhaps to this cause and, in inshore regions, through the upwelling of lower water on the crests of internal waves. The clarity of the pelagic waters of Lakes Tanganyika and Malawi had previously led some to believe that the offshore waters are relatively unproductive (oligotrophic) though this was always difficult to reconcile with the flourishing state of the fisheries. Now however measurements of gross photosynthetic production have shown that the rate per m^2 of surface is comparable with that in Lake Victoria. The depth to which photosynthesis is active is very great (20–

25 m). Production is thus diluted through a great volume of water. Further work has raised the question of sources of organic matter in addition to that synthesised by the phytoplankton. It is certainly not true that the large anoxic mass of lower water forms a permanent trap for nutrients. Further research might even show that, though the deposit of nutrients held in the anoxic 'bank' is very large, the rate of circulation into and out of the upper water may be as great, or perhaps greater, than between mud and water in a lake well stirred to the bottom.

Needless to say, between these two extreme types of lake, the continuously mixed and the permanently stratified, there are many examples in which conditions are intermediate. In some, such as Lake Kariba, stratification with deoxygenation of the lower water develops only in the calm season and is dispersed by the heavy annual winds. Lakes Victoria and Albert are so formed and situated that, though there is some stratification around the turn of the year, it is not maintained for long enough to cause more than a partial reduction of oxygen with some accumulation of nutrients at the bottom. An increased algal production in Lake Victoria results from stirring by the heavy annual southeast trade winds from June to August.

Regular seasonal fluctuations in production and in consequent density of biomasses are thus characteristic of tropical lakes. These fluctuations differ from those in temperate climate lakes for which the main determinant is the wide seasonal swing in temperature and illumination. During the winter the plankton biomasses are often reduced almost to zero, but in summer they may rise to levels above the maximum found in some apparently comparable tropical lakes. The amplitude of the seasonal fluctuations in tropical lakes is characteristically less and the biomasses do not normally fall to such low levels as in winter at high latitudes. It must however be admitted that the information so far available on seasonal fluctuations and latitude is still inadequate for a full understanding of the relationship.

Of all the lakes so far examined sufficiently thoroughly Lake George is exceptional in the stability of the rate of production and of the density of its biomasses. This seems to be due to a fortuitous conjunction of circumstances: (a) its shallowness and open exposure to wind, so that complete stirring is repeated almost daily throughout the year, (b) the relative constancy of the inflows derived from the Rwenzori Mountains, so that there is no very great seasonal inrush of nutrient-rich water, and (c) with the 'buffered' outflow, the normal seasonal changes in water level are very slight and very little peripheral land is flooded. This would appear to be an unusual, perhaps unique, situation.

One of the characteristic features of the tropical climate is the occasional and unpredictable storm with abnormally violent winds, and more rarely, an exceptionally long calm period. The former is often very local, but most lakes have a history of occasional catastrophes (once every year or two), due to such aberrations in the normal regular wind regime, causing the destruction of very large quantities of fish. It is to be hoped that more opportunities will arise for studying the chemical and biological effects of these disturbances which would surely throw further light on the normal balance of processes in the lakes.

The water of some lakes, that have no surface outlet, are nevertheless fresh and support a normal freshwater fauna and flora in spite of a previous great contraction through evaporation. This poses the problem of the fate of the salts that must have accumulated. Most of such lakes have become hypersaline or even saltpans. There is clear evidence of a subterranean seepage of water from Lake Chad and it is most likely that this also accounts for the fresh condition of Lake Naivasha. Lake Bosumtwi has not yet been investigated in this respect. Lake Turkana is especially interesting because we have the kind of information from which to solve the problem, but unfortunately some of the data relating to the past is too uncertain. Apart from underground seepage of water, precipitation and binding of ions to minerals in the sediments also contribute to maintain the unexpectedly low salinity of these lakes.

Lake Turkana is unique among the large lakes in that its salinity, though not excessive, seems to have reached a level which excludes some species of invertebrate animals.

Apart from high salinity there are other difficult situations in the tropical African inland waters which some organisms have managed to colonise by means of special structural and physiological adaptations. Occasional or frequent desiccation, very high salinity, high temperatures, torrential flow and severe battering by waves on rocky shores – these hazards are common to inland waters in many parts of the world, though tropical Africa provides good opportunities for observing and studying them. The very great areas of heavily shaded swamp are however less common elsewhere and are of particular interest because of the frequent depletion of dissolved oxygen due to organic decomposition. This demands special physiological and biochemical adaptations on the part of the many organisms living in them. Since the swamps are not quite stagnant, and often have a considerable flow-through of water, they apparently do not accumulate toxic products of decomposition and may be said to provide an environment which is 'healthy' apart from the low level of oxygen, the effects of which can thus be studied in apparent isolation. In this respect swamps differ from the more nearly, though not completely, stagnant anoxic deep water in permanently stratified tropical lakes, where there is usually an accumulation of hydrogen sulphide. The latter is also worthy of further study as an environment that has been invaded by at least a few animals. The case of the cichlid fish in the West African crater lake Barombi Mbo suggests that there may be similar adaptations in lakes such as Lake Tanganyika. The ability of some chaoborid and chironomid fly larvae to live for long periods in the oxygen-free lower water and mud of lakes has been known for many years, but little is yet understood of the biochemical mechanisms that make this possible. Further research on organisms adapted to this kind of environment would surely increase our understanding of basic respiratory processes and of the manner in which they are regulated in difficult situations.

This leads us on to raise a few general points concerning objectives and methods of investigation. In order to approach an understanding of the workings of an ecosystem as a whole it is necessary to know its 'structure' (its biological and physical components) and the functional relations between the components, i.e. the processes at work, the most fundamental of which is the flow of energy through the system. Knowledge of the basic processes such as the movements, density changes and evaporation of water, photosynthesis and respiration, etc., and of the factors that control them is needed by an ecologist for interpreting his observations. This knowledge has been gained over the years by scientists from laboratory experiments in which the separate components were studied in artificial isolation. Further special experiments usually have to be done by the ecologist to discover which of the factors known to be essential for a certain process is actually controlling it at a given moment in an ecosystem such as lake. For example, the immediate cause of a seasonal decrease in the rate of primary production may prove by experiment to be due to the exhaustion of the supply of one of the nutrients.* The special needs and tolerances of different species of organisms with respect to certain environmental components (e.g. calcium, oxygen, temperature, etc.) may have to be defined by experiment before observations on their abundance, distribution, reproductive and growth cycles can be properly interpreted. There are several examples in this book of doubts remaining owing to inadequate information on such points. On the other hand, since the object is to understand the workings of the natural ecosystem, the component process to be studied in isolation must be carefully chosen and controlled for its relevance to the natural situation where it reacts with others not represented in the experiment. Rigler (1975) discussed the 'holist' and 'reductionist' approaches to the study of lake ecosystems and traced the historical development of dynamic from static concepts in limnology.

The experimental method can to some extent be applied directly to a natural or semi-natural ecosystem. For example, Beyers (1963) studied the overall energy flow, by measurements of photosynthesis and respiration, in a series of small, balanced aquaria containing a few selected macrophytes, some algae and the microorganisms in the mud. The climatic factors were rigorously controlled. He measured the effects of changes in day-length, and of the addition of snails which grazed on the plants. This kind of experiment can give indications of events in fully natural waters. 'Outdoor' artificial ponds, such as fishponds, with uncontrolled but monitored climate, can be used for experiments (e.g. by addition of nutrients, addition or subtraction of species, etc.) under conditions more nearly approaching the natural and similar in principle to the experimental plots in agricultural research. With most natural ecosystems, such as lakes, experiments of this kind are hampered by the large size and heterogeneity of the system, by the lack of other identical lakes for 'controls', and by the danger of irreparable damage to a valuable natural resource. An important technical advance for experiments on lakes under near-natural conditions is the 'Lund tube' (Lund, 1978) whereby a portion of the lake is isolated in a plastic tube about 45 m diameter supported by floats at the surface and sunk into the sediment at the bottom. In such an isolated section of the open water conditions are natural except that there is no exchange of nutrients or of organisms with the peripheral regions and inflows. But chemical changes and fluctuations in the biota can be measured and experiments can be done involving quantitative addition of nutrients or organisms. Even this has obvious limitations as a means of understanding the workings of the whole ecosystem. 'Experiments' have been done on some of the large African lakes, not always, it must be admitted, with due regard for ecological princi-

---

* The experimental use of algal cultures in limnology was the subject of a symposium in 1976 (Skulberg, 1977).

ples, and some even accidentally. Such are the introduction of the Nile perch and tilapias into Lake Victoria, introduction of the Lake Tanganyika 'sardine' into Lake Kivu and Lake Kariba, and the disastrous accidental introduction of the water hyacinth into the Zaïre and Nile basins and of the water fern *Salvinia* into Lake Kariba. An increase of available nutrients in a lake due to artificial fertilisers on the surrounding land or to urban organic pollution may be regarded as unintentional experiments. It cannot be denied that, intentional or not, they have provided some useful ecological information.

Some might agree to stretch the definition of 'experiment' to include observations on the effects of definable natural events – experiments set up by Nature herself. Such have in fact provided the best opportunities for ecological research. An alteration of climate or some geological event may cause changes in a lake ecosystem which can be analysed and recorded. The most familiar examples are the seasonal events such as fluctuations in production and biomasses which have the advantage for an investigator of frequent repetition. Less frequent and spasmodic events, such as contraction, salinification and even total desiccation through evaporation (e.g. Lake Chilwa), the destruction by a volcanic lava flow of life and its regeneration as in the Bay of Kabuno in Lake Kivu, and the occasional 'fish kills' caused by sudden very violent windstorms, provide opportunities, though less easily taken, for the study of some important aspects of lake ecosystems.

Each set of observations or 'experiments' is aimed at establishing the separate links in a chain of events, such as those between a windstorm and an outburst of algal production, and each link must be examined separately to avoid confusion. But in a natural ecosystem every organism is influenced simultaneously by more than one of the physical features of the environment and at the same time is interacting, directly or indirectly, with most of the other organisms at least in its own section of the ecosystem, and these may be numbered in hundreds of species. To complicate matters further the system is heterogeneous and is composed of partially separate though interdependent regions, e.g. littoral, benthic, pelagic, etc., which differ in ecological structure. An apparent state of stability within certain seasonal limits is illusory, though the overall long-term drift may be too slow to be of immediate practical interest. But Lake Chilwa is an example of a lake whose main ecological feature is a periodic fluctuation in the basic character of the whole system over a period of a few years – a matter of great practical importance.

How can we hope to master the intricacies of such a complex system sufficiently to predict its future history and to manipulate it for our greater benefit? The technique of 'systems analysis' is at present being applied to ecology with these objectives in view. This is a mathematical procedure involving computers and has been developed especially for military and industrial operational research. It is designed to deal with a mass of simultaneously interacting components with a view to predicting the future course of events from present information and to devise means of leading them in desired directions. In applied ecology the latter usually means the control of the production of certain component species (resource management), the prevention of diseases and pests, and combating pollution. It has also been applied to such mysterious problems as that presented by bird migration which will certainly be understood only in terms of several simultaneously operating controls.*

There is no cause for alarm among biologists. The mathematical techniques and electronic machines are designed only for a final stage in the ecological interpretation, usually for practical purposes, of information got from observation and experiment. The full extent to which these techniques will succeed in giving useful answers to important questons in applied ecology has yet to be demonstrated, but no good will come from them unless all the data supplied by ecologists and experimenters are sufficiently comprehensive, accurate and relevant to the questions to which answers are sought. There is a danger that, like some other sophisticated techniques and complicated equipment, they may so fascinate the operator that he is driven to apply them to problems for which the data are inadequate or of doubtful relevance and that could better be approached by more antiquated methods.

The management of resources is, however, not the only object of research in African limnology. The scientific value of much of the work needs no further emphasis, and some of it can be of practical value in other fields. In answer to the question, 'In what important aspect of limnology can tropical Africa offer better opportunities for research than any other region of the world?' I should reply, 'The origin and evolution of lake ecosystems'. There are two reasons for this. First, all of the Great Lakes, except Chad which may conveniently be regarded as a kind of 'control', originated from the post-Miocene tectonic disturbances. Much of the geological and palaeontological features of the origins and history of the lakes and their ecology are now known. We can confidently expect more of such information from which to reconstruct the story in greater detail. Second, the post-Miocene history is now being repeated in some important respects by the construction of artificial lakes of a size comparable with that of the Great Lakes. The initial chemical and biological readjustments of these ecosystems during the first few years are being observed, and we await with interest the first signs of the genetic changes which we must ultimately expect. These 'experiments' have important implications much beyond the confines of limnology.

---

* For introductions on the use of these techniques in ecology see the set of essays edited by Watt (1966), the review by Dale (1970) and Chapter 10 by C. J. Walters in Odum (1971). For a critical and provocative assessment see Rigler (1976).

Appendix A

Figs. A1–3 illustrate for the non-specialist a few examples of common types of algae and invertebrate animals that form the plankton communities of African lakes. The species shown are not necessarily the most common, but are chosen because suitable illustrations are available. They are not intended as a means of species identification. Some other groups of animals that are often found in the plankton are not included, e.g. insects other than Diptera, Hydrachnida and Protozoa.

There is no comprehensive illustrated book on the invertebrate fauna of African inland waters. Since, however, all the main groups of invertebrates, very many genera, and several species are cosmopolitan, such works as Ward and Whipple (1959) and the annexes (pp. 620–676) in Dussart (1966) are useful to the beginner. For the large aquatic vegetation (macrophytes) see Lind and Morrison (1974), which deals with the flowering plants of Uganda, but most of the genera and many of the species of aquatic macrophytes are widespread over tropical Africa. Some of the many publications dealing with particular groups of organisms in certain areas are mentioned in the text.

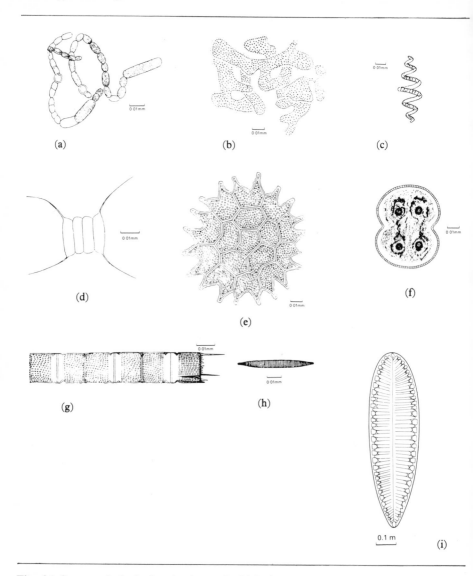

Fig. A1 Some typical planktonic algae, of which there are several hundreds of species in the tropical African Lakes (all reproduced from van Meel, 1954)

Cyanophyceae (blue-green algae)

(*a*) *Anabaena flos aquae*
(*b*) *Microcystic aeruginosa*
(*c*) *Spirulina platensis*

Chlorophyceae (green algae)

(*d*) *Scenedesmus quadricauda*
(*e*) *Pediastrum boryanum*
(*f*) *Cosmarium conatum*

Bacillariophyceae (diatoms)

(*g*) *Melosira granulata*
(*h*) *Nitzchia palea*
(*i*) *Surirella tenera*

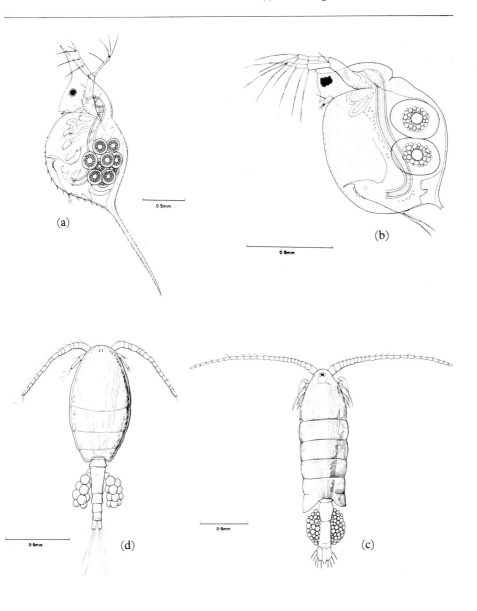

Fig. A2 Typical representatives of the many planktonic Crustacea and Rotifera. Some species of these two groups feed on small organic particles and algea, others prey on other small animals including members of their own group. They are themselves an important item in the food of many fish.

Crustacea Cladocera (redrawn from Sars, 1915, 1925)

(a) *Daphnia lumholtzi*

(b) *Ceriodaphnia caudata*

Crustacea Copepoda (redrawn from Sars, 1925)

(c) *Lovenula falcifera*

(d) *Mesocyclops neglectus*

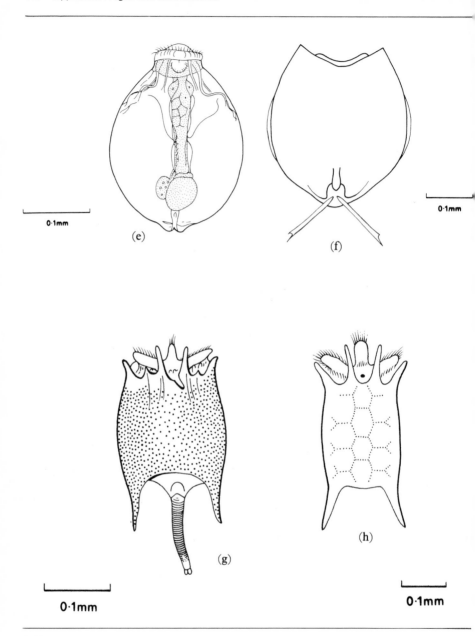

Rotifera (redrawn from Ward and Whipple, 1959)
(e) *Horaella brehmi*
(f) *Lecane luna.* Outline of lorica (outer shell) only
(g) *Brachionus quadridentatus*
(h) *Keratella quadrata*

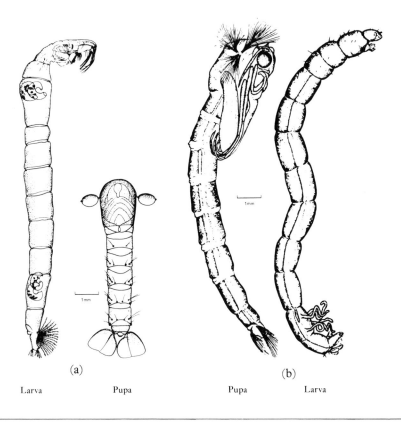

(a)   (b)

Larva   Pupa   Pupa   Larva

Fig. A3 Insecta Diptera (flies). Representatives of immature stages (larvae and pupae) of chaoborid and chironomid flies which are important in lake ecosystems. *Chaoborus* larvae are predacious and feed largely on Crustacea and Rotifera and migrate between the mud and the upper water. Chironomid larvae are often important as feeders on the bottom sediments. Their pupae rise to the surface and become temporarily planktonic. Both are consumed by fish.

(*a*) *Chaoborus anomalus* (from Verbeke, 1957b)
Larva (left) and Pupa right)
(*b*) *Chironomus* sp. (from Wesenberg-Lund, 1943)
Larva (right) and Pupa (left).

Appendix B

Some representative freshwater fish of tropical Africa.

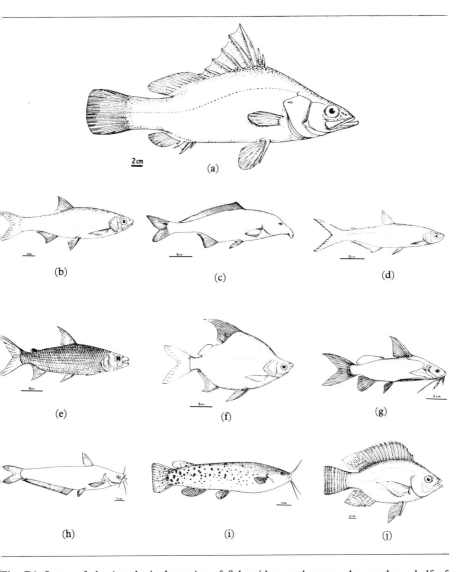

Fig. B1 Some of the 'soudanian' species of fish widespread across the northern half of tropical Africa (all from Greenwood, 1966).

a) *Lates niloticus* (Centropomidae) 'Nile perch'. Food: fish.
b) *Barilius niloticus* (Cyprinidae). Food: fish.
c) *Mormyrus kannume* (Mormyridae) 'elephant snout fish'. Food: mainly insect larvae, especially lake flies.
d) *Alestes baremose* (Characidae). Food: mainly small animals, zooplankton.
e) *Hydrocynus vittatus* (Characidae) 'tiger fish'. Food: fish.
f) *Citharinus latus* (Citharinidae). Food: plankton and small organisms on mud surface.
g) *Synodontis schall* (Mochokidae). Food: molluscs, insect larvae, small fish.
h) *Schilbe mystus* (Schilbeidae) 'butterfish'. Food: small fish, insect larvae.
i) *Malapterurus electricus* (Malapteruridae) 'electric catfish'. Food: mainly fish.
j) *Sarotherodon niloticus* (Cichlidae). Food: mainly phytoplankton.

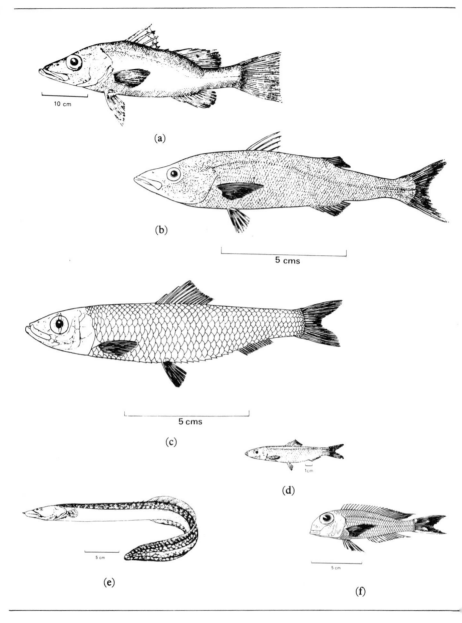

Fig. B2 Some of the endemic species of fish in Lake Tanganyika. The number of endemic species so far known is 176 out of a total of 214 (from Poll, 1952b).

(*a*) *Lates mariae* (Centropomidae).   Food: fish,
(*b*) *Luciolates stappersii* (Centropomidae) (endemic genus). including 'sardines'.
(*c*) *Limnothrissa miodon* (Clupeidae).   'Sardines' or *ndagala*. Food:
(*d*) *Stolothrissa tanganicae* (Clupeidae).   zooplankton.
(*e*) *Mastacembelus moorii* (Mastacembalidae) 'spiny eel'. Food: small fish and molluscs.
(*f*) *Xenotilapia sima* (Cichlidae) (endemic genus). Food: insect larvae, small molluscs.

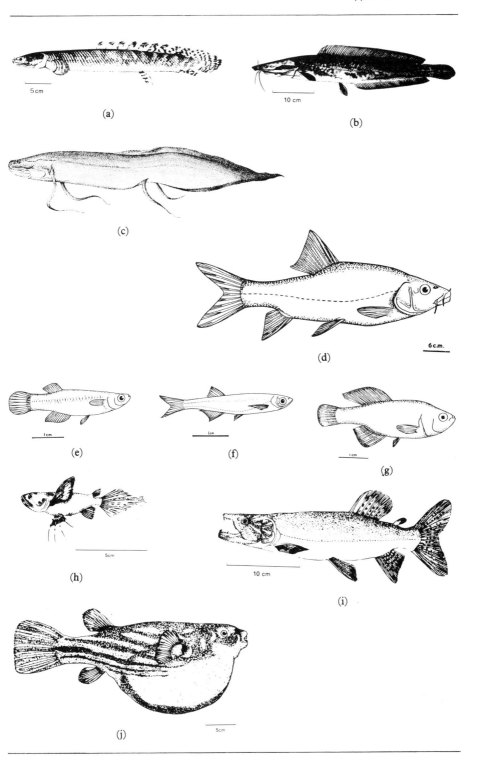

5 cm

(a)

10 cm

(b)

(c)

6 c.m.

(d)

1 cm

(e)

1 cm

(f)

1 cm

(g)

5cm

(h)

10 cm

(i)

5cm

(j)

Fig. B3 Some other fish mentioned in text: *a* and *b* from Poll, 1967; *c* to *g* from Green-wood, 1966; *h* to *j* from Reed, 1967.

(*a*) *Polypterus ornatipinnis* (Polypteridae). Congo basin. Food: probably small fish.

(*b*) *Clarias mossambicus* (Clariidae) 'airbreathing catfish)'. Lake Victoria, East African rivers, Lake Tsana, Lake Tanganyika. Food: omnivorous.

(*c*) *Protopterus aethiopicus* (Lepidosirenidae) 'lungfish'. Eastern half of tropical Africa. Food: molluscs, crabs, small fish.

(*d*) *Barbus altianalis* (Cyprinidae) 'barbel'. Lakes Victoria, Edward, George and Kivu. Food: aquatic vegetation, small fish, molluscs, etc.

(*e*) *Aplocheilichthys pumilus* (Cyprinodontidae). Lakes Victoria, Edward, George and Tanganyika. Food: zooplankton.

(*f*) *Engraulicypris argenteus* (Cyprinidae). Lake Victoria and Victoria Nile. Food: zooplankton.

(*g*) *Nothobranchius taeniopygus* (Cyprinodontidae) 'mosquito fish'. Affluents to Lake Victoria, widespread in Tanzania. Food: insect larvae, zooplankton.

(*h*) *Pantodon buchholzi* (Pantodontidae) 'butterfly fish' or 'freshwater flying fish'. Niger and Zaïre basin. Food: insects.

(*i*) *Hepsetus odoe* (Characidae) 'African river pike'. Niger and Zaïre basins, upper Zambezi. Food: small fish.

(*j*) *Tetraodon faka* (Tetraodontidae) 'puffer fish'. River Niger. Food: probably (like *T. mbu* of the Zaïre basin) molluscs and insects.

References

I.F.A.N. denotes Institut Francais d'Afrique Noire, Dakar, to 1965; Institut Fondamental d'Afrique Noire, Dakar, 1966 onwards.

O.R.S.T.O.M. denotes Office de la Recherche Scientifique et Technique Outre-Mer, Paris.

Those authors to whose names asterisks are attached have assisted with information and comments, for which I am very grateful.

Abid-Magid, A. M., 1971. 'The ability of *Clarias lazera* (Pisces) to survive without air breathing', *J. Zool.*, **163**: 63–72.

Abu-Gideira, Y. B. and **Ali, M. T.**, 1975. 'A preliminary biological survey of 'Lake Nubia', *Hydrobiologia*, **46**: 535–41.

Ackermann, W. C., White, G. F. and **Worthington, E. B.** (eds), 1973. *Manmade lakes: their problems and environmental effects*, Amer. Geophysical Union monogr. **17**: 847 pp., Washington, D.C.

Adam, W., 1957. 'Mollusques quarternaires de la région du lac Edouard', *Expl. Parc. Nat. Albert*, Mission J. de Heinzelin de Braucourt (1950), Fasc. 3.

Adam. W., 1959. 'Mollusques pleistocènes de la région du lac Albert et de la Semliki', *Ann. Mus. Roy. Roy. Belge.*, Vol. 25.

Adeniji, H. A., 1975. 'Some aspects of the limnology and the fishery development of Kainji Lake, Nigeria', *Arch. Hydrobiol.*, **75**: 253–62.

Aleem, A. A. and **Samaan, A. A.,** 1969. 'Productivity of Lake Mariut, Egypt. Part I. Physical and chemical aspects. Part II. Primary production'. *Int. Rev. ges. Hydrobiol.*, **54**: 313–55, 491–527.

Alimen, H., 1955. *Préhistoire d' Afrique*, N. Boubée et Cie, Paris.

Allanson, B. R. and **Gieskes, J. M. T. M.,** 1961. 'Investigations into the ecology of polluted inland waters in the Transvaal. Part II. An introduction to the limnology of Haartebeeespoort Dam with special reference to the effects of industrial and domestic pollution', *Hydrobiologia*, **18**: 77–94.

Ambroggi, R. P., 1966. 'Water under the Sahara'. *Scientific American*, **214**(5): 21–25.

Appleton, C. C., 1978. 'Review of literature on abiotic factors influencing the distribution and life cycles of bilharziasis intermediate snails', *Malacological Review*, **11**: 1–26.

Arad, A. and **Morton, W. H.,** 1969, 'Mineral springs and saline lakes of the Western Rift Valley, Uganda', *Geochim. et Cosmochim. Acta*, **33**: 1169–81.

Arambourg, C., 1935–1948. Mission Scientifique de l'Omo (1932–33), *Mus. Nat. d'Histoire Naturelle*, Paris. Fasc. 1 (1935), Fasc. 2 (1944), Fasc. 3 (1948).

Aubenton, F. d', 1955. 'Etude de l'appareil branchiospinal et de l'organes suprabranchial d'*Heterotis niloticus*', *Bull. I.F.A.N.* **17**: 1179–1201.

★ **Bailey, R. G.**, 1972. 'Observations on the biology of *Nothobranchius guentheri* (Preffer) (Cyprinodentidae), an annual fish from the coastal region of East Africa' *Afr. J. Hydrobiol. Fish.* **2**: 33–43.

Baker, B. H., Mohr, P. A. and **Williams, L. A. J.**, 1972. 'Geology of the Eastern Rift system of Africa', *Geol. Soc. Amer. special paper* **136**.

Baker, Samuel, W., 1866. *The Albert Nyanza, Great Basin of the Nile*, 2 vols. Macmillan, London, and later editions.

Bakker, E. M. van Z., 1962. 'Botanical evidence for quaternary climates in Africa', *Ann. Cape Prov. Mus.*, **2**: 16–31.

Bakker, E. M. van Z., 1964. 'A pollen diagram for Equatorial Africa, Cherangani, Kenya', *Geol. Mijnb.*, **43**: 123–8.

Balinsky, B. I., 1962. 'Patterns of animal distribution on the African Continent', *Ann. Cape Prov. Mus.*, **2**: 299–310.

Balinsky, B. I. and **James, G. V.**, 1960. 'Explosive reproduction of organisms in Kariba Lake', *S. Afr. J. Sci.*, **56**: 101–4.

Ball, J., 1939. *Contributions to the Geography of Egypt*, 308 pp., Cairo, Govt Press.

Balon, E. K., 1974. *Fishes of Lake Kariba*, 144 pp., T.F.H. Publications Inc., Neptune City, N.J.

Balon, E. K. and **Coche, A. G. (eds.), 1974.** *Lake Kariba: a manmade tropical ecosystem in Central Africa*, 767 pp., Junk, The Hague.

Balss, H., 1936. 'Beiträge zur Kenntnis der Potamonidae (Süsswasserkrabben) des Kongogebietes', *Rev. Zool. Bot. Afr.*, **28**: 165–204.

★ **Banage, W.**, 1966. 'Survival of a swamp nematode (*Dorylaimus* sp.) under anaerobic conditions', *Oikos*, **17**: 113–20.

★ **Banister, K. E.**, 1973. 'A revision of the large *Barbus* of East and Central Africa', *Bull., Brit. Mus. Nat. Hist. (Zool.)*, **26**: 1–148.

Banister, K. E. and **Bailey, R. G.**, 1979. 'Fishes collected by the Zaïre River Expedition 1974–75', *Zool. J. Linnaean Soc.*, **66**: 205–49.

Barber, W., 1965. 'Pressure water in the Chad, formation of Bornu and Dikwa Emirates, north-eastern Nigeria', *Geol. Survey of Nigeria*, Bull. No. 35: pp. 138, Lagos.

Barbosa, F. S. and **Barbosa, I.**, 1958. 'Dormancy during the larval stages of the trematode *Schistosoma mansoni* in snails aestivating on the soil of dry natural habitats', *Ecology*, **39**: 763–4.

★ **Barnard, K. H.**, 1927. 'A study of the freshwater isopodan and amphipodan Crustacea of South Africa', *Trans. Roy. Soc. S. Afr.*, **14**: 139–218.

Barnard, K. H., 1940. 'Contributions to the crustacean fauna of South Africa, XII Further additions to the Tanaidacea, Isopoda and Amphipoda, together with keys for the identification of hitherto recorded marine and freshwater species', *Ann. S. Afr. Mus.*, **32**: 381–543.

Barnard, K. H., 1943. 'Revision of the indigenous freshwater fishes of the S.W. Cape region', *Ann. S. Afr. Mus.*, **36**: 101–262.

Barnard, K. H., 1948: 'Report on a collection of fishes from the Okavango River, with notes on the Zambezi fishes', *Ann. S. Afr. Mus.*, **36**: 407–58.

Baxter, R. M. and **Wood, R. B.**, 1965. 'Studies on stratification in the Bishoftu Crater Lakes', *J. Appl. Ecol.* **2**: 416.

Baxter, R. M., Prosser, M. V., Talling, J. F. and **Wood, R. B.**, 1965. 'Stratification in tropical African lakes at moderate altitudes (1 500 to 2 000 m)', *Limnol. Oceanogr.*, **10**: 510–20.

★ **Bayly, I. A. E.** and **Williams, W. D.**, 1966. 'Chemical and biological studies on some saline lakes of South East Australia', *Aust. J. Mar. Freshwat. Res.*, **17**: 177–228.

Bayly, I. A. E. and **Williams, W. D.**, 1973. *Inland Waters and their Ecology*, 315 pp., Longman (Australia).

Beadle, L. C., 1932a. 'Scientific results of the Cambridge Expedition to the East African Lakes 1930–31. 3. Observations on the bionomics of some East Africa Swamps', *J. Linn. Soc. (Zool)*, **38**: 135–55.

Beadle, L. C., 1932b. 'Scientific results of the Cambridge Expedition to the East African Lakes 1930–31. 4. The waters of some East African Lakes in relation to their fauna and flora', *J. Linn. Soc. (Zool.)*, **38**: 157–211.

Beadle, L. C., 1932c. 'Scientific results of the Cambridge Expedition to the East African Lakes 1930–31. 13. Adaptation to aerial respiration in *Alma emini* Mich., an oligochaet from East African Swamps', *J. Linn. Soc. (Zool.)*, **38**: 347–50.

Beadle, L. C., 1943a. 'An ecological survey of some inland saline waters of Algeria', *J. Linn. Soc. (Zool.)*, **41**: 218–42.

Beadle, L. C., 1943b. 'Osmotic Regulation and the faunas of inland waters', *Biol. Revs.*, **18**: 172–83.

Beadle, L. C., 1957. 'Respiration in the African swamp worm *Alma emini*, Mich', *J. Exp. Biol.*, **34**: 1–10.

Beadle, L.C., 1958. 'The measurement of oxygen in swampwater. Further modification of the Winkler method', *J. Exp. Biol.*, **35**: 556–566.

Beadle, L. C., 1959. 'Osmotic regulation in relation to the classification of brackish and inland saline waters', *Arch. Oceanogr. Limnol. Roma* 11 (Suppl.): 143–51.

Beadle, L. C., 1961. 'Adaptations of some aquatic animals to low oxygen levels and to anaerobic conditions', *Sympos. Soc. Exp. Biol.*, **15**: 120–31.

Beadle, L. C., 1962. 'The evolution of species in the lakes of East Africa', *Uganda J.*, **26**: 44–54.

Beadle, L. C., 1963. 'Anaerobic life in a tropical crater lake', *Nature*, **200**: 1223–4.

Beadle, L. C., 1966. 'Prolonged stratification and deoxygenation in tropical lakes. I. Crater Lake Nkugute, Uganda, compared with lakes Bunyoni and Edward', *Limnol. Oceanogr.*, **11**: 152–63.

Beadle, L. C., 1968. 'East African Lakes', in Fairbridge, R. W. (ed.) *The Encyclopedia of Geomorphology*: 303–8. Reinhold.

Beadle, L. C., 1969. 'Osmotic regulation and the adaptation of freshwater animals to inland saline waters', *Verh. int. Ver. Limnol.*, **17**: 421–9.

Beadle, L. C. and **Lind, E. M.**, 1960. 'Research on the Swamps of Uganda', *Uganda J.*, **24**: 84–97.

Beadle, L. C. and **Nilsson, J. R.**, 1959. 'The effect of anaerobic conditions on two heterotrich Ciliate Protozoa from papyrus swamps', *J. Exp. Biol.*, **36**: 583–9.

Beadle, L. C., **Thomas, I. F.** and **Poole, D. F. G.**, 1960. 'Early development of *Limnocnida victoriae*. Gunther (Limnomedusae)', *Proc. Zool. Soc. Lond.*, **134**: 217–19.

*Beauchamp, R. S. A., 1939. 'Hydrology of Lake Tanganyika', *Int. Rev. Hydrobiol.*, **39**: 316–53.

Beauchamp, R. S. A., 1940. 'Chemistry and hydrography of Lakes Tanganyika and Nyasa', *Nature*, **146**: 253–6.

Beauchamp, R. S. A., 1946. 'Lake Tanganyika', *Nature*, **157**: 183–4.

Beauchamp, R. S. A., 1953a. 'Sulphates in African inland waters', *Nature*, **171**: 769–71.

Beauchamp, R. S. A., 1953b. 'Hydrological data from Lake Nyasa', *J. Ecol.*, **41**: 226–39.

Beauchamp, R. S. A., 1956. 'The electrical conductivity of the head waters of the White Nile', *Nature*, **178**: 616–19.

Beauchamp, R. S. A., 1964. 'The Rift Valley Lakes of Afrika', *Verh. int. Ver. Limnol.*, 15: 91–9.

*Becquerel, P., 1950. 'La suspension de la vie des spores des bactéries et de moisissures déséchées dans le vide, vers le zero absolu. Ses conséquences pour la dissémination et la conservation de la vie dans l'Univers', *C. r. Séanc. Soc. Acad. Sci.* Paris, 231: 1392–4.

Begg, G. W., 1976. 'The relationship between the diurnal movements of some zooplankton and the sardine *Limnothrissa miodon* in Lake Kariba, Rhodesia. *Limnol. Oceanogr.*, 21: 529–539.

*Bell-Cross, G., 1965. 'Movement of fish across the Congo-Zambezi watershed in the Mwinilunga District of Northern Rhodesia', *Proc. Centr. Afr. Sci. Med. Congr.*, Lusaka (August 1963): 415–24.

Bell-Cross, G., 1968. 'The distribution of fishes in Central Africa', *Fish. Res. Bull. Zambia*, 4: 1–20.

Bell-Cross, G., 1972. 'The fish of the Zambezi River System', *Arnoldia (Rhodesia)*, 5: 1–19.

Ben, D. van der, 1959. 'La végétation des rives des lacs Kivu, Édouard et Albert', *Explor. Hydrobiol. Lacs Kivu, Édouard et Albert (1952–1954), Résultats Scientifiques*, IV, Fasc. I, 191 pp., Inst. Roy. Sci. Nat. Belg. Bruxelles.

Benech, V., Lemoalle, J. and Quensiere, J., 1976. 'Mortalités des poissons et conditions de milieu dans le Lac Tchad au cours d'une periode de 'sécheresse', *Cah. O.R.S.T.O.M.*, sér. *Hydrobiol.*, X: 119–30.

Bennett, M. V. L., 1971. 'Electric organs', in Hoar, W. S. and Randall, D. J. (eds) *Fish Physiology*, Vol. 4: 347–491. Acad. Press.

Bentley, P. S., 1966. 'Adaptations of Amphibia to arid environments', *Science*, 152: 619–23.

Bergstrand, E. and Cordone, A. J., 1971. 'Exploratory bottom trawling in Lake Victoria', *Afr. J. Hydrobiol. Fish.*, 1: 13–23.

Berns, S. and Peters, H. M., 1969. 'On the reproductive behaviour of *Ctenopoma muriei* and *Ctenopoma damasi* (Anabantidae)', *Ann. Rep. E. Afr. Freshw. Fish. Res. Org.* (1968): 44–9.

*Berrie, A. D., 1970. 'Snail problems in African schistosomiasis', *Adv. Parasitol.*, 8: 43–96.

Berry, L., 1976. 'The Nile in the Sudan, a geomorphological history', in Rzóska (1976a): 11–19.

Berry, L. and Whiteman, A. J., 1968. 'The Nile in the Sudan', *Geogr. J.*, 134: 1–36.

Bertin, L., 1958. 'Organes de la respiration aérienne', in P-P. Grassé (ed.), *Traité de Zoologie*. T. XIII. Fasc. 2: 1363–98. Masson Paris.

Bertram, C. K., Borley, H. J. H. and Trewavas, E., 1942. *Report on the fish and fisheries of Lake Nyasa*. Crown Agents, London.

Beyers, R. J., 1963. 'The metabolism of twelve aquatic laboratory microecosystems', *Ecol. Monogr.*, 33: 281–306.

Bidwell, A., 1976. 'The effect of water level fluctuations on the biology of Lake Kainji, Nigeria', *Nigerian Field*, 41: 156–65.

Bidwell, A. and Clarke, N. V., 1977. 'The invertebrate fauna of Lake Kainji, Nigeria', *Nigerian Field*, 42: 104–10.

Bini, G., 1940a. 'Richerche chimiche nelle acque del Lago Tana', in *Missione di studio al Lago Tana*. III(2): 7–52.

Bini, G., 1940b. 'I pesce del Lago Tana'. in *Missione di studio al Lago Tana*, III(2): 135–206.

Bishai, H. M., 1962. 'The water characteristics of the Nile in the Sudan with a note on the effect of *Eichhornia crassipes* on the hydrobiology of the Nile', *Hydrobiologia*, **19**: 357–82.

*Bishop, W. W., 1965. 'Quaternary Geology and Geomorphology in the Albertine Rift Valley, Uganda', *Geol. Soc. Am. spec. paper* No. **84**: 293–321.

Bishop, W. W., 1969. *Pleistocene Stratigraphy in Uganda*, Geol. Survey of Uganda, Mem. No. X, 128 pp., Govt. Printer, Entebbe.

Bishop, W. W., 1971. The late cenozoic history of East Africa in relation to hominid evolution', Ch. 19 in Turekian K. K. (ed.) *The Late Cenozoic Glacial Ages*, pp. 493–527, Yale Univ. Press.

Bishop, W. W. (ed.), 1978. *Geological Background to Fossil Man*, 564 pp, Scottish Academic Press, Edinburgh.

Bishop, W. W. and Clark, J. D., 1967. *Background to Evolution in Africa*, Univ. of Chicago Press.

Bishop, W. W. and Posnansky, M., 1960. 'Pleistocene environments and early man in Uganda', *Uganda J.*, **24**: 44–61.

Bishop, W. W. and Trendall, A. F., 1967. 'Erosion surfaces, tectonics and volcanic activity in Uganda', *Q. J. Geol. Soc. London*, **122**: 385–420.

Bishop, W. W., Miller, J. A. and Fitch, F. J., 1969. 'New potassium-argon age determinations relevant to the Miocene fossil mammal sequence in East Africa', *Am. J. Sci.*, **276**: 669–99.

Biswas, S., 1966a. 'Oxygen and phytoplankton changes in the newly forming Volta Lake in Ghana', *Nature*, **209**: 218–19.

Biswas, S., 1966b. 'Ecological Studies of Phytoplankton in the newly forming Volta Lake of Ghana', *J.W. Afr. Sci. Assoc.*, **11**: 14–19.

Biswas, S., 1968. 'Hydrobiology of the Volta River and some of its tributaries before the formation of the Volta Lake', *Ghana J. Sci.*, **8**: 152–66.

Biswas, S., 1969. 'The Volta Lake: some ecological observations on the phytoplankton', *Verh. int. Ver. Limnol.*, **17**: 259–72.

Biswas, S., 1973. 'Limnological observations during the early formation of Volta Lake in Ghana', in Ackermann *et al.*

Blache, J., 1964. *Les Poissons du Bassin du Tchad et du Bassin adjacent du Mayo Kebbi. Etude systématique et biologique*, Mém. O.R.S.T.O.M. 4, Tome II, 483 pp.

Blache, J. and Miton, F., 1963. *Première Contribution à la Connaissance de la Pêche dans le Bassin Hydrographique Logone – Chari – Lac Tchad*, Mém. O.R.S.T.O.M. 4, Tome I, 144 pp.

Blake, B. F., 1977a. 'The effect of the impoundment of Lake Kainji, Nigeria, on the indigenous species of mormyrid fishes', *Freshwater Biol.*, **7**: 37–42.

Blake, B. F., 1977b. 'Lake Kainji, Nigeria: a summary of the changes within the fish population since the impoundment of the Niger in 1968', *Hydrobiologia*, **53**: 131–7.

Blanc, M., 1954. 'La répartition des poissons d'eau douce africaine', *Bull. I.F.A.N. (A)* **16**: 599–628.

Blanc, M. and Daget, J., 1957. *Les Eaux et les Poissons du Haute-Volta*, Mém. I.F.A.N. 70 pp. Dakar.

Blanc, M., Daget, J. and Aubenton, F. d', 1955a. 'Recherches hydrobiologiques dans le bassin du Moyen Niger', *Bull. I.F.A.N. (A)*, **17**: 680–746.

Blanc, M., Daget, J. and Aubenton, F. d', 1955b. 'L'Exploitation des eaux douces dans le bassin du Moyen Niger', *Bull, I.F.A.N. (A)*, **17**: 1157–73.

Blanton, J. O., 1973. 'Vertical entrainment into the epilimnia of stratified lakes', *Limnol. Oceanogr*, **18**: 697–704.

Bond, W. J. and Roberts, M.G., 1978. 'The colonisation of Cabora Bassa, Moçambique, a new manmade lake, by floating vegetation', *Hydrobiologia*, **60**: 243–59.

Bond, W. J., Coe, N., Jackson, P. B. N. and **Rogers, K. H.**, 1978. 'The limnology of Cabora Bassa, Moçambique, during its first year', *Freshwater Biology*, **8**: 433–47.

Bont, A. F. de, 1969. 'The status of limnological knowledge of the Congo Basin', *Rep. Regional Meeting of Hydrobiologists in Tropical Africa, Kampala*, 1968: 15–31. Nairobi.

Bott, R., 1955. 'Die Süsswasserkrabben von Afrika (Crustacea Decapoda) und ihre Stammesgeschichte', *Ann. Mus. Roy. Cong. Belge.*, Sect. C. Zool. (3) **1**: 209–352.

Bouchardeau, A., 1958. 'Le lac Tchad', *Ann. Hydrol. France Outre Mer.*, 1956 O.R.S.T.O.M., Paris: 9–26.

Boughey, A. S., 1963. 'The explosive development of a floating weed vegetation on Lake Kariba', *Adamsonia*, **3**: 49–61.

Bouillon, J., 1957. 'Etude monographique de genre *Limnocnida* (Limnomedusae)', *Ann. R. Soc. Zool. Belg.*, **87**: 254–500.

Boulenger, G. A., 1902. 'Descriptions of new fishes from the collection made by Mr E. Degen in Abyssinia', *Ann. Nat. Hist.*, **10**: 421–39.

Boulenger, G. A., 1907. 'The fishes of the Nile', in Anderson's *Zoology of Egypt*, 2 vols, Hugh Rees, London.

Boulenger, G. A., 1909–1916. *Catalogue of the freshwater fishes of Africa in the British Museum (N.H.)*, London, Vol. I–IV.

Bourn, D. M., 1974. 'The feeding of three commercially important fish species in Lake Chilwa, Malawi', *Afr. J. Hydrobiol. Fish.*, **3**: 135–45.

Boutakoff, N., 1937. 'Sur l'ecoulement vers le Nord du lac Tanganika au Pleistocène', *Ac. R. Belg. Bull. Cl. Sci.*, (5), **23**: 703–15.

Bovill, E. W., 1968a. *The Golden Trade of the Moors*, 2nd edn, 281 pp. Oxford Univ. Press.

Bovill, E. W., 1968b. *The Niger Explored*. 263 pp. Oxford Univ. Press.

★ **Bowmaker, A. P.**, 1968. 'Some Upper Congo fish which offer a means of biological control of the snail vectors of bilharziasis', *Proc. Trans. Rhod. Assoc.*, **52**: 28–37.

Bowmaker, A. P., 1969. 'Contribution to knowledge of the biology of *Alestes macrophthalmus* Gunther (Pisces: Characidae)', *Hydrobiologia*, **33**: 302–41.

Bowmaker, A. P., Jackson, P. B. N. and **Jubb, R. A.**, 1978. 'Freshwater fishes', Ch. 36 in M. J. A. Werger (ed.).

Brachi, R. M., 1960. 'Excavation of a rock shelter at Hippo Bay, Entebbe', *Uganda J.*, **24**: 62–70.

Bradford, G. R., Bair, F. L. and **Hunsaker, V.**, 1968. 'Trace and major element content of 170 High Sierra lakes in California', *Limnol. Oceanogr.*, **13**: 526–9.

Braestrup, F. W., 1947. 'Remarks on faunal exchange through the Sahara', *Vidensk. Medd. naturh. Foren.*, **110**: 1–15.

Brand, T. Von, 1946. 'Anaerobiosis in invertebrates', *Biodynamica Monogr.*, **4**: 1–328.

Brien, P. and **Bouillon, J.**, 1959. 'Ethologie des larves de *Protopterus dolloi* BLGR et étude de leur organes respiratoires', *Ann. Mus. R. Congo Belge. Sci. Zool.*, **71**: 25–70.

Brien, P., Poll, M. and **Bouillon, J.**, 1959. 'Ethologie de la réproduction de *Protopterus dolloi* BLGR', *Ann. Mus. R. Congo Belge. Sci. Zool.*, **71**: 3–23.

Brock, T. D., 1967. 'Micro organisms adapted to high temperatures', *Nature*, **214**: 882.

Brook, A. J., 1952. 'A systematic account of the phytoplankton of the Blue and White Nile', *Ann. Mag. Nat. Hist.*, Ser. 12, **7**: 648–56.

Brook, A. J. and **Rzóska, J.**, 1954. 'The influence of the Gebel Aulyia Dam on the development of Nile plankton', *J. Anim. Ecol.*, **23**: 101–14.

★ **Brooks, J. L.**, 1950. 'Speciation in ancient lakes', *Q. Rev. Biol.*, **25**: 30–60, 131–76.

Brooks, J. L., 1957. 'The species problem in freshwater animals', in E. Mayr (ed.) *The Species Problem*. Am. Assoc. Adv. Sci. Washington D. C. **50**: 81–123.

Brooks, J. L., 1965. 'Predation and relative helmet size in cyclomorphic *Daphnia*', *Proc. Nat. Acad. Sci. U.S.A.*, **53**: 119–26.

*Brown, D. S., 1965. 'Freshwater gastropod Mollusca from Ethiopia', *Bull. Brit. Mus. Nat. Hist.*, **12**: 39–94.

Brown, G. W. (ed.), 1969, 1974. *Desert Biology*. Vol. I (1969), Vol. II (1974), Academic Press, N. Y.

Brunelli, G., 1940. 'Le mutazione del genere *Barbus* del Lago Tana', in *Missione di studio al Lago Tana*, III: 209–12.

Brunelli, G. and Cannicci, G., 1940. 'Le caratteristiche biologiche del Lago Tana', in *Missione di studio al Lago Tana*, III: 69–1134.

Burgis, M. J., 1969. 'A preliminary study of the ecology of zooplankton in Lake George, Uganda', *Verh. int. Ver. Limnol.*, **17**: 297–302.

Burgis, M. J., 1970. 'The effect of temperature on the development time of eggs of *Thermocyclops* sp., a tropical cyclopoid copepod from Lake George, Uganda', *Limnol. Oceanogr.*, **15**: 742–7.

Burgis, M. J., 1971a. 'An ecological study of zooplankton of Lake George, Uganda', Ph.D. Thesis, Univ. London.

Burgis, M. J., 1971b. 'The ecology and production of copepods, particularly *Thermocyclops hyalinus*, in the tropical Lake George, Uganda', *Freshwat. Biol.*, **1**: 169–92.

Burgis, M. J., 1974. 'Revised estimates for the biomass and production of zooplankton in Lake George, Uganda, *Freshwater Biol.*, **4**: 535–41.

Burgis, M. J., 1978. 'The Lake George ecosystem', *Verh. int. Ver. Limnol.*, **20**: 1139–52.

Burton, R. F., 1860. *The Lake Regions of Central Africa*. 2 Vols. Longmans Green, London.

Butcher, D. A. P., 1973. 'Sociological aspects of fishery development on Volta Lake', in Ackermann *et al.* (eds), 108–13.

Butzer, K. W., 1966. 'Climatic changes in the arid zones of Africa during early to mid-Holocene times', in *Proc. Int. Sympos. World Climate from 8000–0* B.C. *Roy. Met. Soc.*, 72–83.

Butzer, K. W., 1971a. 'The lower Omo Basin: Geology, fauna and hominids of Plio-Pleistocene formations', *Nature*, **58**: 7–16.

Butzer, K. W., 1971b. *Recent History of an Ethiopean Delta*, 184 pp. Univ. Chicago, Dept. of Geography, Res. Paper No. 136.

Butzer, K. W., 1972. *Environment and Archaeology*, 2nd edn, 703 pp. Methuen.

Butzer, K. W. and Thurber, D. L., 1969. 'Some late coenozoic sedimentary formations of the lower Omo basin', *Nature*, **222**: 1138–42.

Butzer, K. W., Isaac, G. L., Richardson, J. L. and Washbourn-Kamau, C., 1972. 'Radiocarbon dating of East African lake levels', *Science*, **175**: 1069–76.

Cadwalladr, D. A. and Stoneman, J., 1968. 'The siting and relative importance of the fish landings of Lake Albert', *Uganda J.*, **32**: 39–52.

Cahen, L., 1954. *Géologie du Congo Belge*, 577 pp. Vaillant Carmanne, Liège.

Cain, A. J., 1969. 'Speciation in tropical environments: summing up', *Biol. J. Linn. Soc.*, **1**: 233–6.

Cambridge Expedition 1932–33. 'Scientific results of the Cambridge Expedition to the East African Lakes, 1930–31. Nos. 1–14', *Journ. Linn. Soc. Lond.*, **38**: 99–362.

Campbell, A. C., 1977a. 'The Okavango Delta', *African Wildlife*, **31**: 13–27.

Campbell, A. C., 1977b. 'Traditional utilisation of the Okavango Delta', in Nteta and Hermans, 163–74.

Camus, C., 1965. 'Fluctuations du niveau du lac Tanganika', *Bull. Acad. roy. S. Outre-Mer*, nouv. sér. **11**: 1242–56.

*Capart, A., 1949. 'Sondages et carte bathymétrique', *Expl. hydrobiol. Lac Tanganika (1946–47)*, **II**: 1–16.

Capart, A., 1952a. 'Le Lac Tanganika et sa faune', *Bull. Soc. Zool. France*, **77**: 245–51.

Capart, A., 1952b. 'Le milieu géographique et geophysique', *Explor. Hydrobiol. Lac Tanganika*, **I**: 3–27.

Capot-Rey, R., 1953. *Le Sahara Français*, Presses Universitaires de France, Paris.

Carlisle, D. B., 1968. '*Triops* (Entomostraca) eggs killed only by boiling', *Science*, **161**: 279–80.

Carmouze, J-P., 1969. 'La salure globale et les salures spécifiques des eaux du lac Tchad', *Cah. O.R.S.T.O.M. Sér. Hydrobiol.*, **3**: 3–14.

Carmouze, J-P., 1970. 'Salures globales et spécifiques des eaux du Lac Tchad', *Cah. O.R.S.T.O.M. Sér. Géol.*, **2**: 61–5.

Carmouze, J-P., 1971. 'Circulation générale des eaux dans le Lac Tchad', *Cah. O.R.S.T.O.M. Sér. Hydrobiol.*, **5**: 191–212.

Carmouze, J-P., 1973. 'Régulation hydrique et saline du Lac Tchad', *Cah. O.R.S.T.O.M. Sér. Hydrobiol.*, **7**: 17–23.

Carmouze, J-P., 1976. 'Les grands traits de l'hydrologie et de hydrochimie du Lac Tchad', *Cah. O.R.S.T.O.M. Sér. Hydrobiol.*, **10**: 33–56.

Carmouze, J-P., Dejoux, C., Durand, J. R., Gras, R., Iltis, A., Lauzanne, L., Lemoalle, J., Lévêque, C., Loubens, G., Saint-Jean, L., 1972. 'Contribution à la connaissance du Bassin Tchadien. Sommaire. Grandes zones écologiques du Lac Tchad', *Cah. O.R.S.T.O.M. Sér. Hydrobiol.*, **6**: 103–69.

Carmouze, J-P., Arce, C. and Quintanilla, J., 1977. 'Larégulation hydrique des lacs Titicaca et Poopó'. *Cah. O.R.S.T.O.M. Sér. Hydrobiol.* **XI**: 269–283.

*Carter, G. S., 1955. *The Papyrus Swamps of Uganda*. 15 pp., Heffer, Cambridge.

Carter, G. S., 1957. 'Air breathing', in M. E. Brown (ed.) *The Physiology of Fishes*, **I**: 65–79, Academic Press, New York.

Carter, G. S. and Beadle, L. C., 1930–31. 'The fauna of the swamps of the Paraguayan Chaco in relation to its environment. I. Physico-chemical nature of the environment. II. Respiratory adaptations in the fishes. III. Respiratory adaptations in the Oligochaeta', *J. Linn. Soc. (Zool.)*, **37**: 205–58, 327–68, 379–86.

Chaisemartin, C., 1965. 'La teneur en calcium de l'eau et l'équilibre calciques chez l'Ecrevisse *Astacus pallipes* (Lereboullet) en période d'intermue', *C. r. Séanc. Soc. Biol.*, **159**: 1214–17.

Chambers, R., 1970. *The Volta Resettlement Scheme*, 286 pp. Pall Mall.

Chappuis, P. A., 1939. 'Le peuplement du Lac Rodolphe et la répartition des Mormyridae dans le nord-est de l'Afrique', *C. R. somm. Soc. Biogeograph.*, **16**: 55–7.

Chouret, A., 1978. 'La persistance des effets de la sécheresse sur le Lac Tchad', in R. L. Welcomme, 74–91.

Clark, J. D., 1962. 'The Kalambo Falls prehistoric site', *Actes IVe Congr. Panafr. Préhist. (Leopoldville 1959)*: 195–202.

Clements, A. N., 1963. *The Physiology of Mosquitoes*. 303 pp. Pergamon, Oxford.

Cloudesley-Thompson, J. L., 1967. 'Diurnal rhythm, temperature and water relations of the African Toad *Bufo vulgaris*', *J. Zool.*, **152**: 43–54.

Coche, A. G., 1968. 'Description of physico-chemical aspects of Lake Kariba, an impoundment in Zambia–Rhodesia', *Fish, Res. Bull. Zambia*, **5**: 200–67.

Coche, A. G., 1974. 'Limnological study of a tropical reservoir', in E. K. Balon and A. G. Coche.

Cochrane, N. J., 1957. 'Lake Nyasa and the Shire River', *Proc. Inst. Civ. Eng.*, **8**: 363–382.

Cockson, A., 1971. 'Histochemical and chromatographic studies on the kidney of

Aspatharia sp. (Mollusca; Bivalvia) and its associated "kidney stones"', *Rev. Ciênc. Biol.*, **4**, Sér. A: 1–8.

Coe, M. J., 1966. 'The Biology of *Tilapia grahami* in Lake Magadi, Kenya', *Acta Tropica*, **23**: 146–177.

Coe, M. J., 1969. 'Observations on *Tilapia alcalica* in Lake Natron on the Kenya–Tanzania border', *Rev. Zool. Bot. Afr.*, **80**: 1–14.

Coetzee, J. A., 1964. 'Evidence for a considerable depression of vegetation belts during the Upper Pleistocene on the East Africa Mountains', *Nature*, **204**: 564–6.

Cole, G. A., 1969. 'Desert Limnology', in G. W. Brown (ed.) *Desert Biology*, Vol. 1: 424–87.

Cole, G. A. and Brown, R. J., 1967. 'The chemistry of *Artemia* habitats', *Ecology*, **48**: 858–61.

Coles, G. C., 1967. 'Modified carbohydrate metabolism in the tropical swampworm *Alma emini*', *Nature*, **216**: 685–6.

Coles, G. C., 1969a. 'The termination of aestivation in the large freshwater snail *Pila ovata* (Ampullariidae). I. Changes in oxygen uptake. II. *In vitro* studies', *Comp. Biochem. Physiol.*, **25**: 517–22; **29**: 373–81.

Coles, G. C., 1969b. 'Observations on weight loss and oxygen uptake of aestivating *Bulinus nasutus*, an intermediate host of *Schistosoma haematobium*', *Ann. Trop. Med. Parasitol.*, **63**: 393–8.

Coles, G.C., 1970a. 'Snail "metabolic hormone" and snail parasite metabolism', *Comp. Biochem. Physiol.*, **34**: 213–19.

Coles, G. C., 1970b. 'Some biochemical adaptations of the Swampworm *Alma emini* to low oxygen levels in tropical swamps', *Comp. Biochem. Physiol.*, **34**: 481–9.

Coles, S., 1954. *The Prehistory of East Africa*, 382 pp. Weidenfeld and Nicolson.

Compère, P., 1970. 'Contributions a l'étude des eaux douces de l'Ennedi. VI. Algues', *Bull. I.F.A.N. (A)*, **31**: 18–64.

Compère, P., 1977. 'Algues de la region du Lac Tchad. VII. Chlorophycophytes (3ᵉpartie: Desmidées)', *Cah. O.R.S.T.O.M. sér. Hydrobiol.*, **11**: 77–177.

Cooke, H. J., 1977. 'The palaeogeography of the middle Kalahari of northern Botswana and adjacent areas', in Nteta and Hermans, 21–8.

Cooke, M.B.S., 1958. 'Observations relating to quaternary environments in East and Southern Africa', *Trans. Geol. Soc. S. Afr.*, **20** (annex 74): 1–73

Coppens, Y., Howell, F., Clark, I., Glynn, L. and Leakey, R. E. F. (eds), 1976. *Early man and environments in the Lake Rudolf basin*, 615 pp., Univ. of Chicago Press.

Corbet, P. S., 1958. 'Lunar periodicity of aquatic insects in Lake Victoria', *Nature*, **82**: 330–1.

Corbet, P. S., 1961. 'The food of non-cichlid fishes in the Lake Victoria basin, with remarks on their evolution and adaptation to lacustrine conditions', *Proc. Zool. Soc. Lond.*, **136**: 1–101.

Corbet, P. S., 1964. 'Temporal patterns of emergence in aquatic insects', *Canadian Entomologist*, **96**: 264–79.

Corbet, S. A., Sellick, R. D. and Willoughby, N. G., 1974. 'Notes on the biology of the mayfly *Povilla adusta* in West Africa', *J. Zool. London*, **172**: 491–502.

*Cott, H. B., 1963. 'Scientific results of an enquiry into the ecology and economic status of the Nile Crocodile (*Crocodilus niloticus*) in Uganda and Northern Rhodesia', *Trans. Zool. Soc. Lond.*, **29**: 211–337.

*Coulter, G. W., 1963. 'Hydrological changes in relation to biological production in Southern Lake Tanganyika', *Limnol. Oceanog.*, **8**: 463–77.

Coulter, G. W., 1966. 'Hydrological processes and the deep-water fish community in Lake Tanganyika', thesis for Ph.D. Queen's Univ. of Belfast, 204 pp.

Coulter, G. W., 1967a 'Low apparent oxygen requirements of deep-water fishes in Lake Tanganyika', *Nature*, **215**: 317–18.

Coulter, G. W., 1967b. 'What's happening at Kariba?' *New Scientist*, **36**: 750–2.

Coulter, G. W., 1968. 'Thermal stratification in the deep hypolimnion of Lake Tanganyika', *Limnol-Oceanogr.*, **13**: 385–7.

Coulter, G. W., 1976. 'The biology of *Lates* species(Nile perch) in Lake Tanganyika, and the status of the pelagic fishery for *Lates* species and *Luciolates stappersi* (Blgr.)', *J. Fish. Biol.*, **9**: 235–59.

Couvering, J. A. van and **Miller, J. A.**, 1969. 'Miocene stratigraphy and age determinations, Rusinga Island, Kenya', *Nature*, **221**: 628–32.

Crawford, O. G. S., 1949. 'Some medieval theories about the Nile', *Geogr. J.*, **114**: 6–29.

Crowe, J. H., 1971. 'Anhydrobiosis: an unsolved problem, *Am. Nat.*, **105**: 563–73.

Crowley, T. E., Pain, T. and **Woodward, F. R.**, 1964. 'A monographic review of the Mollusca of Lake Nyasa', *Ann. Mus. Roy. de l'Afrique Centr. Tervuren, Belgique*, **131**: 1–58.

Cunnington, W. A., 1920. 'The fauna of the African lakes: a study in comparative limnology with special reference to Tanganyika', *Proc. Zool. Soc. London*, 507–622.

★ **Daget, J.**, 1954. 'Les Poissons du Niger Supérieur', *Mem. I.F.A.N.*, No. 36: 1–391.

Daget, J., 1957. 'Données récentes sur la biologie des poissons dans le delta central du Niger', *Hydrobiologia*, **9**: 321–47.

Daget, J., 1958. 'Sur la presence de *Porcus* cf *docmac* (Poisson Siluriforme) dans le gisement neolithique saharien de Faya', *Bull. I.F.A.N. (A)*, **20**: 1379–86.

Daget, J., 1959a. 'Note sur les poissons du Borkon–Ennedi–Tibesti', *Trav. Inst. Rech. Sahar. Alger.*, **18**: 173–81.

Daget, J., 1959b. 'Restes de *Lates niloticus* (Poissons Centropomidae) du Quaternaire saharien', *Bull. I.F.A.N.*, **21**: 1105–11.

Daget, J., 1960a. 'Contribution à la connaissance de la faune du fleuve Sénégal. Poissons du Baoulé et du Bakoy', *Bull. Mus. d' Hist. Nat.*, **32**: 506–12.

Daget, J., 1960b. 'Poissons de la Volta Noire et de la Haute Comoe', *Bull. Mus. Hist. Nat.*, 2ᵉ Sér. No. 4: 320–30.

Daget, J., 1960c. 'La faune ichtyologique du bassin de la Gambie', *Bull. I.F.A.N.*, **22**: 610–19.

Daget, J., 1961. 'Restes de poissons du Quaternaire Saharien', *Bull. I.F.A.N.*, **23**: 182–91.

Daget, J., 1962. 'Les poisson du Fouta Dialon et de la Basse Guinée', *Mém. I.F.A.N.* **65**: 1–207.

Daget, J., 1968. 'The distribution of freshwater fish in Western Africa' (in French and English) in *International Atlas of W. Africa*, 1st instalment, Org. Afr. Unity.

Daget, J. and **Iltis, A.**, 1965. 'Poissons de Côte d'Ivoire (eaux douces et saumâtres)', *Mém. I.F.A.N.*, **74**: 1–387.

Dale, M. B., 1970. 'Systems analysis and ecology', *Ecology*, **51**: 2–16.

Damas, H., 1937. 'Recherches hydrobiologiques dans les lacs Kivu, Edouard, et Ndalaga', *Exploration du Parc National Albert, Mission H. Damas (1935–1936)*, fasc. I., 128 pp.

Damas, H., 1938. 'La stratification thermique et chimique des lacs Kivu, Edouard et Ndalaga (Congo Belge)', *Verh. int. Ver. Limnol.*, **8**: 51–68.

Damas, H., 1954–55. 'Etude limnologique de quelques lacs ruandais. I. Le cadre géographique. II. Etude thermique et chimique. III. Le plancton', *Mém. Acad. roy. Sci. colon. Belge, Sect. Sci. nat. et med.*, **24**(2): 1–92; **24**(4): 1–116; N.S. **1**(3): 1–66.

Damas, H., 1964. 'Le plancton de quelques lacs d'Afrique Central', *Verh. int. Ver. Limnol.*, **15**: 128–38.

Darlington, J. P. E. C., 1977. 'Temporal and spatial variation in the benthic invertebrate fauna of Lake George, Uganda', *J. Zool. London*, **181**: 95–111.

Darlington, P. T., 1957. *Zoogeography*, 675 pp., Wiley, N. Y.

Darnell, R. M., 1964. 'Organic detritus in relation to secondary production in aquatic communities', *Verh. int. Ver. Limnol.*, **15**: 462–70.

Darteville, E. and **Swetz, J.**, 1947. 'Contribution a l'étude de la faune malacologique des grands lacs Africains. Les lacs Albert, Edouard et Kivu', *Mém. Int. Col. Belge*, **15**: 85–142.

David, L., 1935. 'Die Entwicklung der Clariidae und ihre Verbreitung', *Rev. Zool. Bot. Afr.*, **28**: 77–147.

David, L. and **Poll, M.**, 1937. 'Contribution à la faune ichthyologique du Congo Belge. Collections du Dr Schouteden et d'autres récolteurs', *Ann. Mus. Congo Belge, Tervuren*, Zool. Sér. I, t. III. fasc. 5.

Davidson, B., 1959. *Old Africa Rediscovered*, 287 pp., Gollancz.

Davies, B. R., Hall, A. and **Jackson, P. B. N.**, 1975. 'Some ecological aspects of the Cabora Bassa Dam', *Biol. Conservation*, **8**: 189–201.

Dawson, J. B., 1962. 'Sodium carbonate lavas from Oldoinyo Lengai, Tanganyika', *Nature*, **195**: 1075–6.

Debenham, F., 1947. 'The Bangweulu Swamp of Central Africa', *Geogr. Rev.*, **37**: 351–68.

Deevey, E. S., 1972. 'Biogeochemistry of lakes: major substances', in Likens, 14–20.

⋆**Degens, E. T., von Herzen, R. P.** and **Wong, H. K.**, 1971a. 'Lake Tanganyika: Water chemistry, sediments, geological structure', *Naturw.*, **58**: 229–41.

Degens, E. T., Deuser, W. G., von Herzen, R. P., Wong, H. K., Wooding, F. B., Jannasch, H. W. and **Kanwisher, J. W.**, 1971b. 'Lake Kivu Expedition: Geophysics, Hydrography, Sedimentology' (preliminary report), *Woods Hole Oceanogr. Inst.*

Dejoux, C., 1969. 'Les insects aquatiques du lac Tchad – Aperçu systématique et bioécologique', *Verh. int. Ver. Limnol.*, **17**: 900–6.

Dejoux, C., Lauzanne, L. and **Lévêque, C.**, 1969. 'Evolution qualitative et quantitative de la faune benthique dans la partie est du lac Tchad', *Chad. O.R.S.T.O.M. sér. Hydrobiol.*, **3**: 3–58.

Dejoux, C., Lauzanne, L. and **Lévêque, C.**, 1971.'Nature des fonds et repartition des organismes benthiques dans la region de Bol (Archipel est du lac Tchad)', *Cah. O.R.S.T.O.M. sér. Hydrobiol.*, **5**: 213–24.

Denny, P., 1972. 'Lakes of Southwestern Uganda. I. Physical and chemical studies on Lake Bunyoni'. *Freshw. Biol.*, **2**: 143–58.

Deuser, W. G., Degens, E.T. and **Harvey, G. R.**, 1973. 'Methane in Lake Kivu: new data bearing on its origin', *Science*, **181**: 51–4.

Devroey, E., 1938. 'Le problème de la Lukuga', *Inst. Roy. Col. Belge. Techn.*, Mém. I: 1–127.

Dieleman, P. J., and **de Ridder, N. A.**, 1964. 'Studies of salt and water movement in the Bol Guini Polder, Chad Republic', *Int. Inst. for Land Reclamation and Improvement, Wageningen.*, Bull. 5: 1–40.

Dixey, F., 1924. 'Variations in the level of Lake Nyasa', *Nature*, **114**: 659–61.

Dixey, F., 1938. 'The Nyasa-Shire Rift', *Geogr. J.*, **91**: 51–6.

Dobzhansky, T., 1959. 'Evolution in the tropics', *Am. Sci.*, **38**: 209–21.

Donelly, B. G., 1969. 'A preliminary survey of *Tilapia* nurseries on Lake Kariba during 1967/68', *Hydrobiologia*, **34**: 195–206.

Doornkamp, J. C. and **Temple, P. H.**, 1966. 'Surface drainage and tectonic instability of part of Southern Uganda', *Geogr. J.*, **132**: 238–51.

Douglas, E. L., Friedl, W. A. and **Pickwell, G. V.**, 1976. 'Fishes in oxygen-minimum zones: blood oxygen characteristics', *Science*, **191**: 957–9.

Dubois, J. T., 1955. 'Etude hydrobiologique d'un lac africain d'altitude: Le Lungwe. Étude physique et chimique' *Bull. Centre. Belge étude et de documentation des eaux*, **53**: 79–82.

Dubois, J. T., 1958a. 'Evolution de la température de l'oxygene dissous et de la transparence dans la Baie Nord du lac Tanganika', *Hydrobiologia*, **10**: 215–40

Dubois, J. T., 1958b. 'Composition chimique des affluents du nord du lac Tanganika', *Bull. Séances Acad. Roy. Sci. Colon.*, **4**: 1226–37.

Dubois, J. T., 1959. 'Note sur la chimie des eaux du lac Tumba', *Bull. Séances Acad. Roy. Sci. outremer*, **5**: 1321–34.

Dufalla, H., 1975. *The Nubian Exodus*, 342 pp., C. Hurst, London.

*Dumont, H. J., 1978. 'Neolithic hyperarid period preceding the present climate of the central Sahara', *Nature*, **274**: 356–8.

Dumont, H. J. and Decraemer, W., 1974. '*Nitocrella ioneli* n. sp. (Crustacea: Copepoda), a new phreatic harpacticoid copepod from the pre-Sahara in Morocco', *Biol. Jb. Dodonaea*, **42**: 105–11.

Dumont, H. J. and Decraemer, W., 1977. 'On the continental copepod fauna of Morocco', *Hydrobiologia*, **52**: 257–78.

Dumont, H. J. and Van Der Velde, I., 1975. 'Anostraca, Cladocera and Copepoda from Rio de Oro (Northwestern Sahara)', *Biol. Jb. Dodonaea*, **43**: 137–45.

Dumont, H. J., Miron, I., Dallaster, U., Decraemer, W., Claus, C. and Somers, D., 1973. 'Limnological aspects of some Moroccan Atlas lakes', *Int. Rev. ges. Hydrobiol.*, **58**: 33–60.

*Dunn, I. G., 1965. 'Notes on mass fish death following drought in South Malaya', *Malaysian Afr. J.*, **45**: 204–11.

Dunn, I. G., 1972. 'Ecological studies on the fish of Lake George, Uganda with particular reference to the cichlid genus *Haplochromis*', thesis for Ph.D., University of London, 240 pp.

Dunn, I. G., Burgis, M. J., Ganf, G. G., McGowan, L. M. and Viner, A. B., 1969. 'Lake George, Uganda: a limnological survey', *Verh. int. Ver. Limnol.* **17**: 284–288.

Dupont, B. and Lévêque, C., 1968. 'Biomasses en mollusques et nature des fonds dans la zone est du Lac Tchad, *Cah. O.R.S.T.O.M. sér Hydrobiol.*, **2**: 113–126.

Dupont, B., 1970. 'Distribution et nature des fonds du Lac Tchad (nouvelles données)', *Cah. O.R.S.T.O.M. sér. Géol.*, **2**: 9–42.

Dupuis, C. E., 1936. 'Lake Tana and the Nile', *J. Roy. Afr. Soc.*, **35**: 18–25.

Durand, J. R., 1970. 'Les peuplements ichthyologiques de l'El Beid. Première note: Presentation du milieu et résultats généraux', *Cah. O.R.S.T.O.M. sér. Hydrobiol.*, **4**: 1–26.

Durand, J. R. and Loubens, G., 1969. 'Croissance en longueur d'*Alestes baremoze* dans le bas Chari et le lac Tchad', *Cah. O.R.S.T.O.M. sér. Hydrobiol.*, **3**: 59–105.

*Dussart, B., 1966. *Limnologie. L'étude des Eaux Continentales*, 676 pp. Gauthier-Villars, Paris.

Dussart, B. and Gras, R., 1966. 'Faune planctonique du Lac Tchad', *Cah. O.R.S.T.O.M. sér. Océanogr.*, **4**: 77–92.

*Eccles, D. H., 1962. 'An internal wave in Lake Nyasa and its probable significance in the nutrient cycle', *Nature*, **194**: 832–3.

Eccles, D. H., 1974. 'An outline of the physical limnology of Lake Malawi', *Limnol. Oceanogr.*, **19**: 730–42.

Edmondson, W. T. and Winberg, G. G., 1971. *Secondary Productivity in Freshwater*, I.B.P. Handbook No. 17, 358 pp., Blackwell.

Edwards, P. (ed.), 1967. *Equiano's Travels*, 196 pp., Heinemann Educational.

Eggeling, W. J., 1935. 'The vegetation of Namanve Swamp, Uganda', *J. Ecol.*, **23**: 422–35.

Ellis, C. M. A., 1971. 'The size at maturity maturity and breeding seasons of sardines in southern Lake Tanganyika', *Afr. J. Hydrobiol. Fish.*, **1**: 59–66.

Elton, C. S., 1958. *The Ecology of Invasions by Animals and Plants*, 181 pp., Methuen.

Endler, J. A., 1977. *Geographic variation, speciation and clines*, 246 pp., Princeton Univ. Press.

Entz, B., 1969. 'Limnological conditions in Volta Lake, the greatest man-made Lake of Africa', *U.N.E.S.C.O. Nature and Resources*, **5**: 9–16.

Entz, B., 1972. 'Comparison of the physical and chemical environments of the Volta Lake (Ghana) and Lake Nasser (U.A.R.)', in *Productivity Problems in Freshwaters. I.B.P./U.N.E.S.C.O. Symposium, Warsaw/Krakow, Poland, May 1970*: 883–91.

Entz, B., 1976. 'Lake Nasser and Nubia', in Rzóska, 271–98.

Entz, B., 1978. 'Sedimentation processes above the Aswan High Dam in Lake Nasser-Nubia (Egypt-Sudan)', *Verh. int. Ver. Limnol.*, **20**: 1667–71.

Esaki, T. and **China, W. E.**, 1927. A new family of aquatic Heteroptera', *Trans. Ent. Soc. Lond.*, **73**: 279–95.

Estève, R., 1949. 'Poissons du Sahara Central', *Bull. Soc. Zool. France*, **74**: 19–20.

Evans, J. H., 1958–59. 'The survival of freshwater algae during dry periods. Part I. An investigation of the algae of five small ponds. Part II. Dry experiments. Part III. Stratification of algae in pond margins, litter and mud', *J. Ecol.*, **46**: 149–67; **47**: 55–81.

Evans, J. H., 1962. 'The distribution of phytoplankton in some Central East African waters', *Hydrobiologia*, **19**: 299–315.

Evans, W. A. and **Vanderpuye, J.**, 1973. 'Early development of the fish populations and fisheries of Volta Lake', in Ackermann *et al.*: 115–20.

Ewer, D. W., 1966. 'Biological investigations on the Volta Lake, May 1964 to May 1965', in R. Lowe-McConnell: 21–31.

Ewer, R. F., 1952. 'The effects of posterior pituitary extracts on the water balance in *Bufo carens* and *Xenopus laevis*, together with some general considerations of anuran water economy', *J. Exp. Biol.*, **29**: 429–39.

Eydoux, H. P., 1938. *L'Exploration du Sahara*, 240 pp., Gallimard, Paris.

Fage, J. D., 1958. *An Atlas of African History*, 64 pp., Arnold.

Fage, J. D., 1978. *A History of Africa*, 534 pp., Hutchinson, London.

Fairbridge, R. W., 1962. 'New radiocarbon dating of Nile sediments', *Nature*, **196**: 108–10.

Fairbridge, R. W., 1963. 'Nile sedimentation above Wadi Halfa during the last 20 000 years', *Kush.*, **11**: 96–107.

Fairhead, J. D. and ***Girdler, R. W.**, 1969. 'How far does the Rift System extend through Africa?' *Nature*, **221**: 1018–20.

Falconer, A. C., **Marshall, B. E.** and **Mitchell, D. S.**, 1970. 'Hydrobiological studies of Lake McIlwaine in relation to its pollution', *Rep. to Min. of Water Dev. and Salisbury City Council* (cyclostyled), 49 pp.

Farid, M. A., 1975. 'The Aswan High Dam Development Project', in Stanley and Alpers, 89–102.

Faure, H., 1966. 'Evolution des grands lacs sahariens à l'Holocène', *Quaternaria*, **8**: 167–75.

Faure, H., 1967. 'Une importante periode humide du Quaternaire supérieur du Sahara', *Bull. I.F.A.N. (A)*, **29**: 851–2.

Faure, H., 1969. 'Lacs quaternaires du Sahara', *Mitt. int. Ver. Limnol.*, **17**: 131–46.

*★**Ferguson, A. J. D.**, 1978a. 'Studies on the zooplankton of Lake Turkana', in A. J. Hopson, 1978a.

Ferguson, A. J. D., 1978b, 'An annotated checklist of invertebrates recorded in Lake Turkana', in A. J. Hopson, 1978a.

Ferguson, A. J. D. and **Harbott, B. J.**, 1978. 'Geographical, physical and chemical aspects of Lake Turkana', in A. J. Hopson 1978a.

Fessard, A., 1958. 'Les organes électriques', in P-P. Grassé (ed.), *Traité de Zoologie.* T.XIII. Fasc. 2: 1143–233.

Findenegg, I., 1964. 'Produktionsbiologische Planktonuntersuchungen an Ostalpenseen', *Int. Rev. ges. Hydrobiol.*, **49**: 381–416.

Fischer, A. G., 1960. 'Latitudinal variations in organic diversity', *Evolution*, **14**: 64–81.

Fischer-Piette, E., 1948. 'Sur quelques mollusques fluviatiles du Sahara (Aïr, Itchouma, Fezzan)', *Bull. Mus. d'Hist. Nat.*, 2e sér. 3: 180–2.

*★**Fish, G. R.**, 1952. 'Digestion in *Tilapia esculenta*', *Nature*, **167**: 900.

Fish, G. R., 1955a. 'The food of *Tilapia* in East Africa', *Uganda J.*, **19**: 85–9.

Fish, G. R., 1955b. 'Some aspects of the respiration of six species of fish from Uganda', *J. Exp. Biol.*, **33**: 186–95.

Fish, G. R., 1956. 'Chemical factors limiting growth of phytoplankton in Lake Victoria', *E. Afr. Agric. J.*, **21**: 152–8.

Fish, G. R., 1957. *A Seiche Movement and its effect on the Hydrology of Lake Victoria*, Col. Off. Fish. Publ., 10, 68 pp. H.H.S.O.

Fittkau, E. J., 1973. 'Crocodiles and the nutrient metabolism of Amazonian waters', *Amazoniana*, **4**: 103–33.

Flint, R. F. and **Dorsey, H. G.**, 1945. 'Glaciation of Siberia', *Bull. Geol. Soc. Am*, **59**: 89–106.

Fogg, G. E., 1959. 'Nitrogen nutrition and metabolic products in algae', *Sympos. Soc. exp. Biol.*, **13**: 106–25.

Fogg, G. E., 1968. *Photosynthesis*. 116 pp., English Univ. Press.

Fogg, G. E., 1971. 'Extracellular products of algae in freshwaters', *Ergebn. Limnol.*, **5**: 1–25.

Fogg, G. E. and **Wolf, M.**, 1955. 'The nitrogen metabolism of the blue-green algae', *Sympos. Soc. Gen. Microbiol.*, **4**: 99–125.

Fogg, G. E., **Stewart, W. D. P.**, **Fay, P.** and **Walsby, A. E.**, 1973. *The Blue-green Algae*. 459 pp., Academic Press.

Ford, V. C. R., 1958. *The Trade of Lake Victoria*. East African Studies No. 3, 66 pp. East African Literature Bureau, Nairobi.

Fotius, G. and **Lemoalle, J.**, 1976. 'Reconnaisance de l'evolution de la végétation du Lac Tchad entre Janvier 1974 et Juin 1976', *Rapport de Mission, Centre O.R.S.T.O.M.* N'jamena, 13 pp.

Fox, P. J., 1977. 'Preliminary observations on fish communities of the Okavango Delta', in Nteta and Hermans, 125–30.

*★**Fryer, G.**, 1959a. 'The trophic relationships and ecology of some littoral communities of Lake Nyasa with special reference to the fishes; and an account of the evolution and ecology of rock-frequenting cichlidae', *Proc. Zool. Soc. Lond.*, **132**: 153–281.

Fryer, G., 1959b. 'Some aspects of evolution in Lake Nyasa', *Evolution*, **13**: 440–51.

Fryer, G., 1959c. 'A report on the parasitic Copepods and Branchiura of the fishes of Lake Bangweulu (northern Rhodesia)', *Proc. Zool. Soc. London*, **132**: 517–50.

Fryer, G., 1959d. 'Lunar rhythm of emergence, differential behaviour of sexes and other phenomena in the African midge *Chironomus brevibucca* (Kief)', *Bull. Ent. Res.*, **50**: Pt. I.

Fryer, G., 1960a. 'Evolution of fishes in Lake Nyasa', *Evolution*, **14**: 396–9.

Fryer, G., 1960b. 'Some controversial aspects of speciation of African cichlid fishes', *Proc. Zool. Soc. Lond.*, **135**: 569–78.

Fryer, G., 1960c. 'Concerning the introduction of the Nile perch into Lake Victoria', *E. Afr. Agric. J.*, **26**: 1–5.

Fryer, G., 1961. 'Observations on the biology of the cichlid fish *Tilapia variabilis*, Boulenger in the northern waters of Lake Victoria', *Rev. Zool. Bot. Afr.*, **64**: 1–33.

Fryer, G., 1964. 'Further studies on the parasitic Crustacea of African freshwater fishes', *Proc. Zool. Soc. Lond.*, **143**: 79.

Fryer, G., 1965. 'Predation and its effects on migration and speciation in African fishes: a comment, with further comments by P. H. Greenwood and a reply by P. B. N. Jackson', *Proc. Zool. Soc. Lond.*, **144**: 301–22.

Fryer, G., 1967. 'Speciation and adaptive radiation in African Lakes', *Verh. int. Ver. Limnol.*, **17**: 303–13.

Fryer, G., 1968a. 'A new parasitic isopod of the family Cymothoidae from clupeid fishes of Lake Tanganyika – a further Lake Tanganyika enigma', *J. Zool.*, **156**: 35–44.

Fryer, G., 1968b. 'The parastic crustacea of African freshwater fishes; their biology and distribution', *J. Zool.*, **156**: 45–96.

Fryer, G. and Iles, T. D., 1955. 'Predation pressure and evolution in Lake Nyasa', *Nature*, **176**: 470.

Fryer, G. and Iles, T. D., 1969. 'Alternative routes to evolutionary success by African cichlid fishes of the genus *Tilapia* and the species flocks of the Great Lakes', *Evolution*, **23**: 359–69.

Fryer, G. and Iles, T. D., 1972. *The Cichlid Fishes of the Great Lakes of Africa: their Biology and Evolution*, 641 pp., Oliver and Boyd.

Fryer, G. and Whitehead, P. J. P., 1959. 'The breeding habits, embryology and larval development of *Labeo victorianus* Boulenger', *Rev. Zool. Bot. Afr.*, **59**: 33–49.

Fryer, G., Greenwood, P. H. and Trewavas, E., 1955. 'Scale eating habits of African cichlid fishes.', *Nature*, **175**: 1089.

Fuchs, V. E., 1937. 'Extinct Pleistocene Mollusca from Lake Edward and their bearing on the Tanganyika problem', *J. Linn. Soc. (Zool.)*, **40**: 93–106.

Fuchs, V. E., 1939. 'The geological history of the Lake Rudolph basin, Kenya Colony', *Phil. Trans. Roy. Soc. (B)*, **299**: 219–74.

Fuchs, V. E., 1950. 'Pleistocene events in the Baringo Basin, Kenya Colony', *Geol. Mag.*, **87**: 149–74.

Fuchs, V. E., and Paterson, T. T., 1947. 'The relation of volcanicity and orogeny to climatic change', *Geol. Mag.*, **84**: 321–33.

Gallais, J., 1967. 'Le Delta intérieur du Niger. Étude de géographie régionale', *Mém. 79*, 619 pp., I.F.A.N.

★**Ganf, G. G.,** 1969. 'Physiological and ecological aspects of the phytoplankton of Lake George, Uganda', Ph.D. Thesis, Univ. of Lancaster, 164 pp.

Ganf, G. G., 1974. 'Rates of oxygen uptake by the planktonic community of a shallow equatorial lake (Lake George, Uganda)', *Oecologia (Berl.)*, **15**: 17–32.

Ganf, G. G., 1975. 'Photosynthetic production and irradiance – photosynthesis relationships of the phytoplankton from a shallow equatorial lake (Lake George, Uganda)', *Oecologia (Berl.)*, **18**: 165–83.

Ganf, G. G. and Blazka, P., 1974. 'Oxygen uptake, ammonia and phosphate, excretion by zooplankton of a shallow equatorial lake (Lake George, Uganda)', *Limnol. Oceanogr.*, **19**: 313–25.

Ganf, G. G. and Viner, A. B., 1973. 'Ecological stability in a shallow equatorial lake (Lake George, Uganda)', in Greenwood and Lund, 321–46.

Garrod, D. J., 1961. 'The history of the fishing industry of Lake Victoria, East Africa, in relation to the expansion of marketing facilities', *E. Afr. Agric. For. J.*, **24**(2): 111–20.

Garstin, W. E., 1911. 'Lake Tsana', *Encyclopedia Britannica* 11th edit., **27** 347

Gasith, A. and **Hasler, A. D.**, 1976 'Airborne litterfall as a source of organic matter in lakes', *Limnol. Oceanogr.*, **21**: 253–8

Gasse, F. and **Street, F. A.**, 1978. 'The main stages of the quaternary evolution of the northern Rift Valley and Afar lakes (Ethiopia and Djibouti)'. *Pol. Arch. Hydrobiol.*, **25**: 145–50.

* **Gaudet, J. J.**, 1975. 'Mineral concentrations in papyrus in various African swamps', *J. Ecol.*, **63**: 483–91.

Gaudet, J. J., 1976. 'Nutrient relationships in the detritus of a tropical swamp', *Arch. Hydrobiol.*, **78**: 213–39.

Gaudet, J. J., 1977. 'Uptake, accumulation and loss of nutrients by papyrus in tropical swamps', *Ecology*, **58**: 415–22.

Gauthier, H., 1928. *Recherches sur la Faune des Eaux Continentales de l' Algérie et de la Tunisie.* Alger.

Gauthier, H., 1929. 'Sur la faune aquatique du Sahara centrale', *C.R. Acad. Sci. Paris*, **189**: 201–3.

Gauthier, H., 1938. 'Le vie aquatique dans le déserts subtropicaux', in 'La Vie dans la region désertique nord-tropicale de l'ancien monde', *Societe de Biogeographie*, Mém. VI: 107–20.

Gauthier, H., 1939. 'Contributions a l'étude de la faune dulcaquicole de la region du Tchad et particulièrement des Branchiopodes et des Ostracodes', *Bull. I.F.A.N.*, **1**: 110–244.

Gauthier-Lièvre, L., 1931. *Recherches sur la Flore des Eaux Continentales de l'Afrique du Nord.* Alger.

Gautier, A., 1970. 'Fossil freshwater Mollusca from the Lake Albert–Lake Edward Rift', *Ann. Mus. r. Afr. Centr. Ser. 8Vo. Sciences Geol.*, **67**: 1–144.

Gautier, E. F., 1928. *Le Sahara*, 232 pp. Payot. Paris; English translation (1935) *Sahara: The Great Desert*, Columbia Univ. Press. N.Y.

Gay, P. A., 1960. 'Ecological studies of *Eichhornia crassipes* in the Sudan', *J. Ecol.*, **48**: 180–99.

Gayral, P., 1954. 'Recherches phytolimnologiques au Maroc', *Trav. de l'Inst. Sci. Cherifien. Sér. Bot.*, No. 4: 1–306.

Gee, J. M., 1968. 'Trawling results within northern waters of Lake Victoria', *Ann. Rep. E. Afric. Fish. Res. Org.*, 1968: 15–27.

Gee, J. M., 1969. 'A comparison of certain aspects of the biology of *Lates niloticus* in some East African Lakes', *Rev. Zool. Bot. Afr.*, **80**: 244–62.

Gee, J. M., 1975. 'The genus *Xenoclarias* (Pisces, Siluriformes) in Lake Victoria with a redescription of *Xenoclarias eupogon* and notes on the biology of the species', *J. Zool.*, **175**: 201–17.

Gentile, J. H., 1971. 'Blue-green algae toxins', in S. Kadis and A. Ciegler (eds), *Microbial Toxins*, Vol. VII, 27–66, Academic Press, New York.

George, C. G., 1972. 'Role of the Aswan High Dam in changing the fisheries of the southeastern Mediterranean'., in Farvar and Milton (eds), *The careless technology. Ecology and national development*, Doubleday, New York.

Gerking, S. D. (ed.), 1967. *The biological basis of freshwater fish production*, 495 pp., Blackwell, Oxford.

Germain, M. L., 1932. 'Contribution à la faune malacologique de l'Afrique équatoriale', *Bull. Mus. d'Hist. Nat. 2e Sér.*, **4**: 890–4.

Gessner, F , 1957. 'Van Golu Zur Limnologie des grossen Soda-Sees in Ostanatolien (Türkei)', *Arch. Hydrobiol.*, **53** 1–22

Gillman, C., 1933. 'The hydrology of Lake Tanganyika', *Geol. Surv. Dept Tanganyika, Bull.*, **5**: 1–25.

* **Gilson, H. C.**, 1964. 'Lake Titicaca', *Verh. int. Ver. theor. angew. Limnol.*, **15**: 112–27

Goldman, C. R., 1960. 'Molybdenum as a factor limiting primary productivity in Castle Lake, California', *Science*, **132**: 1016–17.

Goldman, C. R., 1972. 'The role of minor nutrients in limiting the productivity of aquatic ecosystems', in Likens, 1972: 21–33.

Golterman, H. L., 1960. 'Studies in the cycle of elements in freshwater', *Acta. Bot. Neerl.*, **9**: 1–58.

Golterman, H. L., 1964. 'Mineralization of algae under sterile conditions or by bacterial breakdown', *Verh. int. Ver. theor. angew. Limnol.*, **15**: 544–8.

Golterman, H. L., 1971. 'The determination of mineralisation losses in correlation with the estimation of net primary production with the oxygyen method and chemical inhibitors', *Freshw. Biol.*, **1**: 249–56.

Golterman, H. L. and **Clymo, R. S.**, 1969. *Methods for Chemical Analysis of freshwaters.* Int. Biol. Prog. Handbook No. 8, 188 pp., Blackwell, Oxford.

* **Goma, L. K. H.**, 1959. 'The Productivity of various mosquito breeding places in the swamps of Uganda', *Bull. Ent. Res.*, **49**: 437–8.

Goma, L. K. H., 1960a. 'The swamp breeding mosquitoes of Uganda: records of larvae and their habitats', *Bull. Ent. Res.*, **51**: 77–94.

Goma, L. K. H., 1960b. 'Experimental breeding of *Anopheles gambiae* Giles in papyrus swamps', *Nature*, **187**: 1137–8.

Goma, L. K. H., 1961a. 'The Swamp Environment and the breeding of mosquitoes in Uganda', *Proc. Roy. Ent. Soc. London*, **36**: 27–36.

Goma, L. K. H., 1961b. 'The influence of man's activities on swamp-breeding mosquitoes in Uganda', *J. Ent. Soc. S. Africa*, **24**: 221–47.

Goodland, R. J. A. and **Irwin, H. S.**, 1975. *Amazon jungle: green hell to red desert*, 185 pp., Elsevier, Amsterdam.

Gordon, I. and **Monod, Th.**, 1968. 'Sur quelques Crustacées des eaux douces de Zanzibar', *Bull. I.F.A.N. (A)*, **30**: 497–517.

Gorham, P. R., 1960. 'Toxic water blooms of blue-green algae', *Can. Vet. J.*, **1**: 235–45.

Gosse, J. P., 1963. 'Le milieu aquatique et l'écologie des poissons dans la region de Yangambi', *Ann. Mus. Roy. Afr. Centr. Sci. Zool.*, **116**: 113–270.

Gourou, P., 1970. *L'Afrique*, 88 pp., Hachette, Paris.

Graham, M., 1929. *The Victoria Nyanza and its Fisheries*, Crown Agents for the Colonies, London.

Gras, R., Iltis, A. and **Lévêque-Duwat, S.**, 1967. 'La plancton du Bas Chari et de la partie est du lac Tchad', *Cah. O.R.S.T.O.M. sér. Hydrobiol.*, **1**: 25–100.

* **Green, J.**, 1960a. 'Zooplankton of the River Sokoto. The Rotifera, *Proc. Zool. Soc. Lond.*, **135**: 491–523.

Green, J., 1960b. 'Zooplankton of the River Sokoto. The freshwater medusa *Limnocnida*', *Proc. Zool. Soc. Lond.*, **135**: 613–18.

Green, J., 1962. 'Zooplankton of the River Sokoto. The Crustacea', *Proc. Zool. Soc. Lond.*, **138**: 415–53.

Green, J., 1963. 'Zooplankton of the River Sokoto. The Rhizopoda Testacea', *Proc. Zool. Soc. Lond.*, **141**: 497–514.

Green, J., 1967a. 'The distribution and variation of *Daphnia lumholtzi* (Crustacea: Cladocera) in relation to fish predation in Lake Albert, East Africa', *J. Zool. Lond.*, **151**: 181–97.

Green, J., 1967b. 'Associations of Rotifera in the zooplankton of the lake sources of the White Nile', *J. Zool. Lond.*, **151**: 343–78.

Green, J., 1971. 'Associations of Cladocera in the zooplankton of the lake sources of the White Nile', *J. Zool.*, **165**: 373–414.

Green, J., 1972. 'Ecological studies on crater lakes in West Cameroon. Zooplankton of Barombi Mbo, Mboandong, Lake Kotto and Lake Soden', *J. Zool. Lond.*, **166**: 283–301.

Green, J., 1976. 'Changes in the zooplankton of Lakes Mutanda, Bunyoni and Mulehe (Uganda)', *Freshwater Biol.*, **6**: 433–6.

Green, J., Corbet, S. A. and **Betney, E.**, 1973. 'Ecological studies on crater lakes in West Cameroon. The blood of endemic cichlids in Barombi Mbo in relation to stratification and vertical distribution of the zooplankton', *J. Zool.*, **170**: 299–358.

Green, J., Corbet, S. A., Watts, E. and **Oey Biauw Lan**, 1976. 'Ecological studies on Indonesian lakes. Overturn and restratification of Ranu Lamongan', *J. Zool.*, **180**: 315–54.

Greenaway, P., 1970. 'Sodium regulation in the freshwater mollusc *Lymnaea stagnalis* (L) (Gastropoda: Pulmonata)', *J. exp. Biol.*, **53**: 147–63.

Greenaway, P., 1971. 'Calcium regulation in the freshwater mollusc *Lymnaea stagnalis* (L) (Gastropoda: Pulmonata)', *J. exp. Biol.*, **54**: 199–214.

*★**Greenwood, P. H.**, 1951a. 'Evolution of the African cichlid fishes: the *Haplochromis* species flock in Lake Victoria', *Nature*, **167**: 19–20.

Greenwood, P. H., 1951b. 'Fish remains from Miocene deposits of Rusinga Island and Kavirondo Province, Kenya', *Ann. Mag. Nat. Hist.*, (12)**4**: 1192–201.

Greenwood, P. H., 1953. 'The feeding mechanism of the cichlid fish *Tilapia esculenta*', *Nature*, **172**: 207.

Greenwood, P. H., 1955. 'Reproduction in the catfish *Clarias mossambica* Peters', *Nature*, **176**: 516.

Greenwood, P. H., 1958. 'Reproduction in the East African lungfish *Protopterus aethiopicus*, Heckel', *Proc. Zool. Soc. Lond.*, **130**: 547–67.

Greenwood, P. H., 1959. 'Quaternary fish fossils', *Explor. du Parc. Nat. Albert*, Fasc. **4**. No. 1.

Greenwood, P. H., 1961. 'A revision of the genus *Dinotopterus* Blqr. (Pisces, Clariidae) with notes on the comparative anatomy of the superbranchial organs in the Clariidae', *Bull. Brit. Mus. (Nat. Hist.) Zool.*, **7**: 217–41.

Greenwood, P. H., 1963. 'The swimbladder in African Notopteridae (Pisces) and its bearing on the taxonomy of the family', *Bull. Brit. Mus. (Nat. Hist.)*, **11**: 379–412.

Greenwood, P. H., 1965a. 'Explosive speciation in African Lakes', *Proc. Roy. Inst.*, **40**: 256–69.

Greenwood, P. H., 1965b. 'The cichlid fishes of Lake Nabugabo, Uganda', *Bull. Brit. Mus. (Nat. Hist.) Zool.*, **12**: 315–57.

Greenwood, P. H., 1965c. 'Environmental effects on the pharyngeal mill of a cichlid fish, *Astatoreochromis alluardi*, and their taxonomic implications', *Proc. Linn. Soc. Zool.*, **176**: 1–10.

Greenwood, P. H., 1965d. 'Further comments on Fryer's comments on predation and its effects on migration and speciation in African lakes', *Proc. Zool. Soc. Lond.*, **144**: 310–22.

Greenwood, P. H., 1966. *The Fishes of Uganda.*, 131 pp., The Uganda Society, Kampala.

Greenwood, P. H., 1970. 'A revision of the cyprinid species *Barbus (Enteromius) radiatus* Peters, 1853, with a note on the synonomy of the subgenera *Beirabarbus* and *Enteromius*', *Rev. Zool. Bot. Afr.*, **82**: 1–13.

Greenwood, P. H., 1973. 'A revision of the *Haplochromis* and related species (Pisces: Cichlidae) from Lake George, Uganda', *Bull. Brit. Mus. (Nat. Hist.) Zool.*, **25**: 141–242.

Greenwood, P. H., 1974a. 'The cichlid fishes of Lake Victoria, East Africa: the biology and evolution of a species flock', *Bull. Brit. Mus. Nat. Hist. (Zool.)* suppl., **26**: 1–134.

Greenwood, P. H., 1974b. 'The *Haplochromis* species (Pisces, Cichlidae) of Lake Rudolf, East Africa. *Bull. Brit. Mus. Nat. Hist. (Zool.)* suppl., **27**: 139–65.

Greenwood, P. H. and Lund, J. W. G. (Organisers), 1973. 'A discussion on the biology of an equatorial lake: Lake George, Uganda, *Proc. R. Soc. London. B.*, **184**: 227–346.

Greenwood, P. H. and Thomson, K. S., 1960. 'The pectoral anatomy of *Pantodon buchholzi*, and the related Osteoglossidae', *Proc. Zool. Soc. London.*, **135**: 283–301.

Gregory, J. W., 1896. *The Great Rift Valley*, 422 pp., Murray, London.

Gregory, J. W., 1921. *Rift Valleys and Geology of East Africa*. Seeley Service, London.

Griffiths, J. F. (ed.), 1972. *Climates of Africa*, 'World Survey of Climatology', Vol. 10, 504 pp., Elsevier.

Grindley, D. N., 1952. 'The composition of the body fat of small green chironomids', *J. exp. Biol.*, **29**: 440–4.

Grossowicz, N., Hestrin, S. and Keynan, A. (eds), 1961. *Cryptobiotic Stages in Biological Systems*, 232 pp., Elsevier.

Groth, P., 1971. 'Untersuchungen über einige Spurenelemente in Seen', *Arch. Hydrobiol.*, 68(3): 305–75.

*Grove, A. T., 1969. 'Land forms and climatic change in the Kalahari and Ngamiland', *Geogr. J.*, **135**: 191–202.

Grove, A. T., 1970a. *Africa South of the Sahara*. 2nd edition, 280 pp., Oxford Univ. Press.

Grove, A. T., 1970b. 'Rise and fall of Lake Chad', *Geogr. Mag.*, **42**: 432–9.

Grove, A. T., 1972. 'The dissolved and solid load carried by some West African Rivers: Senegal, Niger, Benue and Shari', *J. Hydrol.*, **16**: 277–300.

Grove, A. T., and Goudie, A., 1971. 'Late quaternary lake levels in the Rift Valley of Southern Ethiopia and elsewhere in tropical Africa', *Nature*, **234**: 403–5.

Grove, A. T. and Pullan, R. H., 1963. 'Some aspects of the Pleistocene Palaeogeography of the Chad Basin', *Vicking Publications in Anthropology*, **36**: 230–45.

Grove, A. T. and Warren. A., 1968. 'Quaternary landforms and climate on the south side of the Sahara', *Geogr. J.*, **134**: 194–208.

Grove, A. T., Street, F. A. and Goudie, A. S., 1975. 'Former lake levels and climatic change in the Rift Valley of southern Ethiopia', *Geogr. J.*, **141**: 177–202.

Gunn, D. L., 1973. 'Consequences of cycles in East African climate', *Nature.*, **242**: 457.

Günther, R. T., 1898. 'A further contribution to the anatomy of *Limnocnida tanganyikae*', *Quart. J. Microsc. Soc.*, **36**: 271.

Gwahaba, J. J., 1974. 'Effects of fishing on the *Tilapia nilotica* population in Lake George, Uganda over the past 20 years', *E. Afr. Wildlife J.*, **11**: 311–28.

Gwahaba, J. J., 1975. 'The distribution, population density and biomass of fish in an equatorial lake, Lake George, Uganda', *Proc. R. Soc. London B.*, **190**: 393–414.

Gwahaba, J. J., 1978. 'The biology of cichlid fishes (Teleostei) in an equatorial lake (Lake George, Uganda)', *Arch. Hydrobiol.*, **83**: 538–51.

Hall, A., Davies, B. R. and Valente, I., 1976. 'Cabora Bassa: some preliminary physico-chemical and zooplankton pre-impoundment survey results', *Hydrobiologia*, **50**: 17–25.

Hall, J. B. and Pople, W., 1968. 'Recent vegetational changes in the Lower Volta River', *Ghana J. Sci.*, **8**: 24–9.

Hallett, R. (ed.), 1964. *Records of the African Association*, 318 pp. Nelson.

Hallett, R., 1965a. *The Penetration of Africa*. Vol. I. *To* **1815**, 458 pp. Routledge & Kegan Paul.

Hallett, R. (ed.), 1965b. *The Niger. Journal of Richard and John Lander*. 317 pp. Routledge & Kegan Paul.

Hamblyn, E. L., 1966. 'The food and feeding habits of Nile perch (*Lates niloticus*. Pisces: Centropomidae)', *Rev. Zool. Bot. Afr.*, **74**: 1–28.

*★***Hammerton, D.**, 1972. 'The Nile River – A case history' in R. T. Oglesby, C. A. Carlson and J. A. McCann (eds), *River Ecology and Man*, 171–214. Academic Press. N.Y.

*★***Harbott, B. J.**, 1978. 'Studies on algal dynamics and primary productivity in Lake Turkana', in Hopson, 1978a.

Harder, W., Scheif, A. and **Uhlemann, H.**, 1964. 'Zum Funktion des elektrischen Organs von *Gnathonemus petersi* (Gthr. 1864).' *Zeitschr. Vergleich Physiol.*, **48**: 302–31.

*★***Harding, D.**, 1964. 'Hydrology and fisheries of Lake Kariba'. *Verh. int. Ver. Limnol.* **15**: 139–49.

Harding, D., 1966. 'Lake Kariba. The Hydrology and development of Fisheries' in R. Lowe-McConnell, 7–19.

Harding, J. P., 1957. 'Crustacea: Cladocera', *Explor. Hydrobiol. L. Tanganika* (1946–47), III. **6**: 55–89.

Harris, N., Pallister, J. W. and **Brown, J. M.**, 1956. *Oil in Uganda*. Geol. Survey Uganda, Mem. No. 9. Entebbe, Govt. Printer, 33 pp.

*★***Harrison, A. D.**, 1964. 'An ecological survey of the Great Berg River', in D. H. S. (ed.), *Ecological Studies in Southern Africa*, 415 pp., Junk, The Hague.

Harrison, A. D., 1965. 'Geographical distribution of riverine invertebrates in Southern Africa', *Arch. Hydrobiol.*, **61**: 387–94.

Harrison, A. D., 1966. 'Recolonisation of a Rhodesian stream after drought', *Arch. Hydrobiol.*, **62**: 405–21.

Harrison, A. D., 1978. 'Freshwater invertebrates (except molluscs)', Ch. 34 in M. J. A. Werger.

*★***Hartland-Rowe, R. C. B.**, 1955. 'Lunar rhythm in the emergence of an ephemeropteran' *Nature*, **176**: 657.

Hartland-Rowe, R. C. B., 1958. 'The biology of the tropical mayfly *Povilla adusta* Navas (Ephemeroptera, Polymitarcidae) with special reference to the lunar rhythm of emergence', *Rev. Zool. Bot. Afr.*, **58**: 185–202.

Harvey, T. J., 1976. 'The palaeolimnology of Lake Mobutu Sese Seko, Uganda-Zaïre: the last 28 000 years', Thesis for Ph.D., Duke Univ., N. Carolina, 104 pp.

Haworth, E. Y., 1977. 'The sediments of Lake George (Uganda). V. The diatoms in relation to the ecological history', *Arch. Hydrobiol.*, **80**: 200–215.

Hayes, F. R., 1964. 'The mud-water interface', *Oceanogr. Mar. Biol. Ann. Rev.*, **2**: 121–45.

Hayes, F. R. and **Beckett, N. R.**, 1956. 'The flow of minerals through the thermocline of a lake', *Arch. Hydrobiol.*, **51**: 391–409.

Hecky, R. E. and **Kilham, P.**, 1973. 'Diatoms in alkaline, saline lakes: ecology and geochemical implications', *Limnol. Oceanogr.*, **18**: 53–71.

Hecky, R. E., Klee, E. J. and **Kling, H.**, 1977. 'Studies on the planktonic ecology of Lake Tanganyika', *Freshw. Inst. Winnipeg, Canada*, multigr.

Heinzelin, J. de, 1955. *Le Fossé Tectonique sous le Parallèle d'Ishango*, Explor. du Parc. Nat. Albert, Mission J. de Heinzelin de Braucourt (1950), fasc. 1.

Heinzelin, J. de, 1957. *Les Fouilles d'Ishango*, Explor. du Parc Nat. Albert, fasc 2, Inst. des Parcs Nat. du Congo Belge.

Heinzelin, J. de, 1963. 'Palaeoecological conditions of the Lake Albert – Lake Edward Rift' in *Africa Ecology and Human Evolution*, F. C. Howell and F. Bourlière (eds), *Viking Publications in Anthropology* **36**: 276–84.

Henderson, F., 1973. 'Stratification and circulation in Kainji Lake', in Ackermann *et al.*, 489–94.

Hepher, B., 1962. 'Primary production in fishponds and its application to fertilisation experiments', *Limnol. Oceanogr.*, **7**: 131–6.

** Hepper, F. N.**, 1970. 'Hovercraft on Lake Chad 1. Plant life on sandbanks and papyrus swamps', *Geogr. Mag.*, **42**: 577–82.

Herre, W., 1933. 'The fishes of Lake Lanao. A problem in evolution', *American Nat.*, **67**: 154–62.

Hickling, C. F., 1961. *Tropical Inland Fisheries*, 287 pp., Longman.

Hickling, C. F., 1962. *Fish Culture*, 295 pp., Faber & Faber.

Hickling, C. F., 1968. *The Farming of Fish*, 88 pp., Pergamon.

Hicks, R. M., 1975. 'The mammalian urinary bladder: an accomodating organ', *Biol. Revs.*, **50**: 215–46.

Hinton, H. E., 1954. 'Resistance of the dry eggs of *Artemia salina* to high temperatures', *Ann. Mag. Nat. Hist.*, **7**: 158–60.

Hinton, H. E., 1960a. 'Cryptobiosis in the larva of *Polypedilum vanderplanki* Hint. (Chironomidae)', *J. Insect. Physiol.*, **5**: 286–300.

Hinton, H. E., 1960b. 'A fly that tolerates dehydration and temperatures from −270 to + 102°C', *Nature*, **186**: 336–7.

Hinton, H. E., 1968. 'Reversible suspension of metabolism and the origin of life', *Proc. Roy. Soc. B.*, **171**: 43–57.

Hira, P. R., 1969. 'Transmission of schistosomiasis in Lake Kariba, Zambia', *Nature*, **244**: 670–2.

Hira, P. R. and Muller, R., 1966. 'Studies on the ecology of snails transmitting urinary schistosomiasis in Western Nigeria', *Ann. Trop. Med. Parasit.*, **60**: 198.

Höhnel, L. R., Von, 1894. *Discovery of Lakes Rudolf and Stephanie*, 2 vols. (Transl. from German), Bell, London.

Höhnel, L. R., Von, 1938. 'The Lake Rudolf region, its discovery and subsequent exploration. 1888–1909', *J. Roy. Afr. Soc. London*, **37**: 21–45, 206–26.

Holden, M. J., 1963. 'The populations of fish in the dry season pools of the River Sokoto', *Colon. Off. Fish. Publ.*, **19**: 1–58, H.M.S.O.

Holden, M. J., 1967. 'The systematics of the genus *Lates* (Telesotei: Centropomidae) in Lake Albert, East Africa'. *J. Zool.* **151**: 329–42.

Holden, M. J., 1970. 'The feeding habits of *Alestes baremose* and *Hydrocynus forskali* (Pisces) in Lake Albert, East Africa', *J. Zool.*, **161**: 137–44.

Holden, M. J. and Green, J., 1960. 'The hydrology and plankton of the River Sokoto', *J. Anim. Ecol.*, **29**: 65–84.

Holdship, S. A., 1976. 'The palaeolimnology of Lake Manyara, Tanzania: a diatom analysis of a 56 meter sediment core', Thesis for PhD., Duke Univ., N. Carolina, 122 p.

Holmes, A., 1965. *Principles of Physical Geology*. 1288 pp., Nelson.

Hopkins, G. H. E., 1952. *Mosquitos of the Ethiopian Region I. Larval Bionomics of Mosquitos and Taxonomy of culicine Larvae*, 2nd ed. with notes and addenda by P. F. Mattingly, 355 pp., Brit. Mus. Nat. Hist.

** Hopson, A. J.**, 1967. 'Fisheries of Lake Chad', in W. Reed, 188–200.

Hopson, A. J., 1968. 'Winter scale rings in *Lates niloticus* (Pisces, Centropomidae) from Lake Chad', *Nature*, **208**: 1013–14.

Hopson, A. J., 1969a. 'Seasonal change with pattern of salinity distribution in the northern basin of Lake Chad', *Ann. Rep. 1966–67. Fed. Fish. Services, Republic of Niger-*

ia, L. Chad Research Station, Appendix 2: 13–26.

Hopson, A. J., 1969b. 'A preliminary survey of bottom deposits and benthos in the northern basin of Lake Chad', *Ann. Rep. Fed. Fish. Services, Republic of Nigeria, L. Chad Research Station*, Appendix 5: 43–50.

Hopson, A. J., 1969c. 'A description of the pelagic embryos and larval stages of *Lates niloticus* (L.) (Pisces: Centropomidae) from Lake Chad, with a review of early development in lower period fishes', *Zool. J. Linn. Soc.*, **48**: 117–34.

Hopson, A. J., 1972. *A Study of the Nile Perch in Lake Chad*, Overseas Development Administration, Overseas Res. Publ. No. 19, 93 pp. H.M.S.O.

Hopson, A. J. (ed.), 1978a. *A survey of the fisheries of Lake Turkana, Kenya, and of the lacustrine environment*. Report to Overseas Development Administration, London. (To be published).

Hopson, A. J., 1978b. 'Conclusions and recommendations' in Hopson 1978a:

Hopson, A. J. and Hopson, J., 1978. 'The fishes of Lake Turkana' in Hopson 1978a.

Hopson, A. J., Ferguson, A. J. D., Harbott, B. J. and McLeod, A. A. Q. R., 1978. 'Shore survey' in Hopson 1978a.

Hopson, J., 1972. 'Breeding and growth in two populations of *Alestes baremose* (Johannis) (Pisces, Characidae) from the northern basin of Lake Chad', Overseas Res. Publ. No. 20. 50 pp. H.M.S.O.

Horne, A. J. and Viner, A. B., 1971. 'Nitrogen fixation and its significance in tropical Lake George, Uganda', *Nature*, **232**: 417–18.

Howell, F. C. and Bourlière, F. (eds), 1964. *African Ecology and Human Evolution*, 666 pp., Methuen.

Howell, P. P. 1953. 'The Equatorial Nile project and its effects in the Sudan', *Geogr. J.*, **119**: 33–48.

Hubbs, C. L., 1961. 'Isolating mechanisms with speciation of fishes' in W. F. Blair (ed.), *Vertebrate Speciation*, 5–23 pp. Univ. Texas Press, Austin.

Hubendick, B., 1952. 'On the evolution of the so-called thalassoid molluscs of Lake Tanganyika', *Arch. Zool. Stockholm* (ser. 2), **3**: 319–23.

Hughes, G. M., 1966. 'Evolution between air and water', in *Development of the Lung*, Ciba Foundation Symposium: 64–80.

Hugot, M. M., 1974. *Le Sahara avant le désert*, 343 pp., Hespérides, Toulouse.

Hulot, A., 1956. 'Apercu sur la question de la pêche industrielle aux Lacs Kivu, Edouard et Albert', *Bull. Agr. Congo Belge*, **47**: 815–82.

Hustedt, F., 1939. 'Systematische und ökologische untersuchungen über die Diaton-flora von Java, Bali und Sumatra III'., *Arch. Hydrobiol.*, **16**: 274–394.

Hurst, H. E., 1957. *The Nile*, 2nd rev. edn, 331 pp., Constable.

Hurst, H. E. and Phillips, P., 1931. *The Nile Basin*, Vol. I. *General description of the Basin*, Physical Dept. Papers, Vol. 1, Cairo.

Hutchinson, G. E., 1930. 'On the chemical ecology of Lake Tanganyika', *Science*, **121**: 616.

Hutchinson, G. E., 1933. 'Experimental Studies in Ecology I. The magnesium tolerance of Daphniidae and its ecological significance', *Int. Rev. ges. Hydrobiol. u. Hydrogr.*, **28**: 90–108.

Hutchinson, G. E., 1957, 1967. *A Treatise on Limnology*, Vol. I, *Geography Physics and Chemistry*, 1016 pp.; Vol. II, *Introduction to Lake Biology and the Limnoplankton*, 1115 ppp., Wiley, N. Y.

Hyder, M., 1969. 'Gonadal development and reproductive activity of the cichlid fish *Tilapia leucosticta* (Trewavas) in an equatorial lake', *Nature*, **224**: 1112.

Hynes, H. B. N., 1960. *The Biology of Polluted Waters*, 202 pp., Liverpool Univ. Press.

Iles, T. D., 1959. 'A group of zooplankton feeders of the genus *Haplochromis* (Cichlidae)',

Ann. Mag. Nat. Hist., **2**. (13): 257–80.

Iltis, A., 1968. 'Tolérance de salinité de *Spirulina platensis* (Gom.) Geitl., (Cyanophyta) dans les mares natronées du Kanem (Tchad)', *Cah. O.R.S.T.O.M. sér. Hydrobiol.*, **2**: 119–25.

Iltis, A., 1969–75. Le phytoplancton des eaux natronées du Kanem (Tchad). I. Les lacs permanents à Spirulines, II. Les mares temporaires, III. Variations annuelles du plancton d'une mare temporaire, IV. Note sur les espèces du gênre *Oscillatoria* sous genre *Spirulina* (Cyanophyta.), V. Les lacs mésohalins, X. *Conclusions*, *Cah. O.R.S.T.O.M. sér Hydrobiol.* **3**: 29–44, **3**: 3–19, **4**: 53–60, **3/4**: 129–34, **5**: 73–84, **9**: 13–18.

Iltis, A., 1971. 'Note sur *Oscillatoria* (sans-genre *Spirulina*) *platensis* (nordst.) Bourrelly (Cyanophyta) au Tchad', *Cah. O.R.S.T.O.M. sér. Hydrobiol.*, **5**: 53–72.

Iltis, A., 1977. 'Peuplements phytoplanktoniques du Lac. Tchad. I. Stade Tchad normal (fevrier 1971 et janvier 1972). II. Stade petit Tchad (avril 1974, novembre 1974 et février 1975). III. Remarques générales, *Cah. O.R.S.T.O.M. Sér Hydrobiol.*, **11**: 32–5, 53–72, 189–99.

Iltis, A. and **Riou-Duwat, S.**, 1971. 'Variations saisonnières du peuplement en Rotifères des eaux natronées du Kanem (Tchad)', *Cah. O.R.S.T.O.M. sér. Hydrobiol.*, **5**: 101–12.

* **Imevbore, A. M. A.**, 1967. 'Hydrology and plankton of Eleiyele reservoir, Nigeria', *Hydrobiologia*, **30**: 154–76.

Imevbore, A. M. A., 1970. 'The chemistry of the River Niger in the Kainji Reservoir area', *Arch. Hydrobiol.*, **67**: 412–31.

Imevbore, A. M. A., 1971. 'The first symposium on Lake Kainji, Nigeria's man-made lake', *Afr. J. Trop. Hydrobiol. Fish.*, **1**: 67–8.

Imevbore, A. M. A. and **Adegoke, O. S.** (eds), (1975). '*The Ecology of Lake Kainji, the transition from River to Lake*,' 210 pp., Ife University Press, Nigeria.

Institute of Aquatic Biology (Ghana), 1969–70. Annual Report, 109 pp., Achimota, Ghana.

Jackson, P. B. N., 1959. 'Revision of the clariid catfishes of Nyasaland, with a description of a new genus and seven new species'. *Proc. Zool. Soc. Lond.*, **132**: 109–28.

Jackson, P. B. N., 1961a *The Fishes of Northern Rhodesia*, 140 pp., Govt Printer, Lusaka.

Jackson, P. B. N., 1961b 'The impact of predation especially by the tiger-fish (*Hydrocyon vittatus*) on the African freshwater fishes', *Proc. Zool. Soc. Lond.*, **136**: 603–22.

Jackson, P. B. N., 1971. 'The African Great Lakes fisheries: past, present and future', *Afr. J. Trop. Hydrobiol. Fish.*, **1**: 35–49.

Jackson, P. B. N. and **Davies, B. R.**, 1976. 'Cabora Bassa in its first year: some ecological aspects and comparisons', *Rhodesia Sci. News*, **10**: 128–33.

Jackson, P. B. N. and **Rogers, K. H.**, 1976. 'Cabora Bassa fish populations before and during the filling period', *Zoologica Africana*, **11**: 373–97.

Jackson, P. B. N., Iles, T. D., Harding, D. and **Fryer, G.**, 1963. *Report on a Survey of Northern Lake Nyasa by the Joint Fisheries Research Organization*, 1953–55, Govt Printer, Zomba, Malawi.

Jannasch, H. W., 1975. 'Methane oxidation in Lake Kivu (Central Africa)', *Limnol. Oceanogr.*, **20**: 860–64.

Jaworski, G. and **Lund, J. W. G.**, 1970. 'Drought resistance and dispersal of *Asterionella formosa* Hass.', *Nova Hedwigia, Beiheft*, **31**: 37–48.

Jenkin, P. M., 1936. 'Reports on the Percy Sladen Expedition to some Rift Valley lakes in Kenya in 1929. VII. Summary of the Ecological results with special reference to the alkaline lakes', *Ann. Mag. Nat. Hist., Ser.* 101, **18**: 133–81.

Jenkin, P. M., 1957. The filter-feeding and food of flamingoes (Phoenicopteri)', *Phil. Trans. Roy. Soc. B.*, **240**: 401–93.

Johansen, K., 1970. 'Airbreathing in fishes', in W. S. Hoar and D. J. Randall, *Fish Physiology*, Vol. IV: 361–413, Academic Press, N.Y.

Johansen, K., Maloiy, G. M. O. and **Lykkeboe, G.**, 1975. 'A fish in extreme alkalinity', *Respiration Physiol.*, **24**: 159–62.

Johnels, A. G., 1954. 'Notes on fishes from the Gambia River', *Ark. Zool. Ser. 2, VI*, **17**: 327–411.

Johnels, A. G. and **Svensson, G. S. O.**, 1954. On the biology of *Protopterus annectens* (Owen)', *Ark. Zool.* **7**: 131–64.

Jónassen, P. M. and **Mathiesen, H.**, 1959. 'Measurements of primary production in two Danish eutrophic lakes, Esrom Sø and Furesø', *Oikos*, **10**: 137–67.

*Jones, **J. D.**, 1964. 'Respiratory gas exchange in the aquatic pulmonate *Biomphalaria sudanica*', *Comp. Biochem. Physiol.*, **12**: 297–310.

Jones, J. R. E., 1964. *Fish and River Pollution*, 203 pp., Butterworth, London.

Jonglei Investigation Team, 1954. *The Equatorial Nile Project and its Effects in the Anglo-Egyptian Sudan*, 3 vols, Sudan Govt. Publication.

Jordan, P. and **Webbe, G.**, 1969. *Human Schistosomiasis*, 212 pp., Heinemann.

Jørgensen, C. B., 1976. August Pütter, August Krogh, and modern ideas on the role of dissolved organic matter in aquatic environments', *Biol. Revs*, **51**: 291–328.

*Jubb, **R. A.**, 1961. *An illustrated guide to the freshwater fishes of the Zambezi River, Lake Kariba, Pungwe, Sabi, Lundi and Limpopo Rivers*, 171 pp. Stuart Manning, Bulawayo.

Jubb, R. A., 1967. *The Freshwater Fishes of Southern Africa*, 248 pp., Balkema, Capetown/Amsterdam.

Junk, W., 1970, 1973. 'Investigations on the ecology and production-biology of the 'floating meadows' (Paspalo-Echinochloetum) on the Middle Amazon. Part I. The floating vegetation and its ecology. Part II. The aquatic fauna in the root zone of the floating vegetation', *Amazoniana*, II: 393–400, **IV**: 9–102.

Kahan, D., 1969. 'The fauna of hot springs', *Verh. int. Ver. Limnol.*, **17**: 811–16.

Kalitsi, E. A. K., 1973. 'Volta Lake in relation to the human population and some issues in economics and management' in Ackermann *et al.* 77–85.

Kalk, M., 1969. 'A report on the Lake Chilwa limnological programme in Malawi. Regional meeting of hydrobiologists, in tropical Africa (1968)', *Publ. Unesco Regional Centre for Sci. & Techn. Afr.*, pp. 88–91, Nairobi.

Kalk, M. (ed.), 1970. *Decline and Recovery of a Lake*, 60 pp., Govt Printer, Zomba, Malawi.

Kalk, M. (ed.), 1979. *Lake Chilwa – Studies of change in a tropical ecosystem*, 462 pp., Junk, The Hague.

Keilin, D., 1959. 'The problem of anabiosis or latent life: history and current concept', *Proc. Roy. Soc. B.*, **150**: 149–91.

Kempner, E. S., 1963. 'Upper temperature limit of life', *Science*, **142**: 1318.

Kendall, R. L., 1969. 'An ecological history of the Lake Victoria Basin', *Ecol. Monogr.*, **39**: 121–76.

Kenmuir, D. H. S., 1975. 'Sardines in Cabora Bassa Lake?', *New Scientist*, **65**: 379–80.

Kenmuir, D. H. S., 1978. *A Wilderness called Kariba: the Wildlife and Natural History of Lake Kariba*, Wilderness Publications, Salisbury, Zimbabwe-Rhodesia.

Kent, P. E., 1942. 'The Pleistocene beds of Kanam and Kanjera', *Geol. Mag.*, **79**: 117–132.

Kent, P. E., 1944. 'The Miocene beds of Kavirondo, Kenya', *Q.J. Geol. Soc. Lond.* **100**: 85–118.

Khalil, L. F., 1969. 'The helminth parasites of the freshwater fishes of the Sudan', *J. Zool.*, **158**: 143–70.

Kilham, P., 1971a. 'Biogeochemistry of African lakes and rivers', Thesis for Ph.D., Duke Univ., N. Carolina, 199 pp.

Kilham, P., 1971b. 'A hypothesis concerning silica and the freshwater planktonic diatoms', *Limnol. Oceanogr.*, **16**: 10–18.

Kilham, P. and **Hecky, K. E.**, 1973. 'Fluoride: geological and ecological significance in East African waters and sediments', *Limnol. Oceanogr.*, **18**: 932–45.

Kimpe, P. de, 1964. 'Contribution à l'étude hydrobiologique du Luapula-Moéro', *Ann. Mus. Roy. Afr. Centr. Tervuren*, **128**, 238 pp.

King, L. C., 1951. *South African Scenery*, 2nd edn., 379 pp., Oliver and Boyd, Edinburgh.

Kirk, R. G., 1967. 'The zoogeographical affinities of the fishes of the Chilwa-Chiuta depression of Malawi', *Rev. Zool. Bot. Afr.*, **76**: 295–312.

Kirwan, L. P., 1957. 'Rome beyond the southern Egyptian frontier', *Geogr. J.*, **123**: 13–19.

Kiss, R., 1959. 'Analyse quantitative du zooplancton du Lac Kivu', *Folia Sci. Afr. Centralis*, **4**: 78–80.

Kitaka, G. E. B., 1971. 'An instance of cyclonic upwelling in the southern offshore waters of Lake Victoria', *Afr. J. Trop. Hydrobiol. Fish*, **1**: 85–92.

Klinge, H. and **Ohle, W.**, 1964. 'Chemical properties of rivers in the Amazonian area in relation to soil conditions', *Verh. int. Ver. Limnol.*, **15**: 1067–76.

Klitsch, E., 1967. 'Über den Grundwasserhaushalt der Sahara', *Afrika Spektrum, Hamburg*, **3**: 25–37.

Klitsch, E., 1971. 'Das Wasser im Untergrund der Sahara', in H. Schiffers, Vol. I: 417–28.

Kosswig, C., 1947. 'Selective mating as a factor for speciation in cichlid fish of East African Lakes', *Nature*, **159**: 604–5.

Kosswig, C., 1963. 'Ways of speciation in fishes', *Copeia*, **2**: 238–44.

Kozhov, M., 1963. *Lake Baikal and its Life*, 344 pp. Junk, The Hague.

Kraus, E. B. and **Turner, J. S.**, 1967. 'A one-dimensional model of the seasonal thermocline', *Tellus*, **19**: 98–105.

Krogh, A., 1939. *Osmotic Regulation in Aquatic Animals*, 242 pp., Cambridge Univ. Press.

Kufferath, J., 1952. 'Le milieu biochimique', *Explor. Hydrobiol. du Lac Tanganika (1946–47)*, I: 31–47.

Kusnezov, S. I., 1968. 'Recent studies on the role of microorganisms in the cycling of substances in lakes'. *Limnol. Oceanogr.*, **13**: 211–24.

Kuzoe, F. A. S., 1973. 'Entomological aspects of trypanosomiasis at Volta Lake' in Ac-Kermann *et al.*, 129–31.

Labarbera, M. C. and **Kilham, P.**, 1974. 'The chemical ecology of copepod distribution in the lakes of East and Central Africa.' *Limnol. Oceanogr.*, **19**: 459–65.

Lajoux, J-D., 1963. *The rock paintings of the Tassili*, London.

Lamb, H. H., 1966. 'Climate in the 1960's', *Geogr. J.*, **132**: 182–212.

Langlands, B. W., 1962. 'Concepts of the Nile', *Uganda J.*, **26**: 1–22.

Larson, D. W. and **Donaldson, J. R.**, 1970. 'Waldo Lake, Oregon: a special study', *Oregon State Univ. Water Resources Inst.*, **2**: 1–22.

Latif, A. F. A., 1976. 'Fishes and fisheries of Lake Nasser' in J. Rzóska, 299–307.

Lauzanne, L., 1968. 'Inventaire préliminaire des oligochètes du lac Tchad', *Cah. O.R.S.T.O.M. sér. Hydrobiol.*, **11**: 88–110.

Lauzanne, L., 1969. 'Étude quantitative de la nutrition des *Alestes baremoze* (Pisc. Charac.)', *Cah. O.R.S.T.O.M. sér. Hydrobiol.*, **3**: 15–27.

Lawson, G. W., 1967. '"Sudd" formation on the Volta Lake', *Bull. I.F.A.N.*, A, **29**: 1–4.

Lawson, G. W., Petr, T., Biswas, S., Biswas, E. R. I. and **Reynolds, J. D.**, 1969. 'Hydrobiological work of the Volta Basin Research Project', *Bull. I.F.A.N.*, A, **31**: 965–1003.

Lawson, R. M., 1963. 'The economic organisation of the *Egeria* fishing industry on the River Volta', *Proc. Malacol. Soc. Lond.*, **35**: 273–87.

Lelek, A., 1973. 'Sequence of changes in fish populations of the new tropical manmade lake, Kainji, Nigeria, West Africa', *Archiv. f. Hydrobiol.*, **71**: 381–420.

★**Leloup, E.**, 1950a. 'Gasteropodes', *Explor. Hydrobiol. L. Tanganika*, III, fasc. 4: 1–273.

Leloup, E., 1950b. 'Lamellibranches', *Explor. Hydrobiol. L. Tanganika*, III, fasc. 2: 1–153.

Leloup, E., 1952. 'Les Invertébrés', *Explor. Hydrobiol. L. Tanganika*, I, 71–100.

★**Lemoalle, J.**, 1969. 'Premières données sur la production primaire dans la région de Bol (Avril–Octobre 1968)', *Cah. O.R.S.T.O.M. sér. Hydrobiol.*, **3**: 107–19.

Lemoalle, J., 1975. 'L'activité photosynthétique du phytoplankton en relation avec le niveau des eaux du Lac Tchad (Afrique)', *Verh. int. Ver. Limnol.*, **19**: 1398–403.

Lenfant, C. and **Johansen, K.**, 1968. 'Respiration in the African Lungfish *Protopterus aethiopicus* I. Respiratory properties of blood and normal patterns of breathing and gas exchange. II. Control of breathing', *J. Exp. Biol.*, **49**: 437–52, 453–68.

Léonard, J. and **Compère, P.**, 1967. '*Spirulina platensis* (Gom.) Geitl., algue bleue de grande valeur alimentaire par sa richesse en proteines'. *Bull, Jard. Bot. Nat. Belge.*, **37** (1) Suppl., 23 pp.

Lévêque, C., 1967a. 'Mollusques aquatiques de la zone est du lac Tchad', *Bull. I.F.A.N. A.*, **4**: 1494–53.

Lévêque, C., 1967b. 'Biologie de *Bulinus forskali* dans les mares temporaires de la région de Fort-Lamy', Rapport O.R.S.T.O.M., 8 pp.

Lévêque, C., 1972a. 'Mollusques benthiques du Lac Tchad: écologie, études des peuplements et estimation des biomasses', *Cah. O.R.S.T.O.M. sér. Hydrobiol.*, **6**: 3–46.

Lévêque, C., 1972b. *Mollusques Benthiques de Lac Tchad: Écologie, production et bilans energétique, Thèse présentée à l'Université de Paris*, 225 pp., O.R.S.T.O.M., Paris.

Lévêque, C. and **Gaborit, M.**, 1972. 'Utilisation de l'analyse factorielle des correspondances pour l'étude des peuplements en mollusques benthiques du Lac Tchad', *Cah. O.R.S.T.O.M.*, **6**: 47–66.

Lewis, D. J., 1966. 'Nile Control and its effects on insects of medical importance', in R. Lowe McConnell.

Lewis, D. S. C., 1974. 'The effects of the formation of Lake Kainji (Nigeria) upon the indigenous fish population', *Hydrobiologia*, **45**: 281–301.

Lhote, H., 1958, *A la Découverte des Fresques du Tassili*, 268 pp., Arthaud, Grenoble.

Lhote, H., 1959. *The Search for the Tassili Frescoes*, trans. from the French, 237 pp., Hutchinson.

Likens, G. E. (ed.), 1972. *Nutrients and Eutrophication*, Amer. Soc. Limnol. Oceanogr. Spec. Sympos. **1**, 328 pp.

★**Lind, E. M.**, 1956. 'Studies in Uganda Swamps', *Uganda J.*, **20**: 166–76.

Lind, E. M. and **Morrison, M. E. S.**, 1974. *East African Vegetation*, 257 pp. Longman.

Lind, E. M. and **Visser, S. A.**, 1963. 'A study of a swamp at the north end of Lake Victoria', *J. Ecol.*, **50**: 599–613.

Lindberg, K., 1951. 'Cyclopides (Crustacés, Copépodes),' *Explor. hydrobiol. L'Tanganika*, **III**: fasc. 2, 47–78.

Lindberg, K., 1956. 'Cyclopides (Crustacés, Copépodes) de l'Ouganda', *Kungl. Fysiogr. Sallsk. Lund Forhandl.*, **26**: No. 3: 1–14.

Linder, F., 1941., 'Contribution to the morphology and the taxonomy of the Branchiopoda Anostraca', *Zool. Bidrag. f. Uppsala*, **20**: 101–302.

*Lissmann, H. W., 1958. 'On the function and evolution of electric organs in fish', *J. Exper. Biol.*, 35: 156–91.

Livingstone, David, 1865. *Narrative of an Expedition to the Zambezi (1858–64)*, 608 pp., Murray (and later Editions).

*Livingstone, D. A., 1962. 'Age of deglaciation in the Ruwenzori Range, Uganda', *Nature*, 194: 859–60.

Livingstone, D. A., 1963. *Chemical composition of Rivers and Lakes*, Ch. G., Data of Geochemistry, M. Fleischer, (ed.), 64 pp., U.S. Govt Printing Office, Washington D.C.

Livingstone, D. A., 1965. 'Sedimentation and the history of water level change in Lake Tanganyika', *Limnol. Oceanogr.*, 10: 607–9.

Livingstone, D. A., 1967. 'Postglacial vegetation of the Ruwenzori Mountains in Equatorial Africa', *Ecol. Mongr.*, 1: 25–52.

Livingstone, D. A., 1975. 'Late quaternary climatic change in Africa', *Ann. Rev. Ecol. and Systematics*, 6: 249–80.

Livingstone, D. A., 1976. 'The Nile – Palaeolimnology of headwaters' in Rzóska, 21–30.

Livingstone, D. A. and Kendall, R. L., 1969. 'Stratigraphic studies of East African lakes', *Mitt. Int. Ver. Limnol.*, 17: 147–53.

Lewis, W. M., 1973. 'The thermal regime of Lake Lanao (Philippines) and its theoretical implications for tropical lakes', *Limnol. Oceanogr.*, 18: 200–217.

Lock, J. M., 1973. 'The aquatic vegetation of Lake George, Uganda', *Phytocoenologia*, 1: 250–62.

Löffler, H., 1953. 'Limnologische Ergebnisse der Oesterreichischen Iran Expedition 1949 – 50', *Naturwiss. Rdsch. (Stuttgart)*, 6: 64.

Löffler, H., 1964. 'Zur Limnologie der tropischen Hochgebirge', *Verh. int. Ver. Limnol.*, 15: 176–93.

Löffler, H., 1968. 'Tropical high mountain lakes. Their distribution, ecology and zoogeographical importance', *Colloq. Geogr.*, 9: 57–76, Dümmlers, Bonn.

Lombard, M., 1975. *The Golden Age of Islam*, 259 pp., N. Holland Publ. Co., Amsterdam, Trans from the French *L'Islam dans sa première grandear*, Flammarion, Paris, 1971.

Lomholt, J. P., Johansen, K. and Maloiy, G. M. O., 1975. 'Is the aestivating lungfish the first vertebrate with suctorial breathing?', *Nature*, 257: 787–8.

Loveridge, J. P., 1970. 'Observations on nitrogenous excretion and water relations of *Chiromantis xerampelina* (Amphibia, Anura)', *Arnoldia (Rhodesia)*, 5: 1–6.

*Lowe, R. H., 1952. *Report on the Tilapia and other fish and fisheries of Lake Nyasa*, Col. Off. Fish. Publ. Vol. 1. No. 2, 126 pp., H.M.S.O.

Lowe, R. H., 1953. 'Notes on the ecology and evolution of Nyasa fishes of the genus *Tilapia*, with a description of *T. saka*', *Proc. Zool. Soc. Lond.*, 122: 1035–41.

Lowe, R. H., 1956. 'Observations on the biology of *Tilapia* (Pisces: Cichlidae) in Lake Victoria', *E. Afr. Fish. Res. Org.*, Suppl. Publ. No. 1, 72 pp.

Lowe, R. H., 1957. 'Observations on the diagnosis and biology of *Tilapia leucosticta* Trewavas in E. Africa', *Rev. Zool. Bot. Afr.*, 55: 353.

Lowe, R. H., 1958. 'Observations on the biology of *Tilapia nilotica*. Linne. in E. African waters', *Rev. Zool. Bot. Afr.*, 57: 129.

Lowe-McConnell, R. H., 1959. 'Breeding behaviour patterns and ecological differences between *Tilapia* species and their significance for evolution within the genus *Tilapia* (Pisces: Cichlidae)', *Proc. Zool. Soc. London*, 132: 1–30.

Lowe-McConnell, R. H. (ed.), 1966. *Man-Made Lakes*. Sympos. Inst. Biol. No. 15 (1965), 218 pp. Academic Press, London.

Lowe-McConnell, R. H., 1969. 'Speciation in tropical freshwater fishes', *Biol. J. Linn. Soc.*, 1: 51–75.

Lowe-McConnell, R. H., 1975. *Fish communities in tropical freshwaters*, 337 pp., Longman, London.

Lowe-McConnell, R. H., 1977. *Ecology of fishes in tropical waters*, 64 pp., Inst. Biol. Studies in Biology No. 76.

Lowndes, A. G., 1936. 'Scientific results of the Cambridge Expedition to the East African Lakes 1930–31. No. 16. The smaller Crustacea', *J. Linn. Soc. Zool.*, **40**: 1–31.

Lund, J. W. G., 1965. 'The ecology of the freshwater phytoplankton', *Biol. Rev.*, **40**: 231–93.

Lund, J. W. G., 1978. 'Experiments with lake phytoplankton in large enclosures', *Freshw. Biol. Assoc. Ann. Rep.*, **46**: 32–9.

Lund, J. W. G., Mackereth, F. J. H. and **Mortimer, C. H.**, 1963. Changes in depth and time of certain chemical and physical conditions and of the standing crop of *Asterionella formosa* Hass. in the North Basin of Windermere in 1947', *Phil. Trans. Roy. Soc. B.*, **246**: 255–90.

Luther, H. and **Rzóska, J.**, 1971. *Project Aqua: A Source Book of Inland Waters Proposed for Conservation.*, Int. Biol. Programme Handbook No. 21, 239 pp., Blackwell, Oxford.

Macan, T. T., 1963. *Freshwater Ecology*, 338 pp., Longman.

McBurney, C. B. M., 1960. *The Stone Age of Northern Africa*, 288 pp., Penguin, Harmondsworth.

McCarthy, J., 1962a. 'The colonisation of a swamp forest clearing (with special reference to *Mitragyna stipulosa*)', *E. Afr. Agric. & For. J.*, **28**: 22–8.

McCarthy, J., 1962b. 'The form and development of knee roots in *Mitragyna stipulosa*,' *Phytomorphology*, **12**: 20–30.

McConnell, R. B., 1967. 'The East African Rift System', *Nature*, **215**: 578–81.

Macdonald, W. W., 1953. 'Lake flies', *Uganda J.*, **17**: 124–34.

Macdonald, W. W., 1956. 'Observations on the biology of chaoborids and chironomids in Lake Victoria and on the feeding habits of the "elephant-snout fish" *Mormyrus kannume*', *J. Anim. Ecol.*, **25**: 36–53.

McDougall, I., Morton, W. H. and **Williams, M. A. J.**, 1975. 'Age and rates of denudation of trap series basalts at Blue Nile gorge, Ethiopia', *Nature*. **254**: 207–9.

Macfadyen, A., 1948. 'The meaning of productivity in biological systems', *J. Anim. Ecol.*, **17**: 75–418.

McGowan, J. M., 1974. 'Ecological studies on *Chaoborus* (Diptera, Chaoboridae) in Lake George, Uganda', *Freshwater Biol.*, **4**: 483–505.

McGregor Reid, G. and **Sydenham, H.**, 1979. 'A checklist of lower Benue fishes and an icthyogeographical review of the Benue River (West Africa)', *J. Nat. Hist.*, **13**: 41–67.

Mackereth, F. J. H., 1966. 'Some chemical observations on post-glacial lake sediments', *Phil. Trans. Roy. Soc. B.*, **250**: 165–213.

* **McLachlan, A. J.**, 1968. 'A study of the bottom fauna of Lake Kariba', thesis for Ph.D., University of London.

McLachlan, A. J., 1969a. 'The effect of aquatic macrophytes on the variety and abundance of benthic fauna in a newly created lake in the tropics (Lake Kariba)', *Arch. Hydrobiol.*, **66**: 212–31.

McLachlan, A. J., 1969b. 'Substrate preferences and invasion behaviour exhibited by larvae of *Nilodorum brevibucca* Freeman (Chironomidae) under experimental conditions', *Hydrobiologia*, **33**: 237–49.

McLachlan, A. J., 1970a. 'Some effects of annual fluctuations in water level on the larval chironomid communities of Lake Kariba', *J. Anim. Ecol.*, **39**: 70–90.

McLachlan, A. J., 1970b. 'Submerged trees as a substrate for benthic fauna in the recently created Lake Kariba (Central Africa)', *J. Appl. Ecol.*, **7**: 253–66.

McLachlan, A. J., 1974. 'Development of some lake ecosystems in tropical Africa with special reference to the invertebrates', *Biol. Revs.*, **49**: 365–97.

McLachlan, A. J. and McLachlan, S. M., 1969. 'The bottom fauna and sediments in a drying phase of a saline African lake (Lake Chilwa, Malawi)', *Hydrobiologia*, **34**: 401–13.

McLachlan, A. J. and McLachlan, S. M., 1971. 'Benthic fauna and sediments in the newly created Lake Kariba (Central Africa)', *Ecology*, **52**: 800–809.

McLachlan, A. J., Morgan, P. R., Howard-Williams, C., McLachlan, S. M. and Bourn, D., 1972. 'Aspects of the recovery of a saline African lake following a dry period', *Arch. Hydrobiol.*, **70**: 325–40.

Mclachlan, S. M., 1970. 'The influence of lake level fluctuation and the thermocline on water chemistry in two gradually shelving areas in Lake Kariba, Central Africa', *Archiv. f. Hydrobiol.*, **66**: 499–510.

McLeod, A. A. Q. R., 1978. 'A study of *Caridina nilotica* (Roux) Atyidae and *Macrobrachium niloticum* (Roux) Palaemonidae (Decapoda: Crustacea) in Lake Turkana' in A. J. Hopson, 1978a:

McMahon, B. R., 1970. 'The relative efficiency of gaseous exchange across the lungs and gills of an African lungfish *Protopterus aethiopicus*', *J. Exp. Biol.*, **52**: 1–16.

Maetz, J. and de Renzis, G., 1978. 'Aspects of the adaptation of fish to high external salinity: a comparison of *Tilapia grahami* and *T. mossambic*' in Schmidt–Nielsen *et al.*, 213–28.

Magis, N., 1962. 'Etude limnologique des lacs artificiels de la Lufira et du Lualaba (Haut Katanga). I. Le régime hydraulique, less variations saisonières de la température', *Int. Rev. Hydrobiol.*, **47**: 33–84.

Maglione, G., 1969. 'Premières données sur le régime hydrogéochimique des lacs permanent du Kanem (Tchad)', *Cah. O.R.S.T.O.M. sér. Hydrobiol.*, **3**: 121–41.

Maguire, B., 1963. 'The passive dispersal of small aquatic organisms and their colonisation of small isolated bodies of water'. *Ecol. Monogr.*, **33**: 161–85.

Maire, R., 1928. 'La végétation et la flore du Hoggar', *C. R. Ac. Sci. Paris*, **186**: 1680–82.

Makerere Expedition Report, 1961. *Expedition to Lake Manyara April–July 1961*, 22 pp., Makerere Univ. Library.

Maloiy, G. M. O., Lykkeboe, G., Johansen, K. and Bamford, O. S., 1978. 'Osmoregulation in *Tilapia grahami*: a fish in extreme alkalinity' in Schmidt-Nielsen *et al.*, 229–38.

Mandahl-Barth, G., 1954. *The Freshwater Molluscs of Uganda and Adjacent Territories*, Ann. Mus. Congo Belge., Sér. 8^0, Sci. Zool., 32, 206 pp.

Mandahl-Barth, G., 1972. 'The freshwater molluscs of Lake Malawi', *Rev. Zool. Bot. Afr.*, **56**: 257–88.

Mangum, C. P., Lykkeboe, G. and Johansen, K., 1975. 'Oxygen uptake and the role of haemoglobin in the East African swampworm *Alma emini*', *Comp. Biochem. Physiol.* **52A**: 477–82.

Mann, K. H., 1969. 'The dynamics of aquatic ecosystems' in J. B. Cragg (ed.), *Advances in Ecological Research*, Vol. **6**: 1–71, Academic Press, London.

* Mann, M. J., 1969. 'A résumé of the evolution of the *Tilapia* fisheries of Lake Victoria up to the year 1960', *Ann. Rep. E. Afr. freshw. Fish Res. Org.*, 1969: 21–7.

Mann, M. J. and Ssentongo, G. W., 1968. 'A note on the *Tilapia* populations of Lake Naivasha in 1968', *Ann. Rep. E. Afr. Fish. Res. Org.*, 1968: 28–31.

* Marlier, G., 1938. 'Considérations sur les organes accessoires servant à la respiration aerienne chex les Téléostéens', *Ann. Soc. zool. Belg.*, **69**: 163–85.

Marlier, G., 1951–1954. 'Recherches hydrobiologiques dans les rivières du Congo Oriental. I. La conductivité électrique. II. Études écologiques.' *Hydrobiologia*, **3**: 1961, **6**:
225–64.

Marlier, G., 1953. 'Etude biographique du Bassin de la Ruzizi, basée sur la distribution
des poissons', *Ann. Soc. Roy. Zool. de Belgique*, Fasc. 1. **84**: 175–224.

Marlier, G., 1955. 'Un Trichoptère pelagique nouveau du lac Tanganika', *Rev. Zool.
Bot. Afr.*, **52**: 150–55.

Marlier, G., 1958a. 'Recherches hydrobiologiques au lac Tumba', *Hydrobiologia*, **10**: 352–
85.

Marlier, G., 1958b. 'Réflexions sur l'origine du lac Kivu', *Bull. Séances Acad. Roy. Sci.
Colon. Bruxelles* **4**: 1001–14.

Marlier, G., 1959. 'Observations sur la biologie littorale du lac Tanganika', *Rev. Zool.
Bot. Afr.*, **59**: 164–83.

Marlier, G., 1962. 'Genera des Trichoptères de l'Afrique', *Ann. Mus. Roy. Afr. Centr.
Sciences Zoologiques*, **109**, 261 pp.

Marlier, G., 1967. 'Ecological studies on some lakes of the Amazon Valley', *Amazoniana*,
1: 91–116.

Marlier, G., 1973. 'Limnology of the Congo and Amazon Rivers' in E. S. Meggers, E. S.
Eyensu and W. D. Duckworthy, *Tropical forest ecosystems in Africa and South America*: 223–38, Smithsonian Inst. Press, Washington, D.C.

Marlier, G., Bouillon, J., Dubois, T. and Leleup, N., 1955. 'Le lac Lungwe', *Acad. Roy. Sci.
Col. Bull. Sci.*, **I**: 665–76.

Marshall, B. E., 1978. 'Aspects of the ecology of the benthic fauna in Lake McIlwaine,
Rhodesia', *Freshw. Biol.*, **8**: 241–9.

Marshall, B. E., 1979. 'Fish populations and fisheries potential of Lake Kariba', *S. Afr.
J. Sci.*, **75**: 485–8.

Marshall, B. E. and Falconer, A. C., 1973a. 'Physico-chemical aspects of Lake
McIlwaine (Rhodesia), a eutrophic tropical empoundment', *Hydrobiologia*, **42**: 45–62.

Marshall, B. E. and Falconer, A. C., 1973b. 'Eutrophication of a Tropical African impoundment (Lake McIlwaine, Rhodesia)', *Hydrobiologia*, **43**: 109–23.

Matthes, H., 1962. 'Poissons nouveaux ou intéressants du lac Tanganika et du Ruanda',
Ann. Mus. Roy. Afr. Cent., **111**: 27–88.

Matthes, H., 1964. 'Les poissons du lac Tumba et de la région d'Ikela', *Ann. Mus. Roy.
Afr. Centr.*, **126**, 204 pp.

Matthes, B., 1978. 'The problem of the rice-eating fish in the central Niger delta, Mali',
in R. L. Welcomme, 227–52.

Mauny, R., 1956. 'Préhistoire et zoologie: la grande faune éthiopéene du Nord-ouest
Africain du palaeolithique à nos jours', *Bull. I.F.A.N. (A)*, **18**: 264–79.

Mauny, R., 1957. 'Repartition de la grande "faune éthiopéenne" du Nord-ouest Africain
du palaeolithique à nos jours', *Proc. 3rd Pan-Afr. Congr. Préhist.* (Livingstone 1955),
18: 102–5.

Mauny, R., 1961. 'Tableau géographique de l'Quest Africain au moyen âge', *Mém.
I.F.A.N.*, **61**, 587 pp.

Mauny, R., 1978. 'Trans-Saharan contacts and the Iron Age in West Africa', Ch. 5. in
J. D. Fage (ed.), *Cambridge History of Africa, Vol. 2. From c. 500 BC to 1050 AD*,
Cambridge Univ. Press.

Maurette, F., 1938. *Afrique Equatoriale, Orientale et Australe*, in T. Vidal de la Blache et
L.. Gallois (eds), *Geographie Universelle*, T. XII, 398 pp., Armand Colin, Paris.

Mayhew, W. W., 1969. 'Biology of desert Amphibians and Reptiles', in G. W. Brown,
Vol. **I**: 196–356.

Mayr, E., 1963. *Animal Species and Evolution*, 797 pp., Harvard Univ. Press, Cambridge, Mass.

Mayr, E., 1970. *Populations, Species and Evolution* (Abridgement of *Animal Species and Evolution*), 453 pp., Harvard Univ. Press, Cambridge, Mass.

Meckelein, W., 1959. *Forschungen in der zentralen Sahara. I. Klimageomorphologie*, 181 pp. Geog Westermann, Braunschweig.

Meel, L. van, 1952. 'Le milieu végétal', *Explor. Hydrobiol. Lac Tanganika (1946–47)*, 1: 51–68.

Meel, L. van, 1953. 'Contribution à l'étude du Lac Upemba. A. Le milieu physico-chimique', *Explor. Parc. Nat. Upemba*, Fasc. 9, 190 pp. Inst. Parcs. Nat. Congo Belge, Bruxelles.

Meel, L. van, 1954. 'Le Phytoplankton', *Explor. Hydrobiol. Lac Tanganika (1946–1947)*, 4:1–681.

Meijering, M. P.D., 1970. 'Süsswasser Cladocera unter dem Einfluss mariner Sturm-fluten', *Arch. Hydrobiol.*, 67: 1–31.

Melack, J. M., 1976. 'Limnology and dynamics of phytoplankton in equatorial African lakes', Thesis for Ph.D., Duke Univ., N. Carolina, 453 pp.

Melack, J. M., 1978. 'Morphometric, physical and chemical factors of the volcanic crater lakes of western Uganda', *Arch. Hydrobiol.*, 84: 30–53.

Melack, J. M., 1979. 'Photosynthesis and growth of *Spirulina platensis* (Cyanophyta) in an equatorial lake (Lake Simbi, Kenya)', *Limnol. Oceanogr.*, 24: 753–60.

Melack, J. M. and Kilham, P., 1974 'Photosynthetic rates of phythoplankton in East African alkaline, saline lakes', *Limnol. Oceanogr.*, 19: 743–55.

Methuen, P. A., 1911. 'On an amphipod from the Transvaal', *Proc. Zool. Soc. Land.*, 4: 948–57.

Milburn, T. R. and Beadle, L. C., 1960. 'The determination of total carbon dioxide in water', *J. Exp. Biol.*, 37: 444–60.

Ministry of Agriculture, Botswana, 1977. 'Agricultural development in the Okavango region' in Nteta and Hermans, 187–92.

Missione Di Studio Al Lago Tana, 1938–40. 3 vols, *Reale Acad. d'Ital. Centro Studi Afr. Orient. Ital.*

Mitchell, D.S., 1969, 'The Ecology of vascular hydrophytes on Lake Kariba', *Hydrobiologia*, 34: 448–64.

Mitchell, D. S., 1973. 'Supply of plant nutrient chemicals in Lake Kariba' in Acker-mann *et al.*, 165–9.

Mitchell, D. S., 1978. 'Freshwater plants' Ch. 33. in M.J.A. Werger.

Modha, M. L., 1967. 'The ecology of the Nile Crocodile (*Crocodilus niloticus* Laurent) on Central Island, Lake Rudolf', *E. Afr. Wildlife J.*, 5: 74–95.

Moghraby, A. I., 1977. 'A study on diapause of zooplankton in a tropical river – the Blue Nile', *Freshw. Biol.*, 7: 202–12.

Mohr, P. A., 1962. *The geology of Ethiopa*, University of Addis Ababa Press.

Mohr, P. A., 1966. 'Geographical report on the Lake Langano and adjacent plateau re-gions', *Bull. geophys. Obs. Addis Ababa*, 9: 59–75.

Monod, Th., 1945a. 'Sur la lecture des écailles de quelques poissons du Niger Moyen', C. R. Acad. Sc. 220: 629–30.

Monod, Th., 1945b. 'Un nouveaux *Stenasellus* Ouest-Africain', *Bull, I.F.A.N.*, 7: 101–14.

Monod, Th., 1951. 'Contribution à l'étude du peuplement de la Mauritanie. Poissons d'eau douce', *Bull. I.F.A.N.*, 13: 802–12.

Monod, Th., 1954. 'Contributions à l'étude du peuplement de la Mauritanie. Poissons d'eau douce (2me note)', *Bull. I.F.A.N.*, **16**: 295–9.

Monod, Th., 1958. *Majâbat al. Koubrâ. Contribution à l'étude de l''Empty Quarter' Ouest-Saharien*. Mém. No. 52. I.F.A.N., 406 pp.

Monod, Th., 1963. 'The late Tertiary and Pleistocene in the Sahara' in F. C. Howell and F. Bourlière (eds), *African Ecology and Human Evolution, Viking Publ. Anthropol.*, **36**: 117–229.

Monod, Th., 1968a. 'Rapport sur une mission exécutée dans le Nord-Est du Tchad en decembre 1966 et janvier 1967', *Études et documents tchadiens.*, Sér. A, No. 3, 65 pp

Monod, Th., 1968b. 'Contribution à l'étude des eaux douces de l'Ennedi III. Crustacés Décapodes', *Bull. I.F.A.N.*, **30**: 1350–3.

Monod, Th., 1969. 'A propos du Lac des Vers ou Bahr ed-Dûd (Libye)', *Bill. I.F.A.N.*, **31**: 25–41.

Monod, Th., 1972. 'Contribution á l'étude de la Grotte de Sof Omar (Ethiopie Méridionale). No. 3 – Sur une espèc nouvelle de Cirolanidé cavernicole, *Skotobaena mortoni* (Crust. Isopoda)', *Ann. Spéléol.*, **27**: 205–20.

Monod, Th. and **Mauny, R.**, 1957. 'Découverte de nouveaux instruments en os de l'Ouest africain', *Actes 3e Pan-Afr. Congr. Préhist.*, (Livingstone 1955): 242–7.

Moore, J. E. S., 1903. *The Tanganyika Problem*, 356 pp., Hurst and Blacket.

Moorhead, A., 1960. *The White Nile*, 385 pp., Hamish Hamilton.

Moorhead, A., 1962. *The Blue Nile*, 308 pp., Hamish Hamilton.

Morandini, G., 1940. 'Richerche limnologiche, geographia fisica' in *Missione di studio al Lago Tana*, III(1), 319 pp.

★ **Moreau, R. E.**, 1961. 'Problems of Mediterranean–Sahara migration', *Ibis*, **103a**: 393–427, 580–623.

Moreau, R. E., 1963a. 'The distribution of tropical African birds as an indicator of past climatic changes', *Viking Publ. in Anthropol.*, **36**: 28–42.

Moreau, R. E., 1963b. 'Vicissitudes of the African biomes in the late Pleistocene'. *Proc. Zool. Soc. Lond.*, **141**: 395–421.

Moreau, R. E., 1966. *The Bird Faunas of Africa and Its Islands*, 424 pp., Academic Press, London.

Moreau, R. E., 1967. 'Water birds over the Sahara', *Ibis*, **109**: 232–59.

Morgan, A. and **Kalk, M.**, 1970. 'Seasonal changes in the waters of Lake Chilwa (Malawi) in a drying phase, 1966–68', *Hydrobiologia*, **36**: 81–103.

Morgan, P. R., 1970. 'The Lake Chilwa *Tilapia* and its fishery', *Afr. J. Trop. Hydrobiol. Fish.*, **1**: 51–8.

Moriarty, C. M. and **Moriarty, D. J. W.**, 1973a. 'Quantitative estimation of the daily ingestion by *Tilapia nilotica* and *Haplochromis nigripinnis* in Lake George, Uganda', *J. Zool.*, **171**: 15–23.

' **Moriarty, D. J. W.**, 1973. 'The physiology of digestion of blue-green algae in the cichlid fish *Tilapia nilotica*', *J. Zool.*, **171**: 25–39.

Moriarty, D. J. W. and **Moriarty, C. M.**, 1973b. 'The assimilation of carbon from phytoplankton by two herbivorous fishes: *Tilapia nilotica* and *Haplochromis nigripinnis*', *J. Zool.*, **171**: 41–55.

Moriarty, D. J. W., Darlington, J. P. E. C., Dunn, I. G., Moriarty, C. M. and **Tevlin, M. P.**, 1973. 'Feeding and grazing in Lake George, Uganda' in Greenwood and Lund, 299–320.

Morris, J. E., 1971. 'Hydration, its reversibility, and the beginning of development in the brine-shrimp, *Artemia salina*', *Comp. Biochem. Physiol.*, **39A**: 843–57.

Morris, P., Largen, M. J. and **Yalden, D. W.**, 1976. 'Notes on the biogeography of the Blue Nile (Great Abbai) Gorge in Ethiopia', in Rzóska, 233–42.

★ **Morrison, P. E. S.**, 1968. 'Vegetation and climate in the uplands of south-western Uganda during the later Pleistocene period. I. Muchoya Swamps, Kigezi District', *J. Ecol.*, **56**: 363–84.

Mortimer, C. H., 1941–1942. 'Exchange of dissolved substances between water and mud in lakes, I–IV', *J. Ecol.*, **29**: 280–329; **30**: 147–201.

Mortimer, C. H., 1952. 'Water movements in lakes during summer stratification; evidence from the distribution of temperature in Windermere', *Phil. Trans. Roy. Soc. B.*, **236**: 355–404.

Mortimer, C. H., 1953. 'The resonant response of stratified lakes to wind', *Schweiz. Z. Hydrol.*, **15**: 94–151.

Mortimer, C. H., 1959. 'Motion in thermoclines', *Verh. int. Ver. Limnol.*, **14**: 79–83.

Mortimer, C. H., 1969. 'Physical factors with bearing on eutrophication in Lakes', in G. A. Rohlich, 340–68.

Moss, B., 1969. 'Limitation of algae growth in some Central African waters', *Limnol. Oceanogr*, **14**: 591–601.

Moss, B., 1972–73. 'The influence of environmental factors on the distribution of freshwater algae: an experimental study. I. Introduction and the influence of calcium. II. The role of pH and the carbon dioxide-bicarbonate system. III. Effects of temperature, vitamin requirements and inorganic nitrogen compounds on growth. IV. Growth of test species in natural lake waters, and conclusion', *J. Ecol.*, **60**: 917–32, **61**: 157–77, **61**: 179–92, **61**: 193–211.

Moss, B. and Moss, J., 1969. 'Aspects of the limnology of an endorheic African lake (Lake Chilwa, Malawi)', *Ecology*, **50**: 109–18.

Munro, J. L., 1966. 'A limnological survey of Lake McIlwaine, Rhodesia', *Hydrobiologia*, **28**: 281–308.

Munro, J. L., 1967. 'The food of a community of East African freshwater fishes', *Proc. Zool. Soc. Lond.*, **151**: 389–415.

Myers, G. S., 1949. 'Usage of anadromous, catadromous and allied terms for migratory fishes', *Copeia*, **2**: 89–97.

Myers, G. S., 1952. 'Annual Fishes', *Aquarium J.*, **23**: No. 7.

Myers, G. S., 1960. 'The endemic fish fauna of Lake Lanao, and the evolution of higher taxonomic categories', *Evolution*, **14**: 323–33.

Mzumara, A. J. P., 1967. 'The Lake Chilwa Fisheries', *Soc. Malawi J.*, **20**: 58–68.

Nawar, G., 1959. 'Observations on the breeding of six members of the Nile Mormyridae', *Ann. Mag. Nat. Hist.*, **13**: 493–504.

Nelson, G. S., 1970. 'Onchocerciasis', *Adv. Parasitol.*, **8**: 173–224.

Nemenz, H., 1970. 'Ionenverhältnisse und die Besiedlung hyperhaliner Gewässer, besonders durch Insekten', *Acta Biotheoretica*, **19**: 148–70.

Newell, B. S., 1960. 'The hydrology of Lake Victoria', *Hydrobiologia*, **15**: 363–83.

Nillson, E., 1940. 'Ancient changes of climate in British East Africa and Abbyssinia. A study of ancient lakes and glaciers', *Geograph. Annalr.*, **22**: 1–78.

Nilsson, E., 1963. 'Pluvial lakes and glaciers in East Africa', *Stockholm Contrib. Geol.*, **11**(2): 21–57.

Njogu, A. P. and *Kinoti, G. K., 1971. 'Observations on the breeding sites of mosquitoes in Lake Manyara, a saline lake in the East African Rift Valley', *Bull. Ent. Res.*, **60**: 473–9.

Nteta, D. N. and Hermans, J. (eds), 1977. *Proceedings of the Symposium on the Okavango Delta and its future utilisation*, 350 pp., Botswana Society, National Museum, Gaborone.

Obeng, L. (ed.), 1969. *International Symposium on Man-Made Lakes, Accra (1966)*, Ghana Univ. Press.

Odum, E. P., 1971. *Fundamentals of Ecology* (3rd edn), 574 pp., W. B. Saunders.

*Okedi, J., 1965. 'The biology and habits of the Mormyrid fishes, *Gnathonemus longibarbis*, G. *nigricans*, *Marcusenius grahami*, M. *nigricans*, and *Petrocephalus catastoma*', *J. Appl. Ecol.*, **2**(2): 408–9.

Okedi, J., 1969. 'Observations on the breeding and growth of certain mormyrid fishes of the Lake Victoria Basin', *Rev. Zool. Bot. Afr.*, **79**: 34–65.

Okedi, J., 1970. 'The food and feeding habits of the small mormyrid fishes of Lake Victoria, E. Africa', *Afr. J. Trop. Hydrobiol. Fish.*, **1**: 1–12.

Okedi, J. (ed.), 1973. *Symposium on evaluation of fishery resources in the development of inland fisheries*, Afr. J. Hydrobiol. Fish. Special issues I and II. 143 pp.

Oliver, R. and Fage, J. D., 1966. *A Short History of Africa* (2nd edn.), 284 pp., Penguin, Harmondsworth.

Omer, S. M. and Cloudesley-Thompson, J. L., 1968. 'Dry season biology of *Anopheles gambiae* in the Sudan', *Nature*, **217**: 879–80.

Osmaston, H. A., 1967. 'The sequence of glaciations in the Ruwenzoris and their correlation with glaciations of other mountains in East Africa and Ethiopia', in Bakker, E. M. von. Z. (ed.) *Palaeoecology of Africa*, **2**: 26–8.

Otobo, F. O., 1974. 'The potential for the clupeid fishery in Lake Kainji, Nigeria', *Afr. J. Hydrobiol. Fish.*, **3**: 123–34.

Overbeck, J., 1974. 'Microbiology and biochemistry', *Mitt. int. ver. Limnol.* **20**: 198–228.

Padgett, D. E., 1976. 'Leaf decomposition by fungi in a tropical rainforest stream', *Biotropica*, **8**: 166–78.

Paperna, I., 1969. 'Evolution of the shoreline, aquatic weeds, snails and bilharzia transmission in the newly formed Volta Lake' (Abstract), *Verh. int. Ver. Limnol.*, **17**: 282–3.

Paperna, I., 1970. 'Study of an outbreak of schistosomiasis in the newly formed Volta Lake, Ghana', *Zeitschr. Tropenmed. Parasitol.*, **221**: 411–25.

Parenzan, P., 1939. 'I pesce del bacino del Lago Regina Margherita nel Galla e Sidama', *Boll. Pesca Piscic. Idrobiol.*, **15**: 146–7.

Parsons, T. R. and Seki, H., 1970. 'Importance and general implications of organic matter in aquatic environments' in D. W. Hood (ed.), *Organic matter in natural waters*, Univ. Alaska. Inst. Mar. Sci. occas. publ. No. 1.

Patterson, L., 1977. 'An introduction to the ecology and zoogeography of the Okavango Delta' in Nteta and Hermans, 55–60.

Pellegrin, J., 1911a. 'Les Vertébrés aquatiques du Sahara', *C. R. Ac. Sci. (Paris)*, **153**: 972–4.

Pellegrin, J., 1911b. 'La distribution geographique des poissons d'eau douce en Afrique', *C. R. Ac. Sci. (Paris)*, **153**: 297–9.

Pellegrin, J., 1921. 'Les poissons des eaux douces de l'Afrique du Nord Française', *Mém. Soc. Sci. Nat. Maroc.*, **1**(2): 1–216.

Pellegrin, J., 1929. 'Mission Saharienne Angiéras-Draper 1927–28. Poissons', *Bull. Mus. Nat. d'Hist. Nat.*, 2ᵉ Sér. **I**: 134–9.

Pellegrin, J., 1931. 'Réptiles, Batraciens et Poissons du Sahara Central recueillis par le Prof. Seurat', *Bull. Mus. Hist. Nat.*, 2ᵉ Sér. **3**: 216–18.

Penfound, W. T. and Earle, T. T., 1948. 'The biology of the water hiacynth', *Ecol. Monogr.*, **18**: 447–72.

*Pesce, A., 1968. *Gemini Space Photographs of Libya and Tibesti*, A. S. Campbell (ed.), 81 pp., Petroleum Explor. Soc. of Libya, Tripoli.

*Peters, H. M., 1947. 'Über Bau, Entwicklung und Funktion eines eigenartigen hydrostatischen Apparates larvaler Labyrinthfische', *Biol. Zentralbl.*, **66**: 304–29.

Peters, N., 1963. 'Embryonale Anpassungen oviparer Zahnkarpfen aus periodisch austrocknenden Gewässern', *Int. Rev. Hydrobiol.*, **48**: 257–313.

* Petr, T., 1968a. 'Population changes in aquatic invertebrates living on two water plants in a tropical man-made lake', *Hydrobiologia*, 32: 3–4, 449–85.

Petr, T., 1968b. 'The establishment of the lacustrine fish fauna of Volta Lake in Ghana during 1964–65', *Bull. I.F.A.N.*, 30: 259–69.

Petr, T., 1968c. 'Distribution, abundance and food of commercial fish in the black Volta and the Volta man-made lake in Ghana during its first period of filling (1964–1966). I. Mormyridae', *Hydrobiologia*, 32: 417–48.

Petr, T., 1969. 'The development of the bottom fauna in the man-made Volta Lake in Ghana', *Verh. int. Ver. Limnol.*, 17: 273–81.

Petr, T., 1970a. 'Chironomidae (Diptera) from light catches in the man-made Volta Lake in Ghana', *Hydrobiologia*, 35: 449–68.

Petr, T., 1970b. 'Macro-invertebrates of flooded trees in the man-made Volta Lake (Ghana) with special reference to the burrowing mayfly *Povilla adusta* Navas', *Hydrobiologia*, 36: 399–418.

Petr, T., 1971. 'Establishment of chironomids in a large tropical man-made lake', *Canadian Entomologist*, 103: 380–5.

Petr, T., 1972. 'On some factors determining the quantitative changes in chaoborids in the Volta man-made lake during the filling period', *Rev. Zool. Bot. Afr.*, 85: 147–59.

Petr, T., 1975. 'On some factors associated with the initial high fish catches in new African man-made lakes', *Arch. Hydrobiol.*, 75: 32–49.

Phillipson, J., 1966. *Ecological Energetics*, 57 pp., Arnold.

Pianka, E. R., 1966. 'Latitudinal gradients in species diversity: a review of concepts', *Amer. Natur.*, 100: 33–46.

Pias, J., 1958. 'Transgressions et régressions du Lac Tchad à la fin du Tertiaire et au Quaternaire', *C. R. Acad. Sci. (Paris)*, 246: 800–3.

Pike, J. G. and Rimmington, G. T., 1965. *Malawi. A Geographical Study*, 229 pp., Oxford Univ. Press.

Pilsbry, H. A. and Bequaert, J., 1927. 'The aquatic molluscs of the Belgian Congo with a geographical and ecological account of Congo malacology', *Bull. Amer. Mus. Nat. Hist.*, 53: 69–602.

* Poll, M., 1939a. 'Poissons', *Explor. Parc. Nat. Albert. Mission G. F. de Witte (1933–35)*. Fasc. 24, Inst. Parcs Nat. Congo Belge. Bruxelles.

Poll, M., 1939b. 'Poissons', *Explor. Parc. Nat. Albert. Mission H. Damas (1935–36)*, Fasc. 6, Inst. Parcs Nat. Congo Belge. Bruxelles.

Poll, M., 1942. 'Les poissons du Lac Tumba', *Bull. Mus. R. Hist. Nat. Belg.*, 18 (No. 36): 1–25.

Poll, M., 1946. 'Révision de la faune ichthyologique du Lac Tanganika', *Ann. Mus. Roy. Congo Belge. C. Zool.*, 4: 141–364.

Poll, M., 1950. 'Histoire du peuplement et origine des espèces de la faune ichthyologique du Lac Tanganika', *Ann. Soc. Roy. Zool. Belg.*, 81: 111–40.

Poll, M., 1952a. 'Ségrégation géographique et formation des espèces', *Ann. Soc. Roy. Zool. Belg.*, 83: 211–24.

Poll, M., 1952b. 'Les Vertébrés. Explor. Hydrobiol. Lac Tanganika', *Inst. Roy. Soc. Nat. Belg.*, 1: 103–65.

Poll, M., 1953. 'Poissons non-cichlidae. Explor. Hydrobiol. Lac. Tanganika (1946–47)', *Inst. Roy. Sci. Nat. Belg.*, III, Fasc. 5A, 251 pp.

Poll, M., 1954. 'Zoogéographie des Protoptères et Polytères', *Bull. Soc. Zool. France*, 79: 282–9.

Poll, M., 1956. 'Poissons cichlidae', *Explor. Hydrobiol. Lac Tanganika (1946–47)*, III Fasc. 5: 1–619.

Poll, M., 1957. 'Les genres des poissons d'eau douce de l'Afrique', *Ann. Mus. Roy. Congo Belge.*, **54**: 1–191.

Poll, M., 1959. 'Aspects nouveaux de la faune ichthyologique du Congo Belge', *Bull. Soc. Zool. France*, **84**: 259–71.

Poll, M., 1963. 'Zoogéographie ichthyologique du cours superieur du Lualaba', *Publ. Univ. Elizabethville*, **6**: 95–106.

Poll, M., 1966. 'Géographie ichthyologique de l'Angola', *Bull. Séances Ac. Roy. Sci. Outre-mer*, **2**: 355–65.

Poll, M., 1967. 'Contributions à la faune ichthyologique de l'Angola', Publ. Cult. No. 75, *Museo do Dundo. Lisboa*, 381 pp.

Poll, M., 1973. 'Nombre et distribution géographique des poissons d'eau douce africaine', *Bull. Mus. Hist. Nat. Paris*, **150**: 113–28.

Poll, M., 1976. 'Poissons', *Explor. du Parc National de l'Upemba. Mission G. F. de Witte (1946–49)*, Fasc. 73, Bruxelles.

Poll, M. and **Gosse, J. P.**, 1963. 'Contribution à l'étude systématique de la fauna ichthyologique du Congo Central', *Ann. Mus. Roy. Afr. Centr. Sci. Zool.*, **116**: 45–101.

Poll, M. and **Matthes, H.**, 1962. 'Trois poissons remarkables du Lac Tanganika', *Ann. Mus. Roy. Afr. Centr.*, **111**: 1–26.

Poll, M. and **Nysten, M.**, 1962. 'Vessie natatoire pulmonoïde et pneumatisation des vertébrés chez *Pantodon buchholzi*', *Ac. Roy. Sci. d'Outre-mer. Bull. Séances.*, **8**: 434–54.

Potts, W. T. W. and **Parry, G.**, 1964. *Osmotic and Ionic Regulation in Animals*, 423 pp., Pergamon Press, Oxford.

Pourriot, R., 1968. 'Rotifères du Lac Tchad', *Bull. I.F.A.N.(A)*, **30**: 471–96.

Pourriot, R., **Iltis, A.** and **Lévêque-Duwat, S.**, 1967. 'Le plancton des mares natronées du Tchad', *Int. Rev. ges. Hydrobiol.*, **52**: 535–43.

Poynton, J. C., 1964. 'The amphibia of southern Africa', *Ann. Natal Mus.*, **17**: 1–334.

Prescott, G. W., 1948. 'Objectionable algae with reference to the killing of fish and other animals', *Hydrobiologia*, **1**: 1–13.

* **Prosser, M. W.**, **Wood, R. B.** and **Baxter, R. M.**, 1969. 'The Bishoftu Crater Lakes: a bathymetric and chemical study', *Arch. Hydrobiol.*, **65**: 309–24.

Provasoli, L., **Mclauchlin, J. J. A.** and **Pinter, I. J.**, 'Relative and limiting concentrations of major mineral constituents for the growth of algal flagellates', *Trans. N. Y. Acad. Sci.*, Ser. 11, **15**: 412–17.

Prowse, G. A. and **Talling, J. F.**, 1958. 'The seasonal growth and succession of plankton algae in the White Nile', *Limnol. and Oceanog.*, **3**: 222–38.

Pullan, R. A., 1964. 'Recent geomorphological evolution of the Chad Basin', *J. W. Afr. Sci. Assoc.*, **9**: 115–39.

Purchon, R. D., 1963. 'A note on the biology of *Egeria radiata* (Bivalvia, Donacidae)', *Proc. Malacol Soc. Lond.*, **35**: 251–71.

Quensière, J., 1976. 'Influence de la sécheresse sur les pêcheries du delta du Chari (1971–73)', *Cah. O.R.S.T.O.M. sér. Hydrobiol.*, **X**: 3–18.

Raheja, P. C., 1973. 'Lake Nasser' in Ackermann *et al.* 1973: 234–245.

Ramamurthi, R., 1965. 'Salinity tolerance and chloride regulation in a holeurysaline teleost, *Tilapia mossambria*', *Current Science*, **34**: 694–5.

Ransford, O., 1966. *Livingstone's Lake*. 313 pp. John Murray.

Rawson, D. S., 1955. 'Morphometry as a dominant factor in the productivity of large lakes', *Proc. Int. Assoc. theor. appl. Limnol.*, **12**: 164–75.

Reed, W., 1967. *Fish and Fisheries of Northern Nigeria*, 225 pp. Min. Agric., N. Nigeria.

Regier, H. A., 1970. 'Current problems in assessing Lake Victoria's stocks', Working paper No. 1. *Fish stock assessment on African Inland Waters.*, F.A.O., Rome.

Regier, H. A., (ed.), 1971. 'Evaluation of fisheries resources in African fresh waters', *Afr. J. Trop. Hydrobiol. Fish. Notes and comments*, **1**: 69–83.

Reite, O. B., Maloiy, G. M. O. and **Aasehaug, B.**, 1974. 'pH, salinity and temperature tolerance of Lake Magadi Tilapia', *Nature*, **247**: 315.

Rey, J. and **Saint-Jean, L.**, 1968. 'Les Cladocères (Crustacés, Branchiopodes) du Tchad' (1re note), *Cah. O.R.S.T.O.M. sér Hydrobiol.*, **2**: 79–118.

Rey, J. and **Saint-Jean, L.**, 1969. 'Les Cladocères (Crustacés, Branchiopodes) du Tchad' (2me note), *Cah. O.R.S.T.O.M. sér. Hydrobiol.*, **3**: 21–42.

Reynolds, C. S. and **Walsby, A. E.**, 1975. 'Water blooms', *Biol. Revs.*, **50**: 437–81.

Reynolds, J. D., 1969. 'The biology of the Clupeids in the Volta Lake, Ghana', in L. Obeng, 195–203.

Ricardo, C. K., 1939a. *Report on the Fish and Fisheries of Lake Rukwa in Tanganyika Territory and the Bangweulu Region in Northern Rhodesia*, 78 pp. Crown Agents for the Colonies, London.

Ricardo, C. K., 1939b. 'The fishes of Lake Rukwa', *J. Linn. Soc. Zool.*, **40**: 625–57.

Ricardo, C. K., 1943. 'The fishes of the Bangweulu region', *J. Linn. Soc. Zool.*, **41**: 183–217.

Richards, C. S., 1967. 'Aestivation of *Biomphalaria glabrata* (Basommatophora: Planorbidae): associated characteristics and relation to infection with *Schistosoma mansoni*', *Am. J. trop. Med. Hyg.*, **16**: 797.

* **Richardson, J. L.**, 1966. 'Changes in level of Lake Naivasha, Kenya during post-glacial times', *Nature*, **209**: 290–1.

Richardson, J. L., 1968. 'Diatoms and lake typology in East and Central Africa', *Int. Rev. ges. Hydrobiol.*, **53**: 299–338.

Richardson, J. L., 1969. 'Former lake-level fluctuations – their recognition and interpretation', *Mitt. int. Ver. Limnol.*, **17**: 78–93.

Richardson, J. L. and **Richardson, A. E.**, 1972. 'History of an African Rift lake and its climatic implications', *Ecol. Monogr.*, **42**: 499–534.

Richerson, P. J., Widmer, C., Kittel, T. and **Landa, C. A.**, 1975. 'A survey of the physical and chemical limnology of Lake Titicaca', *Verh. int. Ver. Limnol.*, **19**: 1948–503.

Ricker, W. E. (ed.), 1968. *Methods for the Assessment of Fish Production in Fresh Waters*, I.B.P. Handbook, No. 3, 313 pp., Blackwell, Oxford.

Rigler, F. H., 1975. 'Nutrient kinetics and the new typology', *Verh. int. Ver. Limnol.*, **19**: 197–210.

Rigler, F. H., 1976. Review of B. C. Patten (ed.), 'Systems analysis and simulation in ecology', *Limnol. Oceanogr.*, **21**: 481–3.

Rijks, D. A., 1969. 'Evaporation from a papyrus swamp', *Q. J. Roy. Met. Soc.*, **95**: 643–9.

Robarts, R. D., 1979. 'Under water light penetration, chlorophyll *a* and primary production in a tropical African lake (Lake McIlwaine, Rhodesia), *Arch. Hydrobiol.*, **86**: 423–444.

Robert, M., 1946. *Le Congo Physique* (3rd edn.), 449 pp., Vaillant Carmanne, Liège.

Roberts, R. D. and **Southall, G. C.**, 1977. 'Nutrient limitation of phytoplankton in seven tropical manmade lakes, with special reference to Lake McIlwaine, Rhodesia', *Arch. Hydrobiol.*, **79**: 1–35.

Roberts, T. R., 1972. 'Ecology of fishes in the Amazon and Congo basins', *Bull. Mus. Comp. Zool. Harvard*, **143**: 117–47.

Robinson, A. H., 1969. 'Notes on diurnal and seasonal changes in temperature and oxygen regimes in Lake Chad', *Ann. Rep.* (1966–67). *Fed. Fish Services, Republic of*

Nigeria, Lake Chad Res. Stn, Appendix 3: 26–34.

Robinson, A. H. and **Robinson, P. K.**, 1969. 'A comparative study of the food habits of *Microalestes acutiderns* and *Alestes dageti* (Pisces: Characidae) from the northern basin of Lake Chad', *Bull. I.F.A.N. (A)*, **31**: 951–64.

Robinson, A. H. and **Robinson, P. K.**, 1971. 'Seasonal distribution of zooplankton in the northern basin of Lake Chad', *J. Zool. Lond.*, **163**: 25–61.

Roche, M. A., 1970. 'Evaluation des pertes du Lac Tchad par abandon superficiel et infiltrations marginales', *Cah. O.R.S.T.O.M.*, *sér. Géol.*, **11**: 67–80.

Roger, J., 1943. 'Mollusques fossiles et subfossiles du Bassin du Lac Rudolphe', *Mus. Nat. Hist. Nat. Mission Scièntifique de l'Omo (1932–33)*, T.I. Fasc. 1.

Rogers, A. S., Imevbore, A. M. A. and **Adegoke, O. S.**, 1969. 'Physical and chemical properties of the Ikogosi Warm Spring, Western Nigeria', *J. Min. Geol. (Nigeria)*, **4**: 69–82.

Rohde W., 1974. 'Limology turns to warm lakes', *Arch. Hydrobiol.*, **73**: 527–48.

Rohlich, G. A. (ed.), 1969. *Eutrophication: Causes, Consequences and Corrections*, Proc. Sympos. Nat. Acad. Sci. (1967). Publ. No. 1700, 661 pp., Washington, D.C.

Roncière, C. de la, 1924–27. *La Découverte de l'Afrique au Moyen Age*, Mém. Soc. Roy. Géograph. d'Egypte, 3 vols.

Rubin, N. and **Warren, W. N.** (eds), 1968. *Dams in Africa: An interdisciplinary study of man-made Lakes in Africa*, 188 pp., Cass.

Russell-Hunter, W. D., 1970. *Aquatic Productivity*, 306 pp., Macmillan.

Ruttner, F., 1931. 'Hydrographische und hydrochemische Beobachtungen auf Java, Sumatra und Bali', *Arch. f. Hydrobiol.*, Suppl. 8.

Ruttner, F., 1953. 'Die Kohlenstoffquellen für die Kohlensaureassimilation submerser Wasserpflanzen', *Scientia*, 6 Ser., **47**: 1–8.

Ruttner, F., 1963. *Fundamentals of Limnology* (3rd edn), Trans. from German by D. G. Frey and F. E. J. Frey, 295 pp., Univ. of Toronto Press.

Ryder, R. A., 1965. 'A method of estimating the potential fish production of north-temperate lakes', *Trans. Am. Fish. Soc.*, **94**(3): 214–18.

Ryther, J. H., 1963. 'Geographic variations in productivity', in M. N. Hill (ed.), *The Sea*, Vol. 2, Interscience.

*Rzóska, J.**, 1961. 'Observations on tropical rainpools and general remarks on temporary waters', *Hydrobiologia*, **17**: 268–86.

Rzóska, J., 1964. 'Mass outbreaks of insects in the Sudanese Nile Basin', *Verh. int. Ver Limnol.*, **15**: 194–200.

Rzóska, J., 1968. 'Observations on zooplankton distribution in a tropical river dam-basin (Gebel Aulyia, White Nile, Sudan)', *J. Amin. Ecol.*, **37**: 185–98.

Rzóska, J., 1974. 'The Upper Nile Swamps: A Tropical Wetland Study', *Freshw. Biol.*, **4**: 1–30.

Rzóska, J. (ed.), 1976a. *The Nile. Biology of an ancient River*, 417 pp., Junk, The Hague.

Rzóska, J., 1976b. 'A controversy reviewed', *Nature*, **261**: 444–5.

Rzóska, J., Brook, A. J. and **Prowse, G. A.**, 1955. 'Seasonal plankton development in the White and Blue Nile near Khartoum', *Verh. int. Ver. Limnol.*, **12**: 327–34.

Sandford, K. S., 1933. 'Past climate and early man in the Southern Lybian Desert', *Geogr. J.*, **82**: 219–22.

Sandford, K. S., 1936. 'Observations on the distribution of land and freshwater Mollusca in the southern Libyan Desert', *Quart. J. Geol. Soc.*, **92**: 201–20.

Sandon, H., 1950. 'Illustrated guide to the freshwater fishes of the Sudan', *Sudan Notes and Records*, **25**: 1–61.

Sars, G. O., 1915, 1925. 'The freshwater Entomostraca of Cape Province. I. Cladocera. III. Copepoda', *Ann. S. Afr. Mus.*, **15**: 303–51, **25**: 85–149.

Savage, R. J. G. and **Williamson, P. G.**, 1978. 'The early history of the Turkana depression' in W. W. Bishop, 375–94.

Schelpe, E. A. C. L. E., 1961. 'The ecology of *Salvinia auriculata* and associated vegetation on Kariba Lake', *J. S. Afr. Bot.*, **27**: 181–7.

Schiffers, H., 1950. *Die Sahara*, 254 pp., Franckh'sche Verlagshandlung, Stuttgart.

Schiffers, H., 1967. 'Das Wasser der Sahara', *Bild der Wissenschaft*, Stuttgart, **9**: 748–58.

Schiffers, H. (ed.), 1971–1973. *Die Sahara und ihre Randgebiete.* 3 vols. Afrika-Studien 60, 61 and 62, I.F.O. Inst. f. Wirtschaftsforschung, Weltforum Verlag, Munchen.

Schmidt, G. W., 1976. 'Primary production in the three types of Amazonian waters. IV. On the primary productivity of phytoplankton in a bay of the lower Rio Negro (Amazonas, Brazil)', *Amazoniana*, **V**: 517–28.

Schmidt-Nielsen, B., 1964. *Desert Animals. Physiological Problems of Heat and Water*, 277 pp., Oxford Univ. Press.

Schmidt-Nielsen, K., Bolis, L. and **Maddrell, S. H. P.** (eds), 1978. *Comparative Physiology: water, ions and fluid mechanisms*, 360 pp., Cambridge Univ. Press.

Scientific American, 1977. 'Continents adrift and continents aground', Readings from the Scientific American, 230 pp., W. H. Freeman, N.Y.

S.C.O.P.E., 1972. *Man-made Lakes as Modified Ecosystems*, Sci. Comm. on Problems of the Environment, Report No. 2, 77 pp., Int. Council. Sci. Unions, Rome.

Scudder, T., 1966. 'Man-made lakes and population resettlement in Africa', in R. Lowe-McConnell, 99–108.

Scudder, T., 1972. 'Ecological bottlenecks and the development of the Kariba Lake basin' in M. T. Farrar and J. P. Milton (eds), *The careless technology. Ecology and international development*, Doubleday, N.Y.

Servant, M. and **Servant, S.**, 1970. 'Les formations lacustres et les diatomées du quaternaire recent du fond de la cuvette tchadienne', *Rev. Géogr. phys. Géol. dynam.*, **13**: 63–76.

Servant, M. and **Servant, S.**, 1973. 'Le plio-quaternaire du Bassin du Tchad', *Gème Congrés Int. de l'INQUA* décembre 1973: 169–75.

Seurat, L. G., 1934. 'Etudes zoologiques sur le Sahara Central', *Mém. Soc. d'Hist. Nat. Afr. du Nord*, No. 4. Mission du Hoggar III.

* **Shaw, J.**, 1959a. 'Salt and water balance in the East African fresh water crab *Potamon niloticus* (M. Edev.)', *J. Exp. Biol.*, **36**: 157–76.

Shaw, J., 1959b. 'The absorbtion of sodium ions by the crayfish *Astacus pallipes* (Lereboullet). I. The effect of external and internal sodium concentrations', *J. exp. Biol.*, **36**: 126–44.

Shinnie, M., 1965. *Ancient African Kingdoms*, 126 pp. Arnold.

* **Sioli, H.**, 1955. Beiträge zur regionalen Limnologie des Amazonasgebietes. III. Über einige Gewässer des oberen Rio Negro-Gebietes, *Arch. f. Hydrobiol.*, **50**: 1–32.

Sioli, H., 1964. 'General features of the Limnology of Amazonia', *Verh. int. Ver. Limnol.*, **15**: 1053–8.

Skulberg, O. M., 1977. 'Experimental use of algal cultures in limnology, Mitt. No. 21, *Int. Assoc. Theor. and Appl. Limnol.*, 607 pp.

Smart, J., 1943. *Insects of Medical Importance*, 269 pp., British Museum (Natural History).

Smith, H. W., 1956. *Kamongo or the Lungfish*, Viking.

Smith, H. W., 1961. *From Fish to Philosopher*, 293 pp., Doubleday, N.Y.

* **Smith, I. R.**, 1975. 'Turbulence in Lakes and Rivers', *Freshw. Biol. Assoc. Sci. Publ.* No. 29, 79 pp.

Smith, I. R. and **Sinclair, I. J.**, 1972. 'Deep-water waves in lakes', *Freshwat. Biol.*, **2**: 387–99.

Smith, M. H., 1969. 'Do intestinal parasites require oxygen?', *Nature*, **223**: 1129–32.

Smith, P. A., 1977. 'An outline of the vegetation of the Okavango drainage system' in Nteta and Hermans, 93–112.

Sollaud, E. and **Tilho, J.**, 1911. 'Sur la presence dans le Lac Tchad du *Palaemon niloticus* Roux', *C. R. Ac. Sci.*, **152**: 1868–71.

Sorokin, J. I., 1965. 'On the trophic role of chemosynthesis and bacterial biosynthesis in water bodies', *Mem. 1st. Ital. Idrobiol.*, **18** Suppl.: 187–205.

Sparks, B. W. and **Grove, A. T.**, 1961. 'Some quarternary fossil non-marine mollusca from the Central Sahara', *J. Linn. Soc. (Zool.)*, **44**: 355–64.

Speke, J. H., 1863, *What led to the discovery of the source of the Nile*, Blackwood.

Stankovic, S., 1960. *The Balkan Lake Ohrid and its living World*, 357 pp., Junk, The Hague.

Stanley, H. M., 1878. *Through the Dark Continent*, 2 vols., Sampson Low, Marston, Searle and Rivington, London.

Stanley, N. F. and **Alpers, M. P.**, 1975. *Man-made lakes and human health*, 515 pp., Academic Press, London.

Stark, N., 1970., 'Nutrient cycling II. Nutrient distribution in Amazonian vegetation', *Trop. Ecol.*, **12**: 177–201.

Stauch, A., 1977. 'Fish statistics in the Lake Chad basin during the drought', *Cah. O.R.S.T.O.M. sér. Hydrobiol.*, **XI**: 201–15.

Stebbins, G. L., 1966. *Processes of Organic Evolution*, 191 pp., Prentice-Hall, Englewood Cliffs, N. J.

Steeman–Nielsen, E., 1947. 'Photosynthesis of aquatic plants with special reference to the carbon sources', *Dansk Bott. Ark.*, **12**: 1–71.

Steeman–Nielsen, E., 1959. 'Untersuchungen über Primärproduktion des Planktons in einigen Alpenseen Österreichs', *Oikos*, **10**: 24–37.

Steeman–Nielsen, E., 1962. 'On the maximum quantity of plankton chlorophyll per surface unit of a lake or the sea', *Int. Rev. ges. Hydrobiol.*, **47**: 333–8.

Steeman–Nielsen, E., 1963. 'Productivity, definition and measurement', in M. N. Hill (ed.), *The Sea*, **2**: 129–59, Interscience. N.Y.

Steeman–Nielsen, E., 1965. 'On the terminology concerning production in aquatic ecology with a note about excess production', *Arch. Hydrobiol.*, **61**: 184–9.

Steeman–Nielsen, E. and **Hansen, V. K.**, 1959. 'Light adaptation in marine phytoplankton populations and its interrelation with temperature', *Physiol. Plant.*, **12**: 353–70.

Stobbart, R. H., 1965. 'The effect of some anions and cations upon fluxes and net uptake of sodium in the larva of *Aedes aegypti* (L)', *J. exp. Biol.*, **42**: 29–43.

Stross, R. G., 1966. 'Light and temperature requirement for diapause release and development', *Ecology*, **47**: 368–74.

Sutton, J. E. G., 1974. 'The aquatic civilisation of Middle Africa', *J. Afr. Hist.*, **15**: 527–46.

Svensson, G. S. O., 1933. 'Freshwater fishes from the Gambia River. Results of the Swedish Expedition', *Kungl. Svenska. Vet. Handl.*, **12**: 102.

⋆ **Symoens, J. J.**, 1963. 'Un siècle belge de recherches sur la floristique et l'écologie des algues', *Bull. Soc. Roy. Bot. Belg.*, **95**: 153–91.

Symoens, J. J., 1968. 'La minéralisation des eaux naturelles. Explor. hydrobiol. du Bassin du lac Bangweolo et du Luapula', *Cercle hydrobiol. Bruxelles*, II, **1**: 1–99.

⋆ **Talling, J. F.**, 1957a. 'The longitudinal succession of water characteristics in the White Nile', *Hydrobiologia*, **11**: 73–89.

Talling, J. F., 1957b. 'Diurnal changes of stratification and photosynthesis in some tropical African waters', *Proc. Roy. Soc. B*, **147**: 57–83.

Talling, J. F., 1961. 'Photosynthesis under natural conditions', *Ann. Rev. Plant Physiol.*, **12**: 133–54.

Talling, J. F., 1963. 'Origin of stratification in an African Rift lake', *Limnol. Oceanog.*, **8**: 68–78.

Talling, J. F., 1964. 'The annual cycle of stratification and primary production in Lake Victoria (E. Africa)', *Verh. int. Ver. Limnol.*, **15**: 384–5.

Talling, J. F., 1965a. 'The photosynthetic activity of phytoplankton in East African lakes', *Int. Rev. ges. Hydrobiol.*, **50**: 1–32.

Talling, J. F., 1965b. 'Comparative problems of phytoplankton production and photosynthetic productivity in a tropical and a temperate lake', *Mem. Ist. Ital. Idrobiol.*, 18 Suppl.: 399–424.

Talling, J. F., 1966. 'The annual cycle of stratification and phytoplankton growth in Lake Victoria (East Africa)', *Int. Rev. ges. Hydrobiol.*, **51**: 545–621.

Talling, J. F., 1969. 'The incidence of vertical mixing and some biological and chemical consequences in tropical African lakes', *Verh. int. Ver. Limnol.*, **17**: 998–1012.

Talling, J. F., 1976. 'Water characteristics', 'Phytoplankton: composition development and productivity' in J. Rzóska, 357–84, 385–402.

Talling, J. F. and **Rzóska, J.**, 1967. 'The development of plankton in relation to hydrological regime in the Blue Nile', *J. Ecol.*, **55**: 637–62.

Talling, J. F. and **Talling, I. B.**, 1965. 'The chemical composition of African lake waters', *Int. Rev. ges. Hydrobiol.*, **50**: 421–63.

Talling, J. F., **Wood, R. B.**, **Prosser, M. V.** and **Baxter, R. M.**, 1973. 'The upper limit of photosynthetic productivity by phytoplankton: evidence from Ethiopian soda lakes', *Freshw. Biol.*, **3**: 57–76.

Tarling, D. H. and **Tarling, M. P.**, 1971. *Continental Drift*, 112 pp., Bell.

Taylor, B. W., 1973. 'People in a rapidly changing environment: the first six years of Volta Lake' in Ackermann *et al.*, 99–113.

* **Temple, P. H.**, 1969. 'Some biological implications of a revised geological history of Lake Victoria', *Biol. J. Linn. Soc.*, **1**: 363–71.

Temple-Perkins, E. A., 1951. 'The first finding of a live pigmy crocodile in Uganda', *Uganda J.*, **15**: 182–6.

Thienemann, A., 'Tropische Seen und Seetypenlehre', *Arch. Hydrobiol.*, Suppl. 9: 205–31.

Thines, G., 1955. 'Les poissons aveugles', *Ann. Soc. Roy. Zool. Belg.*, **86**: 1–128.

* **Thomas, I. F.**, 1961. 'The Cladocera of the swamps of Uganda', *Crustaceana*, **2**: 108–25.

Thomas, J. D., 1966. 'Some preliminary observations on the fauna and flora of a small man-made lake in the West African savanna', *Bull. I.F.A.N. (A)*, **28**: 542–62.

Thompson, B. W., 1965. *Climate of Africa*, 132 pp., Oxford Univ. Press.

Thompson, K., 1976. 'Swamp development in the head waters of the White Nile' in J. Rzóska, 177–96.

Thompson, K., 1977. 'The primary productivity of African wetlands with particular reference to the Okavango Delta' in Nteta and Hermans, 67–80.

Thornton, I., 1965. 'Nutrient content of rainwater in the Gambia', *Nature*, **205**: 1025.

Thys van den Audenaerde, D. F. E., 1959. 'Existence d'une vessie natatoire pulmonoïde chez *Phractolaemus ansorgei* Blgr.', *Rev. Zool. Bot. Afr.*, **59**: 364–6.

Thys van den Audenaerde, D. F. E., 1963. 'Description d'une espèce nouvelle d'*Haplochromis* (Pisces, Cichlidae) avec observations sur les *Haplochromis* réophiles du Congo Oriental', *Rev. Zool. Bot. Afr.*, **68**: 140–52.

Thys van den Audenaerde, D. F. E., 1968. 'An annotated bibliography of *Tilapia*', *Mus, Roy. Afr. Centr. Document zool.*, No. 14: 1–406.

Tilho, J., 1911. *Documents scientifiques de la Mission Tilho*. 3 vols. Paris.

Tilho, J., 1928. 'Variations et disparition possible du Lac Tchad', *Ann. Géographie*, 37: 238–60.

Tjønneland, A., 1958a. 'Observations on *Chaoborus edulis* Edwards (Diptera, Culicidae)', *Univ. Bergen. Arb. r.*, 16: 1–12.

Tjønneland, A., 1958b. 'Observations on three species of East African Chironomidae (Diptera)', *Univ. Bergen. Arb. r.*, 17: 1–20.

*Trewavas, E., 1933. 'The Cichlid Fishes. Scientific Results of the Cambridge Expedition to the East African Lakes (1930–31), No. 11', *J. Linn. Soc. Zool.*, 38: 309–41.

Trewavas, E., 1935. 'A synopsis of the cichlid fishes of Lake Nyasa', *Ann. Mag. Nat. Hist.*, 16: 65–118.

Trewavas, E., 1937. 'Fossil cichlids from Dr L. S. B. Leakey's Expedition to Kenya 1934–35', *Ann. Mag. Nat. Hist.*, 19: 381–6.

Trewavas, E., 1947. 'Speciation in cichlid fishes of E. African lakes', *Nature*, 160: 96.

Trewavas, E., 1962. 'Fishes of the Crater Lakes of the Northwestern Cameroons', *Bonn. Zool. Beitr.*, 1/3: 146–92.

Trewavas, E., 1964. 'A revision of the Genus *Serranochromis*', *Ann. Mus. Roy. Afr. Centr.*, 125: 1–58.

Trewavas, E., 1966a. 'Fishes of the genus *Tilapia* with four anal spines in Malawi, Rhodesia, Mozambique and Southern Tanzania', *Rev. Zool. Bot. Afr.*, 74: 1–2.

Trewavas, E., 1966b. 'A preliminary review of fishes of the genus *Tilapia* in the eastward flowing rivers of Africa, with proposals for two new specific names', *Rev. Zool. Bot. Afr.*, 74: 394–424.

Trewavas, E., 1973. 'On the cichlid fishes of the genus *Pelmatochromis* with proposal for a new genus for *P. congicus*; on the relationship between *Pelmatochromis* and *Tilapia* and the recognition of *Sarotherodon* as a distinct genus', *Bull. Brit. Mus. Nat. Hist.*, 25: 1–26.

Trewavas, E., 1978. 'A discussion on *Tilapia* and *Sarotherodon*', *Cichlidae (Brit. Cichlid Assoc.)*, 3: 127–31.

Trewavas, E., Green, J. and Corbet, S. A., 1972. 'Eclogical studies on crater lakes in West Cameroon. Fishes of Barombi Mbo', *J. Zool.*, 166: 15–30.

Urvoy, Y., 1942. 'Les bassins du Niger. Étude de géographique physique et de paléogéographie', *Mém. I.F.A.N.*, No. 4, 135 pp.

Van Der Ben, D., 1959. 'La végétation des rives des lacs Kivu, Edouard et Albert', Explor. hydrobiol. des Lacs Kivu, Edouard et Albert (1952–54), 4(1), 191 pp., *Inst. Roy. Sci. Nat. Belge, Bruxelles*.

Van Der Borght, O., 1962. 'Absorbtion directe du calcium et du strontium en solution dans de milieu ambiant par un gastéropode dulcicole: *Lymnaea stagnalis* L.', *Arch. Int. Physiol. Biochim.*, 70: 611–23.

Vanderplank, F. L., 1941. '*Nothobranchius* and *Barbus* species: indigenous antimalaria fish in East Africa', *E. Af. Med. J.*, 17: 431–6.

*Vareschi, E., 1978. 'The ecology of Lake Nakuru (Kenya) I. Abundance and feeding of the lesser flamingo', *Oecologia (Berl.)*, 32: 11–35.

Vareschi, E., 1979. 'The ecology of Lake Nakuru (Kenya) II. Biomass and spatial distribution of fish (*Tilapia grahami* = *Sarotherodon alcalicum grahami*. Boulenger)', *Oecologia*, 37: 321–35.

Verbeke, J., 1957a. 'Recherches écologiques sur la faune des grands lacs de l'Est Congo Belge', Explor. Hydrobiol. Lacs Kivu, Edouard et Albert (1952–54), 3(1), 177 pp., *Inst. Roy. Sci. Nat. Belg., Bruxelles*.

Verbeke, J., 1957b. 'Chaoboridae (Diptera Nematocera). Stades immatures et adultes', Explor. Hydrobiol. Lacs Kivu, Edouard et Albert (1952–54), 3(2): 185–202. *Inst.*

Roy. Sci. Nat. Belg. Bruxelles.

Verbeke, J., 1959a. 'Le régime alimentaire des poissons du Lac Kivu et exploitation des ressources naturelles du lac', Expl. Hydrobiol. des Lacs Kivu, Edouard et Albert (1952–54) 3(2), 66 pp., *Inst. Roy. Sci. Nat. Belge.*, Bruxelles.

Verbeke, J., 1959b. 'Le régime alimentaire des poissons du lacs Edouard et Albert', Expl. Hydrobiol. des lacs Kivu, Edouard et Albert (1952–54), 3(3), 66 pp., *Inst. Roy. Sci. Nat. Belge.*, Bruxelles.

Verduin, J., 1956. 'Primary production in lakes', *Limnol. Oceanogr.*, 1: 85–91.

Vincent, M., 1963. 'Le calcium total chez *Gammarus pulex pulex* (L.) et la teneur en calcium de l'eau', *C. r. Séanc. Soc. Biol.*, 157: 1274–7.

Vincent, M., 1969. 'Teneur en calcium de l'eau et récupération du calcium de la carapace après la nue chez *Gammarus pulex pulex* (L.)', *C. r. Séanc. Soc. Biol.*, 163: 736–9.

*Viner, A. B., 1969. 'Observation of the hydrobiology of the Volta Lake April 1965–April 1966', in L. Obeng, 195–203.

Viner, A. B., 1970a. 'Hydrobiology of Lake Volta, Ghana. I. Stratification and circulation of water', *Hydrobiologia*, 35: 209–29.

Viner, A. B., 1970b. 'Hydrobiology of Lake Volta. Ghana. II. Some observations on biological features associated with the morphology and water stratification', *Hydrobiologia*, 35: 230–48.

Viner, A. B., 1970c. 'Ecological Chemistry of a Tropical African Lake', 192 pp., Thesis for Ph.D. of London University.

Viner, A. B., 1975. 'The supply of minerals to tropical rivers and lakes (Uganda)', in Olson, G. (ed.) *An introduction to land-water relationships*, Springer Verlag N. Y., 227–61.

Viner, A. B., 1977. 'The sediments of Lake George IV. Vertical distribution of chemical features in relation to ecological history and nutrient recycling', *Arch. Hydrobiol.*, 80: 40–69.

Viner, A. B. and Smith, I. R., 1973. 'Geographical, historical and physical aspects of Lake George' in Greenwood and Lund, 235–70.

*Visser, S. A., 1961. 'Chemical composition of rainwater in Kampala, Uganda and its relation to meteorology and topographical conditions', *J. Geophys. Res.*, 66: 3759–65.

Visser, S. A., 1962. 'Chemical investigations into a system of lakes, rivers and swamps in S. W. Kigezi, Uganda', *E. Afr. Agric. For. J.*, 28: 81–6.

Visser, S. A., 1963. 'Gas production in the decomposition of *Cyperus papyrus*', *J. Water Pollution Control Federation*, 35: 973–88.

Visser, S. A., 1964a. 'A study in the decomposition of *Cyperus papyrus* in the swamps of Uganda, in natural peat deposits as well as in the presence of various additives', *E. Afr. Agr. For. J.*, 29: 268–87.

Visser, S. A., 1964b. 'Origin of nitrates in tropical rainwater', *Nature*, 201: 35–6.

Visser, S. A., 1965. 'A study of the metabolism during aestivation of the amphibious snail *Pila ovata*', *W. Afr. J. Biol. Chem.*, 8: 41–7.

Visser, S. A. (ed.), 1970. *Kainji, a Nigerian man-made lake*, 126 pp., Niger. Inst. Soc. Econ. Res., Ibadan.

Visser, S. A., 1974. 'Composition of waters of lakes and rivers in East and West Africa', *Afr. J. Trop. Hydrobiol. Fish.*, 3: 43–60.

Vollenweider, R. A. (ed.), 1969. *A Manual on Methods for Measuring Primary Production in Aquatic Environments*, Int. Biol. Prog. Handbook, No. 12, 224 pp., Blackwell, Oxford.

Voute, C., 1962. 'Geological and morphological evolution of the Niger and Benue Valleys', *Ann. Mus. Roy. Congo Belge*, Sér. 8. 40(1): 189–205.

Waddy, B. B., 1975. 'Research into the health problems of man-made lakes, with special reference to Africa', *Trans. Roy. Soc. Trop. Med.*, **69**: 39–50.

Wafa, T. A. and **Labib, A. H.**, 1973. 'Seepage losses from Lake Nasser' in Ackermann *et al.*, 287–91.

Walker, H. O., 1956. 'Evaporation from Lake Victoria', *Weather*, **11**: 384.

Warburg, M. R., 1972. 'Water economy and thermal balance of Israeli and Australian Amphibia from xeric habitats' in G. M. O. Maoiy, (ed.), *Comparative Physiology of Desert Animals*, Sympos. Zool. Soc. London: 79–112, Academic Press.

Ward, H. B. and **Whipple, G. C.**, 1959. *Freshwater Biology* (2nd edn), W. T. Edmondson (ed.), 1248 pp., Wiley, N.Y.

*****Wasawo, D. P. S.**, 1959. 'A dry season burrow of *Protopterus aethiopicus*', *Rev. Zool. Bot. Afr.*, **60**: 65–70.

Wasawo, D. P. S. and **Visser, S. A.**, 1959. 'Swamp worms and tussock mounds in the swamps of Teso, Uganda', *E. Af. Agric. J.*, **25**: 86–90.

Watt, K. E. F. (ed.), 1966. *Systems Analysis in Ecology*, 276 pp., Academic Press.

Wayland, E. J., 1925. *Petroleum in Uganda*. Geol. Survey of Uganda. Mem. No. 1. Crown Agents for the Colonies, London.

Wayland, E. J., 1931. 'The Rift Valley and Uganda Waterways', *Summary of Progress of the Geol. Survey of Uganda (1919 to 1929)*, pp. 40–44, Govt Printer, Entebbe.

Wayland, E. J., 1934. 'Katwe', *Uganda J.*, **1**: 96–106.

*****Webbe, G.**, 1962. 'The transmission of *Schistosoma haematobium* in an area of Lake Province, Tanganyika', *Bull. Wld. Hlth. Org.*, **27**: 59–85.

*****Welcomme, R. L.**, 1964. 'The habitats and habitat preferences of the young of the Lake Victoria *Tilapia*', *Rev. Zool. Bot. Afr.*, **70**: 1–28.

Welcomme, R. L., 1966. 'Recent changes in the stocks of *Tilapia* in Lake Victoria', *Nature*, **212**: 52–4.

Welcomme, R. L., 1968. 'Observations on the biology of the introduced species of *Tilapia* in Lake Victoria', *Rev. Zool. Bot. Afr.*, **76**: 249–76.

Welcomme, R. L., 1969. 'The effect of rapidly changing water level in Lake Victoria upon commercial catches of *Tilapia* (Pisces: Cichlidae)', in L. E. Obeng.

Welcomme, R. L., 1970. 'Studies on the effects of abnormally high water levels and the ecology of fish in certain shallow regions of Lake Victoria', *J. Zool. Lond.*, **160**: 405–36.

Welcomme, R. L., 1976. 'Some general and, theoretical considerations on the fish yield of African rivers', *J. Fish. Biol.*, **8**: 351–64.

Welcomme, R. L. (ed.), 1978. *Symposium on river and flood-plain fisheries in Africa*, Bujumbura 1977, Food Agric. Org., Rome.

Welcomme, R. L., 1979. *Fisheries Ecology of Floodplain Rivers*, 317 pp., Longman, London.

Wellington, J. H., 1955. *Southern Africa: geographical study. Vol. I. Physical Geography*, 528 pp., Cambridge Univ. Press.

Werger, M. J. A. (ed.), 1978. *Biogeography and Ecology of Southern Africa*, 2 vols. 1439 pp., Junk, The Hague.

Wesenberg-Lund, C., 1943. *Biologie der Süsswasserinsekten*, 682 pp., Springer, Berlin, Gyldendal, Copenhagen.

Wetzel, R. G., 1975. *Limnology*, 743 pp., W. B. Saunders, Philadelphia.

White, E. (ed.), 1965. 'The First Scientific Report of the Kainji Biological Research Team', Liverpool: Kainji Biological Research Team (Mimeographed).

White, M. J. D., 1978. *Modes of speciation*, 455 pp., W. H. Freeman, San Francisco.

Whitehead, P. J. P., 1958. 'Indigenous river fishing methods in Kenya', *E. Afr. Agric. For. J.*, **24**(2): 111–20.

Whitehead, P. J. P., 1959. 'The anadromous fishes of Lake Victoria', *Rev. Zool. Bot. Afr.*, **59**: 329–63.

Whitworth, T., 1965. 'The Pleistocene Lake Beds of Kabua, Northern Kenya', *Durham Univ. J.*, new ser. **26**: 88–100.

★Whyte, S. A., 1975. 'Distribution, trophic relationships and breeding habits of the fish populations in a tropical lake basin (Lake Bosumtwi, Ghana)', *J. Zool.*, **177**: 25–56.

Wickens, G. E., 1975. 'Changes in the climate and vegetation of the Sudan since 20 000 B.P.', *Boissiera*, **24**: 43–65.

Widmer, C., Kittel, T. and Richerson, P. J., 1975. 'A survey of the biological limnology of Lake Titicaca', *Verh. int. Ver. Limnol.*, **19**: 1504–10.

Williams, M. A. J., 1966. 'Age of alluvial clays in the Western Gazira, Republic of the Sudan', *Nature*, **211**: 270–1.

Williams, M. A. J. and Adamson, D. A., 1974. 'Late Pleistocene desiccation along the White Nile', *Nature*, **248**: 584–6.

Williamson, P. G., 1978. 'Evidence for the major features and development of Rift palaeolakes in the Neogene of East Africa from certain aspects of lacustrine mollusc assemblages' in W. W. Bishop, 507–27.

Willis, B., 1936. *East African Plateaus and Rift Valleys*, Carnegie. Inst. Washington, Publ. 470.

Wilson, B. H. and Dincer, T., 1977. 'An introduction to the hydrography of the Okavango Delta' in Nteta and Hermans, 33–48.

Winberg, G. G.(ed.), 1971. *Methods for the Estimation of Production of Aquatic Animals*, Trans. from the Russian by Annie Duncan, 175 pp., Academic Press.

Wolfe, R. S., 1971. 'Microbial formation of methane', *Adv. Microb. Physiol.*, **6**:107–46.

Wolverton, W. and McDonald, R. C., 1976. 'Don't waste waterweeds', *New Scientist*, **71**: 318–20.

★Wood, R. B., Prosser, M. V. and Baxter, R. M., 1969. 'A seasonal study of stratification in a tropical African lake at 1 800 m altitude (Abstract)', *Verh. int. Ver. Limnol.*, **17**: 1050–1.

Wood, R. B., Prosser, M. V. and Baxter, R. M., 1976. 'The seasonal pattern of thermal characteristics of four of the Bishoftu crater lakes, Ethiopia', *Freshw. Biol.*, **6**: 519–30.

Woodward, J. G., 1974. 'The success and failure of the clupeid introduction', in Balon and Coche, 524–41.

★Worthington, E. B., 1929. 'The life of Lake Albert and Lake Kioga', *Geogr. J.*, **74**: 109–32.

Worthington, E. B., 1932. *A Report on the Fisheries of Uganda Investigated by the Cambridge Expedition to the East African Lakes 1932–33*, 88 pp., Crown Agents for the Colonies, London.

Worthington, E. B., 1933. 'The fishes of Lake Nyasa (other than Cichlidae)', *Proc. Zool. Soc. Lond.* Pt. 1: 285–316.

Worthington, E. B., 1954. 'Speciation of fishes in African Lakes', *Nature*, **173**: 1064.

Worthington, E. B. and Ricardo, C. K., 1936a. 'Scientific Results of the Cambridge Expedition to the East African Lakes 1930–1931. No. 15. The fish of Lake Rudolph and Lake Baringo', *J. Linn. Soc. Zool.*, **39**: 353–89.

Worthington, E. B. and Ricardo, C. K., 1936b. 'The fish of Lake Tanganyika (other than Cichlidae)', *Proc. Zool. Soc. Lond.*, Pt. 2: 1061–112.

Worthington, E. B. and Worthington, S., 1933. *Inland Waters of Africa*, 259 pp., Macmillan.

★Wourms, J. P., 1963. 'Naturally occurring developmental arrest in teleost fishes with associated modifications of epiboly and embryogenesis', *Proc. XVI Int. Congr. Zool.*, 2: 264.

Wourms, J. P., 1964. 'Comparative Studies on the early embryology of *Notobranchius taeniopygus* and *Aplocheilichthys pumilus* with especial reference to the problem of naturally occurring diapause in teleost fishes', *Ann. Rep. E. Afr. Freshw. Fish. Inst.*, 68–9.

Wourms, J. P., 1972. 'The developmental biology of annual fishes. III. Pre-embryonic and embryonic diapause of variable duration in the eggs of annual fishes', *J. Exp. Zool.*, 182: 389–414.

Wright, C. A., Klein, J. and **Eccles, D. H.**, 1967. 'Endemic species of *Bulinus* (Mollusca: Planorbidae) in Lake Malawi (Lake Nyasa)', *J. Zool. Lond.*, 151: 199–209.

Yuretich, R. F., 1976. 'Sedimentology, geochemistry and geological significance of modern sediments in Lake Rudolf (Lake Turkana), Eastern Rift Valley, Kenya', Thesis for Ph.D., Princeton, Univ. New Jersey.

Glossary

abyssal region. The water at the bottom of a deep lake.

aerobic organisms. Those that require molecular oxygen as the final acceptor of electrons in their chain of energy releasing reactions, i.e. cannot live in the absence of free oxygen.

aestivation. A resting inactive condition in which some organisms can survive complete drought, involving a great reduction of metabolism and in some cases *cryptobiosis*.

alkalinity of water. The sum of the concentrations of dissolved carbonate and bicarbonate ions, usually expressed as milliequivalents per litre.

allochthonous. Derived from outside the ecosystem.

allopatric species or populations, that occupy two or more separate geographical areas. *a. speciation* – with geographical separation of sections of a population.

anadromous fish. One that periodically ascends rivers, usually for breeding.

anaerobic organisms. Those that can live by releasing energy from a chain of reactions not ending in the reduction of molecular oxygen, i.e. in the absence of free oxygen.

anoxic. Devoid of oxygen, as is much of the water in tropical swamps and in the lower layers of some stratified lakes.

Aufwuchs. A German word used to denote the carpet-like growth of algae and small animals which encrusts the surface of submerged objects such as rocks, plant stems, etc.

autochthonous. Derived from inside the ecosystem.

autotrophic organisms. Those that synthesise their own organic matter, using an external source of energy. e.g. photosynthesising green plants and some bacteria.

benthic fauna and flora. The organisms inhabiting the substratum at the bottom of a body of water. Collectively known as 'benthos'.

biomass. The total mass of a species or group of species occupying a specific part of or the whole of an ecosystem. Usually expressed as weight per unit area or in the whole ecosystem, e.g. a lake.

biota. All the fauna and flora of a specified region.

capture (river). Natural erosion backwards by the upper reaches of a river so that the watershed is breached and the water system on the far side is diverted into it, i.e. 'captured' by the aforesaid river.

chemocline. A region where there is a sharp change in chemical composition between two water layers within each of which the composition is relatively uniform.

clone. A collection of genetically identical descendants derived by asexual reproduction from the same sexually produced individual.

community of organisms. All the species occupying a certain section of an ecosystem, e.g. the planktonic, benthic or littoral communities of a lake.

conductivity (electrical) of water. An approximate measure of total salinity. The reciprocal of the electrical resistance at a stated temperature of a 1-cm sided cube of water, usually expressed for freshwaters as 1/ohms $\times$ 10^{-6} (μmhos at $x°C$, or K_x).

convection. The vertical movements of water masses of different densities, due to differences of temperature or of dissolved or suspended matter.

Coriolis force. Due to the earth's rotation, a force that can deflect the direction of the movements of water masses in large lakes.

cryptobiosis. A condition in which the metabolism of an organism is reversibly reduced to zero.

cyclomorphosis. The periodic seasonal appearance of a well-defined series of forms in a genetically uniform population of a species, e.g. some planktonic rotifers and Cladocera.

detritus. Finely divided particles of organic matter suspended in water or lying on the bottom. An important food for many organisms.

diapause. A temporary interruption of growth and development at an immature stage. Shown by many invertebrates and by a few cyprinodont fish to tide over periods of drought.

dimictic lake. One this is subject to two overturns in the year, as in the north temperate regions.

draw-down. The controlled lowering of the water level in a manmade lake to make room for a seasonal heavy inflow.

dynamic equilibrium (*steady state*). An equilibrium maintained by a balance between active processes.

ecosystem. The entire complex of interdependent organisms and processes that characterise a more or less definable and stable situation such as a lake, river, swamp, etc. No ecosystem however is isolated and its stability is dependent upon exchanges with the outside.

endemic species. A species that exists only in a given region.

epilimnion. The frequently stirred water layer above the thermocline in a temperate climate lake during summer stratification.

epiphytic plant. One that lives on the surface of other plants but gains no nourishment from them.

equatorial. Near the Equator. In this book used to denote a region without about 10° latitude of the Equator.

escarpment. A steep slope or cliff due to a *fault*, well shown in the sides of the Rift Valleys as more or less continuous lines of cliffs, sometimes more than 100 m high.

euphotic zone or *photozone*. The upper section of a water mass penetrated by light of sufficient intensity and of suitable wavelength to promote photosynthesis by aquatic plants. For convenience of quantitative definition, it has been defined as the water down to a depth reached by 1% of the light just under the water surface during the period of maximum illumination.

eutrophic water. Water of relatively high organic productivity.

eutrophication. The process by which water may become more highly productive through an increase of available nutrients, either naturally or by artificial means, leading in extreme cases to organic pollution.

fault (geological). A fracture in a land surface resulting in a vertical displacement to different levels on either side of the crack. Commonly associated with earth tremors and quakes.

fecundity. The potential of an organism for reproduction, usually expressed by the number of germ-cells produced over a certain time.

gene. A unit of inheritance carried on a chromosome and passed on through the germ cells. Controls the basic course of the development of characters.

gene pool. All the genes possessed by a population of a species at one time.

genotype. The whole constellation of genes possessed by a species population.

geochemical. Relating to the composition of and the chemical processes occuring within the earth's crust.

heterotrophic organisms. Those that derive their organic food from substances already synthesised by autotrophs such as green plants.

holomictic lakes. Lakes in which the seasonal overturns cause mixing of the water column down to the bottom.

homothermic. Of uniform temperature throughout. Used here to denote uniform temperature from top to bottom of a water column.

hybridisation. Crossing between two individuals of genetically different populations of the same species which have secondarily come into contact.

hypertonic regulation. The mechanism by which freshwater animals maintain the concentration of the salts in their internal fluids at a higher level than that in the external water.

hypolimnion. The water mass below the thermocline in a temperate climate lake during summer stratification and thus not stirred directly by wind action at water surface.

hypotonic regulation. The mechanism whereby some organisms can live in saline waters by maintaining the concentration of the salts in their internal fluids lower than that in the external water.

ichthyogeographic. Relating to the geographical distribution of fish.

indigenous species. A species that exists in a given region and has not been artificially introduced there. It may or may not be endemic.

ions. The positively and negatively charged particles into which the molecules of salts, acids and bases are dissociated in water.

isopleth. A line joining points that represent the same quantitative value, e.g. concentration of a substance, level of temperature (isotherm), etc.

isotonic condition of two or more fluids is one in which they contain the same concentration of dissolved salts.

limiting factor. A condition in which one of the essentials for the life and growth of an organism or group of organisms is deficient and retards production, e.g. too low temperature or illumination, an insufficient supply of oxygen or of an essential nutrient etc.

littoral of a lake. The shore region where the water is shallow enough for continuous mixing and for photosynthesis to the bottom, and which is affected in various ways by the proximity of the land and inflows.

macrophytes. Vegetation composed of the large higher plants, represented in freshwaters by submerged water weeds, floating and emergent plants.

meromictic lakes. Those in which the seasonal overturns do not stir the entire water column, but leave an undisturbed mass of water below.

metabolism. The entire complex of constructive and destructive processes continuously at work in a living organism. Sometimes applied to an ecosystem as a whole, such as a lake.

metalimnion. The layer between the epi- and hypolimnion of a temperate climate lake during summer stratication in which the steep vertical temperature gradient (thermocline) is situated.

monimnolimnion. The seasonally unstirred layer at the bottom of meromictic lakes in temperate climates.

monomictic lake. One that is subject to one annual overturn. Typical of lakes at lower latitudes of temperate regions wherein the temperature never falls below 4°C.

montane forest. Characteristic of high altitudes below the open alpine zone.

mutation. A change in the structure or arrangement of genes in the genotype of an organism.

niche. The constellation of environmental factors into which a species fits; the outward projection of an organism; its specific way of utilising its environment (Mayr, 1970).

nutrients. Substances dissolved in the water that are required for the growth of aquatic plants, but are not directly concerned (as are carbon dioxide, oxygen and water itself) in the storage and release of energy. Most of the inorganic ions are 'nutrients' in this sense, but the term is applied especially to compounds of nitrogen, phosphorus, sulphur and silicon that may become scarce and limit production (*limiting factors*).

oligotrophic water. Water of relatively low organic productivity.

orogenesis. The upward movement of a land mass due to lateral compression, e.g. of the Alps, Himalayas and the North African Atlas Mountains.

overturn. The destruction of the state of *stratification* whereby the horizontal layers of water are stirred together.

parthenogenesis. Development of unfertilised eggs.

pelagic region. The open deep water region of a large lake.

pH. Denotes the pressure or concentration of free hydrogen ions in water. Expressed as the negative logarithm of the gram equivalent concentration. A balance between H^+ and OH^- ions (neutrality) is reached at pH 7, alkaline conditions above and acid conditions below pH 7.

phenotype. The entire make-up of a developed organism comprising both inherited and acquired characters.

photosynthesis. The building-up of organic compounds from inorganic constituents, including water and carbon dioxide, by autotrophic organisms with the energy derived from sunlight and absorbed by special pigments. Green plants are the commonest and most important photosynthesising organisms.

phytoplankton. See *plankton*.

plankton. The community of small organisms, plants (phytoplankton) and animals (zooplankton), freely moving in the open water.

polymictic lake. One that is frequently overturned, usually at irregular intervals.

polymorphs. Distinct forms regularly produced by some species, which are not reproductively isolated.

population. A group of interbreeding individuals of a species occupying a certain locality and isolated from other populations of the same species.

production. The building up of living matter in the bodies or organisms, *primary* production by autotrophic plants and *secondary* production by *heterotrophic* organisms such as animals and many micro-organisms.

productivity. The rate of production on the part of a species or group of species. *Gross p.*: the rate of initial production, *net p.*: the actual rate of accumulation of living matter after part has been broken down by respiration.

recycling. The physical and chemical processes by which the breakdown products from the metabolism and decomposition of the organisms in an ecosystem are made available for re-use by the photosynthesising plants.

redox potential. A measure of the 'pressure' of electrons whose transfer from one substance to another is the essence of oxidation and reduction. It is measured electrically as millivolts against a standard electrode, and is an important environmental factor in relation to the energy sources available to many organisms, especially micro-organisms.

rift valley. Valley formed by roughly parallel faults, between which a block of land has fallen to form the floor of the valley.

seiche. The simple type of periodic standing wave in a water mass, detectable as an oscillation of water level or, in a stratified lake, of the boundaries between layers of different density or temperature.

soudanian fish fauna. That typical of the 'Soudan' which extends across northern Tropical Africa from the Nile to the West African coast. Representative of an ancient fauna which was more widely distributed before the post-Miocene tectonic upheavals. A word first proposed in this book for this icthyogeographic region.

speciation. The divergence between members of a species that lead to the evolution of new species, involving the development of reproductive isolating mechanisms.

species. A reproductively isolated aggregate of actually or potentially interbreeding populations.

standing crop. The total quantity (usually expressed as weight) of a species or group of species per unit surface area or in a whole ecosystem.

sterile. Not able to produce viable germ cells.

subspecies. Recognisably distinct forms of a species occupying different parts of the range of the species, but still capable of interbreeding, i.e. speciation has not gone far enough to cause reproductive isolation.

sympatric speciation. Speciation without geographical isolation of sections of a population. Its occurrence is not generally accepted.

taxonomy. The interrelations between a group of organisms upon which a classification is based. The study of the criteria on which a classification may be justified.

tectonic movements. Movements of masses of the earth's surface, e.g. upwarping, faulting, etc.

thalassoid features of some freshwater animals that are apparently similar to those characteristic of certain marine animals.

thermocline. The steep vertical temperature gradient between two of the layers in a stratified lake. Typical of temperate climate lakes during the summer, but commonly found under less definable circumstances in many tropical lakes.

topography. The shape, dimension and arrangement of the superficial features on the earth's surface.

total dissolved solids (T.D.S.). Organic and inorganic substances dissolved in water.

travelling waves. Undulating waves that move in a definite direction on the water surface or on the boundary between two water layers of different density.

turbulence in a fluid. Random movements of the constituent particles.

upwarping. The vertical upward movement of a land mass.

vector. An animal that carries the parasitic agent of a disease affecting another organism (and may itself be affected thereby), e.g. mosquitoes (malaria), snails (bilharzia), etc.

volcanic action. The outpouring of molten rock from below onto the earth's surface, affecting water systems by impeding drainage, providing craters in which lakes may form and contaminating waters with soluble materials.

zooplankton. See plankton.

General index

aerial respiratory organs, 319
aestivation, 348–52
Africa Association, 13
Agulhas current, 149
airborne particles, 40
air-breathing fishes, 319–20
allochthonous organic matter, 40
allopatric speciation, 136
Amphibia and water needs, 349–50
anadromous fish, 259
anaerobic decomposition, 48, 324
anaerobic life, 321, 324
annual fish, 349
autochthonous organic matter, 40
autotrophic organisms, 46

barriers to dispersal, 134
bilharzia (schistosomiasis), 65, 351, 376, 385
blue-green algae, consumption by fish, 51

calcium balance and deficiency, 63,
chotts, 195
climatic regions, 25, 33–8
clupeid fish, 145
continental drift, 150
coriolis force, 77
crocodiles, 187, 198, 237
cryptobiosis, 347, 352
cyclomorphosis, 136–7

density–temperature relation of water, 86
desiccation, 347–53
diatoms and water composition, 49
drawdown in man-made lakes, 364, 369, 383
dystrophic, 56

echo sounding, 284

ecosystems, 39
electrical conductivity, 59
electric fish, 186
Emin Pasha, 17
euphotic zone, 34
eutrophic, 56
eutrophication, 57, 392
evaporation from lakes, 28, 30
extracellular products, 41, 44

Ferguson's Gulf, 181
fish culture, 56, 88
fish kills, 88, 109, 230, 244, 358
flamingoes, 342
foggara, 205

Gauthiot Falls, 216
Gazira scheme, 389–90

harmattan wind, 93, 219
heterotrophic organisms, 46
holomictic, 85
hot springs, 141, 345–6
hydroelectric power, 151

Jonglei canal, 327
Jonglei Investigation Team, 327

Kaiso fossil beds, 174, 235
Karuma Falls, 257
Kazinga Channel, 236

Lakes
Abaya, 190, 191
Abiata, 111, 187ff

Index of organisms